Navigating Global Mangroves Management:

Policy and Practice

Hartuti Purnaweni
Diponegoro University, Indonesia

Ali Roziqin
https://orcid.org/0000-0001-8336-3103
Universitas Muhammadiyah Malang, Indonesia

Nilam Adini Rakhma
https://orcid.org/0009-0003-9436-0030
Diponegoro University, Indonesia

Setyo Budi Kurniawan
National Research and Innovation Agency, Indonesia

IGI Global
Scientific Publishing
Publishing Tomorrow's Research Today

Vice President of Editorial Melissa Wagner
Director of Acquisitions Mikaela Felty
Director of Book Development Jocelynn Hessler
Production Manager Mike Brehm
Cover Design Jose Rosado

Published in the United States of America by
IGI Global Scientific Publishing
701 East Chocolate Avenue
Hershey, PA, 17033, USA
Tel: 717-533-8845 | Fax: 717-533-7115
Website: https://www.igi-global.com E-mail: cust@igi-global.com

Library of Congress Cataloging-in-Publication Data

Names: Purnaweni, Hartuti, 1961- editor | Roziqin, Ali editor | Rakhma, Nilam Adini, 1993- editor | Kurniawan, Setyo Budi, 1994- editor
Title: Navigating global mangroves management : policy and practice / edited by Hartuti Purnaweni, Ali Roziqin, Nilam Adini Rakhma, Setyo Budi Kurniawan.
Description: Hershey, PA : IGI Global Scientific Publishing, 2027. | Includes bibliographical references and index. | Summary: "Mangroves are among the most vital coastal ecosystems, providing multiple ecological, social, and economic benefits. They serve as natural shields against coastal erosion and storm surges, act as highly efficient carbon sinks, and sustain biodiversity as breeding grounds for numerous marine species. Beyond ecology, mangroves are integral to community livelihoods, supporting fisheries, eco-tourism, and the blue economy. Despite their importance, mangroves face escalating threats from unsustainable coastal development, aquaculture, deforestation, and climate change. This book navigates the complexity of global mangrove management, highlighting the interplay between policy, governance, and practice. By drawing on interdisciplinary insights and case studies from diverse regions, this volume will provide scholars, policymakers, and practitioners with a comprehensive understanding of current challenges and potential solutions in sustaining mangroves for future generations"-- Provided by publisher.
Identifiers: LCCN 2026019549 (print) | LCCN 2026019550 (ebook) | ISBN 9798337393209 hardcover | ISBN 9798337393216 paperback | ISBN 9798337393223 ebook
Subjects: LCSH: Mangrove forests | Coastal ecology | Climatic changes | Ecotourism | Biodiversity conservation | Natural resources--Management
Classification: LCC SD397.M25 N39 2026 (print) | LCC SD397.M25 (ebook)
LC record available at https://lccn.loc.gov/2026019549
LC ebook record available at https://lccn.loc.gov/2026019550

British Cataloguing in Publication Data
A Cataloguing in Publication record for this book is available from the British Library.

Editorial Advisory Board

List of Reviewers

Table of Contents

Detailed Table of Contents

This paper explores the interlinkages between mangrove ecosystems, coastal livelihoods, and sustainability through a comparative analysis of India & Pakistan. However, these ecosystems face mounting pressures from industrialization, aquaculture expansion, urbanization, and climate change. Drawing on case studies from major mangrove regions in both countries, the chapter examines how policy interventions influence community engagement; and recognizing under international instruments such as Ramsar Convention, CBD, and climate adaptation agendas for their role. It highlights contrasts & convergences in governance approaches, including community-based management, and public–private initiatives. The analysis underscores the importance of integrating livelihood security with ecological sustainability to ensure long-term mangrove conservation. The chapter argues that sustainable mangrove management in the Indo-Pak region requires inclusive governance frameworks that balance ecological protection with socioeconomic needs, thereby enhancing resilience among vulnerable coastal communities.

products. In the context of climate change, mangroves play a critical role as both vulnerable ecosystems and powerful climate mitigation assets.

Uzair Bhatti, Hainan University, China
Yonis Gulzar, Department of Management Information Systems, College of Business Administration, King Faisal University, Saudi Arabia
Yazeed Yasin Ghadi, Al Ain University of Science and Technology, Abu Dhabi, UAE
Dilmurod Turaev, Department of Information Technology and Exact Sciences, Termez University of Economics and Service, Termez, Uzbekistan
Lola Safarova, Samarkand State University of Veterinary Medicine, Livestock, and Biotechnologies, Uzbekistan
Rustam Kudratov, Tashkent University of Information Technologies Named After Muhammad al-Khwarizmi, Samarkand, Uzbekistan
Gulkhan Dadenova, Department of Information Technologies, Kimyo International University in Tashkent, Uzbekistan
Kudratov Alisher Toshpulatovich, Samarkand State University of Veterinary Medicine, Livestock, and Biotechnologies, Uzbekistan

Coastal pollution is a major and increasing hazard to marine life, livelihoods of coastline peoples, and the environment on a worldwide scale. Mangrove forests in the interface of terrestrial and marine environments are important for ameliorating coastal pollution as they trap sediments, cycle nutrients, sequester contaminants and biodegrade substances. Nevertheless, the considerable degradation of mangroves due to urbanization, expansion of aquaculture and inadequate governance has eroded their ecological functions. In this context, grass-roots mangrove conservation has arisen as a positive participatory tool combining local knowledge management and livelihood incentives. This paper provides an overview of the factors responsible for mangroves' ability to mitigate against coastal pollution, and presents case studies of community-based mangrove conservation projects occurring in many different geographical settings along with an analysis of their challenges and policy implications.

challenges in implementing it, such as institutional fragmentation, weak policy enforcement, and limited meaningful participation from the community. It shows that there is a gap between policy and practice, which often becomes a major obstacle to mangrove sustainability. Therefore, studies on the importance of inclusive and adaptive collaborative governance are a prerequisite for sustainable mangrove management.

Mangrove ecosystems are crucial for maintaining the equilibrium of coastal settings. This study examines the enhancement of mangrove management through the implementation of a Community-Based Mangrove Management (CBMM) approach. This approach empowers local communities to oversee the design, execution, and evaluation of the activity. The research employs qualitative approach which data collected through in-depth interviews, field trips, and document analysis from official government report. The locus study is Rembang Regency, Indonesia, particularly in several areas which have mangrove areas. The findings indicate that effective mangrove management is enhanced and sustained when communities are actively engaged, local organizations are strengthened, and collaboration occurs across communities, government, and other entities.The study indicates that community-based mangrove management effectively safeguards the environment, supports local populations, and ensures the long-term resilience of coastal regions.

Mangrove ecosystems play a vital role in maintaining the stability of coastal areas, but they currently face serious threats due to human activities and weak institutional systems. This study analyzes the role of Community based Mangrove Management in reducing poverty and ensuring environmental sustainability along the northern coast of Central Java. Using a Systematic Literature Review and semi structured interviews, this study explores various management models, ranging from participatory and collaborative models to those based on local knowledge. The study's findings indicate that active community involvement, supported by economic incentives and clear management rights, is crucial for the success of ecosystem restoration while

This study examines sustainable marketing practices in mangrove ecotourism by focusing on value creation, green branding, and policy support within the context of coastal management. Using a quantitative approach, data were collected through structured questionnaires administered to 25 managers and 100 visitors at mangrove ecotourism destinations in Indonesia. Descriptive and inferential analyses were conducted to compare perceptions of sustainable marketing implementation across dimensions such as sustainability communication, stakeholder engagement, community participation, environmental education, and destination competitiveness. The findings reveal significant perception gaps between managers and visitors. While managers perceive branding, collaboration, and sustainability communication as effective, visitors report limited information clarity, inadequate environmental education, and low involvement in conservation activities. These results highlight the need for stronger communication, inclusive stakeholder collaboration, and clearer sustainability narratives.

Mangrove ecosystem degradation in coastal areas is not only an ecological problem, but also a governance, institutional, and livelihood sustainability issue. This article examines the transformation of Pandansari Hamlet, Kaliwlingi Village, Brebes Regency, from a coastal area affected by abrasion, tidal flooding, aquaculture decline, and mangrove degradation into a model of community-based mangrove governance. Using a qualitative case study approach, this study shows how coastal crisis generated ecological awareness, collective action, local institutional strengthening, the formation of Satgas Jaga Segara or mangrove police, culturally grounded environmental campaigns, multi-actor collaboration, and mangrove ecotourism development. The findings suggest that Pandansari's success cannot be reduced to mangrove planting alone, but reflects a broader transformation of coastal governance. Sustainable mangrove restoration requires social legitimacy,

Preface

INTRODUCTION

Mangrove ecosystems stand at the intersection of environmental sustainability, climate resilience, biodiversity conservation, and socioeconomic development. Occupying the fragile interface between land and sea, these coastal forests provide invaluable ecosystem services that extend well beyond their geographic boundaries. They protect shorelines from erosion and extreme weather events, sequester significant amounts of atmospheric carbon, support rich biodiversity, and sustain the livelihoods of millions of people through fisheries, tourism, and other components of the blue economy. As global environmental pressures intensify, the need for effective, integrated, and evidence-based approaches to mangrove management has never been more urgent.

While the ecological significance of mangroves has long been recognized, their management remains a complex global challenge. Rapid coastal development, expanding aquaculture, deforestation, pollution, and the accelerating impacts of climate change continue to threaten these ecosystems. At the same time, governance systems must reconcile competing priorities, including environmental protection, economic development, community livelihoods, and climate adaptation. Although international agreements and national policies have increasingly emphasized mangrove conservation and restoration, implementation often varies considerably across regions, highlighting the need for stronger policy integration, cross-sector collaboration, and locally informed management practices.

Navigating Global Mangroves Management: Policy and Practice was conceived to address these challenges by bringing together a diverse collection of scholarly perspectives that examine mangrove governance from multiple disciplinary and geographic viewpoints. The chapters included in this volume explore the evolving relationship between environmental policy, scientific innovation, governance frameworks, community participation, and sustainable development. Through interdisciplinary research and practical case studies, the contributors examine both the successes and the persistent obstacles associated with protecting and restoring mangrove ecosystems in an era of unprecedented environmental change.

The volume demonstrates that successful mangrove management requires more than ecological expertise alone. It demands integrated governance structures, informed policymaking, technological innovation, stakeholder engagement, and adaptive management strategies capable of responding to rapidly changing environmental and socioeconomic conditions. By examining experiences from different regions of the world, this book illustrates how local practices, national policies, and international frameworks can collectively contribute to more resilient and sustainable coastal management systems.

As editors, we believe this volume offers timely and meaningful contributions to the growing body of scholarship on coastal ecosystem governance. It provides readers with both theoretical foundations and practical insights that can inform future research, policy development, conservation initiatives, and management practices. The interdisciplinary perspectives presented throughout the chapters encourage dialogue across environmental science, public administration, sustainability studies, climate policy, and development planning, reflecting the interconnected nature of contemporary mangrove management.

This book is intended to serve a broad audience, including researchers and academics in environmental science, coastal studies, public administration, sustainability, and development studies; policymakers and government officials responsible for environmental and coastal governance; non-governmental organizations and community leaders engaged in conservation and local resource management; as well as international organizations and development agencies working to advance biodiversity conservation, climate resilience, and the sustainable blue economy.

We extend our sincere appreciation to all contributing authors whose expertise and dedication made this volume possible. Their research reflects the diversity of challenges and opportunities facing mangrove ecosystems worldwide while offering thoughtful and practical pathways toward more sustainable management. We also express our gratitude to the reviewers whose constructive feedback strengthened the quality of each chapter, and to the editorial and publishing teams whose guidance and support have been instrumental throughout the development of this book.

It is our hope that *Navigating Global Mangroves Management: Policy and Practice* will serve as a valuable resource for advancing interdisciplinary scholarship, informing evidence-based policy, and inspiring collaborative action to protect and restore one of the world's most valuable coastal ecosystems. As global environmental challenges continue to evolve, the knowledge, experiences, and perspectives presented in this volume seek to contribute to more resilient, equitable, and sustainable approaches to mangrove management for generations to come.

ORGANIZATION OF THE BOOK

Chapter 1, "From Global Frameworks to Regional Realities: Mangrove Management in the Indo-Pak Context," establishes the global context for the volume by examining how international environmental frameworks are translated into regional governance and practice. Through a comparative analysis of India and Pakistan, the chapter demonstrates how mangrove conservation is shaped by differing policy approaches, community engagement strategies, and socioeconomic realities. By exploring the relationship between international agreements and local implementation, it emphasizes the importance of governance systems that integrate ecological sustainability with livelihood security, offering valuable insights into the opportunities and challenges of managing shared coastal resources.

Chapter 2, "Global Mangroves Management as Sustainable Socioeconomic Systems: Biotechnology, Climate Science, and Digital Innovations Needs Policy Integration," explores the transformative potential of emerging technologies in advancing mangrove conservation. The chapter examines how biotechnology, microbial engineering, blue carbon modeling, artificial intelligence, and remote sensing can improve ecosystem monitoring, restoration, and adaptive management. At the same time, it highlights the critical need for policy frameworks that translate scientific innovation into practical governance, ensuring that technological advances contribute to equitable, sustainable socioeconomic development and long-term ecosystem resilience.

Chapter 3, "Environmental Challenges and Climate Change: Mangroves' Role in Carbon Sequestration and Climate Mitigation," highlights the ecological significance of mangrove ecosystems within the broader context of global climate change. The chapter examines the multiple ecosystem services provided by mangroves, including shoreline protection, biodiversity conservation, nutrient cycling, and carbon sequestration. By emphasizing both the vulnerability of mangroves to environmental change and their capacity to mitigate climate impacts, the chapter reinforces the central role these ecosystems play in global climate adaptation and sustainable coastal management strategies.

Chapter 4, "Earth Observation-Driven Mangrove Management with Global Species Mapping and Blue Carbon Assessment," demonstrates the growing importance of community participation alongside scientific understanding in addressing coastal environmental degradation. Drawing on examples of mangrove conservation initiatives from diverse geographic settings, the chapter examines how local knowledge, participatory governance, and livelihood incentives strengthen conservation outcomes while mitigating coastal pollution. The discussion highlights the policy implications of integrating grassroots engagement with broader environmental management strategies to enhance the long-term resilience of mangrove ecosystems.

Chapter 5, "Corporate Social Responsibility as a Tool for Strengthening Blue Carbon and Mangrove Conservation in Malaysia," examines the expanding role of corporate social responsibility in supporting mangrove restoration and blue carbon initiatives. Using Malaysia as a case study, the chapter evaluates how corporate involvement can move beyond symbolic environmental activities toward measurable ecological outcomes supported by scientific monitoring and carbon accounting. The discussion provides practical guidance for designing CSR frameworks that promote accountability, ecological integrity, and long-term sustainability within national conservation strategies.

Chapter 6, "Bridging the Gap Toward a Collaborative Governance Framework for Sustainable Coastal Management in Semarang, Indonesia," addresses the governance challenges underlying coastal vulnerability in one of Indonesia's most environmentally sensitive regions. The chapter argues that fragmented institutions, overlapping authority, and insufficient stakeholder coordination limit effective coastal management. It proposes a collaborative governance framework built upon shared problem-solving, integrated knowledge, multi-actor participation, and adaptive implementation, illustrating how governance reform can strengthen both environmental resilience and community well-being.

Chapter 7, "Reframing Mangrove Governance under Forest Area Enforcement: Policy Responses to Mangrove Degradation," investigates the evolving relationship between environmental enforcement and community participation in mangrove governance. Focusing on Indonesia's Langkat Regency, the chapter explores how state-led forest enforcement both strengthens conservation authority and creates social and institutional challenges. By advocating hybrid governance models that combine regulatory enforcement with community engagement and adaptive management, the chapter contributes important perspectives on achieving equitable and sustainable conservation outcomes.

Chapter 8, "Collaborative Governance for Mangrove Sustainability: Lessons from Global Policy and Local Practice," examines the transition from traditional state-centered governance toward more inclusive, multi-stakeholder approaches to mangrove management. The chapter analyzes the roles of governments, local communities, private organizations, non-governmental organizations, and academic institutions within collaborative governance frameworks. By highlighting persistent gaps between policy design and practical implementation, it emphasizes the necessity of inclusive participation, institutional coordination, and adaptive governance for sustaining mangrove ecosystems.

Chapter 9, "Strengthening Mangroves Management in Local Areas: A Community-Based Mangrove Management," demonstrates how empowering local communities can significantly improve the effectiveness and longevity of mangrove conservation efforts. Drawing on experiences from Rembang Regency, Indonesia, the chapter illustrates how community participation in planning, implementation, and evaluation strengthens local institutions, enhances environmental stewardship, and fosters productive collaboration among government agencies and other stakeholders. The findings reinforce the value of community-based management as a cornerstone of sustainable coastal resilience.

Chapter 10, "Community-Based Mangrove Management to Reduce Poverty in the Coastal Areas of Central Java," explores the relationship between environmental conservation and socioeconomic development through participatory mangrove governance. By examining various community-based management models, the chapter demonstrates how local engagement, secure management rights, and diversified livelihoods—including ecotourism and sustainable use of mangrove resources—can simultaneously restore ecosystems and reduce poverty. The discussion highlights the importance of institutional capacity building and stakeholder coordination in achieving inclusive and sustainable coastal development.

Chapter 11, "Gender Mainstreaming in Mangrove Conservation," brings an important social dimension to the discussion by emphasizing the contributions of women to mangrove conservation and coastal sustainability. The chapter explores women's roles in ecosystem restoration, ecotourism, and sustainable livelihoods while addressing the structural barriers that often limit their participation in environmental decision-making. By advocating gender-responsive governance and the recognition of local knowledge, the chapter demonstrates how greater inclusivity strengthens both ecological conservation and social equity.

Chapter 12, "Mageri Segoro as Local Wisdom in Mangrove Management: Good Environmental Governance Based on Empowerment," examines how indigenous knowledge and community empowerment can strengthen environmental governance. Focusing on Central Java's Mageri Segoro initiative, the chapter analyzes a conservation program grounded in local wisdom and participatory governance. Beyond evaluating large-scale mangrove planting efforts, it emphasizes the importance of long-term stewardship, ecological maintenance, and community ownership as essential components of successful restoration initiatives.

Chapter 13, "Pathways to Sustainable Mangrove Governance: Bridging Policy and Practice in a Changing Climate," broadens the discussion by comparing mangrove governance approaches across several countries facing diverse environmental and institutional challenges. Through international case studies, the chapter examines decentralization, community stewardship, restoration initiatives, disaster risk reduction, and carbon finance while identifying persistent governance barriers. It concludes by advocating adaptive, collaborative, and multi-level governance systems capable of strengthening climate resilience and promoting social equity.

Chapter 14, "Overcoming the Gap Between Global Commitments and Local Reality: Mangrove Governance and Legal Fragmentation," investigates the disconnect between international climate commitments and domestic legal implementation. Using Indonesia as a case study, the chapter demonstrates how fragmented legislation, overlapping institutional authority, and inconsistent enforcement undermine conservation objectives. It further critiques symbolic compliance practices and proposes legal reforms that better align national governance structures with global environmental commitments and meaningful ecological outcomes.

Chapter 15, "Transforming Coastal Governance from the Grassroots: Evidence from Brebes, Indonesia," presents an inspiring example of community-driven environmental transformation. Through the case of Pandansari Hamlet, the chapter illustrates how ecological crises can catalyze local leadership, institutional innovation, collaborative partnerships, environmental education, and sustainable livelihood development. The chapter demonstrates that successful mangrove restoration depends not only on ecological interventions but also on building social legitimacy, community ownership, adaptive governance, and long-term resilience, providing a fitting conclusion to the themes explored throughout this volume.

Chapter 16, "Marketing Sustainable Mangroves Value Creation, Green Branding, and Policy Support in Global Coastal Management," conclude the volume by explaining sustainable marketing practices in mangrove ecotourism as a strategic approach to promoting coastal sustainability. The chapter highlights the importance of effective sustainability communication, community participation, and environmental education in bridging stakeholder expectations and enhancing visitor engagement. This chapter provides valuable recommendations for policymakers, destination managers, and tourism practitioners seeking to foster resilient and environmentally responsible coastal tourism.

CONCLUSION

As editors, we believe that the chapters presented in this volume collectively demonstrate that the future of mangrove management depends upon the successful integration of sound science, effective governance, inclusive policymaking, technological innovation, and meaningful community participation. While the contributors examine diverse geographical regions, governance systems, and management approaches, a common message emerges: sustainable mangrove conservation cannot be achieved through isolated interventions. Rather, it requires coordinated, adaptive, and collaborative strategies that recognize the interconnected ecological, social, economic, and institutional dimensions of coastal ecosystems.

The perspectives presented throughout this book illustrate that mangroves are far more than valuable natural resources. They are fundamental to climate resilience, biodiversity conservation, sustainable livelihoods, disaster risk reduction, and the achievement of broader global sustainability objectives. From international policy frameworks and legal reforms to community-based governance, gender inclusion, digital technologies, corporate responsibility, and indigenous knowledge systems, the chapters offer complementary insights into the many pathways available for strengthening mangrove management in an increasingly complex world.

It is our hope that this volume will encourage continued interdisciplinary collaboration among researchers, policymakers, practitioners, development organizations, and coastal communities. By bridging policy and practice and by sharing experiences from multiple regional contexts, this book seeks to contribute to more informed decision-making, stronger governance frameworks, and innovative management approaches that can be adapted across diverse coastal environments.

We extend our sincere gratitude to all of the contributing authors for sharing their expertise and research, as well as to the reviewers whose thoughtful evaluations enhanced the quality of each chapter. We also express our appreciation to the editorial and publishing teams at IGI Global Scientific Publishing for their guidance and support throughout the development of this volume.

We hope that *Navigating Global Mangroves Management: Policy and Practice* will serve as a valuable scholarly resource and practical reference for those committed to protecting and restoring mangrove ecosystems around the world. As global environmental challenges continue to evolve, we trust that the knowledge presented in this collection will inspire future research, foster innovative partnerships, and support evidence-based policies that ensure the long-term resilience of mangroves and the coastal communities that depend upon them.

Acknowledgment

The editors would like to extend their sincere appreciation to everyone whose dedication, contributions, and support made the publication of this book possible. Our sincere appreciation also goes to our fellow editors and co-editors, whose collaboration and dedication were fundamental to the successful completion of this volume. From the initial concept through the editorial process to its final publication, this book has been shaped by a collective effort characterized by shared responsibility, effective coordination, and unwavering mutual support. We are deeply grateful for the commitment and professionalism that each co-editor brought to this publication.

The editors gratefully acknowledge the Faculty of Social and Political Sciences, Diponegoro University, for its generous support throughout the preparation of this book. We are especially grateful for the support provided through the World Class University Project under Letter No. 879/UN9.F7/TU/IV/2026, which significantly contributed to the successful completion of this volume.

We would also like to express our sincere gratitude to IGI Global for its outstanding technical and editorial support throughout the development and publication of this book. Their professionalism, guidance, and assistance have been invaluable in ensuring the quality and successful completion of this volume.

Chapter 1
From Global Frameworks to Regional Realities:
Mangrove Management in the Indo-Pak Context

Ramesh Kumar Sahu
http://orcid.org/0000-0002-4341-7506
MATS University, Raipur, India

Sumaira Lodhi
http://orcid.org/0009-0006-4711-4111
Forman Christian College University, Pakistan

ABSTRACT

This paper explores the interlinkages between mangrove ecosystems, coastal livelihoods, and sustainability through a comparative analysis of India & Pakistan. However, these ecosystems face mounting pressures from industrialization, aquaculture expansion, urbanization, and climate change. Drawing on case studies from major mangrove regions in both countries, the chapter examines how policy interventions influence community engagement; and recognizing under international instruments such as Ramsar Convention, CBD, and climate adaptation agendas for their role. It highlights contrasts & convergences in governance approaches, including community-based management, and public–private initiatives. The analysis underscores the importance of integrating livelihood security with ecological sustainability to ensure long-term mangrove conservation. The chapter argues that sustainable mangrove management in the Indo-Pak region requires inclusive governance frameworks that balance ecological protection with socioeconomic needs, thereby enhancing resilience among vulnerable coastal communities.

DOI: 10.4018/979-8-3373-9320-9.ch001

1. INTRODUCTION

Mangrove ecosystems are among the most productive and ecologically significant coastal environments in the world. Located in the intertidal zones of tropical and subtropical coastlines, these salt-tolerant forests form a unique interface between terrestrial and marine ecosystems. Mangroves are composed of salt-tolerant tree and shrub species that have evolved specialized adaptations enabling them to survive in saline, waterlogged, and oxygen-deficient conditions. Although they occupy a relatively small proportion of the Earth's forest cover, mangroves provide a wide range of ecological, economic, and social benefits that are essential for biodiversity conservation, climate regulation, and human well-being (Alongi, 2012; FAO, 2020). Their specialized adaptations enable them to thrive in saline and waterlogged environments, making them critical components of coastal landscapes.

Over the past few decades, mangroves have gained increasing recognition within international environmental governance due to the ecosystem services they provide. Beyond their traditional value as sources of timber, fuelwood, and fisheries resources, mangroves are now widely acknowledged for their role in shoreline stabilization, nutrient cycling, habitat provision, carbon sequestration, and disaster risk reduction (Barbier et al., 2011). These functions have elevated mangrove conservation to a global environmental priority and led to its inclusion within major international frameworks such as the Ramsar Convention on Wetlands, the Convention on Biological Diversity (CBD), the Paris Agreement, and the Sustainable Development Goals (SDGs).

The growing importance of mangroves is particularly evident in the context of climate change. Coastal regions across the world are increasingly threatened by sea-level rise, storm surges, coastal erosion, salinity intrusion, and extreme weather events. Mangrove forests act as natural barriers that reduce wave energy, protect coastlines, and enhance the resilience of coastal communities. Evidence from the 2004 Indian Ocean tsunami and subsequent cyclonic events in South Asia demonstrated that areas protected by healthy mangrove forests experienced lower levels of damage and recovered more rapidly than regions where mangroves had been degraded (Kathiresan & Rajendran, 2005). Consequently, mangroves are increasingly promoted as nature-based solutions capable of supporting both climate adaptation and mitigation objectives.

Mangrove ecosystems also play a crucial role in sustaining livelihoods. Millions of people across Asia, Africa, and Latin America depend directly or indirectly on mangrove resources for fisheries, aquaculture, fuelwood, honey collection, fodder, and construction materials (Walters et al., 2008). These ecosystems support food security, generate income opportunities, and function as safety nets during periods of economic hardship or environmental stress. For many coastal communities, particularly those characterized by poverty and limited livelihood alternatives, mangroves represent an indispensable source of economic and social security.

The relationship between mangrove ecosystems and livelihoods is especially significant in South Asia, a region characterized by high population density, extensive coastlines, and increasing environmental vulnerability. South Asia contains some of the world's most important mangrove ecosystems, including the Sundarbans of India and Bangladesh and the Indus Delta of Pakistan. These ecosystems support rich biodiversity and provide essential ecosystem services to millions of coastal residents. However, they face growing pressures from urbanization, industrialization, aquaculture expansion, pollution, overexploitation of resources, and climate change.

India and Pakistan provide particularly important case studies for examining the challenges and opportunities associated with sustainable mangrove management. Both countries possess extensive coastlines and significant mangrove resources, while simultaneously confronting coastal degradation, biodiversity loss, and climate-related risks. Despite being parties to major international environmental agreements, the two countries have adopted different approaches to mangrove governance. India has developed a relatively comprehensive framework based on environmental regulation, protected area management, and community participation. Mangrove ecosystems such as the Sundarbans, Bhitarkanika, and the mangroves of Gujarat have benefited from conservation initiatives emphasizing biodiversity protection and integrated coastal zone management.

Pakistan, in contrast, has focused more strongly on restoration and rehabilitation, particularly within the Indus Delta, where mangrove degradation has been driven by reduced freshwater inflows, industrial pollution, and coastal development. Over the past two decades, large-scale plantation and restoration programs have contributed to the recovery of degraded mangrove habitats and have attracted international attention as examples of ecosystem-based adaptation and climate resilience.

The increasing prominence of mangroves within global environmental discourse reflects a broader shift toward integrated and ecosystem-based approaches to sustainability. Contemporary environmental governance recognizes that ecological conservation, climate resilience, poverty reduction, and sustainable development are closely interconnected. Nevertheless, significant gaps often exist between international commitments and local realities. While global frameworks provide broad conservation objectives, their effectiveness ultimately depends on national institutions, governance mechanisms, and community participation. Understanding how these global commitments are translated into regional and local practices is therefore critical for evaluating conservation outcomes.

This chapter examines these issues through a comparative analysis of mangrove management in India and Pakistan. It explores the interactions between international environmental frameworks, national policies, local governance structures, and coastal livelihood systems. Particular attention is given to the role of community participation, climate resilience, and sustainable development in shaping conservation outcomes. The chapter argues that effective mangrove conservation requires more than ecological protection alone; it demands governance frameworks that integrate environmental sustainability with livelihood security and social equity. By linking global environmental commitments with regional realities, the chapter contributes to broader debates on sustainable development, climate adaptation, and coastal resource governance in South Asia.

1.1 Significance of the Study

The study is significant because it examines the critical intersection of environmental governance, coastal livelihoods, and sustainable development in one of the world's most environmentally vulnerable regions. Mangrove ecosystems are globally recognized for their ecological importance, providing essential ecosystem services such as coastal protection, biodiversity conservation, fisheries support, and carbon sequestration (Alongi, 2012; Donato et al., 2011). Despite their significance, mangroves continue to face increasing threats from urbanization, industrial development, aquaculture expansion, pollution, and climate change, particularly in South Asia.

This study is important because it bridges the gap between international environmental commitments and local implementation practices. Global frameworks such as the Ramsar Convention, the Convention on Biological Diversity (CBD), the Paris Agreement, and the Sustainable Development Goals (SDGs) emphasize the conservation and sustainable use of coastal ecosystems. However, the effectiveness of these frameworks depends on how they are translated into national policies and local management strategies (Ramsar Convention Secretariat, 2018). By examining the experiences of India and Pakistan, the study provides valuable insights into the practical challenges and opportunities associated with implementing international environmental norms in diverse socioeconomic and political contexts.

The study also contributes to understanding the relationship between mangrove ecosystems and coastal livelihoods. Millions of people living in the coastal regions of India and Pakistan depend directly on mangrove resources for fisheries, fuelwood, fodder, honey collection, and other livelihood activities (Walters et al., 2008). The degradation of mangroves can therefore have significant implications for food security, income generation, and community resilience in this area. By highlighting the livelihood dimensions of mangrove management, the study underscores the need for conservation approaches that balance ecological sustainability with socioeconomic well-being. Furthermore, the research is highly relevant in the context of climate change. Mangroves are increasingly recognized as effective nature-based solutions for climate mitigation and adaptation because of their capacity to store large quantities of carbon and protect coastlines from storms, flooding, and erosion (Mcleod et al., 2011). Given the growing vulnerability of coastal communities in South Asia to sea-level rise and extreme weather events, understanding effective mangrove governance is essential for enhancing climate resilience.

Finally, the comparative analysis of India and Pakistan offers important policy lessons for sustainable coastal governance. By identifying best practices, governance challenges, and opportunities for community participation, the study contributes to broader debates on environmental sustainability, ecosystem-based adaptation, and inclusive resource management. The findings are relevant not only to policymakers and environmental practitioners in South Asia but also to scholars and decision-makers seeking effective strategies for balancing conservation and development in coastal regions worldwide.

2. GLOBAL FRAMEWORKS GOVERNING MANGROVE CONSERVATION

2.1 Ramsar Convention and Wetland Governance

The Ramsar Convention on Wetlands, adopted in 1971, remains one of the most influential international agreements promoting wetland conservation and sustainable use. Mangrove ecosystems fall within the Convention's broad definition of wetlands and are therefore subject to its principles and conservation mechanisms (Ramsar Convention Secretariat, 2018). The Ramsar Convention emphasizes three pillars; first, Wise use of wetlands; second, Designation and management of Ramsar sites; and third, International cooperation.

Both India and Pakistan are signatories to the Convention and have designated several mangrove-associated wetlands as Ramsar Sites. The Convention has encouraged national governments to integrate wetland conservation into broader environmental policies and management strategies. The principle of "wise use" is particularly relevant to mangrove ecosystems because it seeks to balance conservation objectives with sustainable human utilization. This approach recognizes that local communities are integral stakeholders in wetland governance rather than merely beneficiaries of conservation outcomes.

2.2 Convention on Biological Diversity (CBD)

The CBD, adopted during the 1992 Rio Earth Summit, has significantly influenced mangrove conservation globally. The Convention's objectives include biodiversity conservation, sustainable use of biological resources, and equitable sharing of benefits arising from genetic resources (CBD, 1992). Mangrove ecosystems contribute directly to multiple CBD targets through: Conservation of threatened species, Maintenance of ecosystem services, Restoration of degraded habitats, and Community participation in biodiversity governance.

The ecosystem approach promoted by the CBD has shaped national mangrove management strategies in both India and Pakistan by encouraging integrated and participatory conservation mechanisms.

2.3 Climate Change Frameworks and Blue Carbon Initiatives

The emergence of climate governance has elevated mangrove conservation to a strategic priority within international environmental policy. Mangroves are increasingly recognized as "blue carbon" ecosystems because of their exceptional capacity to sequester and store carbon (Mcleod et al., 2011). Under the Paris Agreement, countries are encouraged to include nature-based solutions in their climate mitigation and adaptation strategies. Mangrove restoration projects contribute to Carbon sequestration, Coastal resilience, Disaster risk reduction, and Biodiversity conservation.

India has incorporated ecosystem-based adaptation measures within its climate policies, while Pakistan has launched large-scale mangrove restoration initiatives linked to climate resilience objectives.

2.4 Sustainable Development Goals

Mangrove conservation intersects with several UN Sustainable Development Goals (SDGs), including:

- SDG 1 (No Poverty)
- SDG 2 (Zero Hunger)
- SDG 13 (Climate Action)
- SDG 14 (Life Below Water)
- SDG 15 (Life on Land)

The SDGs encourage integrated approaches that simultaneously address environmental sustainability and socioeconomic development. This integrated perspective is particularly important for coastal communities whose livelihoods depend directly on mangrove resources (UNSDG, 2015).

3. MANGROVE ECOSYSTEMS IN SOUTH ASIA: REGIONAL CONTEXT AND ECOLOGICAL SIGNIFICANCE

South Asia hosts some of the world's most ecologically significant and socio-economically important mangrove ecosystems (Alongi, 2008; Giri et al., 2011). Stretching across the coastlines of India, Bangladesh, Pakistan, and Sri Lanka, these ecosystems form an interconnected ecological region that supports rich biodiversity, coastal productivity, and the livelihoods of millions of people (FAO, 2020; Barbier et al., 2011). Although mangrove forests in South Asia share common ecological characteristics, they differ considerably in terms of geographical distribution, species composition, hydrological conditions, governance structures, and conservation challenges (Spalding et al., 2010).

The region occupies a unique position within the global mangrove landscape. South Asian mangroves include both the world's largest contiguous mangrove forest, the Sundarbans shared by India and Bangladesh, and one of the largest arid-climate mangrove systems in the Indus Delta of Pakistan (Gopal & Chauhan, 2006; International Union for Conservation of Nature, 2016). These ecosystems serve as critical habitats for numerous terrestrial and marine species, many of which are globally threatened or endangered (Nagelkerken et al., 2008). Moreover, they provide essential ecosystem services, including shoreline stabilization, fisheries support, nutrient cycling, carbon sequestration, and protection from extreme weather events (Barbier et al., 2011; Donato et al., 2011).

Historically, mangrove ecosystems have played a central role in sustaining coastal economies throughout South Asia (Walters et al., 2008). Fishing communities, small-scale farmers, honey collectors, wood gatherers, and artisanal producers have depended upon mangrove resources for generations. However, rapid economic development, urbanization, industrial expansion, and climate change have significantly altered the ecological and socioeconomic landscapes of these coastal regions (Alongi, 2015; Friess et al., 2019). Consequently, South Asian mangroves have become focal points for debates concerning sustainability, conservation, and climate resilience (DasGupta & Shaw, 2013).

3.1 Ecological Characteristics of South Asian Mangroves

Mangroves in South Asia occur under diverse climatic and hydrological conditions ranging from humid deltaic environments to arid coastal regions (Tomlinson, 2016). The ecological diversity of the region contributes to considerable variation in species richness and forest structure (Kathiresan & Bingham, 2001).

South Asian mangrove forests are dominated by species belonging to genera such as *Avicennia*, *Rhizophora*, *Sonneratia*, *Bruguiera*, *Ceriops*, *Heritiera*, and *Excoecaria*

(Kathiresan & Rajendran, 2005; Spalding et al., 2010). Species composition varies according to salinity levels, freshwater availability, sediment characteristics, and tidal influences. Deltaic regions receiving substantial freshwater inflows generally support greater biodiversity compared to arid coastal environments where salinity levels are higher (Gopal & Chauhan, 2006).

The ecological productivity of mangrove ecosystems is closely linked to river systems such as the Ganges–Brahmaputra–Meghna Delta, the Indus River Delta, and numerous estuarine systems along India's eastern and western coasts (Woodroffe, 1992; Alongi, 2008). These river systems transport sediments and nutrients that sustain mangrove growth and facilitate ecological connectivity between terrestrial and marine environments.

In addition to supporting mangrove vegetation, these ecosystems provide nursery grounds for fish, crustaceans, mollusks, reptiles, birds, and mammals (Nagelkerken et al., 2008). Many commercially important fish species depend on mangrove habitats during critical stages of their life cycles, highlighting the economic significance of mangrove conservation for regional fisheries (Mumby et al., 2004).

3.2 Mangrove Ecosystems in India

India possesses one of the largest mangrove resources in South Asia, covering approximately 4,900–5,000 km^2 along its coastline (Forest Survey of India, 2023). Mangrove forests are distributed across both the eastern and western coasts, with significant concentrations in West Bengal, Odisha, Andhra Pradesh, Tamil Nadu, Maharashtra, Gujarat, Goa, Karnataka, and the Andaman and Nicobar Islands (MoEFCC, 2021).

3.2.1 The Sundarbans

The Indian Sundarbans constitute the western portion of the world's largest contiguous mangrove forest (Gopal & Chauhan, 2006). Recognized as a UNESCO World Heritage Site and Ramsar Wetland of International Importance, the Sundarbans support exceptional biodiversity, including the Royal Bengal Tiger, estuarine crocodiles, fishing cats, and numerous avian species (UNESCO, 2024; Mukherjee et al., 2010).

The ecological significance of the Sundarbans extends beyond biodiversity conservation. The mangrove forests provide natural protection against cyclones and storm surges affecting densely populated coastal areas of West Bengal (Das & Vincent, 2009; Kathiresan & Rajendran, 2005). The region also supports millions of people through fisheries, honey collection, fuelwood extraction, crab harvesting, and ecotourism (Jalais, 2010).

3.2.2 Bhitarkanika Mangroves

Located in Odisha, Bhitarkanika represents India's second-largest mangrove ecosystem and is internationally recognized for its ecological richness. It serves as a habitat for saltwater crocodiles, migratory birds, olive ridley turtles, and numerous aquatic species (Nayak, 2003). The Bhitarkanika mangroves also play a vital role in mitigating cyclone impacts along the Bay of Bengal coast (Badola & Hussain, 2005).

3.2.3 Mangroves of Gujarat and Western India

Gujarat contains some of India's most extensive western coastal mangrove forests, particularly in the Gulf of Kachchh and Gulf of Khambhat (Forest Survey of India, 2023). These mangroves occur under relatively arid conditions and are dominated by *Avicennia marina*. Community participation initiatives and industrial development pressures coexist in these regions, making them important examples of contemporary coastal governance challenges (Kumawat et al., 2026).

India's mangrove ecosystems have generally shown positive trends in recent decades due to afforestation efforts, regulatory measures, and conservation programs (Forest Survey of India, 2023; MoEFCC, 2021). Nevertheless, challenges remain, including industrial expansion, coastal development, pollution, and climate-induced environmental change (Alongi, 2015).

3.3 Mangrove Ecosystems in Bangladesh

Bangladesh contains approximately 60 percent of the Sundarbans ecosystem and possesses one of the largest mangrove forest areas globally (Iftekhar & Islam, 2004). The Sundarbans support more than 300 plant species and hundreds of animal species, including the Royal Bengal Tiger, Irrawaddy dolphin, Gangetic dolphin, estuarine crocodile, and numerous migratory birds (Islam & Wahab, 2005).

Mangrove resources provide employment and subsistence opportunities for fishers, honey collectors, and wood gatherers, benefiting millions of people living in coastal districts (Rahman et al., 2021). Bangladesh faces severe climate-related challenges, including sea-level rise, salinity intrusion, cyclones, and coastal erosion (IPCC, 2022). Cyclones Sidr, Aila, and Amphan highlighted the critical role of mangroves in disaster risk reduction and coastal resilience (Dasgupta et al., 2015; Rahman et al., 2021).

3.4 Mangrove Ecosystems in Pakistan

Pakistan's mangrove ecosystems are concentrated primarily within the Indus Delta, one of the world's largest arid-climate mangrove systems (International Union for Conservation of Nature, 2016; Sanaullah et al., 2025). Historically nourished by freshwater and sediment flows from the Indus River, these ecosystems have experienced significant degradation due to upstream water diversions and dam construction (Qureshi, 1993; Ahmed et al., 2025).

The dominant species is *Avicennia marina*, which exhibits high tolerance to saline conditions (Saenger, 2002). Despite relatively low species diversity, the Indus Delta remains ecologically important for fisheries production, biodiversity conservation, and coastal protection (WWF-Pakistan, 2020). Recent large-scale restoration initiatives have significantly expanded mangrove cover and gained international recognition for their contributions to climate adaptation and carbon sequestration (WWF-Pakistan, 2020).

3.5 Mangrove Ecosystems in Sri Lanka

Although smaller in extent than those of India, Bangladesh, and Pakistan, Sri Lanka's mangrove ecosystems possess considerable ecological and socioeconomic importance (Priyashantha A. K. H. & Taufikurahman T., 2020). Mangroves occur along lagoons, estuaries, and sheltered coastal zones and support diverse communities of fish, crustaceans, reptiles, and birds (Jayatissa et al., 2002).

Following the 2004 Indian Ocean tsunami, Sri Lanka increasingly recognized the role of mangroves in coastal protection and disaster resilience (Dahdouh-Guebas et al., 2005). Community-based conservation programs involving local communities, women's groups, and environmental organizations have become internationally recognized examples of integrated coastal resource management (Wickramasinghe et al., 2022). However, urbanization, tourism expansion, and coastal infrastructure development continue to threaten ecosystem sustainability (Fernando et al., 2020).

4. COASTAL LIVELIHOODS AND DEPENDENCE ON MANGROVE RESOURCES

Mangrove ecosystems are among the most valuable natural resources for coastal communities, providing a broad range of ecosystem services that sustain livelihoods, food security, and socioeconomic development. Across South Asia, particularly in India, Bangladesh, Pakistan, and Sri Lanka, millions of people depend on mangrove forests for fisheries, forest products, aquaculture, tourism, and protection from natural hazards. These ecosystems function not only as biodiversity hotspots but also as livelihood-support systems that sustain vulnerable coastal populations. Their close relationship with local communities underscores the need to integrate livelihood considerations into conservation and management strategies (Walters et al., 2008; Hidayati et al., 2026; Awang et al., 2026).

Mangroves contribute to household economies through both direct and indirect benefits. Direct benefits include harvesting fish, crabs, shrimp, mollusks, honey, fuelwood, timber, medicinal plants, and other non-timber forest products (NTFPs). Indirect benefits arise from ecosystem services such as shoreline stabilization, nutrient cycling, carbon sequestration, and support for marine biodiversity, all of which enhance fisheries productivity and agricultural sustainability (Barbier et al., 2011). For many low-income coastal households, mangrove resources represent vital natural capital, providing diversified income opportunities and reducing vulnerability where alternative employment is limited.

4.2 Fisheries, Forest Products, and Household Resource Use

Mangrove ecosystems are fundamental to fisheries and household resource security. Their intricate root systems provide breeding, spawning, nursery, and feeding habitats for numerous species of fish, crustaceans, and mollusks, thereby supporting commercial and subsistence fisheries (Nagelkerken et al., 2008). The Sundarbans of India and Bangladesh support one of South Asia's most productive fishing sectors, while Pakistan's Indus Delta sustains valuable shrimp fisheries and marine resources that contribute significantly to local livelihoods and export earnings. Similarly, Sri Lankan coastal communities depend heavily on mangrove-associated fisheries for subsistence and small-scale commercial activities.

Beyond fisheries, mangroves provide essential forest products used for household consumption and local trade. Fuelwood remains a primary energy source in many rural coastal communities, while mangrove timber is valued for house construction, boat building, fencing, and fishing equipment because of its durability under saline conditions. Non-timber forest products such as honey, beeswax, medicinal plants, fodder, tannins, and thatching materials also contribute significantly to household welfare. Seasonal honey collection in the Sundarbans, for example, provides supplementary income while preserving traditional ecological knowledge and cultural practices (Walters et al., 2008). However, unsustainable harvesting of timber and fuelwood can accelerate forest degradation, emphasizing the need for regulated extraction and sustainable resource management.

4.3 Aquaculture, Ecotourism, and Emerging Livelihood Opportunities

Aquaculture has become an important component of coastal economies throughout South Asia. Shrimp farming, crab cultivation, and fish culture generate substantial employment and income, although rapid commercial expansion has frequently resulted in mangrove clearance and habitat conversion, particularly in parts of India and Bangladesh. To reconcile economic development with environmental conservation, integrated approaches such as silvofisheries have gained increasing attention. By combining mangrove conservation with aquaculture production, these systems enable communities to generate income while maintaining essential ecosystem functions, demonstrating that sustainable livelihoods and conservation objectives can be mutually reinforcing (Primavera, 2005).

Mangrove ecosystems also support growing ecotourism industries owing to their rich biodiversity, scenic landscapes, and cultural significance. The Sundarbans, for instance, have become major ecotourism destinations, generating employment in guiding services, transportation, hospitality, handicrafts, and wildlife tourism. Well-managed ecotourism offers alternative livelihood opportunities that reduce dependence on extractive resource use while creating financial incentives for conservation. Revenue generated through tourism can support community development and ecosystem restoration initiatives. However, poorly planned tourism may cause habitat disturbance and environmental degradation, highlighting the importance of sustainable tourism planning and active community participation.

4.4 Mangroves, Disaster Risk Reduction, and Social Dimensions of Livelihoods

In addition to their economic contributions, mangroves play a crucial role in protecting coastal communities from environmental hazards. They act as natural barriers against cyclones, storm surges, coastal erosion, and sea-level rise by dissipating wave energy, stabilizing shorelines, and reducing flood impacts (Kathiresan & Rajendran, 2005). Their protective function gained global recognition following the 2004 Indian Ocean tsunami and subsequent cyclones in the Bay of Bengal, where communities protected by intact mangrove belts experienced significantly lower levels of damage than those in degraded areas. Consequently, mangroves serve as natural insurance systems that safeguard lives, livelihoods, and infrastructure while strengthening climate resilience.

The social benefits derived from mangroves are not distributed equally. Women play essential roles in shellfish collection, fish processing, fuelwood gathering, honey production, and small-scale trading, contributing significantly to household incomes and food security. Nevertheless, they often face constraints related to land ownership, access to financial resources, and participation in decision-making. Increasingly, community-based conservation and restoration programs across South Asia are adopting gender-sensitive approaches that promote women's participation in mangrove restoration, sustainable resource management, and livelihood diversification, thereby improving both conservation outcomes and social inclusion.

4.5 Livelihood Vulnerability, Sustainability Challenges, and Conservation Linkages

Despite their significant socioeconomic value, mangrove-dependent communities remain highly vulnerable to environmental degradation, climate change, and economic uncertainty. Rising sea levels, increasing salinity, pollution, declining fish stocks, habitat loss, and reduced freshwater inflows threaten both ecosystem health and livelihood security. For example, reduced freshwater discharge in Pakistan's Indus Delta has adversely affected mangrove regeneration and fisheries productivity, while communities in the Sundarbans increasingly face cyclones, coastal erosion, and salinity intrusion.

Addressing these challenges requires integrated management approaches that combine ecological conservation with poverty reduction, livelihood diversification, environmental education, and participatory governance. Conservation policies that restrict resource access without considering local socioeconomic needs often generate conflicts and undermine conservation objectives. Conversely, community-based management, ecosystem-based adaptation, sustainable fisheries, blue economy initiatives, and benefit-sharing mechanisms encourage local participation while improving conservation outcomes. Recognizing coastal communities as active stakeholders rather than passive beneficiaries is therefore fundamental to achieving sustainable mangrove management. Ultimately, long-term conservation success depends on governance frameworks that simultaneously protect mangrove ecosystems and enhance the resilience, livelihoods, and well-being of the communities that depend upon them.

In conclusion, mangrove ecosystems form the foundation of coastal livelihoods across South Asia. Their contribution to fisheries, forest products, climate resilience, and economic security underscores the need for governance frameworks that balance ecological sustainability with human well-being. Long-term conservation success will depend on policies that simultaneously protect mangrove ecosystems and enhance the resilience and prosperity of the communities that rely upon them.

5. COMPARATIVE GOVERNANCE ANALYSIS OF MANGROVE MANAGEMENT IN INDIA AND PAKISTAN

Mangrove governance in India and Pakistan reflects both shared environmental concerns and distinct policy approaches shaped by their respective institutional and political contexts. As signatories to international environmental agreements such as the Ramsar Convention, the Convention on Biological Diversity (CBD), and the Paris Agreement, both countries have committed to conserving coastal ecosystems and promoting sustainable resource management (Ramsar Convention Secretariat, 2018; CBD, 1992; UNFCCC, 2015). However, the implementation of these commitments has followed different trajectories.

India has established a relatively comprehensive legal and institutional framework for mangrove conservation. Environmental governance is supported by legislation such as the Environment (Protection) Act, 1986, the Coastal Regulation Zone (CRZ) Notification, and the Biological Diversity Act, 2002 (MoEFCC, 2023). Mangrove management is coordinated through multiple agencies, including the Ministry of Environment, Forest and Climate Change (MoEFCC), state forest departments, biodiversity boards, and coastal management authorities (Kathiresan, 2012; Forest Survey of India, 2023). Protected areas such as the Sundarbans Biosphere Reserve and Bhitarkanika National Park play a central role in safeguarding mangrove habitats while promoting biodiversity conservation and sustainable resource use (UNESCO, 2023).

In contrast, Pakistan's mangrove governance is concentrated primarily within the Indus Delta region of Sindh Province. Following the 18th Constitutional Amendment, environmental management responsibilities were increasingly decentralized to provincial governments (Government of Pakistan, 2010). Consequently, provincial forest and environmental departments have assumed greater responsibility for mangrove conservation (WWF-Pakistan, 2021). Pakistan has focused significantly on large-scale restoration and rehabilitation programs to address historical mangrove degradation caused by reduced freshwater inflows, pollution, and coastal development (Sanaullah et al., 2025; Abbas et al., 2013). National climate policies and international partnerships have further strengthened restoration efforts, making mangrove rehabilitation an important component of climate adaptation strategies (Government of Pakistan, 2021; IUCN Pakistan, 2020).

A notable distinction between the two countries lies in their policy orientation. India has traditionally emphasized protection and regulation through legal safeguards, protected area management, and coastal zoning measures (MoEFCC, 2023; Kathiresan & Bingham, 2001). Its conservation strategy focuses on preventing habitat loss and preserving ecological integrity. Pakistan, on the other hand, has increasingly prioritized restoration and afforestation, aiming to recover degraded mangrove ecosystems and enhance coastal resilience (WWF-Pakistan, 2021; Abbas et al., 2013). While these approaches differ, both contribute to the broader objective of sustainable mangrove management (Alongi, 2014).

Community participation has emerged as an important element of conservation efforts in both countries. In India, participatory mechanisms such as Joint Forest Management Committees, Eco-Development Committees, and Biodiversity Management Committees encourage local involvement in resource management and conservation activities (Poffenberger, 2000; MoEFCC, 2023). These initiatives seek to align ecological protection with livelihood needs and community development (Datta et al., 2012). Pakistan has similarly expanded stakeholder engagement through plantation campaigns, awareness programs, and restoration projects involving local fishers, community groups, and civil society organizations (WWF-Pakistan, 2021; IUCN Pakistan, 2020). However, participation often remains dependent on project-based interventions and external support (Sanaullah et al., 2025).

Overall, the experiences of India and Pakistan demonstrate that effective mangrove governance requires a balance between ecological conservation, climate resilience, and livelihood security (Barbier et al., 2011; Alongi, 2012). India's emphasis on regulatory protection and Pakistan's focus on ecosystem restoration represent complementary approaches that offer valuable lessons for sustainable coastal governance. The comparative analysis highlights the importance of strong institutions, adaptive policies, and meaningful community participation in ensuring the long-term sustainability of mangrove ecosystems and the well-being of coastal communities (Ostrom, 1990; Armitage et al., 2009).

6. COMMUNITY-BASED MANGROVE MANAGEMENT AND SUSTAINABLE LIVELIHOODS

Community-based mangrove management has emerged as an effective alternative to centralized conservation models, which often excluded local communities from decision-making and created conflicts over resource access (Agrawal & Gibson, 1999; Berkes, 2004; Purnaweni et al., 2025). Based on Ostrom's (1990) common-pool resource theory, this approach recognizes that local communities can sustainably manage shared resources when they possess clearly defined rights, responsibilities, and incentives. Since coastal livelihoods are closely linked to mangrove ecosystem health, participatory governance that incorporates local ecological knowledge has become a key strategy for balancing conservation objectives with socioeconomic development (Berkes, 2004; Walters et al., 2008).

India has increasingly integrated community participation into mangrove conservation through restoration, nursery development, plantation programmes, ecosystem monitoring, and sustainable resource management (Datta et al., 2012; MoEFCC, 2023). Village institutions in Odisha contribute to mangrove regeneration, while Gujarat has combined conservation with sustainable fisheries, ecotourism, and environmental awareness initiatives (Badola & Hussain, 2005; Gujarat Ecology Commission, 2018). Similarly, in Pakistan, restoration efforts in the Indus Delta actively involve coastal communities through plantation campaigns, ecosystem monitoring, and awareness programmes supported by government agencies, non-governmental organizations, and international partners (Abbas et al., 2013; WWF-Pakistan, 2021; IUCN Pakistan, 2020). These collaborative initiatives have strengthened ecosystem restoration and increased public awareness of the ecological and economic importance of mangroves (Sanaullah et al., 2025). However, challenges such as poverty, insecure land tenure, inequitable benefit-sharing, dependence on external funding, and vulnerability to climate-related hazards continue to constrain the long-term effectiveness of community-based programmes (Armitage et al., 2009; Ghosh et al., 2015).

The success of participatory conservation depends largely on the availability of sustainable livelihood opportunities that provide tangible economic benefits to local communities (Barbier et al., 2011). Livelihood diversification through sustainable aquaculture, crab fattening, honey production, mangrove nurseries, handicrafts, ecotourism, and other nature-based enterprises reduces dependence on destructive resource extraction while improving household incomes (Primavera, 2005). By integrating ecological conservation with livelihood security, community-based management strengthens environmental sustainability, promotes local stewardship, and enhances the adaptive capacity of coastal communities. The experiences of India and Pakistan demonstrate that participatory governance, supported by equitable benefit-sharing and livelihood diversification, provides an effective framework for sustainable mangrove management across South Asia.

7. CHALLENGES IN MANGROVE MANAGEMENT: REGIONAL AND GLOBAL

Despite their ecological, economic, and social significance, mangrove ecosystems continue to face numerous interconnected challenges that threaten their long-term sustainability. At the global level, climate change remains the most significant driver of mangrove degradation through sea-level rise, increasing temperatures, changing precipitation patterns, and the growing frequency of cyclones and storm surges. These impacts are compounded by rapid coastal urbanization, industrial development, tourism expansion, aquaculture, habitat fragmentation, and pollution from industrial effluents, agricultural runoff, plastics, and shipping activities. Although international frameworks such as the Ramsar Convention, the Convention on Biological Diversity (CBD), the Sustainable Development Goals (SDGs), and the Paris Agreement promote mangrove conservation, implementation is frequently constrained by limited financial resources, weak institutional capacity, inadequate monitoring, and uneven policy enforcement.

In the Indo-Pak region, these global pressures are intensified by region-specific environmental and socioeconomic challenges. Reduced freshwater inflows, river regulation, sediment depletion, coastal industrialization, and expanding urban settlements have significantly degraded mangrove ecosystems, particularly in Pakistan's Indus Delta and several coastal regions of India. Increasing dependence on fisheries and other mangrove resources, coupled with overexploitation, further accelerates ecosystem degradation. The Sundarbans, shared by India and Bangladesh, remain especially vulnerable to rising salinity, habitat loss, cyclones, and coastal flooding, threatening both biodiversity and the livelihoods of dependent communities.

Additional constraints include fragmented governance, inadequate funding for restoration programmes, limited institutional coordination, insufficient community participation, and the complexities of managing transboundary ecosystems. Addressing these challenges requires integrated and adaptive governance that combines ecosystem conservation with climate resilience, sustainable livelihoods, and regional cooperation. Strengthening institutional capacity, improving policy implementation, promoting equitable community engagement, and enhancing cross-border collaboration will be essential for ensuring the long-term conservation and sustainable management of mangrove ecosystems in the Indo-Pak region and beyond.

8. DISCUSSION: FROM GLOBAL COMMITMENTS TO REGIONAL REALITIES

The comparative analysis reveals both convergence and divergence in mangrove governance across India and Pakistan. Both countries have embraced international environmental commitments and recognize the importance of mangrove ecosystems for biodiversity conservation, climate adaptation, and livelihood support. However, significant differences exist in implementation strategies and governance arrangements.

One major finding is that global environmental frameworks exert substantial influence on national policy development. The Ramsar Convention, CBD, Paris Agreement, and SDGs have provided normative guidance and policy direction for mangrove conservation. Nevertheless, the effectiveness of these frameworks depends on domestic institutional capacity and local stakeholder engagement.

A second finding concerns the central role of livelihoods in shaping conservation outcomes. Communities are not merely beneficiaries of conservation policies but active participants whose decisions influence ecosystem sustainability. Conservation initiatives that ignore livelihood realities often face resistance and limited effectiveness. Conversely, programs that integrate livelihood enhancement with ecosystem protection tend to achieve more durable outcomes.

Third, the comparative evidence suggests that neither purely state-centered nor entirely community-driven approaches are sufficient. Successful mangrove management typically involves hybrid governance arrangements combining governmental oversight, community participation, scientific expertise, and private-sector engagement.

The experiences of India and Pakistan illustrate the importance of adaptive governance. Environmental conditions, socioeconomic contexts, and institutional capacities vary significantly across regions. Flexible governance systems capable of responding to local conditions are therefore more likely to succeed than rigid, centralized models.

Climate change further reinforces the need for integrated management approaches. Mangroves serve as critical components of climate adaptation strategies while simultaneously supporting biodiversity conservation and poverty reduction. Their multifunctional nature makes them particularly valuable within sustainability frameworks. Future policy directions should focus on:

1. Strengthening community participation.
2. Enhancing livelihood diversification.
3. Expanding ecosystem restoration.
4. Improving inter-agency coordination.
5. Integrating climate adaptation and conservation planning.
6. Promoting transboundary knowledge exchange.
7. Increasing investment in blue carbon initiatives.
8. Developing robust monitoring and evaluation systems.

Given shared environmental challenges, India and Pakistan could benefit from greater scientific collaboration, knowledge sharing, and regional cooperation regarding mangrove conservation and coastal resilience.

9. LIMITATIONS AND FUTURE SCOPE OF THE RESEARCH

This study examined the relationship between international environmental governance, coastal livelihoods, and mangrove conservation through a comparative analysis of India and Pakistan. Although it provides a comprehensive synthesis of existing literature, policy frameworks, and conservation practices, several limitations should be acknowledged. The research is primarily based on secondary sources, including academic publications, government reports, policy documents, and reports from international organizations. While these sources offer valuable insights into mangrove governance and livelihood dependence, the absence of primary field-based research limits a deeper understanding of local perceptions, community experiences, traditional ecological knowledge, and contemporary conservation practices. Consequently, some socio-cultural, institutional, and livelihood dynamics operating at the grassroots level may not be fully reflected in the analysis.

Another limitation concerns the geographical scope and availability of consistent data. The study focuses mainly on India and Pakistan, with only limited discussion of other South Asian countries such as Bangladesh and Sri Lanka. Therefore, it may not fully represent the diversity of governance systems, conservation policies, and community-based management approaches across the region. In addition, variations in methodologies, temporal scales, and monitoring techniques often produce inconsistent estimates of mangrove cover, biodiversity, restoration success, and livelihood dependence, making direct comparisons difficult and affecting the accuracy of regional assessments (Friess et al., 2019). Since environmental governance continues to evolve through new policies, restoration initiatives, climate adaptation strategies, and institutional reforms, some recent developments may not be fully captured in this review.

The study also emphasizes formal governance structures and policy frameworks more than informal institutions, customary resource-use systems, local power dynamics, and traditional ecological knowledge, all of which significantly influence conservation outcomes and community participation. Moreover, uncertainties associated with climate change projections make it difficult to accurately assess the long-term impacts of sea-level rise, salinity intrusion, coastal erosion, and extreme weather events on mangrove ecosystems and the livelihoods of dependent communities (Alongi, 2015). These limitations highlight the need for more comprehensive, interdisciplinary, and field-oriented research.

Future research should therefore prioritize empirical investigations using participatory approaches, household surveys, stakeholder interviews, ethnographic studies, and long-term field observations to better understand community perceptions, livelihood transitions, and the effectiveness of conservation policies. Comparative studies should also be expanded to include Bangladesh, Sri Lanka, and other countries within the Indian Ocean region to generate a broader understanding of governance innovations, conservation challenges, and best practices. Greater attention should also be given to evaluating ecosystem-based adaptation, nature-based solutions, mangrove restoration programmes, blue carbon governance, and climate finance mechanisms, recognizing the significant role of mangroves in carbon sequestration, climate mitigation, and sustainable financing (Donato et al., 2011; Purnaweni et al., 2026; Roziqin, 2018).

Future scholarship should further examine issues of gender, social inclusion, and environmental justice, ensuring that the roles and perspectives of women and marginalized coastal communities are adequately incorporated into conservation planning and decision-making. At the same time, advances in remote sensing, Geographic Information Systems (GIS), drone technology, and artificial intelligence should be integrated with ecological and social research to improve monitoring, evidence-based policymaking, and adaptive management. Strengthening regional cooperation for transboundary ecosystems such as the Sundarbans will also be essential. Overall, future research should adopt more empirical, interdisciplinary, and policy-oriented approaches that integrate ecological, economic, social, and governance perspectives to strengthen mangrove conservation, enhance community resilience, and promote sustainable coastal development across South Asia.

10. CONCLUSION

Mangrove ecosystems occupy a critical position at the intersection of environmental sustainability, climate resilience, and human well-being. As demonstrated through the comparative analysis of India and Pakistan, mangroves provide indispensable ecological services while supporting the livelihoods of millions of coastal residents. Their significance extends beyond local and national contexts, contributing to global biodiversity conservation, climate mitigation, and sustainable development objectives.

International frameworks such as the Ramsar Convention, the Convention on Biological Diversity, the Paris Agreement, and the Sustainable Development Goals have elevated mangrove conservation within global environmental governance. These frameworks have influenced policy development in both India and Pakistan, encouraging greater attention to ecosystem protection, restoration, and sustainable use.

Despite these advances, mangrove ecosystems continue to face substantial threats arising from industrialization, urbanization, aquaculture expansion, pollution, and climate change. The experiences of India and Pakistan illustrate that effective conservation cannot be achieved through ecological protection alone. Sustainable management requires integrating conservation objectives with livelihood security, social equity, and community participation.

The comparative analysis demonstrates that inclusive governance arrangements provide the most promising pathway toward long-term sustainability. Community-based management, participatory decision-making, and livelihood diversification strengthen local stewardship while reducing pressures on mangrove ecosystems.

Simultaneously, state institutions play essential roles in policy development, regulation, enforcement, and coordination.

Climate change further underscores the urgency of protecting and restoring mangrove ecosystems. As natural buffers against coastal hazards and significant carbon sinks, mangroves represent critical nature-based solutions capable of enhancing resilience among vulnerable coastal communities.

Ultimately, sustainable mangrove management in the Indo-Pak region depends on the ability of governments, communities, civil society organizations, and international partners to develop governance frameworks that balance ecological integrity with socioeconomic needs. Such an approach not only safeguards valuable ecosystems but also contributes to broader goals of sustainable development, climate resilience, and human security.

REFERENCES

Abbas, S., Qamer, F. M., Ali, G., Tripathi, N. K., Shehzad, K., Saleem, R., & Gilani, H. (2013). An assessment of status and distribution of mangrove forest cover in Pakistan. *Journal of Biodiversity and Environmental Sciences*, *3*(6), 64–78. https://www.researchgate.net/publication/240475916_An_assessment_of_status_and_distribution_of_mangrove_forest_cover_in_Pakistan

Agrawal, A., & Gibson, C. C. (1999). Enchantment and disenchantment: The role of community in natural resource conservation. *World Development*, *27*(4), 629–649. DOI: 10.1016/S0305-750X(98)00161-2

Ahmed, S., Hu, L., Cheng, L., Jia, M., Zhao, C., Zhang, R., Wang, Z., Marnn, P., & Ali, H. (2025). Substantial rehabilitation of mangrove forests along the Indus Delta coastline of Pakistan: A 33-year review. *Ecological Indicators*, *178*, 113837. DOI: 10.1016/j.ecolind.2025.113837

Alongi, D. M. (2008). Mangrove forests: Resilience, protection from tsunamis, and responses to global climate change. *Estuarine, Coastal and Shelf Science*, *76*(1), 1–13. DOI: 10.1016/j.ecss.2007.08.024

Alongi, D. M. (2012). Carbon sequestration in mangrove forests. *Carbon Management*, *3*(3), 313–322. DOI: 10.4155/cmt.12.20

Alongi, D. M. (2014). Carbon cycling and storage in mangrove forests. *Annual Review of Marine Science*, *6*(1), 195–219. DOI: 10.1146/annurev-marine-010213-135020 PMID: 24405426

Alongi, D. M. (2015). The impact of climate change on mangrove forests. *Current Climate Change Reports*, *1*(1), 30–39. DOI: 10.1007/s40641-015-0002-x

Armitage, D., Plummer, R., Berkes, F., Arthur, R., Charles, A., Davidson-Hunt, I., Diduck, A., Doubleday, N., Johnson, D., Marschke, M., McConney, P., Pinkerton, E., & Wollenberg, E. (2009). Adaptive co-management for social–ecological complexity. *Frontiers in Ecology and the Environment*, *7*(2), 95–102. DOI: 10.1890/070089

Awang, M. Y., Farida, N., Listyorini, S., & Nugraha, H. S. (2026). Revisit Intention in Coastal Tourism: The Interaction of Experience Quality Destination Image, Perceived Value, and Destination Trust. *Institutions and Economies*, *18*(2), 227–253. DOI: 10.22452/IJIE.vol18no2.9

Badola, R., & Hussain, S. A. (2005). Valuing ecosystem functions: An empirical study on the storm protection function of Bhitarkanika mangrove ecosystem, India. *Environmental Conservation*, *32*(1), 85–92. DOI: 10.1017/S0376892905001967

Barbier, E. B., Hacker, S. D., Kennedy, C., Koch, E., Stier, A., & Silliman, B. (2011). The value of estuarine and coastal ecosystem services. *Ecological Monographs*, *81*(2), 169–193. DOI: 10.1890/10-1510.1

Berkes, F. (2004). Rethinking community-based conservation. *Conservation Biology : The Journal of the Society for Conservation Biology*, *18*(3), 621–630. DOI: 10.1111/j.1523-1739.2004.00077.x

CBD. (1992). *Convention on Biological Diversity*. United Nations. Retrieved on 30 May 2026, from https://www.cbd.int/doc/legal/cbd-en.pdf

Dahdouh-Guebas, F., Hettiarachchi, S., Lo Seen, D., Batelaan, O., Sooriyarachchi, S., Jayatissa, L. P., & Koedam, N. (2005). Transitions in ancient inland freshwater resource management in Sri Lanka affect biota and human populations in and around coastal lagoons. *Current Biology : CB*, *15*(6), 579–586. DOI: 10.1016/j.cub.2005.01.053 PMID: 15797030

Das, S., & Vincent, J. R. (2009). Mangroves protected villages and reduced death toll during Indian super cyclone. *Proceedings of the National Academy of Sciences of the United States of America*, *106*(18), 7357–7360. DOI: 10.1073/pnas.0810440106 PMID: 19380735

DasGupta, R., & Shaw, R. (2013). Changing perspectives of mangrove management in India—An analytical overview. *Ocean and Coastal Management*, *80*, 107–118. DOI: 10.1016/j.ocecoaman.2013.04.010

Dasgupta, S., Huq, M., Khan, Z. H., Ahmed, M. M. Z., Mukherjee, N., Khan, M. F., & Pandey, K. (2015). Cyclones in a changing climate: The case of Bangladesh. *Climate and Development*, *7*(2), 96–110. DOI: 10.1080/17565529.2013.868335

Datta, D., Chattopadhyay, R. N., & Guha, P. (2012). Community-based mangrove management: A review on status and sustainability. *Journal of Environmental Management*, *107*, 84–95. DOI: 10.1016/j.jenvman.2012.04.013 PMID: 22595074

Donato, D. C., Kauffman, J. B., Murdiyarso, D., Kurnianto, S., Stidham, M., & Kanninen, M. (2011). Mangroves among the most carbon-rich forests in the tropics. *Nature Geoscience*, *4*(5), 293–297. DOI: 10.1038/ngeo1123

FAO. (2020). *Global forest resources assessment 2020*. Food and Agriculture Organization. Retrieved on 30 May 2026, from https://openknowledge.fao.org/server/api/core/bitstreams/9f24d451-2e56-4ae2-8a4a-1bc511f5e60e/content

Fernando, R. L. S., Herath, H. M. I. U. P., & Rathnayake, R. B. P. M. (2020). Coastal Conservation in Sri Lanka: Problems and Prospects. In Singh, A., Fernando, R. L. S., & Haran, N. P. (Eds.), *Development in Coastal Zones and Disaster Management*. Disaster Research and Management Series on the Global South., DOI: 10.1007/978-981-15-4294-7_2

Forest Survey of India. (2023). India State of Forest Report 2023. Dehradun: Ministry of Environment, Forest and Climate Change, Government of India. Retrieved on 02 June 2026, from https://fsi.nic.in/forest-report-2023

Friess, D. A., Rogers, K., Lovelock, C. E., Krauss, K. W., Hamilton, S. E., Lee, S. Y., Lucas, R., Primavera, J., Rajkaran, A., & Shi, S. (2019). The state of the world's mangrove forests: Past, present, and future. *Annual Review of Environment and Resources*, *44*(1), 89–115. DOI: 10.1146/annurev-environ-101718-033302

Ghosh, A., Schmidt, S., Fickert, T., & Nüsser, M. (2015). The Indian Sundarban mangrove forests: History, utilization, conservation strategies and local perception. *Diversity*, *7*(2), 149–169. DOI: 10.3390/d7020149

Giri, C., Ochieng, E., Tieszen, L. L., Zhu, Z., Singh, A., Loveland, T., Masek, J., & Duke, N. (2011). Status and distribution of mangrove forests of the world using earth observation satellite data. *Global Ecology and Biogeography*, *20*(1), 154–159. DOI: 10.1111/j.1466-8238.2010.00584.x

Gopal, B., & Chauhan, M. (2006). Biodiversity and its conservation in the Sundarban Mangrove Ecosystem. *Aquatic Sciences*, *68*(3), 338–354. DOI: 10.1007/s00027-006-0868-8

Government of Pakistan. (2010). Constitution (Eighteenth Amendment) Act, 2010. Islamabad: Government of Pakistan. Retrieved on 02 June 2026, from https://na.gov.pk/uploads/documents/1302138356_934.pdf

Government of Pakistan. (2021). National Climate Change Policy (Updated Version). Islamabad: Ministry of Climate Change. Retrieved on 02 June 2026, from https://mocc.gov.pk/SiteImage/Policy/NCCP%202021.pdf

Gujarat Ecology Commission. (2018). Mangrove Conservation and Livelihood Enhancement Programme in Gujarat. Gandhinagar: Government of Gujarat. Retrieved on 02 June 2026, from https://www.facebook.com/GEC.Gujarat/

Hidayati, I., Prabawani, B., Rahmadhony, A., Hayati, A. N., & Boddy, J. (2026). The Critical Role of Local Leaders in Managing the Impacts of Climate Change in Demak, Indonesia. In: Nurhidayah, L., Young, A., Handayani, W. (eds) *Climate Change, Labour and Migration in Indonesia*. Springer Climate. Springer, Cham. https://doi.org/DOI: 10.1007/978-3-032-15334-0_8

Iftekhar, M. S., & Islam, M. R. (2004). Managing mangroves in Bangladesh: A strategy analysis. *Journal of Coastal Conservation*, *10*(1–2), 139–146. DOI: 10.1652/1400-0350(2004)010[0139:MMIBAS]2.0.CO;2

Intergovernmental Panel on Climate Change (IPCC). (2022). Climate Change 2022: Impacts, Adaptation and Vulnerability. Cambridge University Press. Retrieved on 02 June 2026, from https://www.ipcc.ch/report/ar6/wg2/downloads/report/IPCC_AR6_WGII_FullReport.pdf

International Union for Conservation of Nature. (2016). Mangrove related policy and institutional frameworks in Pakistan, Thailand and Viet Nam. Retrieved on 02 June 2026, from https://portals.iucn.org/library/sites/library/files/documents/2016-084.pdf

Islam, M. S., & Wahab, M. A. (2005). A review on the present status and management of mangrove wetland habitat resources in Bangladesh with emphasis on mangrove fisheries and aquaculture. *Hydrobiologia*, *542*(1), 165–190. DOI: 10.1007/s10750-004-0756-y

Jalais, A. (2010). *Forest of Tigers: People, Politics and Environment in the Sundarbans*. Routledge., https://www.routledge.com/Forest-of-Tigers-People-Politics-and-Environment-in-the-Sundarbans/Jalais/p/book/9780415690461

Jayatissa, L. P., Dahdouh-Guebas, F., & Koedam, N. (2002). A review of the floral composition and distribution of mangroves in Sri Lanka. *Botanical Journal of the Linnean Society*, *138*(1), 29–43. DOI: 10.1046/j.1095-8339.2002.00002.x

Kathiresan, K. (2012). Importance of mangrove ecosystem. *International Journal of Material Science*, *2*, 1–7. DOI: 10.5376/ijms.2012.02.0010

Kathiresan, K., & Bingham, B. L. (2001). Biology of mangroves and mangrove ecosystems. *Advances in Marine Biology*, *40*, 81–251. DOI: 10.1016/S0065-2881(01)40003-4

Kathiresan, K., & Rajendran, N. (2005). Coastal mangrove forests mitigated tsunami. *Estuarine, Coastal and Shelf Science*, *65*(3), 601–606. DOI: 10.1016/j.ecss.2005.06.022

Kumawat, A. K., Jha, S. K., Kumar, S., & Pathak, B. (2026). Landscape-Driven Genetic Structure of Three Mangrove Species of Western India Inferred from SCoT Markers. *Ecological Genetics and Genomics*, *100469*, 100469. Advance online publication. DOI: 10.1016/j.egg.2026.100469

Mcleod, E., Chmura, G. L., Bouillon, S., Salm, R., Björk, M., Duarte, C. M., Lovelock, C. E., Schlesinger, W. H., & Silliman, B. R. (2011). A blueprint for blue carbon. *Frontiers in Ecology and the Environment*, *9*(10), 552–560. DOI: 10.1890/110004

MoEFCC. (2021). National Coastal Mission and Mangrove Conservation Programme Report. New Delhi: Ministry of Environment, Forest and Climate Change, Government of India. Retrieved on 02 June 2026, from https://moef.gov.in/uploads/2022/03/Annual-report-2021-22-Final.pdf

Mukherjee, N., Sutherland, W. J., Dicks, L., Hugé, J., Koedam, N., Dahdouh-Guebas, F., & Koedam, N. (2010). Ecosystem service valuations of mangrove ecosystems. *Biodiversity and Conservation*, *19*(10), 2865–2887. DOI: 10.1371/journal.pone.0111386

Mumby, P. J., Edwards, A. J., Arias-González, J. E., Lindeman, K. C., Blackwell, P. G., Gall, A., Gorczynska, M. I., Harborne, A. R., Pescod, C. L., Renken, H., Wabnitz, C. C. C., & Llewellyn, G. (2004). Mangroves enhance the biomass of coral reef fish communities in the Caribbean. *Nature*, *427*, 533–536. DOI: 10.1038/nature02286 PMID: 14765193

Nagelkerken, I., Blaber, S. J. M., Bouillon, S., Green, P., Haywood, M., Kirton, L. G., Meynecke, J. O., Pawlik, J., Penrose, H. M., Sasekumar, A., & Somerfield, P. J. (2008). The habitat function of mangroves. *Aquatic Botany*, *89*(2), 155–185. DOI: 10.1016/j.aquabot.2007.12.007

Nayak, D. (2003). Biodiversity of Bhitarkanika Mangrove Forest. *Tiger Paper – FAO Regional Quarterly Bulletin on Wildlife and National Parks Management, Bangkok*, 3(4), Oct-Dec. 2003. https://www.researchgate.net/publication/340779332_Biodiversity_of_bhitarkanika_Mangrove_Forests

Ostrom, E. (1990). *Governing the commons: the Evolution of Institutions for Collective Action.* Cambridge University Press., DOI: 10.1017/CBO9780511807763

Pakistan, I. U. C. N. (2020). Indus Delta Mangrove Ecosystem Restoration Programme. Karachi: IUCN Pakistan. Retrieved on 02 June 2026, from https://iucn.org/story/202606/mangrove-restoration-begins-karachi-coast

Poffenberger, M. (2000). Communities and forest management in South Asia. Gland, Switzerland: IUCN. Retrieved on 02 June 2026, from https://portals.iucn.org/library/sites/library/files/documents/2000-054.pdf

Primavera, J. H. (2005). Mangroves, fishponds, and the quest for sustainability. *Science, 310*(5745), 57–59. DOI: 10.1126/science.1115179 PMID: 16210520

Priyashantha, A. K. H., & Taufikurahman, T. (2020). Mangroves of Sri Lanka: Distribution, status and conservation requirements. *Tropical Plant Research, 7*(3), 654–668. DOI: 10.22271/tpr.2020.v7.i3.083

Purnaweni, H., Hadi, S. P., Soeprobowati, T. R., Roziqin, A., & Rakhma, N. A. (2026). What are the Main Concerns for Global Mangrove Management Re-search? Systematic Literature Review and Bibliometric Analysis. *Research in Ecology, 8*(2), 46-66. *Article, 12149.* Advance online publication. DOI: 10.30564/re.v8i2.12149

Purnaweni, H., Soeprobowati, T. R., Chatterjee, U., Kiswoyo, G. P., Rakhma, N. A., & Manik, J. S. T. (2025). The Role of Stakeholders in Community Based Mangrove Management in Tapak Area of Semarang City. In *E3S Web of Conferences* (Vol. 650, p. 02047). EDP Sciences., DOI: 10.1051/e3sconf/202565002047

Qureshi, M. T. (1993). Rehabilitation and management of mangrove forests of Pakistan. *Towards the rational use of high salinity tolerant plants: Vol. 1 Deliberations about High Salinity Tolerant Plants and Ecosystems*, 89-95. https://doi.org/DOI: 10.1007/978-94-011-1858-3_8

Rahman, M. M., Hossain, M. R., & Islam, M. N. (2021). Climate change impacts on Sundarbans: Challenges for Mitigation Strategies. Md. Jakariya and Ms. N. Islam (eds.), Climate Change in Bangladesh, Spinger Climate. pp. 47-64. https://doi.org/DOI: 10.1007/978-3-030-75825-7_3

Ramsar Convention Secretariat. (2018). *Global wetland outlook*. Ramsar Convention Secretariat. Retrieved on 02 June 2026, from https://www.ramsar.org/sites/default/files/documents/library/gwo_e.pdf

Roziqin, A. (2018). Environmental policy of mangroves management in Rembang Regency. In *E3S Web of Conferences* (Vol. 31, p. 09002). EDP Sciences., DOI: 10.1051/e3sconf/20183109002

Saenger, P. (2002). *Mangrove Ecology, Silviculture and Conservation*. Kluwer Academic Publishers., DOI: 10.1007/978-94-015-9962-7

Sanaullah, S., Yang, D., Zhong, R., Zhao, L., Shafi, M., & Akbar, A. J. (2025). Mangrove dynamics in Pakistan: A long-term study of coastal ecosystem shifts over more than three decades. *Ecological Indicators*, *174*, 113452. DOI: 10.1016/j.ecolind.2025.113452

Spalding, M., Kainuma, M., & Collins, L. (2010). *World atlas of mangroves*. Earthscan., DOI: 10.4324/9781849776608

Tomlinson, P. B. (2016). *The botany of mangroves* (2nd ed.). Cambridge University Press., DOI: 10.1017/CBO9781139946575

UNESCO. (2024). The Sundarbans World Heritage Site. Paris: United Nations Educational, Scientific and Cultural Organization. Retrieved on 02 June 2026, from https://whc.unesco.org/en/list/798/

UNFCCC. (2015). *Paris Agreement*. United Nations. Retrieved on 02 June 2026, from https://unfccc.int/sites/default/files/resource/parisagreement_publication.pdf

UNSDG. (2015). The 17 Goals. Department of Economic and Social Affairs. Retrieved on 01 June 2026, from https://sdgs.un.org/goals

Walters, B. B., Ronnback, P., Kovacs, J. M., Crona, B., Hussain, S. A., Badola, R., Primavera, J. H., Barbier, E., & Dahdouh-Guebas, F. (2008). Ethnobiology, socio-economics and management of mangrove forests: A review. *Aquatic Botany*, *89*(2), 220–236. DOI: 10.1016/j.aquabot.2008.02.009

Wickramasinghe, S., Wijayasinghe, M., & Sarathchandra, C. (2022). Sri Lankan Mangroves: Biodiversity, Livelihoods, and Conservation. In: Das, S.C., Pullaiah, Ashton, E.C. (eds) *Mangroves: Biodiversity, Livelihoods and Conservation* (pp. 297-329). Springer, Singapore. https://doi.org/DOI: 10.1007/978-981-19-0519-3_13

Woodroffe, C. D. (1992). Mangrove sediments and geomorphology. In Robertson, A. I., & Alongi, D. M. (Eds.), *Tropical Mangrove Ecosystems* (pp. 7–41). American Geophysical Union., DOI: 10.1029/CE041p0007

World Wide Fund for Nature (WWF-Pakistan). (2021). Mangrove Restoration and Coastal Resilience in Pakistan. Karachi: WWF-Pakistan. Retrieved on 30 May 2026, from https://wwfasia.awsassets.panda.org/downloads/small_final_annual_report_2021__3_.pdf \

World Wildlife Fund Pakistan (WWF-Pakistan). (2020). Mangrove Restoration and Coastal Resilience in Pakistan. Karachi: WWF-Pakistan. Retrieved on 30 May 2026, from https://wwfasia.awsassets.panda.org/downloads/wwf_annual_report_2020.pdf

Chapter 2
Global Mangroves Management as Sustainable Socioeconomic Systems: Biotechnology, Climate Science, and Digital Innovations Needs Policy Integration

Srushti Kumaragouda Patil
http://orcid.org/0009-0001-8636-1258
Sri Sathya Sai University for Human Excellence, India

Rakshitha N. Ramesh
Sri Sathya Sai University for Human Excellence, India

Gowchenahalli Kavya Krishnappa
Sri Sathya Sai University for Human Excellence, India

Ramya Raghavan
http://orcid.org/0000-0002-9953-543X
Sri Sathya Sai University for Human Excellence, India

ABSTRACT

Mangrove ecosystems are essential for global carbon storage and for maintaining biodiversity and coastal resilience. Genomic and microbial engineering approaches, along with blue carbon modeling and AI-driven remote sensing technologies, are working toward predictive, data-driven, and adaptive ecosystem management

DOI: 10.4018/979-8-3373-9320-9.ch002

frameworks. Emerging innovations enable the identification of stress-resilient genotypes, the enhancement of restoration success through engineered microbiomes, and the real-time monitoring of ecosystem dynamics across spatial and temporal scales. Mangrove-associated microbes produce a variety of bioactive compounds, antibiotics, antioxidants, and antitumor agents. Mangrove forests need policy integration to function as socio-ecological systems for future bioprospecting and biotechnological applications. Despite these advances, significant gaps persist in field-scale validation, policy integration, and policy translation, limiting scalability and equitable implementation toward sustainable socioeconomic systems.

INTRODUCTION

Mangrove ecosystems are critical interfaces between terrestrial and marine environments, providing ecological and socio-economic value. Wetlands, both saline and freshwater, play a role in carbon storage and can act as carbon sequestration sinks. Global evidence on the interconnections among carbon storage, biodiversity, and coastal resilience in mangrove ecosystems shows that they function as major carbon sinks, coastal buffers, and biodiversity reservoirs. Despite recognized global importance, ecosystems face compounding pressures from anthropogenic development, localized land-use changes, and severe climatic events, which threaten to destabilize structural integrity and disrupt long-term carbon storage capacity. Mangroves also remain vulnerable to climate change, anthropogenic disturbance, and fragmented governance. Novel solutions are required to realize the climate mitigation potential of conservation under climate change and stress factors. Biotechnology, climate science, and digital innovation are transforming research and conservation of mangrove ecosystems. (Reznikova et al., 2026) Genomic and microbial tools are enhancing the precision and resilience of restoration, while climate science is quantifying blue carbon and guiding adaptation. Digital technologies are enabling real-time, high-resolution monitoring and participatory management. However, challenges remain in scaling up innovations, integrating socio-economic valuation, and ensuring equitable governance. (Kumar et al., 2026) Mangrove ecosystems need novel solutions to maintain coastal resilience, contributing to carbon sequestration, biodiversity conservation, and socio-economic sustainability. The novelty lies in cross-disciplinary conceptual framing of mangrove ecosystems as bio-digital-climate systems. Rather than treating biotechnology, climate science, and digital innovation as parallel domains, the present synthesis highlights functional convergence and interdependence. Methodologically, it emphasizes integrating multi-omics approaches with AI-driven ecological modeling and remote sensing technologies. From a translational perspective, it underscores the emergence of real-world applications such as drone-assisted restoration, synthetic microbial engineering, and real-time ecological monitoring systems. The one health framework represents a shift toward adaptive, feedback-driven ecosystem management, where biological processes and digital intelligence co-evolve. Future research should focus on field-scale deployment of biotechnologies and on data integration for cross-sectoral collaboration to maximize the ecological and societal benefits of mangrove conservation.

Mangrove ecosystems represent one of Earth's most vital lines of defense against global environmental challenges at the critical nexus of climate science, ecological preservation, and coastal resilience. The dynamic intertidal zones of tropical and subtropical coastlines, with their unique halophytic communities, function as efficient blue carbon sinks. Blue carbon sequester organic carbon within waterlogged sediments for carbon storage, at rates that outpace terrestrial forests. (Das et al., 2026) Current literature extensively documents the baseline capacity of mangroves to store carbon and buffer coastlines, but research gaps remain regarding the precise operational dynamics of ecosystems under climate change stressors. There is a lack of clarity surrounding how localized estuarine gradients, seasonal environmental fluxes, and distinct land-use transitions alter biomass distribution and accelerate carbon release through soil greenhouse gas emissions. Functional microbial mechanisms in a stressed mangrove system acts as a carbon sink or a greenhouse gas source remain largely unmapped. Biotechnologists need to engage with carbon accounting standards, the digital governance of biosafety regulations, and community rights and environmental justice. Likewise, readiness assessments should systematically distinguish between technologies that are already validated at scale to make proportionate governance decisions. Most coastal mangrove forests recover all lost carbon within four years, suggesting ecosystems continue to function as effective carbon sinks. (Ananthakrishnan et al., 2026)

Mangroves provide essential structural and ecological services, mitigating the destructive energy of storm surges, stabilizing coastal geomorphology against erosion, and sustaining marine biodiversity. (Wijerathna et al., 2026) Conservation frameworks frequently fail to bridge fine-scale ecological shifts with macro-level policy targets, leaving a critical disconnect between localized ecosystem management and global sustainability initiatives. However, macro-debris and microplastic pollution in the mangrove ecosystems of Negombo and Rekawa, Sri Lanka, were found to have aerial roots in Negombo, indicating that roots act as debris traps. (Gajre et al., 2025) Recent interdisciplinary research integrating biotechnology, climate science, and digital innovation is transforming mangrove research from descriptive ecology to predictive, adaptive domains. Intervention-driven science also highlights that transboundary conservation efforts are more effective at safeguarding climate-resilient mangroves than individual national plans. (Dabalà et al., 2026) The present work examines how biotechnology, climate science, and digital innovation are revolutionizing research and conservation of mangrove ecosystems. In genomics, microbial engineering, remote sensing, AI, and integrated policy frameworks are enabling more precise, adaptive, and scalable approaches to mangrove restoration and monitoring, with ecological, climate, and socio-economic benefits.

Figure 1. Key research gaps in mangrove innovation science

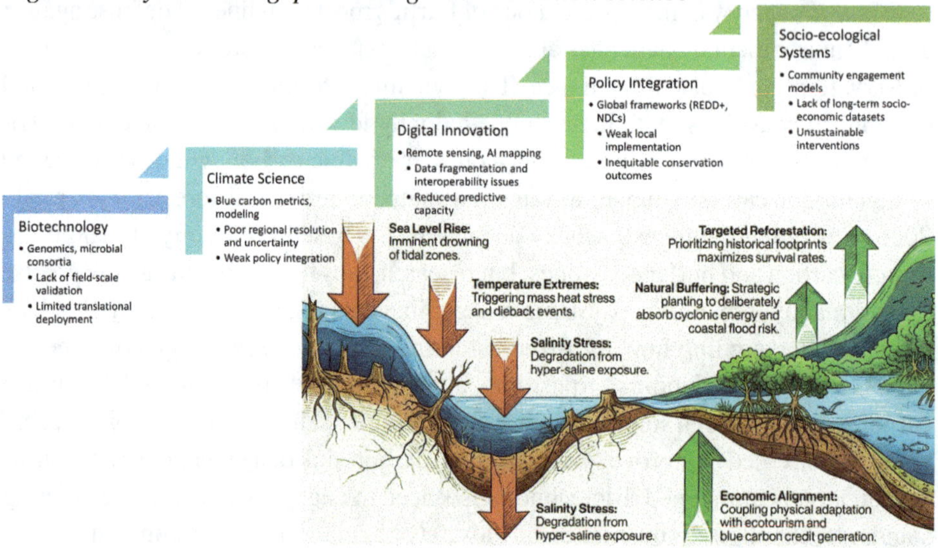

METHODOLOGY

A comprehensive dataset on global blue carbon storage, microbial metagenomics, and technological advancements in mangrove management was compiled through a systematic search across major electronic databases: Scopus, Web of Science, PubMed, and Google Scholar. The search strategy was bounded by publication dates spanning 2025 to 2026 to capture the most recent advancements in high-resolution satellite imagery, multi-omics soil sequencing, and national targets under modern biodiversity frameworks. A narrative synthesis approach, combined with structured tabular mapping, was used to evaluate the compiled data. To reconcile data across varying spatial scales, coastal zones were contextualized within broader global synthesis models. The macro-level policy frameworks, such as the Kunming–Montreal biodiversity targets, are discussed to provide a comprehensive assessment of how modern digital innovations and biotechnological insights address ongoing mangrove degradation challenges worldwide.

BIOTECHNOLOGICAL FRONTIERS

Biotech solutions are enhancing the resilience, restoration success, and ecosystem services of mangrove forests. The genomic studies have identified adaptive traits and genetic diversity across mangrove species, informing targeted breeding and restoration strategies. Whole-genome and chloroplast sequencing are being used to select source populations with enhanced tolerance to salinity and drought and to support a One Health framework for both in situ and ex situ conservation. The greenhouse gas fluxes of saline and freshwater wetlands in Southern Florida, focusing on carbon dioxide (CO2) and methane (CH4) emissions and their variability in response to climate change and water management. The study finds that regional net annual CO2 equivalent uptake in wetlands increased by 18% from around 2003 to 2020. However, approximately 43% of CO2 uptake was offset by CH4 emissions, highlighting the complex dynamics of ecosystems as climate solutions. (Doughty et al., 2026) Key locations for mangrove restoration via natural regeneration include the Amazonia ecoregion, the Tropical Northwestern Atlantic, Indonesia, and tidal systems, providing essential data for effective mangrove restoration strategies and global policy planning.

Biotechnological precision in mangrove conservation and restoration. The rhizosphere is the narrow zone of soil surrounding mangrove roots. Traditional restoration fails because seedlings are planted without native soil bacteria. The microbial-mediated sulfur transformation pathways in mangrove wetlands reveal differences in microbial community composition and sulfur-cycling gene abundance between mangrove and mudflat sediments. The research utilizes publicly available 16S rRNA amplicon datasets and co-occurrence network analysis to identify positively correlated genera and predicted functional roles within the microbial communities. The increase in total ecosystem carbon stocks is primarily attributable to higher soil organic carbon levels resulting from hydrological rehabilitation and reforestation efforts. The ongoing monitoring to support climate mitigation and coastal mangrove research is moving from traditional ecology to high-tech systems science. DNA sequencing technologies, 16S rRNA gene sequencing, 18S-ITS-28S sequencing, and shotgun sequencing are essential for understanding the structure and function of soil microbiomes in ecosystems. (Reznikova et al., 2026) Modern sequencing methods enhance the identification of microbial community diversity and dynamics, thereby aiding the development of sustainable agricultural practices, improving soil fertility, managing plant diseases, and increasing crop yields. The current frontier integrates genomics to engineer resilience, geophysics to map soil carbon at depth, and digital twins to simulate future conservation scenarios. The rhizosphere benefits from this convergence, helping protect existing forests and future-proof restoration efforts against rising sea levels and salinity. (Das et al., 2026)

Rhizosphere

The rhizosphere is one of the most biogeochemically active microenvironments in coastal ecosystems. Steep oxygen gradients, high organic inputs from root exudates, and dynamic interactions between plant, sediment, and microbial communities characterize it. The mangrove rhizosphere mediates processes fundamental to ecosystem function: nitrogen fixation, phosphorus mineralization, sulfur cycling, and organic matter decomposition. Recent metagenomics and stable-isotope probing studies have revealed extraordinary microbial diversity in rhizosphere zones. (Ariadi et al., 2026) Genomics research is testing interspecies variation in buoyancy and salinity responses. For example, genetic diversity and kinship relationships among black mangrove populations *of Rhizophora mucronata in West Sulawesi were assessed* using morphological markers, highlighting the importance of ecosystems for local communities. High genetic diversity was found among black mangrove populations in Polewali Mandar, Majene, and Mamuju, with genetic distances indicating varying levels of kinship among Polewali Mandar and Mamuju. (Wallacea, 2025) Targeted strategies hope to preserve the genetic diversity of *R. mucronata* populations. The application of desulfurization gypsum and cattle manure increased the abundance of nitrogen- and phosphorus-cycling genes. Soil potassium and phosphorus levels influenced microbial communities and nutrient-cycling genes. Biotechnological case study findings provide a scientific foundation for global restoration initiatives, ensuring that conservation programs enhance genetic resilience and adaptability within the mangrove ecosystem.

Microbial Consortia

Genomic tools, whole-genome sequencing, and comparative phylogenetics are enabling the identification of adaptive traits associated with salinity tolerance, thermal resilience, and stress-response pathways. (Mahalakshmi et al., 2026) Insights are facilitating the selection of climate-resilient genotypes for restoration programs. At the same time, microbial biotechnology is gaining prominence through the application of plant growth-promoting bacteria, mycorrhizal fungi, and endophytic consortia that enhance nutrient cycling, stress tolerance, and pollutant degradation. Synthetic microbial consortia assemblages of multiple, functionally complementary microbial strains, plant growth-promoting rhizobacteria (PGPR), arbuscular mycorrhizal fungi, and nitrogen fixers. Experimental biotechnology for mangrove restoration, greenhouse and mesocosm trials in Indonesia, Bangladesh, and Mexico, has demonstrated statistically significant improvements in propagule survival. Indian Ocean case studies show early biomass accumulation following inoculation with consortia containing *Rhizobium, Azospirillum, and Pseudomonas* species. (Wijerathna et al., 2026) Environmental gradients influence microbial functional stratification and sulfur metabolism. Impact of mangrove degradation on greenhouse gas emissions, mainly methane (CH4) and nitrous oxide (N2O), by comparing degraded mangroves and adjacent rice systems in Sundarban, India. Whole-genome metagenomic analysis linked annual CH4 and N2O emissions to soil microbial energy metabolism pathways, methanogenesis, methanotrophy, and sulfur and nitrogen metabolism. The acetoclastic pathway was identified as the dominant pathway for CH4 production in rice systems. At the same time, methane oxidation was more pronounced in degraded mangroves, indicating microbial adaptation to varying salinity and alkalinity conditions. Targeted microbial management in mangrove-rice ecologies could be an effective strategy for mitigating climate change impacts in coastal environments. (Nayak, Bhattacharyya, et al., 2025)

Table 1. Microbial consortia are being engineered to improve mangrove establishment and resilience, especially in degraded or polluted sites

Microbial Species	Pollutant Targeted	Application
Alcanivoracaceae, Rhodospirillaceae	Hydrocarbons (oil)	Oil bioremediation in refinery-impacted mangroves
Actinomycetes (Streptomyces)	Heavy metals, organics	Bioremediation, soil health improvement
Gammaproteobacteria, Deltaproteobacteria	Detoxification processes	Soil detoxification, resilience
Bacillus, Pseudomonas	Nitrogen fixation, hormone production	Enhanced growth, stress resilience
Actinobacteria (Streptomyces)	Biocontrol, pollutant degradation	Bioremediation, disease suppression
Mycorrhizal fungi	Nutrient uptake, salt/drought tolerance	Restoration success, plant establishment
Alcanivoracaceae, Rhodospirillaceae	Hydrocarbons (oil)	Oil bioremediation in refinery-impacted mangroves
Actinomycetes (Streptomyces)	Heavy metals, organics	Bioremediation, soil health improvement
Gammaproteobacteria, Deltaproteobacteria	Detoxification processes	Soil detoxification, resilience

Assisted gene flow (AGF)

Mangrove forests exhibit high resilience to hurricanes, owing to species' evolutionary adaptations that enable them to recover from natural disturbances. Assisted gene flow (AGF) allows deliberate translocation of propagules or genetic material from populations already adapted to projected future climate conditions into populations at risk of maladaptation. AGF is receiving increasing attention as a nature-based adaptation tool. In mangrove restoration contexts, AGF involves sourcing propagules from populations in warmer, more saline, or more drought-prone environments and deploying them in restoration sites expected to experience analogous stressors as climate change progresses. (Ariadi et al., 2026) Trials in Australia and the United States have improved thermal tolerance in AGF-sourced *Avicennia marina and Rhizophora mangle* seedlings. The AGF approach is considered closer to conventional horticultural restoration than to genetic modification and thus faces lower regulatory barriers. Biotechnology is unlocking new frontiers in mangrove conservation, from genetic research and microbial engineering to bioengineering and bioremediation. (Reznikova et al., 2026) The biotechnology study highlights the environmental significance of *actinomycetes* in mangrove soil sediments, emphasizing roles in pollutant biodegradation and nutrient recycling. *Microbacterium alkaliflavum* sp. nov. was derived from mangrove sediments, based on low genetic similarity to existing species. The strain exhibits weak cytotoxicity against nasopharyngeal carcinoma cell lines and harbors gene clusters involved in secondary metabolite biosynthesis. (Byalahunasishetter et al., 2026) The genus *Micromonospora* (Ørskov, 1923) is a Gram-positive, chemoorganotrophic, aerobic actinomycete, the second most abundant genus in the phylum *Actinomycota* in natural environments. The actinomycetes isolated from Bichitrapur mangrove sediments show potential to produce antimicrobial compounds. It also explores the potential to produce novel bioactive compounds, antibiotics, and antioxidants. The superior pharmaceutical potential of *Micromonospora* sp. compared to Streptomyces sp. could be valuable for bioprospecting and biotechnological applications. (Pradhan et al., 2026) Sulfur cycling in mangrove wetlands, on microbial processes and gene-level patterns, responds to environmental gradients. Differences in microbial community composition were observed, with *Escherichia* dominating mangrove sediments and *Salmonella* prevailing in mudflat sediments, indicating spatial variation in microbial populations. The abundance of sulfur-cycling genes decreased with depth, and the distribution of functional genes was influenced by biogeochemical conditions, with oxidative genes more abundant in mudflat profiles and reductive pathways dominating in mangrove sediments—indicating the need to manipulate genetic and microbial pathways for enhanced survival. Six distinct colonies were isolated, and the identities of two strains, *Micromonospora* sp. (BSS-D-04) and Streptomyces sp. (BSS-D-05), were confirmed through genetic sequencing and submitted to the NCBI database. (Pradhan & Das, 2025)

Environmental factors, such as salinity, pH, total nitrogen, and total organic carbon, were negatively correlated with microbial community structure. In contrast, total sulfur, total phosphorus, and sulfate had positive effects on environmental gradients on microbial functional stratification and sulfur metabolic pathways. (Nayak, Bhattacharyya, et al., 2025) Biotechnological interventions are critical for scaling up restoration in challenging environments.

Synthetic Biology

Single-cell omics approaches in synthetic biology, engineered microbial consortia, and hybrid engineered living materials are beginning to offer enhanced stress-tolerance traits in mangrove species. Salinity resistance and heat tolerance enable accelerated pneumatophore development. Synthetic biology approaches envision engineering novel microbial strains for ecological functions. (Korrapati et al., 2026) Bioengineering accelerates sediment accretion by producing extracellular polymeric substances. Dynamic, responsive systems capable of both environmental sensing and remediation. Genome editing technologies such as CRISPR further extend the frontier by enabling targeted manipulation of stress-response pathways, although ecological deployment remains constrained by ethical and regulatory considerations. (Pallavi et al., 2026) Saline-alkali soils exhibit high salinity, low fertility, and poor structure, leading to slow nutrient cycling and reduced microbial activity. Mangrove-associated microbes, plant growth-promoting bacteria (PGPB), actinobacteria, mycorrhizal fungi, and endophytes play vital roles in nutrient cycling, stress tolerance, and bioremediation. The application of different amendments enhances soil nutrient levels and affects nitrogen and phosphorus cycling processes. The cycling and transformation of biogenic elements (C, N, S, Fe) in the stress-response genes. Rhizosphere microbial communities' nitrogen-fixing bacteria, phosphate-solubilizing microorganisms, and sulfur-cycling archaea mediate sediment geochemistry, nutrient cycling, and seedling establishment. Synthetic microbial consortia, engineered to stabilize anaerobic sediments or improve propagule survival under salinity stress, represent an experimental frontier with ecological promise but equally biosafety uncertainties. CRISPR-based gene editing to confer climate stress tolerance in mangrove propagules remains largely speculative, yet it is already attracting research investment. (Ramesh et al., 2026) Bioengineering integrates living plants, engineered microbes, and digital monitoring to stabilize shorelines and enhance ecosystem resilience. AmSOD1 for oxidative stress or AmFer1 for iron storage. Heterotrophic reactions, such as aerobic respiration, denitrification, and nitrification, were dominant in high-flow regions, with microbial reactions contributing to nitrate removal. It paves the way for CRISPR-based editing or marker-assisted selection to breed super-mangroves capable of surviving in hypersaline waters resulting from sea-level rise and aquifer intrusion. Changes in vertical permeability, influenced by mangrove root systems and benthic organisms, affect nutrient cycling by promoting solute exchange and nitrate removal. Next-generation technologies are used in soil microbiome research, highlighting their role in understanding microbial diversity and dynamics, thereby emphasizing the importance of mangrove wetlands in maintaining ecological stability and nutrient cycling. (Reznikova et al., 2026)

Table 2. Biostimulation and bioaugmentation strategies using microbes are improving pollutant degradation and soil detoxification, making restoration feasible in previously inhospitable sites

Technique	Target Species	Application
CRISPR/Cas9	Mangroves, trees	Gene function validation, stress tolerance
Genomic barcoding	Engineered organisms	Traceability, biosafety in restoration
Whole-genome sequencing	Multiple mangrove spp.	Adaptive trait identification, breeding
Approach/Technology	Application/Function	Outcome/Benefit
HELM *Bacillus subtilis* spores	In-situ sensing, pollutant remediation	Enhanced shoreline resilience
Synthetic microbial consortia	Outcompete native populations, remediation	Improved restoration success
IoT/AI integration	Environmental monitoring, data analysis	Adaptive management, predictive modeling

Biotechnology Ethics, Policy and Governance

Scientists are developing bio-inoculants or synthetic consortia of nitrogen-fixing and phosphate-solubilizing bacteria. Vibrio and *Marinomonas* are applied to seedlings in nurseries. A probiotic approach has been shown to increase seedling survival rates by mitigating salt stress and heavy metal toxicity, including in the Sundarban, India, as evidenced by comparisons between degraded mangroves and adjacent rice systems. The impact of mangrove degradation on greenhouse gas emissions, mainly methane and nitrous oxide. Genomic decoding for stress tolerance 2021-2022: researchers fully decoded the genome of Bruguiera gymnorhiza, revealing that jumping genes (transposons) are activated under high-salinity stress. Metagenomic analysis identified key microbial pathways involved in emissions, revealing that rice systems have a higher potential for methane production. At the same time, degraded mangroves exhibit greater methane oxidation capacity. (Nayak, Bhattacharyya, et al., 2025)

The deployment of biotechnology in mangrove ecosystems is shifting from inoculating non-native microbial strains to the hypothetical release of CRISPR-edited organisms, requiring engagement with a layered regulatory landscape. At the international level, the Cartagena Protocol on Biosafety establishes obligations for risk assessment and management of living modified organisms (LMOs). (Jangama et al., 2026). Many mangrove-rich countries are parties to the international protocol, but national biosafety authority capacity varies widely, and regulatory guidance for marine and coastal ecosystems remains underdeveloped. Beyond regulatory compliance, the ethical dimensions of biotechnology deployment in community-managed mangroves are substantial. Many coastal indigenous and customary tenure-holding communities rely on traditional ecological knowledge (TEK) of mangrove systems and bear disproportionate risks from ecological mismanagement. The principle of Free, Prior and Informed Consent (FPIC), embedded in the UN Declaration on the Rights of Indigenous Peoples (UNDRIP) and the Convention on Biological Diversity's Nagoya Protocol, establishes communities' rights to approve, modify, or reject interventions on lands and waters. Rights apply to biotechnological interventions no less than to infrastructure development.

- Risk of competitive displacement of native microbial communities by introduced consortia
- Potential alteration of biogeochemical cycles with unintended consequences for carbon flux
- Intellectual property claims over engineered organisms that may restrict community access
- Inadequate post-deployment monitoring and accountability mechanisms
- Insufficient representation of community knowledge in research design and risk assessment

CLIMATE SCIENCE AND ENVIRONMENT

Climate science provides the quantitative and predictive backbone for understanding mangrove ecosystem dynamics under global change. The concept of blue carbon has elevated mangroves to a central role in climate mitigation strategies, supported by robust estimates of soil organic carbon stocks and sequestration rates. However, climate stressors such as sea-level rise, temperature extremes, salinity fluctuations, and increased storm intensity continue to threaten mangrove stability. Advanced climate models and vulnerability assessments are increasingly used to identify high-risk regions and inform adaptive management strategies. (Girkin et al., 2026) Ecosystem-based adaptation approaches, those emphasizing reforestation over afforestation, have demonstrated superior outcomes in carbon sequestration and ecological resilience. Nevertheless, uncertainties in regional climate projections and limited integration of ecological complexity into models remain challenges. Emerging research examines how mangrove propagule traits influence dispersal potential, focusing on buoyancy and morphological attributes of 13 mangrove species in the central Great Barrier Reef region of Queensland, Australia. Over 90 days, the researchers measured the buoyancy of mangrove propagules. They tracked changes in root and stem growth under varying salinity conditions, revealing interspecies variation in buoyancy and salinity responses. (Thompson-Saud et al., 2025) It emphasizes the importance of microbial community structure and composition for the health and productivity of mangrove ecosystems, which are known for rich biodiversity.

Figure 2. Emerging interdisciplinary innovations

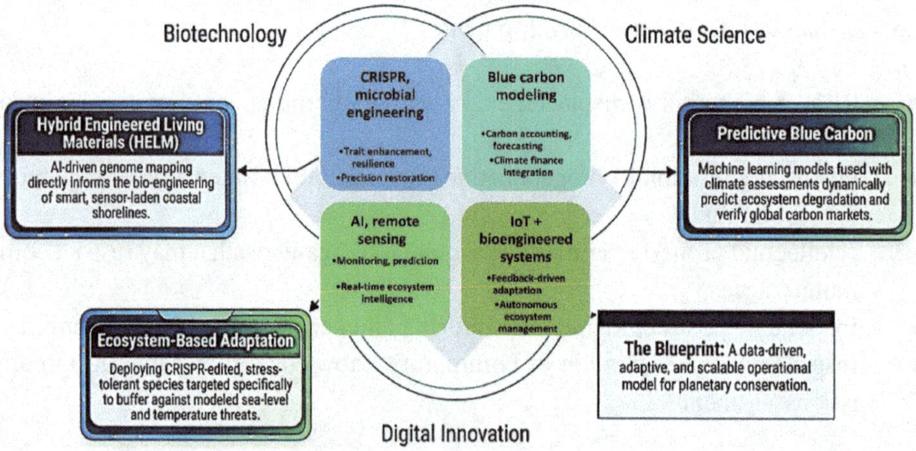

BLUE CARBON

Hurricanes disrupt the carbon cycle by releasing carbon back into the atmosphere, affecting ecosystems at local and regional scales through impacts on vegetation and agricultural areas. The carbon flux interactions during major hurricanes alter net ecosystem exchange (NEE) data from the SMAP satellite and in-situ CO2 flux observations. (Ray et al., 2026) Agriculture 5.0 enhances operational efficiency and environmental sustainability through tools like IoT, AI, and drones. It identifies benefits, such as improved resource use and yield stability, while also addressing challenges like high costs and data privacy concerns. (Kumar et al., 2026) Southern Texas experienced the highest carbon release (0.33 megatons) during Hurricane

Harvey, followed by Florida (0.03 megatons) during Hurricane Irma and Puerto Rico (0.02 megatons) during Hurricane Maria. Land cover reductions were observed across all three regions, with changes in vegetation cover indicating short-term stress. They reduced photosynthetic activity, leading to a temporary transition from carbon sinks to carbon sources in the affected areas. (Ray et al., 2026)

The Verified Carbon Standard (VCS) and Gold Standard now include mangrove methodologies, and a growing number of coastal nations are integrating blue carbon into Nationally Determined Contributions (NDCs). The emergence of high-integrity voluntary carbon markets and Article 6 mechanisms under the Paris Agreement opens new financing pathways for restoration. Beyond Biomass, the research focus has shifted from counting trees to understanding the complex biogeochemistry of soil and water. The Soil Carbon studies highlight that up to 80% of mangrove carbon is stored not in the trees, but in the anaerobic soil layers up to 3 meters deep. The dynamics and drivers of soil organic carbon (SOC) in natural mangroves, restored mangroves, and tidal flats. Natural mangroves have the highest SOC levels and carbon sequestration potential. (Slooten et al., 2026) Restored mangroves have distinct ecological drivers of SOC across ecosystems, which require nutrient and salinity management in mangrove restoration and the inclusion of tidal flats in carbon management strategies. (Nayak, Das et al., 2025)

Lateral Flux carbon washed out of mangroves into the adjacent ocean (outwelling). New models suggest lateral export may account for a portion of the ocean's refractory dissolved organic carbon (DOC), effectively doubling the perceived carbon value of mangroves. Methane-Carbon Trade-offs. As mangroves sequester CO2, degraded or freshwater-impacted mangroves can emit methane (CH4). Precision monitoring is now required to ensure that carbon credit projects are not inadvertently funding methane emitters.

Environmental Justice and Benefit-Sharing in Blue Carbon Projects

Perhaps the most consequential governance challenge in blue carbon finance is the equitable distribution of revenues and the protection of community tenure rights. Carbon projects on community-managed mangroves have, in several cases, been structured as top-down commercial ventures in which external project developers capture the majority of carbon revenues while communities receive only marginal shares. The pattern reproduces longstanding dynamics of conservation-induced displacement and green grabbing. Best practice, as articulated in frameworks such as the Global Alliance for Community Forestry's guidelines and the Equator Principles, requires (1) communities' FPIC before project initiation; (2) transparent benefit-sharing agreements negotiated on equitable terms; (3) community participation in governance structures; (4) legal recognition of customary tenure; and (5) independent grievance mechanisms. Principles are not mere aspirations — they are conditions for the social legitimacy and long-term efficacy of blue carbon projects.

DIGITAL TOOLS FOR MANGROVE ECOSYSTEM

Digital innovation is rapidly reshaping mangrove monitoring and management through the integration of remote sensing, artificial intelligence, and cloud-based data platforms. Deep learning architectures, including convolutional neural networks (CNNs) and transformer models, have rapidly become central to mangrove remote sensing. Applied to multi-spectral, hyperspectral, and SAR imagery, algorithms can classify mangrove species assemblages, estimate canopy height and above-ground biomass, and detect degradation events. Hyperspectral imagery offers particular promise for species discrimination and the detection of stress indicators (chlorophyll fluorescence, water content, phenolic compounds) invisible to conventional multi-spectral sensors. Airborne and increasingly spaceborne hyperspectral platforms are now operationally deployed for research. However, the computational demands of hyperspectral processing and the cost of airborne data acquisition remain barriers to routine national monitoring in lower-income countries, such as die-back from hypersalinity or pest damage, and mapping pneumatophore density as a proxy for ecosystem health. (Lakshmaiah & Raghavan, 2026) Classification accuracies in peer-reviewed studies have consistently exceeded 85 percent, and some approaches using LiDAR-optical fusion have achieved plot-scale above-ground biomass estimation errors of less than 15 percent. (Ray et al., 2026) Satellite systems such as Landsat and Sentinel, combined with LiDAR and hyperspectral imaging, provide high-resolution, multi-temporal data on mangrove extent, structure, and health. Machine learning algorithms, random forests, convolutional neural networks, and gradient boosting models enable automated classification, biomass estimation, and early detection of ecosystem degradation.

Global Mangrove Watch (GMW)

The Global Mangrove Watch (GMW) is a JAXA-led initiative providing global mangrove extent maps at 25-meter resolution. IPCC Wetlands Supplement methodologies provide the technical bridge between national inventory reporting and international climate commitments. GMW enables near-real-time monitoring of changes in mangrove cover at national and sub-national scales. Machine learning algorithms applied to multi-spectral and synthetic aperture radar imagery can distinguish mangrove species assemblages, estimate above-ground biomass, and detect degradation events within weeks of occurrence.

Developed through a partnership led by JAXA and key contributors Aberystwyth University, soloEO, Cinvestav, and USGS, the GMW produces global mangrove extent maps at approximately 25-meter spatial resolution using L-band Synthetic Aperture Radar (SAR) imagery. (Doughty et al., 2026) The dataset provides consistent annual maps from 1996 to the present, enabling change detection at national and sub-national scales. (Li et al., 2026) GMW data underpin mangrove area estimates in global assessments, the FAO Global Forest Resources Assessment, the Global Mangrove Alliance, and IUCN Species Survival Assessments. Critically, the dataset is openly accessible, supporting capacity building in data-constrained nations. The GMW's integration with IPCC Wetlands Supplement methodologies provides the spatial foundation for country-level blue carbon MRV. Complementary data sources airborne LiDAR, hyperspectral imagery, and field inventory are required for higher-tier MRV. (Kumar et al., 2026) The international community is actively working to improve GMW resolution and integrate multi-source data fusion approaches and machine-learning classification of complementary optical imagery.

The Digital Mangrove Projects

A digital twin would integrate satellite imagery, drone surveys, IoT sediment and water quality sensors, microbial community data, climate projections, and hydrological models to simulate ecosystem dynamics and project future states. AI, Remote Sensing, and Digital Twins will drive growth in mangrove research, enabled by the application of Digital Earth technologies, enabling monitoring at scales and resolutions previously impossible. GMW technologies are complemented by platforms such as Google Earth Engine and Global Mangrove Watch, which democratize access to large-scale geospatial data and analytical tools. (Ray et al., 2026) The emergence of citizen science applications further enhances data granularity and community engagement, although issues related to data standardization and interoperability persist. Digital Twins of Ecosystems is a virtual replica of a physical mangrove forest, updated in real-time via IoT sensors (salinity, water level, temperature). Projects such as ManglarIA (Mexico) and various academic pilots enable researchers to run simulations. ManglarIA allows for predictive restoration rather than reactive conservation. (Li et al., 2026) Pilot digital twin frameworks for mangrove systems have been prototyped in Australia, the Netherlands (for deltaic systems broadly), and Indonesia, typically coupling SAR-based extent maps with tidal hydrodynamic models and vegetation growth algorithms. Prototypes demonstrate the architectural feasibility of the approach but reveal substantial challenges.

The GDPR-analogous principles help data minimization, purpose limitation, and the right to explanation.

Table 3. AI and machine learning automate and enhance mangrove monitoring, enabling early warning systems and predictive modeling for proactive conservation

Technology	Application	Outcome
CART Decision Tree	Mangrove classification	Highest accuracy among tested methods
Random Forest, GBM, NN	Extent mapping	overall accuracy
Deep Learning (CNN)	Species identification	Automated, high-accuracy ID
XGBoost	Mapping & monitoring	High classification accuracy
Early Warning ML System	Degradation risk prediction	Early detection of decline
Biomass Estimation DL	Blue carbon stock estimation	Accurate carbon stock mapping

GLOBAL CASE STUDIES ON MANGROVE SOCIO- ECOLOGICAL MODELS

The introductory framework prioritizes regions characterized by high blue carbon potential, critical conservation vulnerability, and severe anthropogenic pressures. By centering the operational analysis on empirical datasets from the Sundarban estuarine gradients in India, Pekalongan and West Sulawesi in Indonesia, and Bac Lieu in Vietnam, the research evaluates regions widely recognized as global priority zones for climate mitigation and habitat recovery. (Das et al., 2026) Mangroves provide livelihoods for coastal communities across South and Southeast Asia, West Africa, and the Caribbean. They harbor extraordinary biodiversity, numerous endemic and threatened species. The Ramsar Convention, the Convention on Biological Diversity's Kunming-Montreal Global Biodiversity Framework (2022), and the UNFCCC Paris Agreement recognize mangrove restoration as a priority action. The degradation of mangroves, therefore, represents climate mitigation, biodiversity, coastal protection, and human wellbeing. The carbon sequestration in mangroves and other blue carbon habitats highlights biogeomorphic feedbacks in carbon burial processes. Revenue from carbon credits is reinvested into community governance and monitoring, ensuring long-term stewardship. Species Protection is critical for the threatened American Crocodile (*Crocodylus acutus*).

The global case studies illustrate the tangible benefits of interdisciplinary innovation, from large-scale restoration and carbon finance to enhanced resilience and community livelihoods. Mangrove ecosystems along the Indian Ocean coast are biodiverse and host a variety of microbial communities, including endophytic fungi which are for producing unique secondary metabolites. The structural diversity and pharmacological activities of 302 secondary metabolites derived from mangrove-associated fungi are primarily classified as polyketides, alkaloids, and terpenoids. A total of 164 compounds were identified, demonstrating diverse biological activities: cytotoxic, antibacterial, antifungal, antioxidant, enzyme-inhibitory, and anti-inflammatory.(Pradhan et al., 2026) Mangrove fungi are valuable sources for new drug discovery and industrial biotechnology, and further exploration along the Indian coast is necessary for novel discoveries. (Wijerathna et al., 2026)The integrative framework, therefore, demands that governance be embedded at each pillar junction, rather than applied as a post hoc oversight layer. Technology developers, carbon project designers, and remote sensing practitioners must co-design governance structures with the communities whose ecosystems and livelihoods are at stake. Environmental justice, equitable distribution of benefits, burdens, and decision-making power, is a normative foundation of the framework. Ethical, legal, and social implications (ELSI) of environmental surveillance data are an emerging area of legal scholarship with direct relevance to mangrove monitoring governance.

Can Gio Biosphere Reserve, Vietnam

Post-War restoration and urban green lung reflect the high biomass density and deep sedimentary profiles of the chosen Indian Ocean. Southeast Asian settings provide an ideal baseline for evaluating deep-soil organic carbon stocks and baseline blue carbon capacities. Locations such as Bac Lieu, Vietnam, directly represent the severe land-use conflicts outlined in the introduction, specifically detailing how commercial expansions like integrated aquaculture can drive localized degradation. Destroyed by herbicides (Agent Orange) during the Vietnam War, the area was replanted and is now a UNESCO Biosphere Reserve. It acts as a critical green lung for nearby Ho Chi Minh City, filtering air pollution and protecting the city from storm surges. Aquaculture expansion is a primary driver of mangrove degradation in Bac Lieu, Vietnam, alongside socio-economic factors such as poverty and limited education. (Van Viet et al., 2026) Key factors influencing mangrove cover include soil water saturation, distance to the sea, increased forestation potential, and sea-level rise. The restoration has seen the return of King Cobras, saltwater crocodiles, and vast bird populations. The increase in carbon storage at restored sites was linked to improved soil organic carbon stocks and better hydrological conditions, emphasizing the importance of hydrological rehabilitation for restoring carbon storage capacity in hurricane-impacted mangroves. It serves as a major eco-tourism hub for the city, generating revenue through entry fees and tours while educating the urban population on coastal resilience.

The ongoing degradation of mangroves in Bac Lieu, Vietnam, is primarily driven by the expansion of small-scale aquaculture and socio-economic factors. Despite the potential of integrated mangrove-shrimp systems for sustainability, weak governance and limited support exacerbate the issue, leading to forest clearing as communities seek to improve income. The findings underscore the urgent need for policy reform and better governance to align economic activities with mangrove conservation efforts. (Van Viet et al., 2026)

Matang Mangrove Forest Reserve, Malaysia

Sustainable Silviculture (Timber) & Fisheries cited as the best-managed mangrove forest in the world, Matang has been managed since 1902. It uses a 30-year rotation cycle for timber harvesting, ensuring that forest cover remains continuous to protect the coastline. The reserve supports a thriving charcoal industry, export quality, and pole production. Long-term patterns in greenhouse gas fluxes varied spatially with wetland management practices, showing higher carbon uptake in mangroves and greater emissions in freshwater sawgrass marshes, with emissions in the latter influenced by hydrological management. (Doughty et al., 2026) The extensive root systems support one of Malaysia's most productive cockle and prawn fisheries. The economic value of the fisheries exceeds that of timber, underscoring the standing forest's value to the blue economy. Evaluating these regions allows for a direct critique of how local conservation efforts match international mandates, such as national targets established under the Kunming–Montreal biodiversity framework.

The Sundarbans in India and Bangladesh

Extreme storm defense generates unique biodiversity among mangrove-derived actinomycetes, which are adapted to harsh environmental conditions. It shows promise for bioremediation due to its ability to produce a variety of bioactive compounds and enzymes. Filamentous bacteria are abundant in mangrove ecosystems and have potential applications in antimicrobial activities, heavy metal removal, and soil decontamination. Above-ground biomass (AGB) in Indian mangroves varies widely, influenced by environmental conditions, with deltaic and island systems identified as hotspots for biomass and carbon. In contrast, stressed west-coast mangroves show lower carbon densities. (Ananthakrishnan et al., 2026)

As the world's largest contiguous mangrove forest, Sundarbans acts as a massive biological shield. Studies show the Sundarbans reduce the energy of cyclone waves by 40-70%, saving billions in potential property damage annually. *Actinomycetes* found in mangrove sediment exhibit unique physiological and ecological characteristics, making them contributors to biodegradation and nutrient recycling in challenging ecosystems. Microbes are crucial for the biodegradation of persistent organic pollutants and the bio-mobilization or immobilization of inorganic contaminants, contributing to environmental health. (Pradhan & Das, 2025) Degraded mangrove sites in the Hooghly and Bidyadhari estuaries exhibit low carbon storage potential due to high anthropogenic activities, limited mangrove species diversity, and proximity to the sea, highlighting the need for conservation efforts in these regions. (Nayak, Das et al., 2025) It supports traditional livelihoods like honey collection (Mouli) and crab farming. The biodiversity of mangrove habitats worldwide supports a tiger population (Royal Bengal Tiger), which is a draw for high-value wildlife tourism. These geographic settings are exposed to severe weather anomalies, making them ideal for monitoring how real-world environmental stress and hurricane-induced changes in precipitation disrupt regional carbon cycles.

Southeast Asia Philippines Disaster Risk Reduction (DRR)

Following the 2004 Indian Ocean Tsunami, regional policy shifted heavily toward Bio-shields. The conservation areas had higher carbon storage than non-conservation sites, with Sonneratia alba being the largest contributor. Mangroves, through their role in climate regulation, support Indonesia's focus on blue carbon ecosystems and coastal resilience. (Ariadi et al., 2026) Mangrove forests demonstrate high resilience to hurricanes, with recovery to pre-disturbance conditions occurring relatively quickly after storms. Villages in Aceh (Indonesia) with intact mangroves suffered fewer casualties than those with converted shrimp ponds. has led to massive government-backed replanting schemes (Indonesia's target to rehabilitate 600,000 hectares). The genetic diversity and kinship relationships among black mangrove populations (Rhizophora mucronata) in West Sulawesi using morphological markers. The findings reveal high genetic diversity among populations, with variation in genetic distance between Polewali Mandar and Mamuju, indicating the need for targeted conservation strategies to preserve ecosystems amidst anthropogenic threats. (Wallacea, 2025) The management frameworks, national policies, and cross-border restoration models have successfully mitigated those stressors on a broader scale. The local case studies illustrate active environmental and human conflicts, while the global models of success demonstrate the scalability of established solutions.

Soil physicochemical properties and vegetation structural variables influenced the variance in carbon stock in Silvo-fishery, a method in which mangroves are planted within aquaculture ponds. maintains 70-80% forest cover while allowing farmers to harvest shrimp and crabs, balancing the shrimp-versus-forest conflict. Afforestation of abandoned ponds with mangroves is an effective approach to global mangrove restoration, but the combined recovery of vegetation and carbon stocks remains limited. Most mangrove species demonstrated potential for long-distance dispersal, remaining buoyant for portions of the experiment and indicating the ability to move through complex coastal habitats. In dispersal modeling, mangrove propagule traits indicate that variability in buoyancy behaviors among species influences dispersal trajectories and connectivity. (Thompson-Saud et al., 2025) The integrated workflows, featuring Google Earth Engine processing, Random Forest machine learning models, and satellite telemetry, are drawn directly from recent advancements in digital agriculture and remote sensing research.

Figure 3. Priority research and innovation directions. mangrove ecosystems could become the prototype for next-generation climate-resilient bioengineered landscapes

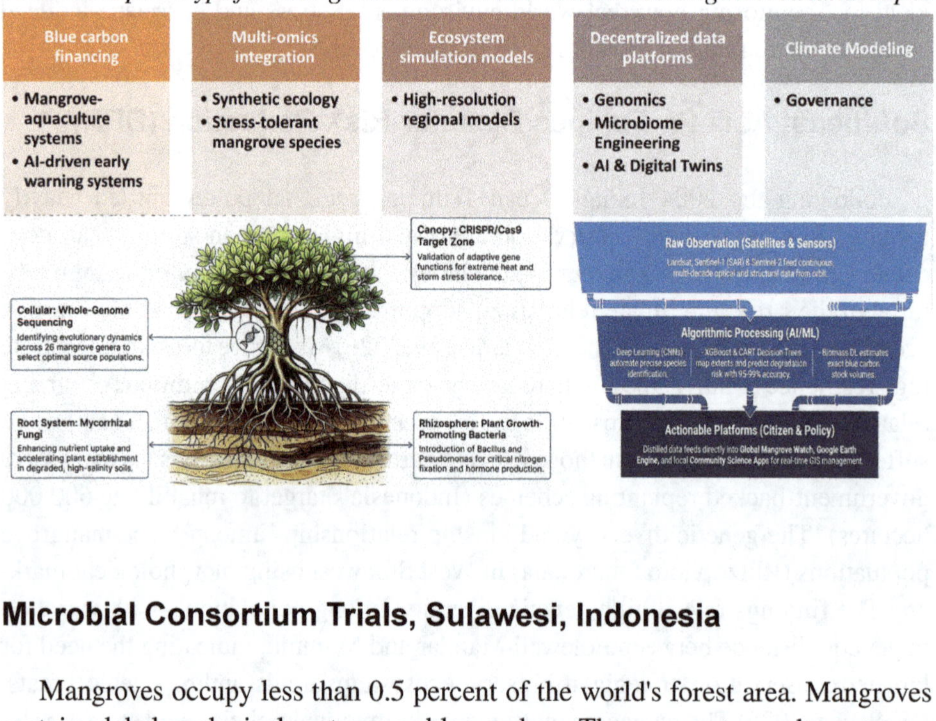

Microbial Consortium Trials, Sulawesi, Indonesia

Mangroves occupy less than 0.5 percent of the world's forest area. Mangroves sustain global ecological systems as blue carbon. They sequester carbon at rates three to five times higher per unit area than tropical upland forests. The most efficient carbon sinks on Earth are below-ground biomass and waterlogged soils, which accumulate organic carbon over millennia in coastal and marine ecosystems. The IPCC Wetlands Supplement (2013) considers mangrove soils and biomass to be carbon pools subject to internationally agreed measurement, reporting, and verification standards.

Indonesia has lost approximately 40% of its mangrove cover since 1980, and restoration efforts under the national Mangrove Restoration Program (BRGM) have had mixed success. A research collaboration involving Institut Pertanian Bogor and international partners initiated contained field trials of phosphate-solubilizing and nitrogen-fixing microbial consortia across six restoration sites in the Sulawesi mangrove belt between 2019 and 2023. GMW-derived baseline maps were used to stratify trial sites by degradation status, and AI-assisted multispectral classification provided quarterly canopy health assessments. Inoculated plots showed statistically significant improvements in *Rhizophora apiculata* establishment rates (23% survival improvement at 12 months) compared to control plots. Crucially, trials were conducted under the oversight of the national biosafety authority, with community monitoring committees co-established at each site in accordance with a participatory action research protocol. Revenue generated by a nested blue carbon pilot project was distributed via a community trust fund under FPIC-validated benefit-sharing agreements. Framework lessons: pillar integration (Biotechnology + Digital monitoring + Blue carbon MRV) was achieved through community governance co-design for social acceptance, while contained trial protocols prevented premature operational scaling.

Table 4. Case studies of innovations on mangrove management

Country	Project	Innovation Technology	Outcome	Policy Framework
UAE (Gulf Cooperation Council)	ADNOC Drone-Led Mangrove Restoration	Drone-based planting, remote sensing, blue carbon finance UAVs, remote sensing, carbon accounting	Large-scale planting, increased sequestration, Net Zero 2050 support	NDCs, COP-28, Blue Carbon Initiatives
China (Zhanjiang)	Integrated Mangrove Aquaculture (IMA)	Coupled aquaculture-mangrove systems, nutrient cycling, digital monitoring Digital sensors, aquaculture-mangrove	Fourfold increase in productivity, improved environmental/economic outcomes	National restoration, co-management
Singapore	Pulau Tekong, Pasir Ris, Pulau Semakau	Ecological engineering, EMR, digital monitoring Engineering, digital monitoring	Enhanced diversity, improved coastal protection, and adaptive management	National adaptation plans
Indonesia (Sulawesi)	Rawa Aopa Watumohai National Park	Vulnerability assessment, blue carbon quantification, community management Remote sensing, blue carbon, community	Lower vulnerability, higher blue carbon storage, improved adaptive capacity	Conservation legislation, REDD+
Western Asia (Regional)	Regional Restoration Programs	Remote sensing, machine learning, and geospatial policy tools Remote sensing, ML, GIS	Improved monitoring, cross-border policy networks, and enhanced enforcement	Ramsar, regional policy networks

Cispatá Bay Vida Manglar Colombia

Mangroves in Central America function as carbon sinks and influence dynamics and resilience in the face of environmental changes. The Central American region is expected to face severe climate change, higher temperatures, more intense cyclones, and more frequent droughts, necessitating an understanding of mangrove resilience to inform effective policy and management strategies for climate solutions and disaster risk reduction. (Girkin et al., 2026)

Blue carbon innovation includes soil carbon and the vegetated zones. It stores more organic carbon than adjacent bare mudflats, despite receiving more sediment input, owing to reduced hydrodynamic disturbance. National targets under the Kunming–Montreal Global Biodiversity Framework work to reduce mangrove loss, highlighting that current strategies are insufficient to halt loss by 2030. The study identifies that 78% of mangrove loss occurred in biodiverse areas outside protected zones, with only a few countries positioned to address the major drivers of loss effectively. It emphasizes the need for policy amendments to incorporate both short-term and long-term strategies for effective mangrove conservation. (Gajre et al., 2025) There is variability in carbon stock estimates for Central American mangroves, influenced by factors such as geomorphology, environmental conditions, and species composition, which differ between the Pacific and Atlantic coastlines. Mangroves within national greenhouse gas inventories and climate finance. Tropical coastal ecosystems are important for blue carbon sequestration, storing high carbon stocks per unit area and sequestering carbon over centuries. The high-resolution benthic habitat maps from the Khaled bin Sultan Living Oceans Foundation's Global Reef Expedition assess blue carbon habitats in remote tropical marine ecosystems. Research on mangrove forests, seagrass meadows, and macroalgal habitats identifies the Solomon Islands, Bahamas, and Colombia as potential blue carbon hotspots. The fragmentation of habitats in French Polynesia underscores the need for globally consistent mapping and stock estimates to improve management and conservation. (Duchâtellier et al., 2025) The project conserves roughly 11,000 hectares because it was among the first to successfully account for the carbon stored in the soil, where up to 80% of mangrove biomass is above ground. The biogeomorphologic factors affect blue carbon accumulation in mangrove sediments. Vegetated zones store more organic carbon than adjacent bare mudflats due to reduced hydrodynamic disturbance and sediment input. Beyond carbon, mangroves provide nursery habitat for commercially vital fish and crustacean species. It shields coastlines from storm surge and erosion and filter terrestrial runoff.

Figure 4. Mangrove microbiomes are harnessed to bioremediate oil, heavy metals, and other pollutants, supporting ecosystem recovery in contaminated areas

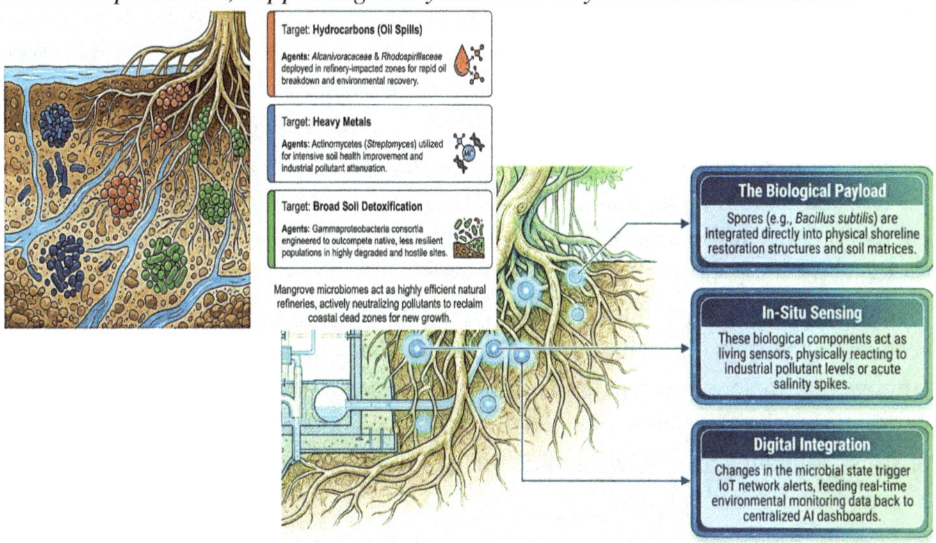

THE CARIBBEAN & LATIN AMERICA RESILIENCE & TOURISM VALUE

A World Bank/TNC report estimated that mangroves in the Caribbean provide hundreds of millions of dollars in annual flood protection benefits to coastal properties.(Das et al., 2026)

Tourism. In places like Bonaire and Mexico (Riviera Maya), mangroves are marketed as essential to the clarity of water on nearby coral reefs. Without the mangroves trapping sediment, the multi-billion-dollar dive tourism industry would collapse due to cloudy water and coral die-offs. (Girkin et al., 2026) Hybrid-engineered living materials (HELM) and synthetic biology approaches are being developed for real-time environmental sensing and targeted remediation, offering new tools for the adaptive management of mangrove shorelines. (Shivaraju et al., 2026)

West Africa (Senegal & Guinea) Food Security & Shellfish

The Sine Saloum Delta in Senegal harbors one of West Africa's largest and best-preserved mangrove systems, managed under decades-old community forestry arrangements. A joint initiative between the Senegalese Ministry of Environment, the NGO Wetlands International, and the voluntary carbon standard body initiated a blue carbon project in 2017, with communities holding formal carbon rights under national forestry law. GMW time-series data (1996–present) served as the baseline deforestation reference level for additionality calculations, and Tier 2 IPCC Wetlands Supplement emission factors, based on regional field inventory data, were applied for MRV. A community benefit-sharing agreement directed 60% of carbon revenues to a village development fund governed by a women-majority committee. AI-assisted annual change detection alerts enabled rapid response to small-scale encroachment. In the Sine Saloum Delta (Senegal), mangrove management is led by women's cooperatives who harvest oysters and clams (Anadara spp.) from the mangrove roots. (Wijerathna et al., 2026) Economic benefits from improved processing of shellfish (smoking/drying) have created a value-added supply chain that empowers female entrepreneurs and incentivizes them to protect mangrove roots from damage during harvest. (Couëdel et al., 2026) Tools accelerate the development of stress-tolerant varieties and enable traceability in restoration programs, ensuring biosafety and genetic integrity.

Mikoko Pamoja (Mangroves Together), Kenya

Mangroves are crucial Blue Carbon ecosystems in the tropics and subtropics, effectively capturing and storing atmospheric carbon dioxide more efficiently than terrestrial forests.

The project protects 117 hectares of mangrove forest in Gazi Bay. 2 trees capture ~2,500 tonnes of CO2 annually (Blue Carbon), while root systems prevent siltation of nearby coral reefs and seagrass beds. Mikoko Pamoja was the first community-based project in the world to sell carbon credits.(Couëdel et al., 2026) The Community-led Blue Carbon Trading revenue funds local needs directly, such as buying textbooks for schools and installing fresh water points for the village. The total ecosystem carbon stocks varied across sites, with soil properties influencing both tree growth and carbon composition. Overall, natural regeneration is highlighted as a cost-effective approach for restoring mangroves while enhancing biodiversity and carbon storage. (Slocum et al., 2026) The project employs local forest guards and supports eco-tourism boardwalks, creating jobs independent of fishing.

Table 5. Summary of benefits across case studies

Benefit Category	Mechanism	Case Study
Coastal Shield	Roots dissipate wave energy and reduce storm-surge height.	Sundarbans (India/ Bangladesh)
Carbon Sink	Sequestration in biomass and deep anaerobic soil layers (4times faster than rainforests).	Mikoko Pamoja (Kenya)
Fisheries	Nursery ground for juvenile fish, crabs, and shrimp.	Matang (Malaysia)
Eco-Tourism	Wildlife viewing (birds, tigers, bioluminescence).	Can Gio (Vietnam)
Livelihoods	Honey, sustainable timber, shellfish, carbon dividends.	Sine Saloum (Senegal)

Digital Twin Prototype for Hydrological Management, Gulf of Fonseca, Honduras

The Gulf of Fonseca, shared among Honduras, El Salvador, and Nicaragua, contains the largest mangrove stand in Central America, under pressure from the expansion of shrimp aquaculture, altered freshwater flows, and sea-level rise. A digital twin prototype, jointly developed by a consortium of national universities and the European Space Agency's Phi-Lab, was piloted from 2021 to 2024 to simulate mangrove hydrological dynamics under various aquaculture and dam management scenarios. The twin integrated SAR-based GMW extent maps, tidal hydrodynamic models, precipitation projections, and sediment accretion rates from field monitoring. Scenario outputs were translated into policy-relevant guidance documents co-developed with national water authorities and indigenous Lenca communities. The project surfaced data sovereignty concerns: drone-derived imagery collected over indigenous fishing grounds required retroactive renegotiation of data access agreements. (Doughty et al., 2026) Framework lessons from the digital twin prototype demonstrated decision-support value and governance of data collection over inhabited territories. It requires advance community consent protocols with cross-border data-sharing agreements formalized before operational deployment; model uncertainty must be explicitly communicated in policy translation documents.

Policy Networks

Climate change and human activities threaten biodiversity and ecosystem services, necessitating effective conservation strategies. The Cartagena Protocol on Biosafety (2003) provides the international legal framework for the risk assessment, handling, transport, and use of living modified organisms (LMOs). Its Advance Informed Agreement (AIA) procedure requires that parties intending to introduce

LMOs provide detailed risk assessment documentation to receiving parties before the first intentional transboundary movement. The climate-smart conservation plans for mangroves found that a moderate 7.3% increase in protected areas can yield a 13.3% benefit in climate resilience. Transboundary conservation plans, which involve cooperation among countries, are more effective at protecting climate-resilient mangroves than nation-based plans. The conservation strategies could be applied to other ecosystems, provided there is sufficient data on distribution and resilience to climate change.(Dabalà et al., 2026) Genomic tools are enabling the identification of stress-resilient genotypes, which can be prioritized for restoration in areas facing climate extremes. A precision breeding approach increases the likelihood of long-term ecosystem stability and function. (Wallacea, 2025) The Nagoya Protocol's benefit-sharing provisions, designed for genetic resources, offer useful analogs for negotiating equitable arrangements around carbon rights derived from community-managed ecosystems.

Technologies such as the Internet of Things, remote sensing, artificial intelligence, and uncrewed aerial vehicles enhance operational efficiency, environmental sustainability, and decision-making in agriculture. A machine learning framework to assess global mangrove forestation potential, revealing 156,682 km^2 of potential under current climate conditions, with the highest capacity in the Tropical Northwestern Atlantic, Amazonia, and Indonesia. (Li et al., 2026) Key factors influencing mangrove cover include soil saturated water content and distance to the sea. At the same time, future climate scenarios indicate a general increase in forestation potential, though sea-level rise could reduce it by 26,820 km^2. The research highlights the carbon storage potential of mangrove forestation in Indonesia, Brazil, Australia, Mexico, and the Philippines. Agriculture 5.0 improves input use efficiency, enables real-time monitoring, optimizes irrigation, and enhances yield stability while reducing environmental impacts. However, challenges include high upfront costs, data privacy concerns, and limited digital literacy among farmers. (Thompson-Saud et al., 2025)

The need for comprehensive policy frameworks, capacity-building efforts, and cooperative innovation ecosystems to overcome barriers and enhance the resilience and adaptability of agroecosystems to environmental and socioeconomic challenges. (Kumar et al., 2026) The following biosafety principles should govern any biotechnology deployment in mangrove restoration contexts

- Proportionality: governance intensity scaled to technology readiness and ecological risk profile
- Precaution burden of proof for safety rests with proponents, not communities or regulators
- Reversibility preference for interventions that can be discontinued or reversed if adverse effects are observed

- Transparency public disclosure of risk assessments, trial protocols, and monitoring data
- Adaptive management mandatory post-deployment monitoring and pre-defined decision triggers for intervention cessation
- Community sovereignty: FPIC exercised before any trial or deployment, with genuine veto rights

Figure 5. Policy frameworks increasingly recognize the multifunctional value of mangroves, linking local action to global climate and biodiversity goals

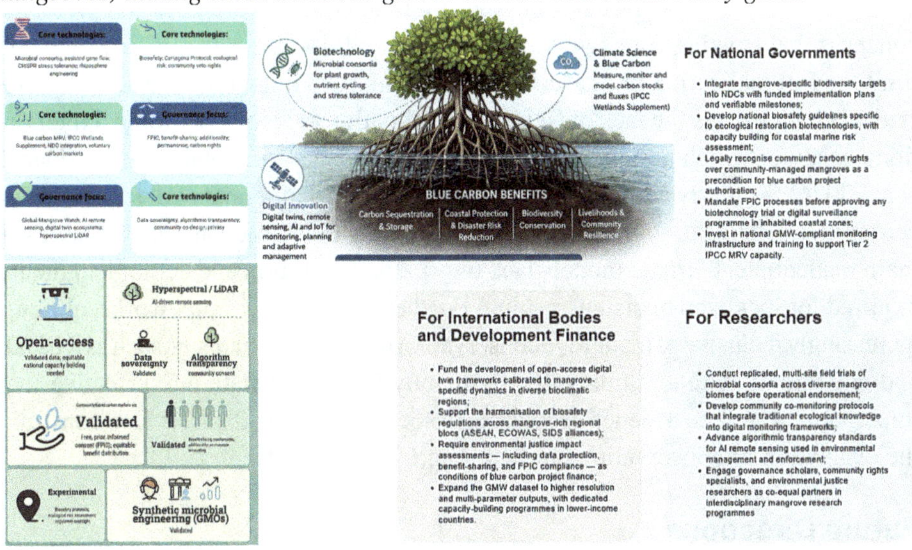

Limitations and Challenges

Mangroves are sensitive to sea-level rise, temperature increases, salinity stress, and storm frequency/intensity. Climate science provides the foundation for quantifying mangrove carbon sequestration, assessing vulnerability, and designing adaptation strategies in the face of global change. The implications of developments are substantial across scientific, environmental, and socio-economic dimensions. Scientifically, they enable a transition toward systems-level ecology and predictive modeling. Marginalization of governance, ethics, and community rights risks in the deployment of biotechnology. (Van Viet et al., 2026) The genetic contamination, disruption of native microbial communities, and unintended ecological cascades demand biosafety frameworks calibrated to mangrove conditions. Environmentally, they enhance the resilience and capacity for restoration of mangrove ecosystems under climate stress. Socio-economically, they support sustainable livelihoods through innovations such as integrated mangrove-aquaculture systems and carbon finance mechanisms.(Dabalà et al., 2026) However, benefits are contingent upon equitable governance, community engagement, and the responsible deployment of emerging technologies.

The promise of blue carbon finance simultaneously raises urgent questions about who owns the carbon rights over community-managed forests, how revenues are distributed, and whether global market mechanisms risk commodifying the commons in ways that dispossess local and indigenous communities. (Kumar et al., 2026) Data generated by AI-driven remote sensing raises data sovereignty and protection concerns that existing frameworks do not fully address. Despite this, research gaps limit the full realization of interdisciplinary mangrove innovation. One of the most critical challenges is the lack of field-scale validation for biotechnological interventions, many of which remain confined to controlled experimental settings. (Dabalà et al., 2026) Similarly, while digital tools generate vast amounts of data, integration across platforms and alignment with ecological and socio-economic variables remain inadequate. Climate models lack the spatial resolution and ecological nuance required for localized decision-making. Furthermore, policy frameworks, although increasingly inclusive of mangrove conservation through mechanisms such as REDD+ and national climate commitments, frequently fail to translate into effective local implementation. The absence of long-term socio-economic datasets also constrains the evaluation of conservation outcomes and sustainability.

Future Directions

Mangrove forests are productive ecosystems located in tropical and subtropical intertidal coastal regions, characterized by a unique blend of terrestrial and marine environments that support rich biodiversity. *Sonneratia apetala* affects aboveground biomass carbon (AGB) and soil carbon stock (SOC) in mangrove ecosystems. ACG accumulates rapidly in younger mangroves but stabilizes early, while SOC increases more gradually and in older mangroves. (Slooten et al., 2026) Remote sensing data indicate that the introduction of *Sonneratia apetala* alters species composition and root structures, affecting carbon dynamics and suggesting that restoration targets for ACG and SOC should be set on different timelines based on stand age and species composition. Biomass, carbon storage, and soil CO2 fluxes in degraded mangroves across different estuarine gradients in Sundarban, India. It finds that total carbon storage varies among sites, with the highest CO2 flux recorded in degraded areas near human habitation, indicating a need for greater intervention to protect ecosystems. (Nayak, Das et al., 2025)

Actinomycetes found in mangrove sediment exhibit unique physiological and ecological characteristics, making them organisms of a specialized ecosystem.

Actinomycetes play a crucial role in the biodegradation of persistent organic pollutants and the bio-mobilization or immobilization of inorganic contaminants, contributing to nutrient recycling in mangrove ecosystems. (Pradhan et al., 2026) Key factors influencing mangrove cover potential were identified as soil saturated water content and distance to the sea. At the same time, future climate scenarios indicate a general increase in mangrove cover potential, despite a potential reduction of 26,820 km^2 due to sea-level rise under SSP5-8.5.

The study highlights that only 19,361 km^2 of current forestation potential aligns with areas projected to increase in future cover, and identifies Indonesia, Brazil, Australia, Mexico, and the Philippines as the top five countries for carbon storage potential from mangrove forestation. Looking ahead, future research must prioritize scaling biotechnological innovations from laboratory to field conditions, alongside developing integrated data architectures that unify biological, environmental, and digital datasets. The concept of digital twin ecosystems, in which real-time data feeds inform predictive simulations and adaptive interventions, represents a promising frontier. In multi-omics, synthetic ecology, and AI-driven modeling, further enhancements in the precision and effectiveness of conservation strategies are expected. (Nayak, Bhattacharyya, et al., 2025) At the same time, strengthening policy-tech interfaces and ensuring inclusive governance will be essential to translating innovation into sustained ecological and societal impact.

In summary, resolving complex environmental challenges requires a multifaceted approach that integrates biotechnology, climate science, and modern digital innovation into contemporary mangrove management. From a biotechnological perspective, leveraging next-generation sequencing and metabolomic profiling enables researchers to uncover hidden functional pathways in sediment-derived microbiomes, track microbial responses to environmental stress, and identify novel biochemical compounds with bioprospecting potential. Concurrently, robust climate science frameworks utilize precise in situ observations and flux measurements to model how coastal wetlands respond to rising sea levels and altered freshwater inputs. When biological and environmental insights are coupled with digital innovations, advanced satellite telemetry, remote sensing arrays, and predictive machine learning, resource managers can achieve precise real-time tracking of canopy health and carbon cycles. (Duchâtellier et al., 2025; Gajre et al., 2025) Integration of molecular biology, environmental monitoring, and predictive technology transforms coastal conservation from a reactive practice into a proactive, data-driven strategy capable of safeguarding blue carbon systems for the future.

Table 6. Community-based and adaptive governance are critical for long-term success

Policy Model	Description	Example	Key Outcomes
Ramsar Convention	International wetland protection and restoration	Western Asia, Southeast Asia	Enhanced legal protection, international cooperation
REDD+	Climate finance for forest/mangrove conservation	Indonesia, Southeast Asia	Carbon credits, funding for restoration
Nationally Determined Contributions (NDCs)	National climate adaptation/ mitigation targets, mangroves	UAE, Singapore	Policy alignment, funding, and international reporting
Community-Based Co-Management	Local stakeholder engagement, shared governance	Indonesia, China, Singapore	Increased project success, socio-economic co-benefits
Adaptive Management	Iterative, evidence-based project design and monitoring	Singapore, Belize	Improved resilience, flexible response to change
Cross-Border Policy Networks	Regional collaboration for shared ecosystem management	Western Asia	Coordinated action, knowledge sharing

CONCLUSION

Mangrove ecosystems are among the world's most productive and valuable coastal habitats, providing critical services such as carbon sequestration, shoreline protection, biodiversity support, and livelihoods for millions. In recent years, a shift has emerged in mangrove research, driven by the convergence of biotechnology, climate science, and digital innovation. Interdisciplinary integration is transforming mangrove science from a largely descriptive discipline into a predictive, intervention-oriented, and technology-enabled field that can support scalable conservation and restoration strategies. However, they face escalating threats from climate change, land conversion, pollution, and unsustainable exploitation. In response, a new wave of interdisciplinary research is emerging at the intersection of biotechnology, climate science, and digital innovation. Mangrove-associated actinomycetes have the potential to produce a variety of bioactive compounds, novel antibiotics, anti-oxidants, and antitumor agents, highlighting promise for future bioprospecting and biotechnological applications.

Mangrove ecosystems are essential for global carbon storage, biodiversity maintenance, and coastal resilience. Mangrove forests are productive ecosystems located in tropical and subtropical intertidal coastal regions, characterized by a unique blend of terrestrial and marine environments that support rich biodiversity. By integrating genomic and microbial engineering approaches with blue carbon modeling and AI-driven remote sensing technologies, the field is transitioning toward predictive, data-driven, and adaptive ecosystem management frameworks.

Innovations enable the identification of stress-resilient genotypes, the enhancement of restoration success through engineered microbiomes, and the real-time monitoring of ecosystem dynamics at spatial and temporal scales. Despite this, gaps persist in field-scale validation, data integration, and policy translation, limiting the scalability and equitable implementation of technologies.

Mangrove-associated actinomycetes have the potential to produce a variety of bioactive compounds, novel antibiotics, antioxidants, and antitumor agents, highlighting promise for future bioprospecting and biotechnological applications. The review synthesizes emerging interdisciplinary research at the intersection of biotechnology, climate science, and digital innovation, thereby transforming our understanding of, and conservation efforts for, mangrove ecosystems.

In conclusion, the convergence of biotechnology, climate science, and digital innovation is redefining the boundaries of mangrove ecosystem research and conservation. While challenges related to scalability, integration, and governance remain, the emerging interdisciplinary paradigm offers a powerful framework for addressing complex environmental problems. If effectively implemented, the approach has the potential to transform mangrove ecosystems into resilient, adaptive, and intelligently managed landscapes that serve as models for global climate solutions.

REFERENCES

Ananthakrishnan, S., Nayak, S. K., Rout, A., Aparna, P., Sahoo, A., Nayak, R. K., & Kandasamy, K. (2026). Mangrove biomass and carbon storage in India: Implications for blue carbon potential. *Biomass Futures*, *1*, 100033. DOI: 10.1016/j.bmf.2026.100033

Ariadi, H., Wafi, A., Madusari, B. D., Syakirin, M. B., Linayati, L., & Mardiana, T. Y. (2026). Assessment of Mangrove Species Composition, Biomass, and Carbon Stock Potential for Climate Change Mitigation in Pekalongan, Indonesia. [Journal of Natural Resources and Environmental Management]. *Jurnal Pengelolaan Sumberdaya Alam Dan Lingkungan*, *16*(1), 36–36. DOI: 10.29244/jpsl.16.1.36

Byalahunasishetter, B. I., Chowdareddy, S., & Raghavan, R. (2026). Cancer Risk Prediction Based on Gut Microbiota Analysis. *Smart Microbiomes for Personalized Medicine*, 307–342. DOI: 10.4018/979-8-3373-8172-5.ch009

Couëdel, A., Falconnier, G. N., Adam, M., Cardinael, R., Six, J., Laub, M., Ruane, A. C., Boote, K., Justes, E., Smith, W. N., Whitbread, A. M., Affholder, F., Akinseye, F. M., Balkovic, J., Basso, B., Bhatia, A., Chakrabarti, B., Chikowo, R., Christina, M., & Corbeels, M. (2026). Beyond Climate Change: The Role of Integrated Soil Fertility Management for Sustaining Future Maize Yield in Sub-Saharan Africa. *Global Change Biology*, *32*(2), e70720. DOI: 10.1111/gcb.70720 PMID: 41631408

Dabalà, A., Brown, C. J., Van der Stocken, T., Buelow, C. A., Schoeman, D. S., Dunn, D. C., Lovelock, C. E., Dahdouh-Guebas, F., Flower, J., Neubert, S., Buenafe, K. C. V., Everett, J. D., Esturas, K. J. T., & Richardson, A. J. (2026). Safeguarding climate-resilient mangroves requires only a moderate increase in the global protected area. *Nature Communications*, *17*(1), 2063. DOI: 10.1038/s41467-026-68877-4 PMID: 41593080

Das, S., Dutta, S., Roy, J., Roy Choudhury, M., Sen, P., Chanda, A., Dutta, H., Nandi, S., & Ghosh, T. (2026). Mangroves in the Anthropocene: A global synthesis of carbon storage, biodiversity, and coastal resilience under climate and anthropogenic stressors. *The Science of the Total Environment*, *1015*, 181405. DOI: 10.1016/j.scitotenv.2026.181405 PMID: 41558381

Doughty, C. L., Ying, Q., Ward, E., Delaria, E., Wolfe, G. M., Malone, S. L., Reed, D. E., Troxler, T., Kominoski, J. S., Castañeda-Moya, E., Shoemaker, W. B., Yannick, D., Starr, G., Oberbauer, S. F., Barenblitt, A., Campbell, A., Charles, S., Fatoyinbo, L., Gewirtzman, J., & Poulter, B. (2026). Compounded effects on wetland greenhouse gas fluxes from climate change and water management along a saline to freshwater gradient. *Proceedings of the National Academy of Sciences of the United States of America, 123*(8), e2513685123. DOI: 10.1073/pnas.2513685123 PMID: 41701819

Duchâtellier, G., Bakker, A., Purkis, S. J., Estevanez, M., Dempsey, A. C., Heuer, R. M., Grosell, M., & Oehlert, A. M. (2025). Organic Carbon Stocks in Earth's Most Remote Blue Carbon Ecosystems. *Environmental Science & Technology*, *59*(45), 24404–24415. DOI: 10.1021/acs.est.5c06377 PMID: 41190867

Gajre, R. B., Barchiesi, S., Friess, D. A., Lang, D., Lee, Y. K., Lim, H. K., Nasry, M., Ochavo, K. G. C., & Peh, S.-H. (2025). Bridging conservation and policy: Evaluating national targets to reduce mangrove loss under the Kunming–Montreal biodiversity framework. *Environmental Research Letters*, *20*(11), 114044. DOI: 10.1088/1748-9326/ae1150

Girkin, N. T., Canty, S. W. J., Rovai, A. S., Morrissette, H. K., Collin, R., Shukla, Y., Romero-González, T. E., Quirós, J., Pineda, J., Wong, J. R., & Cifuentes-Jara, M. (2026). Central American Mangrove Blue Carbon: Distribution, Dynamics and Future Directions. *Wetlands*, *46*(1), 12. DOI: 10.1007/s13157-025-02025-4 PMID: 41567823

J, S. P., Moger, S. S., Byalahunasishetter, B. I., & Raghavan, R. (2026). The One Health Framework and Antimicrobial Resistance: Role of Herbal Medicine. *Herbal Medicine Approaches to Antimicrobial Resistance and One Health*, 1–36. DOI: 10.4018/979-8-3373-7810-7.ch001

Jangama, R., Suresh, C., Korrapati, C., & Raghavan, R. (2026). Hybrid Coating Systems: Bioactive Molecules and Nanoparticles for Cancer Therapy. *Biomedical Applications of Laser-Processed Coatings*, 99–134. DOI: 10.4018/979-8-3373-4104-0.ch003

Korrapati, C., Chowdareddy, S., Suresh, C., & Raghavan, R. (2026). Cancer Cell Membrane-Coated Nanoparticles: Therapeutic Advancements and Hybrid Systems. *Applications of Nanomaterials for Oncology*, 131–160. DOI: 10.4018/979-8-3373-3765-4.ch005

Kumar, A., Thakur, S., & Kumar, R. (2026). Digital convergence in Agri-Tech: Reviewing the role of advanced technologies in enabling precision and sustainability. *Agronomy Journal*, *118*(2), e70310. DOI: 10.1002/agj2.70310

Lakshmaiah, A., & Raghavan, R. (2026). Deep Learning-Enhanced Hyperspectral Imaging: Applications in Oncology Cancer Diagnosis and Tumor Margin Identification. *Biomedical Applications in Deep Learning-Enhanced Hyperspectral Imaging*, 335–368. DOI: 10.4018/979-8-3373-7183-2.ch011

Li, G., Ma, Y., Zhuo, T., Qi, X., & You, X. (2026). A machine learning framework to assess global mangrove forestation potential under current and future climate scenarios. *Environmental Research*, *295*, 123923. DOI: 10.1016/j.envres.2026.123923 PMID: 41621565

Mahalakshmi, K. M., Divya, A. P., & Sai Priyanka, J. (2026). Spatial Multi-Omics in Cancers: Integrating Molecular Landscapes for Precision Oncology. In *Spatially Resolved Single-Cell and Multiple Omics in Human Diseases* (pp. 217-248). IGI Global Scientific Publishing.

Nayak, S. K., Bhattacharyya, P., Pradhan, C., Padhy, S. R., Parida, S. P., Das, A., Moharana, A., Rath, M., & Nayak, A. (2025). Soil functional microbial pathways related to GHGs emissions of coastal degraded mangroves-rice ecologies in Sundarban, India: A soil metagenomic approach. *Journal of Environmental Sciences (China)*. Advance online publication. DOI: 10.1016/j.jes.2025.12.050

Nayak, S. K., Das, A., Bhattacharyya, P., Padhy, S. R., Parida, S. P., Reddy, C. S., & Kandasamy, K. (2025). Comparative assessment of biomass, carbon storage and soil CO2 fluxes in degraded mangroves in the major estuarine gradients of the sundarban, India. *Journal of Environmental Management*, *395*, 127807. DOI: 10.1016/j.jenvman.2025.127807 PMID: 41161277

Ørskov, J. (1923). *Investigations into the Morphology of the Ray Fungi*. Levin & Munksgaard.

Pallavi, V. N., Chowdareddy, S., Byalahunasishetter, B. I., & Raghavan, R. (2026). CRISPR-Cas9 Technology for Precision Cancer Medicine: Target Discovery, Disease Modeling, and Therapeutics. *CRISPR-Cas9 Technology for Precision Medicine*, 35–72. DOI: 10.4018/979-8-3373-4857-5.ch002

Pradhan, D., & Das, M. T. (2025). Antimicrobial potential and extracellular metabolite profiling of bichitrapur mangrove sediment derived Micromonospora sp. BSS-D-04 and Streptomyces sp. BSS-D-05 using LC–MS/MS and GC–MS analysis. *Brazilian Journal of Microbiology : Publication of the Brazilian Society for Microbiology*, *56*(3), 1843–1859. DOI: 10.1007/s42770-025-01710-0 PMID: 40549332

Pradhan, D., Priyadarshini, P., Das, R., & Das, M. T. (2026). Actinomycetes of Mangrove Soil Sediments: Environmental Significance and Their Bio-Prospecting Potential. *Journal of Basic Microbiology*, *66*(1), e70141. DOI: 10.1002/jobm.70141 PMID: 41589713

Ramesh, R. N., Palled, S. B., Shivaraju, A., & Raghavan, R. (2026). Reproductive Toxicities and Transgenerational Cancer Risk From Climate-Related Occupational Hazards. *One Health Perspectives on Climate Change and Emerging Occupational Health Risks*, 191–230. DOI: 10.4018/979-8-3373-8470-2.ch007

Ray, R. L., Kumar, N., Gurau, S., Raza, A., Khan, N., Singh, S. K., Kayode, E. T., & He, Y. (2026). Assessing the impact of hurricanes on the carbon cycle using SMAP satellite and in-situ observations: Role of land cover and precipitation. *Physics and Chemistry of the Earth Parts A/B/C*, *143*, 104381. DOI: 10.1016/j.pce.2026.104381

Reznikova, D. A., Barannikova, M. V., Shnakhova, L. M., Mitkin, N. A., & Vatlin, A. A. (2026). Next-generation sequencing approaches for soil microbiome research. *Frontiers in Soil Science*, *5*, 1706999. Advance online publication. DOI: 10.3389/fsoil.2025.1706999

Shivaraju, A., Vaddi, H., & Yaraladdi, A. B. (2026). *Multiomics Insights into Advanced Carbon and Nutrient Capture: Omics-Driven Strategies for Microalgae Photosynthetic Efficiency and Resource Utilization*. IGI Global Scientific Publishing., DOI: 10.4018/979-8-3373-7630-1.ch005

Slocum, M. P., Barton, R. B., Wagner, S. J., Lehmann, J., Solomon, D., Yeboah, E., & Plante, A. F. (2026). Elevated organic carbon in African Dark Earths is not exclusively attributable to pyrogenic organic matter. *The Science of the Total Environment*, *1016*, 181481. DOI: 10.1016/j.scitotenv.2026.181481 PMID: 41604878

Slooten, E., Raath-Krüger, M. J., Archibald, S., & Siebert, F. (2026). Subtropical grassland root-to-shoot ratios and plant organic carbon estimates: Why global averages fall short. *Annals of Botany*, •••, mcag027. Advance online publication. DOI: 10.1093/aob/mcag027 PMID: 41657197

Thompson-Saud, G., Robertson, A. I., Choukroun, S., Ospina-Alvarez, A., Logan, M., van der Mheen, M., & Grech, A. (2025). Factors influencing the early growth and dispersal potential of mangrove propagules. *Regional Studies in Marine Science*, *91*, 104528. DOI: 10.1016/j.rsma.2025.104528

Van Viet, T., Tsai, W.-P., & Huynh, H. H. (2026). A warning about the double-edged sword of integrated aquaculture: Mangrove degradation ongoing in Bac Lieu, Vietnam, as a case study. *Aquaculture (Amsterdam, Netherlands)*, *615*, 743680. DOI: 10.1016/j.aquaculture.2026.743680

Wallacea, J. P. K. (2025). *Genetic Diversity and Kinship Relationships Among Black Mangrove Populations (Rhizophora mucronata) in West Sulawesi Based on Morphological Markers*. https://journal.unhas.ac.id/index.php/wallacea/article/view/36617

Wijerathna, P., Chen, X., Qiu, R., Wijethilake, P. V. J. S., Chen, Y., Madushanka, N., Perera, I. J. J. U. N., Cai, J., Jayasinghe, L., Liu, Y., Bulugahapitiya, V. P., & Zhou, X. (2026). Advances in Natural Products from Mangrove-Associated Fungi Along the Indian Ocean Coast. *Molecules (Basel, Switzerland)*, *31*(2), 261. DOI: 10.3390/molecules31020261 PMID: 41599310

Chapter 3

Environmental Challenges and Climate Change: Mangroves' Role in Carbon Sequestration and Climate Mitigation

P. Selvakumar
http://orcid.org/0000-0002-3650-4548
Department of Science and Humanities, Nehru Institute of Technology, Coimbatore, India

ABSTRACT

Mangroves are highly specialized coastal ecosystems found in tropical and subtropical regions, characterized by salt-tolerant trees and shrubs that thrive in intertidal zones where land and sea meet. These unique ecosystems provide essential ecological functions, including coastal protection, habitat provision, nutrient cycling, and carbon sequestration. Mangroves act as natural barriers against storm surges, tidal waves, and coastal erosion by stabilizing sediments with their complex root systems, dissipating wave energy, and trapping sediments. They support high biodiversity, serving as nurseries for fish, crustaceans, and other marine organisms, and provide livelihood support to coastal communities through fisheries, timber, and non-timber products. In the context of climate change, mangroves play a critical role as both vulnerable ecosystems and powerful climate mitigation assets.

DOI: 10.4018/979-8-3373-9320-9.ch003

MANGROVES & CLIMATE CHANGE

Mangroves are highly specialized coastal ecosystems found in tropical and subtropical regions, characterized by salt-tolerant trees and shrubs that thrive in intertidal zones where land and sea meet. These unique ecosystems provide essential ecological functions, including coastal protection, habitat provision, nutrient cycling, and carbon sequestration. Mangroves act as natural barriers against storm surges, tidal waves, and coastal erosion by stabilizing sediments with their complex root systems, dissipating wave energy, and trapping sediments. They support high biodiversity, serving as nurseries for fish, crustaceans, and other marine organisms, and provide livelihood support to coastal communities through fisheries, timber, and non-timber products. In the context of climate change, mangroves play a critical role as both vulnerable ecosystems and powerful climate mitigation assets. The impacts of climate change, including sea level rise, increased storm intensity, temperature changes, and altered precipitation patterns, pose significant threats to mangrove ecosystems. Rising sea levels can inundate mangrove habitats, causing loss of land and altering salinity gradients, which may affect mangrove distribution and health. In some regions, mangroves may migrate landward if space is available, but coastal development and land-use changes often prevent such migration, leading to habitat loss. Increased storm intensity and extreme weather events can cause physical damage to mangrove stands, uprooting trees, eroding sediments, and disrupting ecosystem structure (Alongi, 2020).

The vulnerability of mangroves to climate change is further compounded by human activities such as coastal development, aquaculture expansion, pollution, and deforestation. These anthropogenic pressures reduce mangrove resilience by fragmenting habitats, reducing genetic diversity, and altering hydrological regimes. The combined effects of climate change and human disturbance can lead to mangrove degradation, reduced ecosystem services, and loss of biodiversity. Despite their vulnerability, mangroves are recognized as one of the most carbon-rich ecosystems, storing large amounts of carbon in biomass and sediments. Mangrove forests sequester carbon at high rates due to rapid biomass growth and the accumulation of organic matter in waterlogged, anoxic soils that slow decomposition (Alongi, 2022). This makes mangroves a vital component of climate change mitigation strategies. Carbon stored in mangrove sediments can remain sequestered for centuries, contributing to long-term climate regulation. Protecting and restoring mangroves can therefore significantly contribute to reducing greenhouse gas concentrations and meeting climate goals. Mangrove conservation and restoration also offer climate adaptation benefits. By buffering coastlines from storm surges and erosion, mangroves reduce the vulnerability of coastal communities to climate-induced hazards. They help maintain coastal stability and protect infrastructure, livelihoods, and human lives. Restoring degraded mangrove areas can enhance coastal resilience, improve fisheries productivity, and increase biodiversity. Mangrove restoration projects often involve re-establishing natural hydrology, planting native species, and engaging local communities in sustainable management.

Mangrove Carbon Sequestration

Mangrove carbon sequestration refers to the process by which mangrove ecosystems capture atmospheric carbon dioxide (CO_2) through photosynthesis and store it in biomass and sediments, thereby acting as significant carbon sinks that contribute to climate change mitigation. Mangroves are unique coastal forests found in tropical and subtropical intertidal zones, characterized by salt-tolerant trees and shrubs that thrive in saline environments. These ecosystems possess distinctive physiological and ecological traits that enable them to sequester carbon at high rates, often surpassing those of terrestrial forests on a per-unit-area basis. Mangrove carbon sequestration is a critical component of the broader concept of "blue carbon," which includes carbon stored in coastal and marine ecosystems such as mangroves, seagrasses, and salt marshes (Baranzini et al., 2000). Blue carbon ecosystems are recognized for their high carbon density, rapid carbon accumulation, and long-term storage potential. Mangroves, in particular, store carbon in both aboveground biomass (trunks, branches, leaves) and belowground biomass (roots) as well as in organic-rich sediments that accumulate in waterlogged, anaerobic conditions. The combination of high productivity and slow decomposition rates in mangrove soils leads to substantial carbon accumulation over time. The process of carbon sequestration in mangroves begins with photosynthesis, through which mangrove trees absorb CO_2 from the atmosphere and convert it into organic carbon compounds. These compounds are incorporated into plant tissues and contribute to biomass growth.

Hydrology and tidal dynamics play a critical role in sediment deposition, nutrient delivery, and the distribution of mangrove species. Tidal inundation brings sediment and organic matter into mangrove forests, supporting soil formation and carbon burial. The salinity gradient influences species composition and productivity, with certain mangrove species exhibiting higher growth rates and carbon accumulation. Climate factors such as temperature, rainfall, and sea level rise also affect mangrove productivity and carbon storage.(Barbier, 2006) Additionally, the geomorphological setting, including coastal topography and sediment supply, influences the capacity of mangroves to accumulate soils and sequester carbon. Mangrove carbon sequestration has significant implications for climate change mitigation. By capturing and storing carbon, mangroves help reduce atmospheric CO_2 concentrations, contributing to global efforts to limit warming. The carbon sequestration potential of mangroves is particularly important given the urgency of climate action and the need for nature-based solutions. Protecting existing mangrove forests and restoring degraded areas can increase carbon storage and provide cost-effective mitigation benefits. Mangroves offer high carbon sequestration per unit area, making them valuable in climate mitigation strategies, especially in regions where land is limited and coastal ecosystems are abundant.

Climate Mitigation: Mangroves

Mangroves are increasingly recognized as one of the most effective nature-based solutions for climate mitigation, owing to their exceptional capacity to capture and store atmospheric carbon dioxide (CO_2) in biomass and sediments. These coastal ecosystems, found in tropical and subtropical intertidal zones, are characterized by salt-tolerant trees and shrubs that thrive in saline and waterlogged conditions. Mangroves function as carbon sinks through photosynthesis, where carbon is assimilated into plant tissues, and through the accumulation of organic matter in anaerobic soils that slow decomposition. (Callaway et al., 2012)The high productivity of mangrove forests, combined with slow organic matter breakdown, results in substantial carbon sequestration rates that often exceed those of terrestrial forests on a per-unit-area basis. This unique carbon storage potential has led to the inclusion of mangroves in the broader framework of "blue carbon," which encompasses carbon sequestration in coastal and marine ecosystems such as mangroves, seagrasses, and salt marshes. Blue carbon ecosystems are critical for climate mitigation because they store carbon both in living biomass and in deep, carbon-rich sediments that can persist for centuries. Mangroves thus represent a long-term carbon reservoir that can contribute significantly to global efforts to reduce greenhouse gas concentrations. Climate mitigation through mangroves involves both conservation of existing mangrove forests and restoration of degraded areas. Protecting intact mangrove ecosystems is essential because they already store large amounts of carbon, and their destruction can release stored carbon back into the atmosphere. Land-use changes such as conversion to aquaculture, agriculture, urban development, and infrastructure lead to mangrove deforestation, resulting in the oxidation of organic carbon in soils and the emission of CO_2 and methane.

Their complex root systems stabilize shorelines, reduce wave energy, and trap sediments, thereby protecting coastal communities from storm surges, erosion, and sea level rise (Calumpong et al., 2021). These protective services reduce the need for expensive engineered infrastructure, making mangrove conservation a cost-effective mitigation and adaptation strategy. Mangroves also support rich biodiversity, serving as nurseries for fish, crustaceans, and other marine organisms, which contributes to food security and economic livelihoods. Additionally, mangroves filter pollutants and trap sediments, improving water quality and benefiting adjacent ecosystems such as coral reefs and seagrass beds. The integration of mangroves into climate mitigation policy has gained momentum through international frameworks and national commitments (DebRoy & Jayaraman, 2012). The Paris Agreement encourages countries to include nature-based solutions and ecosystem-based approaches in their Nationally Determined Contributions (NDCs). Many countries have incorporated mangrove conservation and restoration into their climate action plans, recognizing the role of blue carbon in meeting emission reduction targets. Mechanisms such as carbon credits, payment for ecosystem services, and conservation finance are increasingly used to support mangrove projects. However, integrating mangroves into climate mitigation policy requires robust governance, clear land tenure, and long-term funding. One of the key challenges in mangrove-based mitigation is measurement, reporting, and verification (MRV) of carbon stocks. Accurate carbon accounting is essential for carbon credit schemes and policy evaluation. MRV involves field measurements of biomass and soil carbon, remote sensing, and modeling to estimate carbon stocks and sequestration rates. Variability in carbon density across mangrove sites, differences in species composition, and environmental conditions pose challenges for standardizing MRV methodologies.

Mangroves: Carbon Sinks

Mangroves are globally recognized as one of the most efficient carbon sinks, offering exceptional capacity for carbon sequestration and long-term storage in coastal ecosystems. These unique forests thrive in tropical and subtropical intertidal zones, where saline water, tidal inundation, and waterlogged soils create specialized conditions for salt-tolerant trees and shrubs. Mangroves are not only important for coastal protection and biodiversity but also play a crucial role in climate regulation through carbon storage. (Dewsbury et al., 2016)Their ability to capture and store carbon has led to their classification as "blue carbon" ecosystems, alongside seagrass meadows and salt marshes, highlighting their significance in global climate mitigation strategies. Carbon sinks are systems that absorb more carbon from the atmosphere than they release, thereby reducing greenhouse gas concentrations and helping to stabilize global climate. Mangroves act as carbon sinks by sequestering carbon through photosynthesis and storing it in biomass and sediments. Their high productivity, coupled with slow decomposition rates in waterlogged, anoxic soils, leads to substantial accumulation of organic carbon over time. This combination of rapid carbon uptake and long-term storage makes mangroves exceptionally valuable as carbon sinks. The carbon sequestration process in mangroves begins with photosynthesis, where mangrove trees absorb CO_2 from the atmosphere and convert it into organic carbon compounds. These compounds are allocated to plant tissues, leading to growth of aboveground biomass such as trunks, branches, and leaves, as well as belowground biomass in complex root systems. Mangrove roots play a vital role in carbon storage by depositing organic matter into soils through root turnover and litter fall. The roots also stabilize sediments, which facilitates the accumulation of organic carbon in the soil. A defining feature of mangrove carbon sinks is the storage of carbon in sediments. Mangrove soils are typically waterlogged and anaerobic, which slows down microbial decomposition of organic matter (Faure & Svatikova, 2012).

Hydrology and tidal dynamics are crucial, as tidal inundation brings sediments and nutrients that support soil formation and carbon burial. Salinity gradients influence species composition and productivity, with certain species exhibiting higher growth rates and carbon accumulation. Climate variables such as temperature, rainfall, and sea level rise also affect mangrove productivity and carbon storage. Geomorphological settings, including coastal topography and sediment supply, determine the capacity of mangroves to build soils and maintain elevation relative to sea level. These factors affect the resilience of mangroves and their ability to continue functioning as carbon sinks under changing environmental conditions. Mangroves' role as carbon sinks has significant implications for climate change mitigation. By capturing and storing carbon, mangroves reduce atmospheric CO_2 concentrations, contributing to global efforts to limit warming.(Favasuli & Sebastian, 2021) The high carbon sequestration potential of mangroves makes them valuable in climate mitigation strategies, especially as nature-based solutions. Protecting existing mangrove forests is essential, as they already contain large carbon stocks, and their destruction can release stored carbon back into the atmosphere. Land-use changes such as conversion to aquaculture, agriculture, urban development, and infrastructure lead to mangrove deforestation, resulting in the oxidation of organic carbon in soils and emissions of CO_2 and methane. Therefore, preventing mangrove loss is a key priority in climate mitigation.

Mangroves in Climate Action

Mangroves play a crucial role in global climate action, serving as one of the most effective nature-based solutions for both climate mitigation and adaptation. Found along tropical and subtropical coastlines, mangrove forests are uniquely adapted to saline and waterlogged environments, with specialized root systems that stabilize sediments, reduce erosion, and support rich biodiversity. These ecosystems are vital for climate action because they sequester and store large quantities of carbon in both aboveground biomass and belowground sediments, a feature that distinguishes them as prominent components of the "blue carbon" framework. The term blue carbon refers to the carbon captured and stored by coastal and marine ecosystems, including mangroves, seagrasses, and salt marshes, which have been increasingly recognized for their high carbon density and long-term storage potential. Mangroves are particularly significant within this framework due to their ability to sequester carbon at high rates and store it in deep, anoxic sediments where decomposition is slow, allowing carbon to remain trapped for centuries or even millennia. Climate action involves both reducing greenhouse gas emissions and enhancing the resilience of communities and ecosystems to climate impacts(Fourqurean et al., 2012). Mangroves contribute to climate mitigation through carbon sequestration, capturing CO_2 through photosynthesis and storing it in biomass and sediments. Their high productivity and slow organic matter breakdown make mangroves highly efficient carbon sinks, often outperforming terrestrial forests on a per-unit-area basis. Protecting intact mangrove forests is essential for maintaining existing carbon stocks, as deforestation and degradation can release stored carbon back into the atmosphere, contributing to greenhouse gas emissions.

Community-based restoration approaches have proven effective, as local communities possess traditional knowledge and rely on mangroves for livelihoods. Engaging communities in restoration promotes sustainable practices, ensures long-term stewardship, and enhances social acceptance of conservation initiatives. In addition to mitigation, mangroves support climate adaptation by providing natural coastal protection. Their dense root networks reduce wave energy, stabilize shorelines, and trap sediments, thereby reducing erosion and buffering coastal areas against storm surges, flooding, and sea level rise.(Gordon et al., 2011) This protective function is increasingly important as climate change intensifies extreme weather events and increases coastal vulnerability. By maintaining coastal integrity, mangroves help protect infrastructure, livelihoods, and human lives, making them valuable components of climate resilience strategies. Mangroves also support biodiversity and ecosystem health, which are integral to climate action. These ecosystems provide habitat and nursery grounds for a diverse array of species, including fish, crustaceans, birds, and reptiles. Healthy mangrove ecosystems contribute to fisheries productivity and food security, particularly for coastal communities. Mangroves also improve water quality by filtering pollutants and trapping sediments, benefiting adjacent ecosystems such as coral reefs and seagrass beds. The co-benefits of mangrove conservation enhance the overall value of climate action initiatives and support sustainable development. Policy and governance mechanisms are essential to integrate mangroves into climate action frameworks. International climate agreements, such as the Paris Agreement, encourage countries to include nature-based solutions and ecosystem-based approaches in their Nationally Determined Contributions (NDCs). Many countries have incorporated mangrove conservation and restoration into their climate action plans, recognizing the role of blue carbon in meeting emission reduction targets. Climate finance mechanisms, including carbon credits, payment for ecosystem services, and conservation funds, support mangrove projects (Hamilton & Casey, 2016).

Environmental Role of Mangroves

Mangroves are among the most ecologically significant coastal ecosystems, performing a wide range of environmental functions that support biodiversity, protect shorelines, and maintain the health of marine and terrestrial systems. Found in tropical and subtropical intertidal zones, mangrove forests are characterized by salt-tolerant trees and shrubs adapted to saline, waterlogged conditions, and fluctuating tidal regimes. Their unique root structures, including prop roots and pneumatophores, enable them to stabilize sediments, reduce coastal erosion, and create complex habitats for diverse flora and fauna. The environmental role of mangroves extends beyond coastal protection, encompassing carbon sequestration, nutrient cycling, water filtration, and the maintenance of ecological connectivity between land and sea. One of the primary environmental functions of mangroves is shoreline stabilization and erosion control. Mangrove roots bind and trap sediments, reducing the impact of wave action and tidal currents. This sediment-trapping function builds and maintains coastal landforms, including mudflats and mangrove islands, which can expand over time through sediment accretion. As a result, mangroves can store large amounts of carbon in their sediments, often at depths of several meters. This long-term carbon storage contributes to climate change mitigation by reducing greenhouse gas concentrations (Harpankar, 2019).

The loss of mangroves can therefore lead to declines in fish stocks and reduced fisheries productivity, affecting food security and livelihoods in coastal communities. Mangroves also contribute to biodiversity conservation by providing habitats for a wide range of species. These ecosystems support unique flora and fauna adapted to saline conditions, including mangrove-specific plants, crustaceans, mollusks, and birds. Mangroves are important stopover sites for migratory birds, offering feeding and resting grounds during long-distance migrations. The structural complexity of mangrove forests supports high species diversity and contributes to the overall health of coastal ecosystems. Mangroves also maintain ecological connectivity between terrestrial and marine environments. They act as ecological corridors that link rivers, estuaries, and coastal waters, facilitating the movement of nutrients, organisms, and sediments. This connectivity supports the functioning of adjacent ecosystems such as coral reefs and seagrass beds. Mangroves filter sediments and pollutants from runoff, improving water quality and protecting downstream ecosystems. By trapping sediments and filtering pollutants such as heavy metals and nutrients, mangroves reduce turbidity and prevent harmful algal blooms, which can damage coral reefs and seagrass meadows (Inoue, 2018).

Mangroves & Carbon Storage

Mangroves are globally recognized as highly efficient carbon storage ecosystems, playing a critical role in climate regulation through their capacity to sequester atmospheric carbon dioxide (CO_2) and store it in biomass and sediment. These unique coastal forests, found in tropical and subtropical intertidal zones, are characterized by salt-tolerant trees and shrubs that thrive in saline and waterlogged environments. Mangrove ecosystems are distinguished by their complex root structures, including prop roots and pneumatophores, which stabilize sediments, support diverse biodiversity, and create conditions favorable for carbon accumulation. Carbon storage in mangroves is a central aspect of the broader concept of "blue carbon," which encompasses carbon captured and stored by coastal and marine ecosystems such as mangroves, seagrasses, and salt marshes. Blue carbon ecosystems are recognized for their high carbon density and long-term storage potential, making them essential components of nature-based climate mitigation strategies. (Islam & Bhuiyan, 2018)The carbon storage capacity of mangroves is rooted in their high primary productivity and the slow decomposition rates in waterlogged, anoxic soils. Through photosynthesis, mangrove trees absorb CO_2 from the atmosphere and convert it into organic carbon compounds that contribute to biomass growth. Carbon is stored in aboveground biomass, including trunks, branches, and leaves, and in belowground biomass such as root systems. Mangrove roots contribute to carbon storage by depositing organic matter into soils through root turnover and litter fall. A defining feature of mangrove carbon storage is the accumulation of organic carbon in sediments.

Figure 1. Importance of Mangroves in Blue Economy

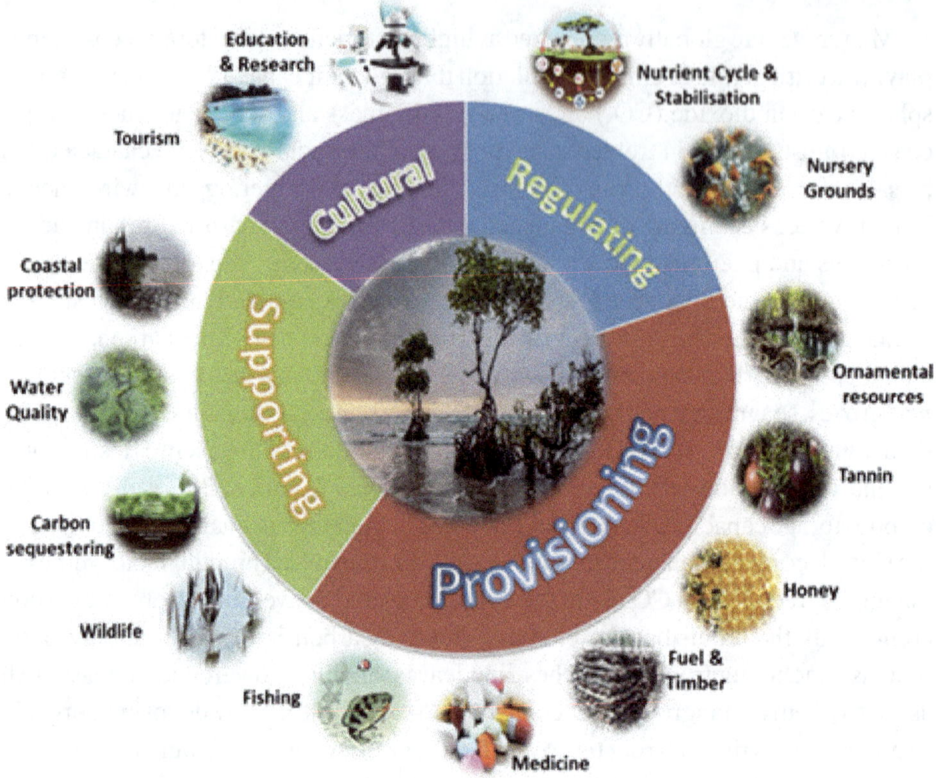

Tidal inundation brings sediments and organic matter into mangrove forests, supporting soil accretion and carbon burial. Salinity gradients influence species composition and productivity, with certain species exhibiting higher growth rates and carbon accumulation. Climate factors such as temperature, rainfall, and sea level rise also affect mangrove productivity and carbon storage. Geomorphological settings, including coastal topography and sediment supply, determine the capacity of mangroves to build soils and maintain elevation relative to sea level. These factors influence the resilience of mangroves to climate change and their ability to continue storing carbon under changing environmental conditions. (Kunnathully et al., 2026). The role of mangroves in carbon storage has significant implications for climate change mitigation. By capturing and storing carbon, mangroves help reduce atmospheric CO_2 concentrations, contributing to global efforts to limit warming. Protecting existing mangrove forests is essential because they already contain large carbon stocks, and their destruction can release stored carbon back into the atmosphere. Land-use changes such as conversion to aquaculture, agriculture, urban development, and infrastructure result in mangrove deforestation, leading to the oxidation of organic carbon in soils and emissions of CO_2 and methane. Therefore, preventing mangrove loss is a key priority in climate mitigation. Restoration of degraded mangroves can enhance carbon storage by re-establishing vegetation and improving sediment dynamics.

Climate Change & Mangroves

Climate change poses both severe threats and significant management opportunities for mangrove ecosystems, which are among the most valuable coastal habitats due to their ecological functions, carbon storage capacity, and socio-economic benefits. Mangroves, found in tropical and subtropical intertidal zones, are uniquely adapted to saline and waterlogged conditions through specialized root systems and physiological mechanisms that allow them to thrive in fluctuating tidal environments. However, the accelerating impacts of climate change—including sea level rise, increased frequency and intensity of storms, altered precipitation patterns, and changing temperature regimes—challenge the resilience of mangrove forests and the services they provide. One of the most direct impacts of climate change on mangroves is sea level rise, which can inundate coastal zones and alter the hydrological and salinity regimes critical to mangrove survival.(Kim et al., 2017) Mangroves depend on a balance between sediment accretion and sea level rise to maintain soil elevation and avoid submergence. In areas where sediment supply is sufficient, mangroves may keep pace with rising seas through sediment deposition and root growth. However, where sediment supply is limited or coastal development prevents natural sediment movement, mangroves may experience coastal squeeze, leading to habitat loss. Landward migration of mangroves is possible if there is space and suitable conditions, but in many coastal regions, hard infrastructure and land-use changes block migration, causing a net loss of mangrove habitat. Increased storm intensity and extreme weather events associated with climate change also pose significant risks to mangroves. Strong winds, storm surges, and heavy rainfall can cause physical damage to mangrove stands, uprooting trees, eroding sediments, and altering forest structure. While mangroves are adapted to withstand periodic disturbances, the increasing frequency and severity of storms can exceed their recovery capacity, leading to long-term degradation.

However, temperature stress can also affect physiological processes and increase susceptibility to pests and diseases. Altered precipitation patterns, including changes in freshwater inflow, can affect salinity gradients, nutrient availability, and sediment dynamics. Reduced freshwater input can increase salinity stress, while increased freshwater may alter sediment deposition and nutrient loads. These changes can influence mangrove productivity and species composition, potentially affecting ecosystem resilience and carbon sequestration capacity. Climate change also interacts with other environmental stressors, amplifying the vulnerability of mangroves. Human activities such as coastal development, aquaculture expansion, pollution, and deforestation have already degraded many mangrove ecosystems. These anthropogenic pressures reduce mangrove resilience by fragmenting habitats, reducing genetic diversity, and disrupting hydrological regimes. When combined with climate change impacts, these stressors can lead to accelerated mangrove loss and diminished ecosystem services. For example, conversion of mangroves to shrimp farms or urban infrastructure not only removes vegetation but also disturbs carbon-rich sediments, leading to greenhouse gas emissions. The vulnerability of mangroves to climate change has profound implications for climate mitigation and adaptation. Mangroves are recognized as significant carbon sinks, storing large amounts of carbon in biomass and sediments(Laffoley et al., 2021). Their high productivity and slow decomposition rates in anaerobic soils allow carbon to accumulate over time, making mangroves important components of the blue carbon framework.

Mangroves for Climate Mitigation

Mangroves are increasingly recognized as critical ecosystems for climate mitigation due to their exceptional capacity to sequester and store carbon, their resilience in coastal environments, and their ability to deliver multiple co-benefits that support sustainable development. Found in tropical and subtropical intertidal zones, mangrove forests consist of salt-tolerant trees and shrubs adapted to saline, waterlogged, and fluctuating tidal conditions. These ecosystems are unique in their ecological structure and function, particularly in their ability to capture atmospheric carbon dioxide (CO_2) through photosynthesis and store it in biomass and sediments. The high productivity of mangroves, combined with slow decomposition rates in anaerobic soils, enables them to accumulate significant amounts of organic carbon over time, making them among the most carbon-dense ecosystems on Earth. This carbon storage capability positions mangroves as valuable components of climate mitigation strategies, especially within the broader framework of "blue carbon," which encompasses carbon sequestration in coastal and marine ecosystems such as mangroves, seagrasses, and salt marshes. Blue carbon ecosystems are distinguished by their ability to sequester carbon in long-lived sediments, where organic matter can be buried and preserved for centuries or even millennia (Lamb et al., 2017). Climate mitigation involves reducing greenhouse gas concentrations in the atmosphere, and mangroves contribute to this goal through both carbon sequestration and avoidance of emissions.

Engaging communities promotes sustainable management, reduces conflicts, and enhances long-term stewardship. The mitigation benefits of mangroves extend beyond carbon storage. Mangroves provide essential ecosystem services that enhance resilience to climate impacts, including coastal protection, erosion control, and storm buffering. (Maulidia et al., 2022)Their complex root systems stabilize sediments, reduce wave energy, and trap sediments, thereby protecting coastal communities from storm surges and sea level rise. These protective services reduce the need for expensive engineered infrastructure, making mangrove conservation a cost-effective adaptation and mitigation strategy. Mangroves also support biodiversity by providing habitat and nursery grounds for fish, crustaceans, birds, and other species. This biodiversity support contributes to fisheries productivity and food security, which are important for coastal communities, particularly in developing regions. Mangroves also improve water quality by filtering pollutants and trapping sediments, benefiting adjacent ecosystems such as coral reefs and seagrass beds. The co-benefits of mangrove conservation enhance the overall value of mitigation efforts and support sustainable development goals. Integrating mangroves into climate mitigation policy has gained momentum through international frameworks and national commitments.

Mangrove Ecosystems & Carbon

Mangrove ecosystems represent a unique intersection of ecological productivity, coastal resilience, and global carbon dynamics, functioning as critical blue carbon reservoirs that sequester and store substantial amounts of carbon in both biomass and sediments. These coastal forests, located in tropical and subtropical intertidal zones, are characterized by salt-tolerant tree and shrub species that have evolved specialized adaptations—such as pneumatophores, prop roots, and salt-excreting leaves—to survive in high salinity, low oxygen, and fluctuating tidal environments. The ecological structure of mangroves enables them to trap sediments, stabilize shorelines, and create complex root networks that support diverse biological communities, including fish, crustaceans, birds, and microorganisms. Mangrove ecosystems are highly productive, with primary productivity often exceeding that of many terrestrial forests. This productivity, combined with the slow decomposition rates in waterlogged, anoxic soils, results in significant accumulation of organic matter in the sediments, where carbon can be buried and stored for long periods. (Nor et al., 2020) The carbon dynamics in mangroves involve both sequestration through photosynthesis and long-term storage in soil and biomass. Mangroves capture atmospheric carbon dioxide (CO_2) via photosynthesis and allocate carbon to aboveground biomass (trunks, branches, leaves) and belowground biomass (roots). A substantial proportion of this carbon is transferred to the soil through leaf litter, root turnover, and organic matter deposition. In the anaerobic conditions of mangrove sediments, decomposition is slow, allowing organic carbon to accumulate and form thick carbon-rich soil layers.

The carbon sequestration potential of mangroves is influenced by various ecological, environmental, and anthropogenic factors. Species composition, forest structure, age, and productivity affect biomass accumulation. Hydrology and tidal inundation influence nutrient availability, salinity gradients, and sediment deposition, which in turn affect carbon accumulation in soils. Geomorphological setting, such as deltaic, estuarine, or fringe mangroves, determines sediment supply and stability. Climate variables like temperature, rainfall, and sea level rise also influence mangrove growth and carbon dynamics. For instance, increased freshwater inflow can reduce salinity stress and enhance productivity, while prolonged droughts can increase salinity and reduce growth. Sea level rise can affect sediment dynamics and soil elevation, impacting carbon storage capacity. Mangrove ecosystems also play a significant role in carbon emissions when they are disturbed or degraded. Conversion of mangroves to aquaculture, agriculture, urban development, or infrastructure leads to the loss of vegetation and disturbance of carbon-rich sediments. (Osland et al., 2012)When mangrove soils are exposed, drained, or disturbed, the stored carbon can be oxidized and released as CO_2 and methane, contributing to greenhouse gas emissions. Studies have shown that mangrove deforestation and degradation can release carbon at rates higher than many terrestrial forests due to the large belowground carbon stocks.

Mangroves: Climate Warriors

Mangroves, often referred to as the "climate warriors" of coastal ecosystems, play a pivotal role in global climate regulation through their extraordinary capacity for carbon sequestration, coastal protection, and ecosystem resilience. These unique intertidal forests, located in tropical and subtropical regions, have evolved specialized adaptations that allow them to thrive in saline, waterlogged, and fluctuating tidal environments. Their complex root systems, including pneumatophores and prop roots, not only support the structural integrity of the trees but also trap sediments and organic matter, creating thick, carbon-rich soils that can store carbon for centuries or millennia. The ecological importance of mangroves extends beyond carbon storage; they serve as biodiversity hotspots, nurseries for fish and crustaceans, and protective barriers against storm surges and coastal erosion. In the context of climate change, mangroves are increasingly recognized for their dual role as both mitigators and adaptors. Their ability to sequester carbon makes them critical for climate mitigation, while their capacity to buffer coastal communities from extreme weather events makes them essential for climate adaptation. The carbon sequestration potential of mangroves is exceptionally high due to their rapid primary productivity and slow decomposition rates in anaerobic soils. Mangroves capture atmospheric carbon dioxide through photosynthesis and allocate it to aboveground biomass—such as trunks, branches, and leaves—and belowground biomass, particularly extensive root systems. (Pedersen, 2019)A significant portion of the captured carbon is transferred to the soil through leaf litter, root turnover, and sediment deposition. Because mangrove soils are often waterlogged and oxygen-poor, microbial decomposition is slow, allowing organic carbon to accumulate over time. This process results in the formation of deep, carbon-dense peat-like soils, which store more carbon per unit area than most terrestrial forests. The concept of "blue carbon" has brought global attention to mangroves as essential carbon sinks within coastal ecosystems. Blue carbon refers to the carbon captured and stored in coastal and marine ecosystems such as mangroves, seagrasses, and salt marshes.

Protecting existing mangrove forests is a key mitigation strategy, as these ecosystems already contain substantial carbon stocks. When mangroves are cleared or degraded, the stored carbon is released back into the atmosphere as carbon dioxide and methane, contributing to greenhouse gas emissions. Conversion of mangroves to aquaculture, agriculture, urban development, and infrastructure not only destroys vegetation but also disturbs carbon-rich sediments, accelerating carbon loss. Consequently, preventing mangrove deforestation is often more cost-effective and environmentally beneficial than restoration alone. Restoration of degraded mangrove areas also contributes to climate mitigation by re-establishing vegetation and enabling carbon accumulation in soils (Pendleton et al., 2012). Restoration efforts typically involve planting native mangrove species, restoring tidal hydrology, and ensuring adequate sediment supply. Successful restoration requires careful planning, including assessment of local environmental conditions, species suitability, and long-term monitoring. Community participation is crucial in restoration projects, as local communities often possess traditional knowledge and depend on mangroves for livelihoods. Engaging communities ensures sustainable management and fosters stewardship, while also addressing social equity and benefit-sharing. Beyond carbon storage, mangroves provide essential ecosystem services that enhance resilience to climate impacts. Their root networks reduce wave energy and stabilize shorelines, protecting coastal communities from storm surges and erosion.

Carbon Sequestration in Mangroves

Carbon sequestration in mangroves represents one of the most effective natural processes for mitigating climate change due to the exceptional ability of these coastal ecosystems to capture, store, and stabilize carbon over long periods. Mangroves, located in tropical and subtropical intertidal zones, are characterized by salt-tolerant trees and shrubs that have evolved unique adaptations such as aerial roots (pneumatophores), prop roots, and salt-excreting leaves, enabling them to thrive in saline, waterlogged, and low-oxygen soils.(Ranjan, 2019). The trapped sediments also contribute to soil building and elevation gain, enabling mangroves to keep pace with sea level rise. This sediment trapping function is crucial for maintaining mangrove resilience and sustaining carbon sequestration over time. Carbon sequestration in mangroves is influenced by multiple factors including species composition, forest structure, age, hydrology, sediment supply, and climate conditions. Different mangrove species vary in growth rates, biomass allocation, and tolerance to salinity and inundation, which affects carbon accumulation. Mature, undisturbed mangrove forests generally have higher carbon stocks due to larger biomass and deeper soils compared to young or degraded stands. Hydrological conditions such as tidal range and frequency of inundation influence nutrient availability, salinity gradients, and soil oxygen levels, impacting productivity and decomposition rates. Sediment supply from rivers and coastal processes affects soil accretion and carbon burial, while climate variables such as temperature, rainfall, and sea level rise influence mangrove growth and carbon dynamics (Ribeiro et al., 2019).

Mangroves & Environmental Challenges

Mangrove ecosystems, renowned for their ecological productivity and multifunctional value, face a growing array of environmental challenges that threaten their survival, compromise their ecosystem services, and undermine their role in climate regulation, coastal protection, and biodiversity conservation. These challenges stem from a combination of natural processes and human-induced pressures, including climate change, coastal development, pollution, hydrological alteration, overexploitation, and invasive species. Mangroves are adapted to harsh intertidal environments characterized by high salinity, waterlogged soils, and tidal fluctuations. Their specialized root systems, such as pneumatophores and prop roots, enable gas exchange and sediment stabilization, while salt-excreting mechanisms allow them to tolerate saline conditions. However, these adaptations are being tested as environmental conditions shift rapidly due to global and local drivers. One of the most significant environmental challenges facing mangroves is climate change, which manifests through sea level rise, changing temperature regimes, altered precipitation patterns, and increasing frequency and intensity of extreme weather events. Sea level rise poses a direct threat to mangroves by inundating coastal zones, altering salinity gradients, and affecting sediment dynamics. Mangroves maintain their elevation relative to sea level through sediment accretion and root growth. In areas with sufficient sediment supply, mangroves may keep pace with rising seas by building soil vertically. However, in regions where sediment supply is limited or where coastal infrastructure disrupts natural sediment flows, mangroves may experience "coastal squeeze," where they are trapped between rising seas and fixed land boundaries, leading to habitat loss.

Pollution is another major environmental challenge for mangroves. Industrial effluents, agricultural runoff, sewage discharge, and oil spills introduce contaminants such as heavy metals, pesticides, excess nutrients, and hydrocarbons into coastal waters. Mangroves can absorb and trap pollutants, but chronic contamination can impair growth, reduce reproductive success, and alter species composition. Nutrient pollution, particularly from agricultural runoff, can lead to eutrophication, algal blooms, and hypoxic conditions, which negatively affect mangrove-associated fauna and microbial communities. Plastic pollution is an emerging threat, as mangroves trap floating debris in their root systems, leading to physical damage, reduced light penetration, and habitat degradation. Hydrological alteration is another challenge that affects mangrove ecosystems. Activities such as dam construction, river channelization, groundwater extraction, and coastal engineering can alter freshwater inflow, sediment supply, and tidal dynamics. Changes in freshwater availability can shift salinity regimes, affecting mangrove species composition and productivity. Reduced sediment supply can limit soil accretion and increase vulnerability to sea level rise. Altered tidal dynamics can disrupt nutrient exchange and sediment deposition, affecting mangrove growth and carbon sequestration.

Mangroves in Climate Resilience

Mangroves are increasingly recognized as foundational ecosystems for climate resilience, providing a suite of ecological services that buffer coastal communities against the impacts of climate change while also supporting biodiversity, livelihoods, and carbon storage. Located in tropical and subtropical intertidal zones, mangrove forests are uniquely adapted to saline, waterlogged, and dynamic tidal environments through specialized root systems, salt filtration mechanisms, and physiological adaptations that allow them to tolerate extreme conditions. Their ecological significance extends beyond biodiversity and habitat provision, encompassing coastal protection, sediment stabilization, nutrient cycling, and blue carbon storage. Climate resilience refers to the capacity of social-ecological systems to anticipate, absorb, adapt to, and recover from climate-related shocks and stresses. In this context, mangroves contribute to resilience by reducing vulnerability to sea level rise, storm surges, coastal erosion, and extreme weather events. Their dense root networks and aboveground biomass dissipate wave energy and stabilize shorelines, thereby reducing the physical impacts of storms and reducing flood risks. By trapping sediments, mangroves also facilitate land building and maintain soil elevation relative to rising sea levels, enhancing the ability of coastlines to adapt to climate change. One of the key resilience functions of mangroves is their role in coastal protection.

The accumulation of sediments not only supports mangrove survival but also creates new land areas that can support additional vegetation and habitat. However, the ability of mangroves to keep pace with sea level rise depends on factors such as sediment availability, hydrological conditions, and coastal development. In areas where sediment supply is reduced due to dam construction or coastal engineering, mangroves may struggle to maintain elevation, leading to habitat loss. Mangroves also enhance climate resilience through their role in biodiversity support and ecosystem productivity. Mangrove forests provide habitat and nursery grounds for a wide range of marine and terrestrial species, including fish, crustaceans, birds, and mammals. The biodiversity supported by mangroves contributes to the resilience of coastal ecosystems by maintaining ecological functions and supporting food webs. Fisheries, which are vital for food security and livelihoods in many coastal communities, depend on mangrove habitats for breeding and juvenile development.(Thomas & Theokritoff, 2021) The resilience of fisheries to climate impacts is therefore linked to the health of mangrove ecosystems. In addition, mangroves support nutrient cycling and water quality improvement by filtering pollutants and trapping sediments. This function is important for maintaining the health of adjacent ecosystems such as coral reefs and seagrass beds, which are also vulnerable to climate change. By improving water clarity and reducing nutrient pollution, mangroves help sustain the broader coastal ecosystem, enhancing resilience at the landscape scale.

Mangrove Forests & Carbon

Mangrove forests are increasingly recognized as pivotal ecosystems in the global carbon cycle due to their extraordinary capacity to sequester and store carbon in both biomass and sediments, making them essential components of climate mitigation and nature-based solutions. Located in tropical and subtropical intertidal zones, mangrove forests consist of salt-tolerant trees and shrubs adapted to high salinity, waterlogged soils, and dynamic tidal regimes. Their distinctive root structures, including pneumatophores, prop roots, and buttress roots, facilitate gas exchange, stabilize sediments, and trap organic matter, enabling mangroves to thrive in conditions that are challenging for most terrestrial plants. These structural adaptations not only support mangrove survival but also enhance their role as carbon sinks. Mangrove forests exhibit high primary productivity, often exceeding that of many terrestrial forests, driven by abundant sunlight, warm temperatures, and nutrient inputs from tidal waters. Through photosynthesis, mangroves convert atmospheric carbon dioxide (CO_2) into organic carbon, which is allocated to aboveground biomass—such as trunks, branches, and leaves—and belowground biomass, particularly extensive root systems.(Twilley et al., 2017) A significant portion of the fixed carbon is transferred to the soil through leaf litter, root turnover, and sediment deposition. Mangrove soils are typically waterlogged and anaerobic, resulting in slow microbial decomposition. This slow decomposition enables organic carbon to accumulate in sediments, forming deep, carbon-rich layers that can store carbon for centuries or millennia.

The carbon storage capacity of mangrove forests is influenced by various ecological and environmental factors. Species composition, forest structure, age, and productivity affect biomass accumulation and carbon allocation. Mature mangrove forests with dense canopy cover and extensive root systems generally have higher carbon stocks compared to young or degraded stands. Hydrological conditions, such as tidal range and frequency of inundation, influence nutrient availability, salinity gradients, and soil oxygen levels, which affect productivity and decomposition rates. Sediment supply from rivers and coastal processes plays a critical role in soil accretion and carbon burial. Climate variables such as temperature, rainfall, and sea level rise also impact mangrove growth and carbon dynamics (Uddin et al., 2013). For instance, increased freshwater inflow can reduce salinity stress and enhance productivity, while prolonged droughts can increase salinity and reduce growth. Sea level rise can affect sediment dynamics and soil elevation, influencing carbon storage capacity. The carbon stored in mangrove forests is vulnerable to disturbance and land-use change. Conversion of mangroves to aquaculture, agriculture, urban development, and infrastructure leads to loss of vegetation and disturbance of carbon-rich sediments. When mangrove soils are drained or exposed, organic carbon is oxidized and released as CO_2 and methane, contributing to greenhouse gas emissions. Mangrove deforestation and degradation are therefore significant sources of emissions, and preventing mangrove loss is essential for climate mitigation. Restoration and conservation of mangrove forests are crucial for enhancing carbon sequestration and maintaining ecosystem services. Restoration efforts often involve planting native mangrove species, restoring hydrological connectivity, and removing barriers to natural regeneration.

Climate Change Impact on Mangroves

Climate change poses a profound and multifaceted threat to mangrove ecosystems, which are among the most productive and ecologically significant coastal habitats in tropical and subtropical regions. Mangroves provide vital ecosystem services, including carbon sequestration, coastal protection, biodiversity support, and livelihood sustenance for millions of people. However, the accelerating impacts of climate change—such as sea level rise, increased frequency and intensity of storms, altered precipitation patterns, and rising temperatures—are reshaping the environmental conditions that sustain mangrove forests, often in ways that exceed their adaptive capacity. One of the most direct and pervasive impacts of climate change on mangroves is sea level rise, which threatens to inundate coastal zones and alter hydrological and salinity regimes essential for mangrove survival. Mangroves maintain their position relative to sea level through sediment accretion and root growth. In areas where sediment supply is sufficient, mangroves may adapt by building soil vertically, thereby keeping pace with rising seas.(Valiela et al., 2001b) However, in regions where sediment supply is reduced—due to upstream damming, coastal engineering, or land-use changes—mangroves may be unable to keep up with the rate of sea level rise, leading to habitat loss.

Rising temperatures can influence mangrove growth rates, reproductive cycles, and species distribution. In some regions, warmer temperatures may allow mangroves to expand poleward into areas previously limited by cold temperatures, potentially increasing their range. However, higher temperatures can also increase heat stress, affect physiological processes, and alter species composition. Temperature changes can influence the timing of flowering and fruiting, affecting reproduction and recruitment. Additionally, warmer conditions can increase the prevalence of pests and diseases, further stressing mangrove populations. Altered precipitation patterns due to climate change also affect mangrove ecosystems. Changes in freshwater inflow from rivers and rainfall influence salinity gradients, nutrient availability, and sediment dynamics. Reduced freshwater input can increase salinity stress, affecting growth and survival, while increased freshwater may alter sediment deposition and nutrient loads. In some regions, prolonged droughts can increase salinity and reduce productivity, while extreme rainfall events can lead to flooding, sediment displacement, and changes in soil chemistry. These hydrological changes can impact mangrove species composition and ecosystem functioning. Climate change also interacts with other environmental stressors, amplifying the vulnerability of mangroves. (Valiela et al., 2001a).

Mangroves: Nature-Based Solution

Mangroves, the unique intertidal forests found along tropical and subtropical coastlines, have emerged as a powerful nature-based solution (NbS) for addressing multiple environmental challenges, including climate change mitigation, coastal protection, biodiversity conservation, and sustainable livelihoods. As nature-based solutions, mangrove ecosystems leverage natural processes and ecological functions to deliver cost-effective, scalable, and sustainable outcomes that align with both environmental and socio-economic objectives. The concept of NbS emphasizes working with nature to restore, manage, and protect ecosystems in ways that provide benefits for both people and the planet, rather than relying solely on engineered or technological interventions. In this context, mangroves represent an exemplary NbS because they simultaneously enhance ecosystem resilience, sequester carbon, reduce disaster risk, support fisheries, and provide cultural and economic value to coastal communities. One of the most significant contributions of mangroves as an NbS is their role in climate change mitigation through carbon sequestration and storage. Mangrove forests are among the most carbon-dense ecosystems globally, storing substantial amounts of carbon in both aboveground biomass and in deep, waterlogged soils. Through high productivity and slow decomposition under anaerobic conditions, mangroves accumulate organic carbon over long periods, effectively removing CO_2 from the atmosphere. (Vanderklift et al., 2019).

Their dense root systems attenuate wave energy, stabilize sediments, and reduce the force of incoming water, thereby diminishing the impact of extreme weather events on coastal communities. Research has shown that mangrove belts can significantly reduce wave height and energy, acting as natural buffers that complement engineered defenses such as seawalls. Unlike hard infrastructure, mangrove ecosystems can adapt and self-repair, growing and expanding in response to sediment deposition and changing conditions. This dynamic nature makes mangroves a resilient and cost-effective alternative or supplement to conventional coastal protection measures. Beyond climate and disaster resilience, mangroves support biodiversity and ecological health as a vital NbS. They serve as nurseries for a wide range of marine and estuarine species, including fish, crustaceans, and mollusks, which are essential for maintaining productive fisheries and supporting food security. Mangroves also provide habitat for numerous bird species, including migratory birds, and support diverse terrestrial and aquatic fauna. The structural complexity of mangrove forests, including canopy layers and root networks, creates niches for various organisms and fosters high levels of biodiversity. By sustaining ecological connectivity between terrestrial and marine ecosystems, mangroves contribute to the health of adjacent coral reefs and seagrass beds, as they trap sediments and filter pollutants, thereby improving water quality and reducing sedimentation impacts on sensitive habitats. The socio-economic benefits of mangroves as an NbS are substantial, particularly for coastal communities that depend on them for livelihoods and cultural identity. Mangroves support fisheries, providing breeding grounds and feeding habitats for commercially important species. They also offer resources such as timber, fuelwood, honey, and medicinal plants, which are integral to local economies and traditional practices. In many regions, mangrove tourism and ecotourism contribute to sustainable income generation, fostering community engagement in conservation.

Mangroves & Blue Carbon

Mangroves represent one of the most critical and effective blue carbon ecosystems, playing a central role in the global carbon cycle by sequestering and storing substantial amounts of carbon in their biomass and, most importantly, in their waterlogged soils. The soil carbon pool in mangroves often exceeds aboveground biomass carbon, highlighting the importance of sediment processes in carbon storage(Yarish et al., 2016). The waterlogged, anoxic environment slows the breakdown of organic matter, while tidal inundation continuously deposits new sediments, creating a stable environment for carbon burial. The sediment-trapping function of mangroves also contributes to soil elevation gain, which can help mangroves adapt to sea level rise. This dual function of carbon storage and coastal stabilization underscores the ecological importance of mangroves as blue carbon ecosystems. Mangrove blue carbon dynamics are influenced by a variety of ecological and environmental factors. Species composition, forest age, productivity, and structural complexity affect carbon accumulation. Mature mangrove forests with dense canopy and well-developed root systems typically store more carbon than younger or degraded forests. Hydrological conditions, such as tidal range, frequency of inundation, and salinity, influence nutrient availability and decomposition rates. Sediment supply from rivers and coastal processes plays a critical role in soil accretion and carbon burial. Climate variables, including temperature, rainfall, and sea level rise, also impact mangrove growth and carbon dynamics.

Mangrove Conservation for Climate

Mangrove conservation is increasingly recognized as a critical climate action strategy, offering a powerful combination of carbon sequestration, coastal protection, biodiversity preservation, and socio-economic benefits. As climate change accelerates, mangroves—unique intertidal forests found along tropical and subtropical coastlines—play a pivotal role in mitigating greenhouse gas emissions and enhancing resilience to climate-related hazards. Their capacity to sequester carbon in both biomass and deep sediments, their ability to buffer storm surges and coastal erosion, and their function as habitats for diverse marine and terrestrial species make mangrove ecosystems indispensable for climate mitigation and adaptation. The conservation of mangroves is essential because these ecosystems store some of the highest amounts of carbon per unit area among terrestrial and coastal habitats. Through photosynthesis, mangrove trees convert atmospheric carbon dioxide into organic carbon, which is stored in trunks, branches, leaves, and extensive root systems. A substantial portion of this carbon is transferred to soils via leaf litter, root turnover, and sediment deposition. The waterlogged, anoxic conditions of mangrove soils slow down decomposition, allowing organic carbon to accumulate over long periods, often for centuries. This long-term carbon storage makes mangroves crucial in climate mitigation, as protecting them prevents the release of stored carbon and maintains their ongoing carbon sequestration.

The conversion of mangroves to shrimp farms, agricultural land, and urban infrastructure disrupts carbon-rich soils, leading to the release of stored carbon as carbon dioxide and methane. Mangrove loss also reduces coastal protection and biodiversity, exacerbating climate vulnerability. Consequently, conserving existing mangrove forests is a cost-effective climate mitigation measure, as it avoids emissions that would result from ecosystem conversion. Conservation strategies must be grounded in ecological science and integrated into broader coastal management frameworks. Effective conservation involves protecting intact mangrove forests through legal mechanisms such as protected areas, conservation easements, and enforcement of regulations against illegal clearing. Strengthening governance and land tenure is crucial to ensure long-term protection, as unclear ownership and competing land uses often drive mangrove loss. In many regions, weak enforcement and lack of resources hinder conservation efforts. Therefore, capacity building, institutional coordination, and community participation are essential to improve governance and ensure sustainable management. Community-based conservation approaches have proven effective in many mangrove regions, as local communities often rely on mangroves for livelihoods and possess traditional ecological knowledge. Engaging communities in conservation fosters stewardship and ensures that benefits are equitably shared. Sustainable livelihood alternatives, such as ecotourism, sustainable fisheries, and non-timber forest products, can reduce pressure on mangroves while supporting local economies.

CONCLUSION

Mangroves are increasingly recognized as one of the most effective natural systems for mitigating climate change, acting as powerful carbon sinks, stabilizing coastal environments, and supporting resilient ecosystems that reduce vulnerability to climate impacts. Located in the intertidal zones of tropical and subtropical coastlines, mangrove forests are uniquely adapted to saline, waterlogged conditions through specialized root systems, such as pneumatophores, stilt roots, and prop roots, which facilitate gas exchange and stabilize sediments. These adaptations not only allow mangroves to thrive in challenging environments but also enable them to capture and store carbon at rates that surpass many terrestrial forests. Mangroves contribute to climate mitigation through several interlinked mechanisms, including carbon sequestration in biomass and soils, carbon burial through sediment accumulation, and protection of coastal areas that reduces climate-induced disasters. Their role in climate mitigation is integral to the concept of "blue carbon," which highlights the capacity of coastal ecosystems to capture and store carbon in sediments for long periods. The process of carbon sequestration in mangroves begins with photosynthesis, whereby mangrove trees convert atmospheric carbon dioxide (CO_2) into organic carbon. This carbon is stored in aboveground biomass, including trunks, branches, and leaves, and belowground biomass, particularly extensive root systems. A substantial portion of the organic carbon is transferred to soils through leaf litter, root turnover, and sediment deposition. Mangrove soils are typically anoxic and waterlogged, leading to slow decomposition of organic matter. As a result, organic carbon accumulates in sediments, creating deep carbon-rich layers that can persist for centuries to millennia. This long-term storage capacity makes mangroves exceptionally effective at removing CO_2 from the atmosphere and storing it in stable forms. Mangroves also contribute to climate mitigation through sediment trapping and carbon burial. Their dense root networks slow water flow and trap sediments carried by tides and rivers. Over time, these sediments build up, leading to soil accretion and vertical land growth, which helps mangroves keep pace with sea level rise. The trapped sediments often contain organic matter from terrestrial and marine sources, further enhancing carbon burial.

Hydrological factors, including tidal range, salinity, and frequency of inundation, affect nutrient availability and decomposition rates, thereby influencing soil carbon accumulation. Climate variables such as temperature, precipitation, and sea level rise also impact mangrove productivity and carbon dynamics. For example, increased freshwater inflow can reduce salinity stress and enhance productivity, while drought conditions can increase salinity and reduce growth. Sea level rise can affect sediment dynamics and soil elevation, influencing carbon storage capacity. Despite their high carbon sequestration potential, mangroves are under threat from anthropogenic activities such as coastal development, aquaculture expansion, deforestation, and pollution. The conversion of mangroves to shrimp farms, agricultural land, and urban infrastructure leads to the loss of vegetation and disturbance of carbon-rich soils. When mangrove soils are drained or exposed, stored carbon is oxidized and released as carbon dioxide and methane, contributing to greenhouse gas emissions. Mangrove loss thus represents a significant source of emissions, and conserving existing mangrove forests is a cost-effective climate mitigation strategy. Restoration and conservation are essential to maintain and enhance mangrove carbon stocks. Restoration efforts often involve planting native species, restoring hydrological connectivity, and removing barriers to natural regeneration. However, restored mangroves may take decades to develop carbon stocks comparable to natural forests, and long-term monitoring is essential to ensure ecological functionality. Natural regeneration, where feasible, is often more successful and cost-effective than planting, as it leverages the ecosystem's inherent recovery capacity. Community-based restoration and conservation can improve outcomes, as local communities often depend on mangroves for livelihoods and possess traditional knowledge about local conditions. Mangroves also support climate mitigation indirectly by protecting coastal ecosystems that are important carbon sinks, such as seagrass beds and coral reefs. By filtering sediments and reducing turbidity, mangroves enhance water quality and support the health of adjacent ecosystems.

REFERENCES

Alongi, D. M. (2020). Carbon balance in salt marsh and mangrove ecosystems: A global synthesis. *Journal of Marine Science and Engineering*, *8*(10), 767. Advance online publication. DOI: 10.3390/jmse8100767

Alongi, D. M. (2022). Impacts of climate change on blue carbon stocks and fluxes in mangrove forests. *Forests*, *13*(2), 149. Advance online publication. DOI: 10.3390/f13020149

Baranzini, A., Goldemberg, J., & Speck, S. (2000). A future for carbon taxes. *Ecological Economics*, *32*(3), 395–412. DOI: 10.1016/S0921-8009(99)00122-6

Barbier, E. B. (2006). Natural barriers to natural disasters: Replanting mangroves after the tsunami. *Frontiers in Ecology and the Environment*, *4*(3), 124–131. DOI: 10.1890/1540-9295(2006)004[0124:NBTNDR]2.0.CO;2

Callaway, J. C., Borgnis, E. L., Turner, R. E., & Milan, C. S. (2012). Carbon sequestration and sediment accretion in San Francisco Bay tidal wetlands. *Estuaries and Coasts*, *35*(5), 1163–1181. DOI: 10.1007/s12237-012-9508-9

Calumpong, H., Kirkman, H., Yokoya, N. Y., Hall-Spencer, J. M., Osman, N. A. R., & Park, C.. (2021). Marine plants and macroalgae. In *The second World Ocean Assessment* (pp. 224–248). United Nations.

DebRoy, P., & Jayaraman, R. (2012). Economic valuation of mangroves for assessing the livelihood of fisherfolk: A case study in India. In A. L. Shriver (Ed.), *Visible possibilities: The economics of sustainable fisheries, aquaculture and seafood trade. Proceedings of the Sixteenth Biennial Conference of the International Institute of Fisheries Economics and Trade, July 16-20, Dar es Salaam, Tanzania*. International Institute of Fisheries Economics and Trade.

Dewsbury, B. M., Bhat, M., & Fourqurean, J. W. (2016). A review of seagrass economic valuations: Gaps and progress in valuation approaches. *Ecosystem Services*, *18*, 68–77. DOI: 10.1016/j.ecoser.2016.02.010

Faure, M. G., & Svatikova, K. (2012). Criminal or administrative law to protect the environment? Evidence from Western Europe. *Journal of Environmental Law*, *24*(2), 253–286. DOI: 10.1093/jel/eqs005

Favasuli, S., & Sebastian, V. (2021, June 10). *Voluntary carbon markets: How they work, how they're priced and who's involved.* S&P Global. https://www.spglobal.com/platts/en/market-insights/blogs/energy-transition/061021-voluntary-carbon-markets-pricing-participants-trading-corsia-credits

Fourqurean, J. W., Duarte, C. M., Kennedy, H., Marbà, N., Holmer, M., Mateo, M. A., Apostolaki, E. T., Kendrick, G. A., Krause-Jensen, D., McGlathery, K. J., & Serrano, O. (2012). Seagrass ecosystems as a globally significant carbon stock. *Nature Geoscience*, *5*(7), 505–509. DOI: 10.1038/ngeo1477

Gordon, D., Murray, B. C., Pendleton, L., & Victor, B. (2011). *Financing options for blue carbon: Opportunities and lessons from the REDD+ experience*. Nicholas Institute for Environmental Policy Solutions, Duke University. https://nicholasinstitute.duke.edu/sites/default/files/publications/financing-options-for-blue-carbon-paper.pdf

Hamilton, S. E., & Casey, D. (2016). Creation of a high spatio-temporal resolution global database of continuous mangrove forest cover for the 21st century (CGMFC-21). *Global Ecology and Biogeography*, *25*(6), 729–738. DOI: 10.1111/geb.12449

Harpankar, K. (2019). Internal carbon pricing: Rationale, promise and limitations. *Carbon Management*, *10*(3), 219–225. DOI: 10.1080/17583004.2019.1577178

Inoue, Y. (2018). Ecosystem carbon stock, atmosphere, and food security in slash-and-burn land use: A geospatial study in mountainous region of Laos. In *Land-atmospheric research applications in South and Southeast Asia* (pp. 641–665). Springer. DOI: 10.1007/978-3-319-67474-2_28

Islam, S. M. D.-U., & Bhuiyan, M. A. H. (2018). Sundarbans mangrove forest of Bangladesh: Causes of degradation and sustainable management options. *Environmental Sustainability*, *1*(2), 113–131. DOI: 10.1007/s42398-018-0018-y

Kim, J. K., Yarish, C., Hwang, E. K., Park, M., & Kim, Y. (2017). Seaweed aquaculture: Cultivation technologies, challenges and its ecosystem services. *Algae - Korean Phycological Society*, *32*(1), 1–13. DOI: 10.4490/algae.2017.32.3.3

Kunnathully, K., Jinesh, M., Satyanarayana, P., Esther, V., Pathak, G. P., & Manjunath, T. C. (2026). Building human rights by design: Ethical AI development for religious freedom. In Chakraborty, S. (Ed.), *Intersections of AI and the freedom of religion or belief* (pp. 81–104). IGI Global Scientific Publishing., DOI: 10.4018/979-8-3373-3830-9.ch004

Laffoley, D., Baxter, J. M., Amon, D. J., Claudet, J., Hall-Spencer, J. M., Grorud-Colvert, K., Levin, L. A., Reid, P. C., Rogers, A. D., Taylor, M. L., Woodall, L. C., & Andersen, N. F. (2021). Evolving the narrative for protecting a rapidly changing ocean, post-COVID-19. *Aquatic Conservation*, *31*(6), 1512–1534. DOI: 10.1002/aqc.3512 PMID: 33362396

Lamb, J. B., van de Water, J. A. J. M., Bourne, D. G., Altier, C., Hein, M. Y., Fiorenza, E. A., Abu, N., Jompa, J., & Harvell, C. D. (2017). Seagrass ecosystems reduce exposure to bacterial pathogens of humans, fishes, and invertebrates. *Science*, *355*(6326), 731–733. DOI: 10.1126/science.aal1956 PMID: 28209895

Maulidia, V., Akbar, A. A., Jumiati, J., Arifin, A., & Sulastri, A. (2022). The value of mangrove ecosystems based on mangrove carbon sequestration in West Kalimantan. *Journal of Wetlands Environmental Management*, *10*(1), 12–25. DOI: 10.20527/jwem.v10i1.279

Nor, A. M., Gray, T. S., Caldwell, G. S., & Stead, S. M. (2020). A value chain analysis of Malaysia's seaweed industry. *Journal of Applied Phycology*, *32*(4), 2161–2171. DOI: 10.1007/s10811-019-02004-3

Osland, M. J., Spivak, A. C., Nestlerode, J. A., Lessmann, J. M., Almario, A. E., Heitmuller, P. T., Russell, M. J., Krauss, K. W., Alvarez, F., Dantin, D. D., Harvey, J. E., From, A. S., Cormier, N., & Stagg, C. L. (2012). Ecosystem development after mangrove wetland creation: Plant–soil change across a 20-year chronosequence. *Ecosystems (New York, N.Y.)*, *15*(5), 848–866. DOI: 10.1007/s10021-012-9551-1

Pedersen, O. (2019, August 19). Punishing the polluters: Why large fines are an important step towards cleaner corporations. *The Conversation*. https://theconversation.com/punishing-the-polluters-why-large-fines-are-an-important-step-towards-cleaner-corporations-115727

Pendleton, L., Donato, D. C., Murray, B. C., Crooks, S., Jenkins, W. A., Sifleet, S., Craft, C., Fourqurean, J. W., Kauffman, J. B., Marbà, N., Megonigal, P., Pidgeon, E., Herr, D., Gordon, D., & Baldera, A. (2012). Estimating global "blue carbon" emissions from conversion and degradation of vegetated coastal ecosystems. *PLoS One*, *7*(9), e43542. Advance online publication. DOI: 10.1371/journal.pone.0043542 PMID: 22962585

Ranjan, R. (2019). Optimal mangrove restoration through community engagement on coastal lands facing climatic risks: The case of Sundarbans region in India. *Land Use Policy*, *81*, 736–749. DOI: 10.1016/j.landusepol.2018.11.047

Ribeiro, R. D. A., Rovai, A. S., Twilley, R. R., & Castañeda-Moya, E. (2019). Spatial variability of mangrove primary productivity in the neotropics. *Ecosphere*, *10*(6), e02841. Advance online publication. DOI: 10.1002/ecs2.2841

Thomas, A., & Theokritoff, E. (2021). Debt-for-climate swaps for small islands. *Nature Climate Change*, *11*(10), 889–891. DOI: 10.1038/s41558-021-01194-4

Twilley, R. R., Castañeda-Moya, E., Rivera-Monroy, V. H., & Rovai, A. (2017). Productivity and carbon dynamics in mangrove wetlands. In *Mangrove ecosystems: A global biogeographic perspective* (pp. 113–162). Springer., DOI: 10.1007/978-3-319-62206-4_5

Uddin, M. S., de Ruyter van Steveninck, E., Stuip, M., & Shah, M. A. R. (2013). Economic valuation of provisioning and cultural services of a protected mangrove ecosystem: A case study on Sundarbans Reserve Forest, Bangladesh. *Ecosystem Services*, *5*, 88–93. DOI: 10.1016/j.ecoser.2013.07.002

Valiela, I., Bowen, J. L., Cole, M. L., Kroeger, K. D., Lawrence, D., Pabich, W. J., Tomasky, G., & Mazzilli, S. (2001b). Following up on a Margalevian concept: Interactions and exchanges among adjacent parcels of coastal landscapes. *Scientia Marina*, *65*(S2), 215–229. DOI: 10.3989/scimar.2001.65s2215

Valiela, I., Bowen, J. L., & York, J. K. (2001a). Mangrove forests: One of the world's threatened major tropical environments. *Bioscience*, *51*(10), 807–815. DOI: 10.1641/0006-3568(2001)051[0807:MFOOTW]2.0.CO;2

Vanderklift, M. A., Marcos-Martinez, R., Butler, J. R. A., Coleman, M., Lawrence, A., Prislan, H., Steven, A. D. L., & Thomas, S. (2019). Constraints and opportunities for market-based finance for the restoration and protection of blue carbon ecosystems. *Marine Policy*, *107*, 103429. Advance online publication. DOI: 10.1016/j.marpol.2019.02.001

Yarish, B., Hansen, B., Valderrama, S., Radulovich, D., Capron, F., & Goudey, H. (2016). *Seaweed aquaculture for food security, income generation and environmental health in tropical developing countries*. World Bank Group.

Chapter 4
Earth Observation-Driven Mangrove Management With Global Species Mapping and Blue Carbon Assessment

Uzair Bhatti
http://orcid.org/0000-0002-8743-2783
Hainan University, China

Yonis Gulzar
http://orcid.org/0000-0002-6515-1569
Department of Management Information Systems, College of Business Administration, King Faisal University, Saudi Arabia

Yazeed Yasin Ghadi
http://orcid.org/0000-0002-7121-495X
Al Ain University of Science and Technology, Abu Dhabi, UAE

Dilmurod Turaev
Department of Information Technology and Exact Sciences, Termez University of Economics and Service, Termez, Uzbekistan

Lola Safarova
Samarkand State University of Veterinary Medicine, Livestock, and Biotechnologies, Uzbekistan

Rustam Kudratov
Tashkent University of Information Technologies Named After Muhammad al-Khwarizmi, Samarkand, Uzbekistan

Gulkhan Dadenova
Department of Information Technologies, Kimyo International University in Tashkent, Uzbekistan

Kudratov Alisher Toshpulatovich
Samarkand State University of Veterinary Medicine, Livestock, and Biotechnologies, Uzbekistan

DOI: 10.4018/979-8-3373-9320-9.ch004

ABSTRACT

Coastal pollution is a major and increasing hazard to marine life, livelihoods of coastline peoples, and the environment on a worldwide scale. Mangrove forests in the interface of terrestrial and marine environments are important for ameliorating coastal pollution as they trap sediments, cycle nutrients, sequester contaminants and biodegrade substances. Nevertheless, the considerable degradation of mangroves due to urbanization, expansion of aquaculture and inadequate governance has eroded their ecological functions. In this context, grass-roots mangrove conservation has arisen as a positive participatory tool combining local knowledge management and livelihood incentives. This paper provides an overview of the factors responsible for mangroves' ability to mitigate against coastal pollution, and presents case studies of community-based mangrove conservation projects occurring in many different geographical settings along with an analysis of their challenges and policy implications.

INTRODUCTION

The anthropogenic pressures affecting the coastal ecosystem on Earth are unprecedented, which means that the strategies of sustainable management need to be designed to ensure the ecological soundness of the ecosystems in balance with the welfare of people (Fatimatuzzahroh et al., 2018). The mangrove forests, particularly, are being realized to have the ability to reduce these threats as well as sustain the lives of people living along the coastlines (Macamo et al., 2024). These special intertidal ecosystems are a source of very important ecosystem services, including carbon sequestration and storm defense, as well as water filtration, thus acting as an important point of contact between the land and the sea (Que et al., 2024; Rahmadi et al., 2023). Nevertheless, these most important ecosystems are becoming more and more degraded by land-based pollutants, such as agricultural runoff and industrial waste, which affect their filtration capacity and biodiversity (Lasaiba, 2023). Therefore, the destruction of mangrove ecosystem results in a lack of their capacity to act as natural filters, which increases the susceptibility of nearby marine environments to pollution (Anu et al., 2024). This has led to a move toward the community-based conservation models in which the local populations can restore and manage mangrove resources because effective environmental stewardship means its active involvement in conservation of these ecosystems (Anu et al., 2024; Sukarmen et al., 2023). Participatory governance frameworks empower local stakeholders to promote ecological sustainability and sustainable socio-economic development, as well as the conservation endeavors are adjusted to the requirements of communities residing along the coastline (Katsanevakis et al., 2025; Sukarmen et al., 2023). The subsequent sections look into the particular dangers of coastal contamination to the mangrove surroundings and the ways community-based management plans can help increase the resilience of these indispensable ecosystems (Hagger et al., 2022; Sukarmen et al., 2023).

The world is witnessing a rise in pollution of coastal ecosystems due to the high rate of urbanization, industrial discharge, agricultural run off and poor waste management. Mangroves are some of these ecosystems that are important natural buffers between the land and the sea and therefore, offer ecosystem services that include: stabilizing shorelines, carbon sequestration, habitat, and water quality maintenance. Nonetheless, human activities are killing the mangrove forests at alarming rates, which affects their ability to prevent pollution of the coast and safeguard the coastal population. The adult capability of such wetlands as effective biofilters depends on the complicated biological and physical processes which allow to trap the land-based contaminants and buffer the coastal hazards against shoreline erosion (Wang and Gu, 2021).

The mangrove is especially efficient in decreasing coastal pollution due to their complex root structure, which entraps the sediments and filters nutrients and captures pollutants, including heavy metals and excess nitrogen and phosphorus. These natural filtration measures will ensure that pollutant does not enter near shore waters and as such, coral reefs, seagrass beds, and fisheries would be preserved. It has been proven through scientific studies that intact mangrove ecosystems can greatly decrease the nutrient loads and enhance the water quality which makes them a vital nature-based remedy in the coastal pollution control. As important ecotonal elements, mangroves can perform purification of the coastline by filtration of roots and sediments as well as adsorbing heavy fees and storing nutrients, which is an excellent option to reduce nearshore eutrophication (Zhao et al., 2025). Moreover, these biogeochemical hotspots are dynamically engaged with groundwater and riverine water inflows to modify the coastal carbonate chemistry, which emphasizes the multifunctional capabilities of these biogeochemical hotspots in remediation of water quality and ecosystem-based adaptation. Nonetheless, these ecosystem services are highly affected by a major influx of nutrients caused by anthropogenic processes, including agriculture and urban development, which can cause eutrophication and change the structure of vegetation communities (Wang and Gu, 2021; Yulianto et al., 2024). Figure 1 shows the flow of this study.

Figure 1. Flowchart of review study

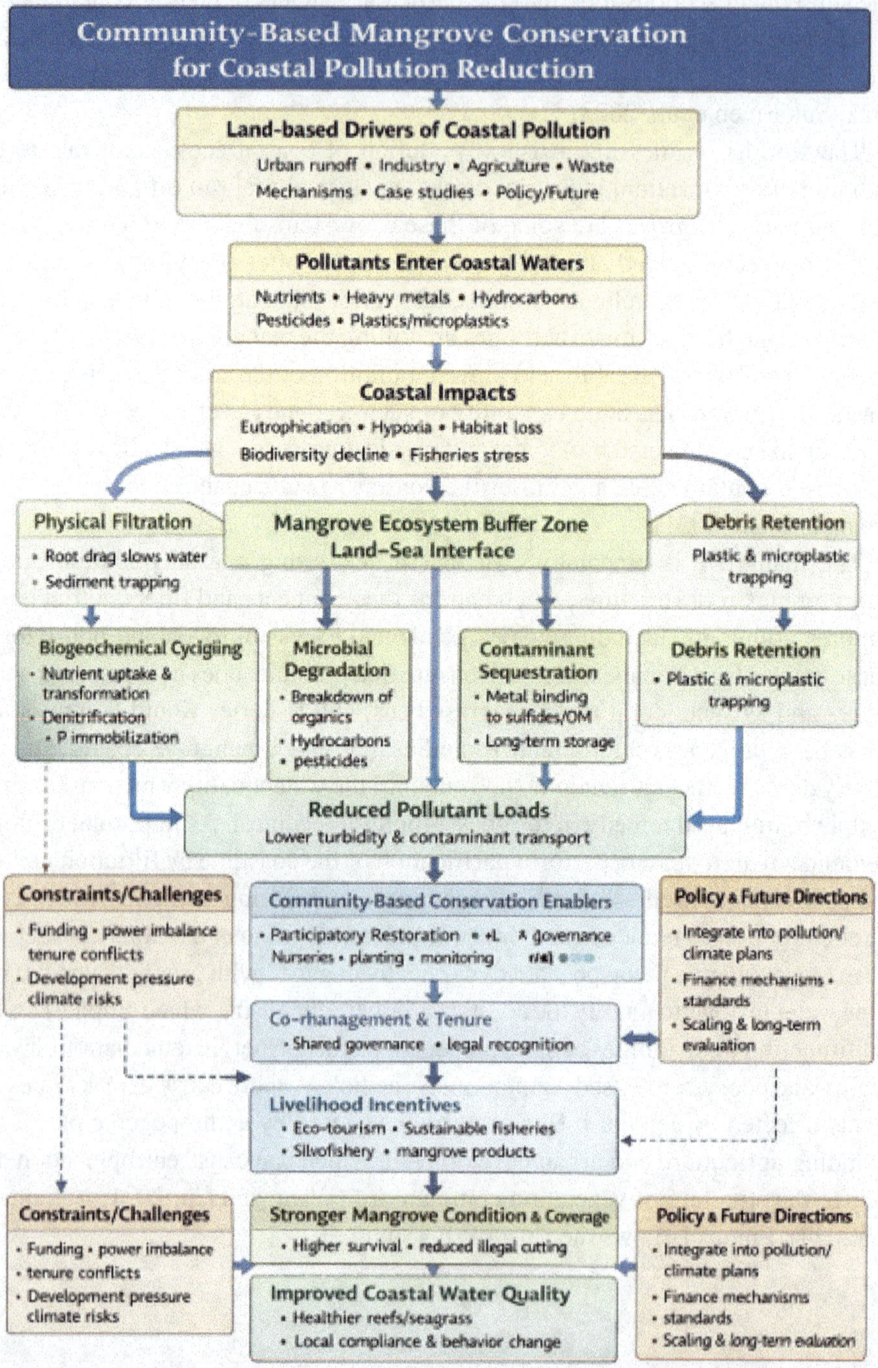

Even though these conservation practices are considered vital in conserving the environment, relying exclusively on the top-down methods of regulation has proved to fail because of the inability to enforce the regulations, the inadequacy of funds and the absence of local participation. Conversely, community-based mangrove conservation has also become a possible alternative whereby local knowledge, livelihood, and stewardship are merged into the management of the environment. Such initiatives by actively engaging communities along the coast in the protection, restoration and monitoring of mangroves will create a sense of ownership and commitment over the long term that is essential in maintaining conservation results. Proper governance mechanisms need both governmental and local input whereby there is collaboration and responsibility to guarantee the sustainable utilization of biological resources and their maintenance of life support networks (Bhagarathi & DaSilva, 2024). This correlation of socio-economic motivation and ecological stewardship forms a sound system of sustainable management, in which local populations are the major beneficiaries of the coastal environment, as well as, the custodians of this same environment. To effectively implement these strategies, it is necessary to address the main determinants, including long-term funding, legislation compliance, and implementation of local ecological knowledge to eliminate conflicts between conservation and competing land use (Friess et al., 2019; Yulianto et al., 2025).

The community-based conservation efforts do not only lead to ecological effectiveness but they also produce co-benefits on social and economic fronts. These are better livelihoods in the form of eco-tourism, sustainable fisheries and mangrove products and awareness of the causes of pollution and mitigation measures. Notably, the mangrove and other resources should be empowered to allow local communities to control the management of this resource because this action will ensure behavioral changes that will limit the source of land-based pollution and hence complement the pollution-filtering role of mangrove habitat. To decrease the number of nutrient loads and turbidity that negatively affect these important habitats, it is necessary to address anthropogenic perturbations of mangrove ecosystems, a crucial element of which is the role of pollution absorbers, including clearing the land and improper wastewater treatment (Trégarot et al., 2023), which is essential to ensure the preservation of the essential role of mangrove ecosystems. Next, the shift toward centralized governance is the less sustainable route to the conservation of the mangroves through the implementation of the more inclusive and community-driven management systems that would help solve the past failures involving the shortage of resources and the lack of appropriate enforcement (Bhagarathi and DaSilva, 2024; Sathiyamoorthy and Sakurai, 2024). In turn, by incorporating local societies into governance systems, it will use their reliance on these ecosystems as a motivating factor to restore, use and protect them through application of indigenous knowledge (Arfan et al., 2024; Elegbede et al., 2023).

This study examines the role of community-based mangrove conservation in reducing coastal pollution, focusing on how participatory governance, local capacity building, and collaborative management can enhance both environmental and social resilience. By synthesizing existing literature and practical case examples, the paper aims to highlight best practices, identify challenges, and propose strategies for integrating community-driven mangrove conservation into broader coastal pollution reduction and sustainable development frameworks.

1. COASTAL POLLUTION AND THE ROLE OF MANGROVES

Coastal pollution, increasingly becoming a worldwide challenge with the growth in population, development of cities and rapid industrialization leading to coastal eutrophication, together with agricultural waste entry and untreated garbage disposal(as shown in Fig 2). Some of the pollutants that find their way into coastal waters from upstream river systems and direct dumping include excess nutrients, heavy metals, hydrocarbons, pesticides and plastic which have led to eutrophication, degradation of habitats and decline of marine biodiversity. The effects are most serious in developing coastal areas with poor sewage treatment and enforcement of environmental regulations. These highly diverse transitional ecosystems, such as mangroves seagrass beds and coral reefs depend largely on the health of the surrounding coastal environment (Atapaththu et al., 2025), making them fragile to the environmental changes which affect the full nature of coastal biome. Nursery A mangrove nursery is situated at the interface between land and sea and receives both terrestrial and marine sources of pollution, but not into other coastal ecosystems such as coral reefs (Friess et al., 2019; Hewindati & Kusumaningrum, 2024).

Mangrove systems are an ecologically important natural treatment system for such CO-estuaries, functioning as biogeochemical filters at the land-sea interface. Their thick underground and above-ground root structures serve to capture fine sediments and pollutants, leading to less transport of contaminants into the nearshore. Fine sediments retained by mangrove roots sequester heavy metals and organic pollutants, and limit their release into marine environments like coral reefs and seagrass beds. Studies have shown that mangroves and saltmarshes are also bioreactors where incoming substrates are retained to be stored in the sediment or exported to the ocean, showing their potential as hotspots for material (Anu et al., 2024). These intricate sediments and the bio-communities that they support also enhance by way of biotransformation and bioaccumulation processes, in reducing toxic pollutants without their outright release into the neighbouring ocean (Maiti & Chowdhury, 2013).

Apart from the physical filtration, mangroves are involved in physiological and biological processes which helps in decreasing pollution load. Microbial populations in mangrove sediments are highly variable and have been implicated in nutrient and contaminant transformation/immobilization (PTI), with processes such as denitrification, sulfate reduction, and metal sequestration. These microbial processes greatly reduce nitrogen and phosphorus inputs, which can lead to eutrophic phenomena such as algal blooms, hypoxia and habitat destruction with severe impacts on coastal water quality and fisheries. Indeed, mangrove systems also dampen organic and plastic pollution adding onto their previous described ability: through the complex roots' grooves, interception is applied to brackish water particles and this feature contributes to lowering particle diffusion in thatd area (Chen, 2022; Earp et al., 2018). Mangroves, with their dense and well-developed root systems reduce wave impact and current flow velocity, which acts to effectively capture or trap floating matter (Zhang et al, 2024).

Figure 2. Pollution sources in coastal system

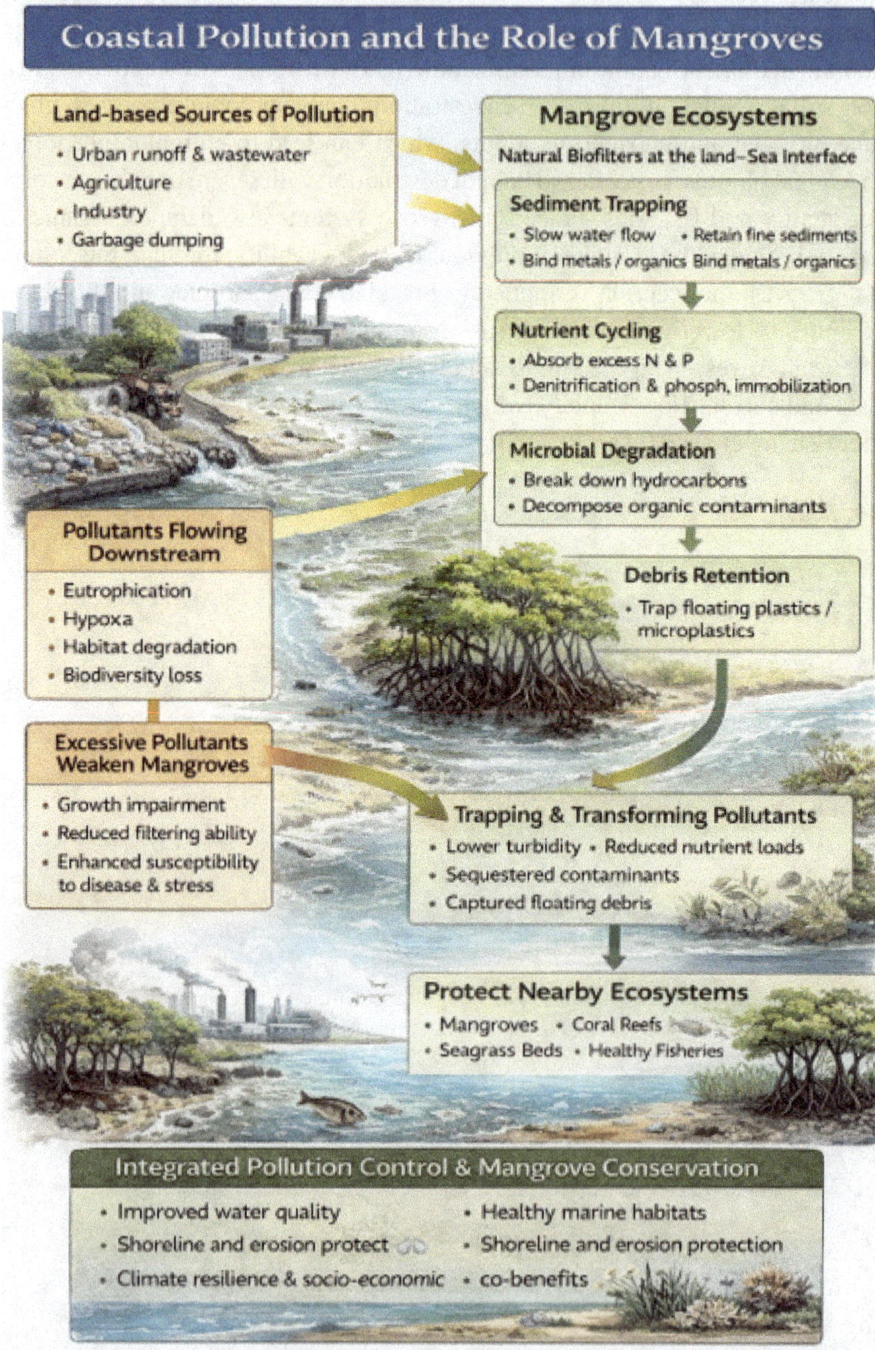

Mangroves also contribute to reducing organic and plastics pollution. The intricate root systems that sustain their structure aid in redirecting water flow, which results in the floating debris and microplastics to settle and be trapped inside the mangrove zone. Although this can result in accumulation of pollutants with mangrove sediments, it also restricts broader dispersion into open waters by demonstrating the retention and containment function of mangroves within the coastal environment. The intricate root-densities above the ground surfaces that are formed by mangrove species have been mediating in soil accumulation and lowered erosion, functioning as a catcher of land-based waste and microplastics to avoid them being moved from the terrestrial landscape into open waters (Chen, 2022; Friess et al., 2020; Li et al., 2025).

But the mangroves' ability to absorb pollution is not infinite. The accumulation of heavy pollutant loads can weaken mangrove ecosystems, causing growth suppression, changes in species composition and enhanced susceptibility to diseases and climate-induced stress. Summary: Continual pollution, especially with heavy metals and hydrocarbons, can damage mangrove physiological functions and diminish their role as natural pollution filters; thus integrated management plans are required for the maintenance of such environments. Mangroves were found to be good barriers against inorganic or organic pollutants, and their intricate root system can trap plastic debris and hinder its migration, thereby reducing the inflow of waste materials into open waters (Nugraheni et al., 2024). Surveys show that large plastic debris accumulates in mangrove wetlands more regularly than on adjacent beaches, becoming land-based dumpsites for waste as well as marine trash dumps that trap copious amounts of litter on the forest floor (Bhagarathi & DaSilva, 2024).

In this vein, preservation and restoration of mangrove habitats are increasingly viewed as efficient nature-based means to mitigate coastal pollution. Conserving mangroves also helps water quality and improves the resilience of both ecosystems, while providing benefits to fisheries, stabilizing shorelines and combating climate change. Inclusion of mangrove conservation into coastal pollution abatement schemes can thus provide diverse environmental and socio-economic co-benefits, further cementing their place as a linchpin of sustainable coastal management.

2. COMMUNITY-BASED MANGROVE CONSERVATION APPROACHES

Community-based mangrove conservation models in turn focus actively on engagement of local stakeholders in the conservation, rehabilitation and management of mangroves with sustainable practices. In contrast to more top-down models of protection, these models acknowledge coastal communities as the primary stewards of mangroves who bring their traditional ecological knowledge, reliance on ecosystem services, and long-term resilience in coastal environments. Community-based approaches instil stewardship and accountability that are essential to long-term success by institutionalizing sustainability within local governance. Equitable benefit-sharing arrangements and recognition of customary tenure are vital building blocks that reinforce local institutional capacity and help maintain the social sustainability that is key to the long-term viability of such projects. Participatory Monitoring Mechanisms Thus, PMFs add value to the local monitoring efforts in a way that strengthen the role of local actors as monitors capable of following ecosystem health, establishing degradation and assessing conservation outcomes using traditional indicators as well as utilizing expert measurements (Khan et al., 2025; Yulianto et al., 2025). These in return use local ecological knowledge to correctly evaluate changes contrast of vegetation which enables well-founded decisions about how to allocate resources and helps bridge the difference between indigenous knowledge-based practices and contemporary, scientific data collection (Souza et al., 2024). Involving local stakeholders to be part of the data collection process, these participatory approaches help establish a sense of ownership and guarantee that conservation strategies are culturally applicable and ecologically relevant for the particular mangrove setting (Yulianto et al., 2025).

Participatory mangrove restoration is one of the most effective community-based approaches in which local community members are involved in seed collection, nursery management, planting and monitoring. This method is more effective in restoration by thinking site specific and creating employment locally. Restoration efforts led by the community frequently produce higher survival compared to externally driven projects, as they integrate local knowledge on tides timing, sediment dynamics, and species compatibility. Rehabilitation guidelines based on scientific studies are essential to substantiate the efforts of these community projects that restore the wild by implementing ecological focus criteria while adapting to local environment (Camacho et al., 2020). Science-based ecological knowledge combined with local ecological knowledge can lead to more successful management in the long-term by enhancing community acceptance and ownership of projects (Marquez & Olavides, 2024). This commitment is enhanced by the inclusion of traditional sociocultural components and norms in management practices, which contributes to the establishment of a symbolic relationship between residents and indigenous mangrove units (Borges et al., 2017; Lawasi et al., 2025). Capacity-building activities such as leadership training and livelihoods, are essential in terms of empowering local populations with the right instruments to effectively play a role in these conservation responsibilities (Camacho et al., 2020; Bhatti et al., 2026).

Co-management models also stand out as significant strategy, through which decision making is shared between communities, government and non-governmental organizations. In co-management, the community has de jure or de facto rights to manage mangroves and external institutions provide technical knowledge, policy advice and money. These participatory governance approaches can resolve conflicts about utilization of resources, increase law enforcement regarding conservation, and promote mangrove protection within the framework of general coastal management. Given that much of the local population relies on the mangrove forest areas themselves to generate their livelihoods and income, models of livelihood development should be based on conceptual agreement with what local communities do and learn from (Pham et al., 2019). In addition, studies such as this reveal that monetary compensation for restoration is a common scenario in mangrove reforestation, which often fails to communicate the long-term trade-offs of having healthy mangroves while maintaining sustainable ecosystem management and may tend to focus on economic gains over the longer term (Macera et al., 2024). To reduce the risk of giving priority to short-term economic gains, examples of integrated mangrove-aquaculture systems (mud crab fattening, mixed shrimp-mangrove-crab systems, or silvofishery in mangroves) were shown as efficient models by which the local people could make a living without deteriorating mangrove ecosystems (Camacho et al., 2020; NGUYEN et al., 2020).

Livelihood-centred conservation interventions are key to maintain community participation in mangrove protection. These linkages between conservation and prosperity – including eco-tourism, sustainable fisheries, honey production and mangrove-friendly aquaculture can encourage communities to protect their mangroves. Such approaches can take the pressure off damaging activities such as illegal logging and land conversion while enhancing economic resilience and social wellbeing. Mangrove management can be well supervised by local fishermen communities and their involvement may enhance the possibility of success for the project by controlling access to resources and direct gain from sustainable management results (Roy & Trivedi, 2023). Good governance in community-based initiatives is crucial, requiring multistakeholder collaboration as well as clear tenure regimes that demarcate rights and responsibilities (Friess et al., 2019). For example, it is important to define clear financial management procedures and accountability mechanisms up front, since trust building mechanisms are vital in order to make sure that local interest in willingness to pay for conser- vation – is maintained (Pham et al., 2019).

Training and education programs are also reinforcing local mangrove conservation. Training activities on environmental care, pollution abatement, surveillance techniques and climate change adaptation assist communities in the decision-making process. Participatory monitoring most notably enables communities to monitor mangrove health and contamination levels, promote adaptive management, and strengthen the link between local actions and environmental condition. Shared management can be effective only if the established coastal local community organizations take part actively for representatives to express ideas, and make decisions on management plans and regulations (Environment and Livelihoods in Tropical Coastal Zones: Managing Agriculture-Fishery-Aquaculture Conflicts., 2006).

However, community-based mangrove conservation initiatives face several challenges such as inadequately funded projects, tenure insecurity, power imbalances and development pressures from outside. To overcome these challenges, supportive policy environments, long-term financial instruments and inclusive governance that ensures meaningful engagement of left-behind-categories are needed. Community-based approaches, when well supported, stand as a scalable and socially equitable route towards mangrove conservation that also increases coastal climate resilience and decreases pollution.

3. MECHANISMS BY WHICH MANGROVES REDUCE COASTAL POLLUTION

Mangroves decrease estuarine pollution by a suite of physical, chemical and biological mechanisms that occur across the land^-sea interface. Their locations contribute to intercepting waste transport by rivers, surface dispersion and tidal motion towards open marine systems. Collectively, these processes render mangroves efficient natural treatments for enhancing the quality of coastal's waters and safeguarding intermingling ecosystems. Detailed root systems form a highly effective physical barrier available to agglomerate fine particles and organic matter combined with related chemical and physical pollutants (Stadtländer, 2025). As water enters this intricate network, its velocity tends to decrease and wetlands can physically trap sediments, heavy metals, or nutrients (Macamo et al., 2024; Singh et al., 2025). In addition to the settling of sediments, this sediment trapping clears not only the water column but also the supply of adsorbed contaminants that might be flowing towards sensitive receiving ecosystems (such as coral reefs and seagrass meadows) elsewhere downstream (Gnansounou et al., 2021; Sabai, 2023). Mangroves, not only trap particulates, are also biological filters which retain and trap sediment and various nutrients from contaminated coastal waters and thus purify it (Mahato, 2024).

One of the major processes is through physical filtering and the retention of sediment. The prop roots and pneumatophores of mangrove species restrict the speed of water flow and help to facilitate the trapping of suspended sediments. Contaminants like heavy metals, hydrocarbons and nutrients are generally associated with fine-sediment particles that become trapped within the mangroves. This operation decreases sediment and pollutant discharges to the coral reefs and seagrass beds, which therefor can keep coastal water clearer and more healthy. The longer water stays in the forest, the more nutrients and heavy metals can be consumed by plants or get absorbed into sediments, improving quality of runoff that flows into nearby marine ecosystems ("A Guide to Forest–Water Management," 2021).

Mangroves also support nutrient absorption and conversion, which is essential to manage eutrophication. Mangroves vegetation sequesters excess nitrogen and phosphorus for growth, and the microbial communities in mangrove sediments play an active role in biogeochemical processes (nitrification, denitrification and phosphorus immobilization). These processes transform biologically reactive nutrients into a less harmful form, and thus reduce algal blooms and oxygen depletion in coastal waters. Moreover, the large root systems and abundant vegetation of mangrove forests act as physical filters that are able to adsorb a wide variety of contaminants from the water column such as heavy metals, hydrocarbons and particulate plastic (Cziesielski et al., 2021). Another significant process is biological treatment of organic contaminants. Mangrove sediments harbor abundant microorganisms that are able to degrade various organic pollutants, such as hydrocarbons and pesticides. Under anaerobic condition prevail on the mangrove soil, specific bacteria are responsible for the degradation or capture of pollutants making them bio-inavailable minimising their toxicity in the coastal environment.

Mangroves are sinks of heavy metals and persistent chemicals, as well. Heavy metals, for example, lead, cadmium and mercury are typically immobilized in mangroves sediments by forming complexes with organic matter and sulfides. While these deposits may present long-term ecological hazards if high levels of pollution occur (such as those from the Exxon Valdez spill), they also effectively retard the immediate transfer of toxins into coastally located waters and food webs. Microbial consortia present in the sediments also contribute to transformation and detoxification of these compounds via multiple metabolic routes (Alsharif et al., 2024; Hasim, 2021). Finally, mangroves reduce pollution by trapping floating rubbish and plastics. The tangled forest slows the tidal flow, leading to plastics and other debris getting stuck in the mangrove zone rather than being swept out into open water. While this leads to localized pollutant accumulation, it also presents the possibility of targeted cleanup and management efforts (building upon the role mangroves provide as part of important buffers in coastal pollution control systems).

4. CASE STUDIES OF COMMUNITY-LED MANGROVE CONSERVATION INITIATIVES

Community-based mangrove protection programs have become effective paradigms towards achieving ecological responsibility, as well as local socioeconomic advantages. In many instances, these cases show that grassroots involvement with proper governance and technical help can lead to the restoration of the mangroves as well as mitigation of coastal pollution. With the help of various examples of geographical contexts, it is evident that with the support of community action, national policies can be improved and the resilience of the ecosystem in the long term can be enhanced.

Local communities in Sundarbans of Bangladesh and India have also been engaged in participatory mangrove restoration programs as a way of managing erosion, fishery habitat and water quality. The villagers have been part of seedling propagation, site selection on the basis of traditional knowledge and continuous monitoring of the restored forests through the collaboration between the community groups, government agencies, and NGOs. These activities not only have promoted the mangrove cover, but also have been associated with the decreases in sediments and nutrient loads into the nearby waterways, hence, contributing to the coastal pollution control and the stabilization of the shoreline areas. Community controlled mangrove forest zones in the Philippines have effectively used the sediment sequestration capacities of the forests to prevent the erosion of coastal aquaculture and farmlands (Bajahmoum & Almaghamsi, 2025). Al Amin et al. (2025) highlighted the importance of climate adaptation, mangrove management, stakeholder engagement, and sustainable coastal development. Mangrove restoration initiatives in the Baros Village of Yogyakarta, Indonesia, have combined coastal management and socio-economic demands and managed to use the reforested lands as a buffer to agricultural areas by abrasion as well as attaining ecological, economic and tourism amenity to the local community (Bhagarathi and DaSilva, 2024).

Another such case is that of the Gulf of California, Mexico, fishing cooperatives joined fishing environmental agencies to conserve and rehabilitate mangroves along important estuarine environments. Having realized that water quality was declining, both marine biodiversity and fisheries productivity were being compromised, local fishers were involved in community mapping, replanting and patrol to stop illegal logging. Increased mangrove density as a result enhanced sediment retention and juvenile fish habitat which led to healthier near shore ecosystems and improved livelihoods due to increased catches. Community-based mangrove management has proven to effectively enhance the protection of forests and the local livelihoods of people in the southern parts of Thailand, especially in the Trang province, with secure user rights and external non-governmental support (Hagger et al., 2022). A comparative study of an NGO-controlled restoration site and an unmanaged reserve in the Indian Sundarbans found that participatory management practices, such as monitoring, replanting and protection against illegal logging, were more effective at regenerating forests than non-participatory protection in the face of population interferences (Costa and Ribeiro, 2017).

Community-based conservation programs in the Mida Creek of Kenya have been directed toward ecological restoration and sustainable use. The management team was established with local stakeholders such as fishers, women groups, and youth associations where reforestation activities, pollution monitoring, and development of eco-tourism were under the watch of the management committee. These initiatives were supported by capacity-building workshops as well as by combining indigenous ecological understanding with scientific methods of measuring water quality and the health of mangroves. The project has produced quantifiable results regarding water quality and fewer cases of harmful algal bloodshed, which proves the overall environmental worth of community stewardship. The Mikoko Pamoja project in Gazi Bay, Kenya, and managed as a co-management regime with communities, government agencies, and NGOs to preserve natural and planted mangroves is one of the brightest examples of such financial and ecological integration (Evans et al., 2023).

Sitangkai municipality in Tawi-Tawi in the Philippines has set a good example of community-based mangrove management coupled with pollution control. With un-treated sewage and commercial-waste causing massive coastal erosion, residents formed clean-up movements and established bylaws regulating waste disposal and mangrove conservation. Community enforcement and school-based education have resulted in a substantial reduction of solid waste entering into the mangrove habitat, mangrove regeneration and a positive change in the near-shore water quality. Fig 3 highlights the key sources of mangroves.

Figure 3. Mangroves study areas

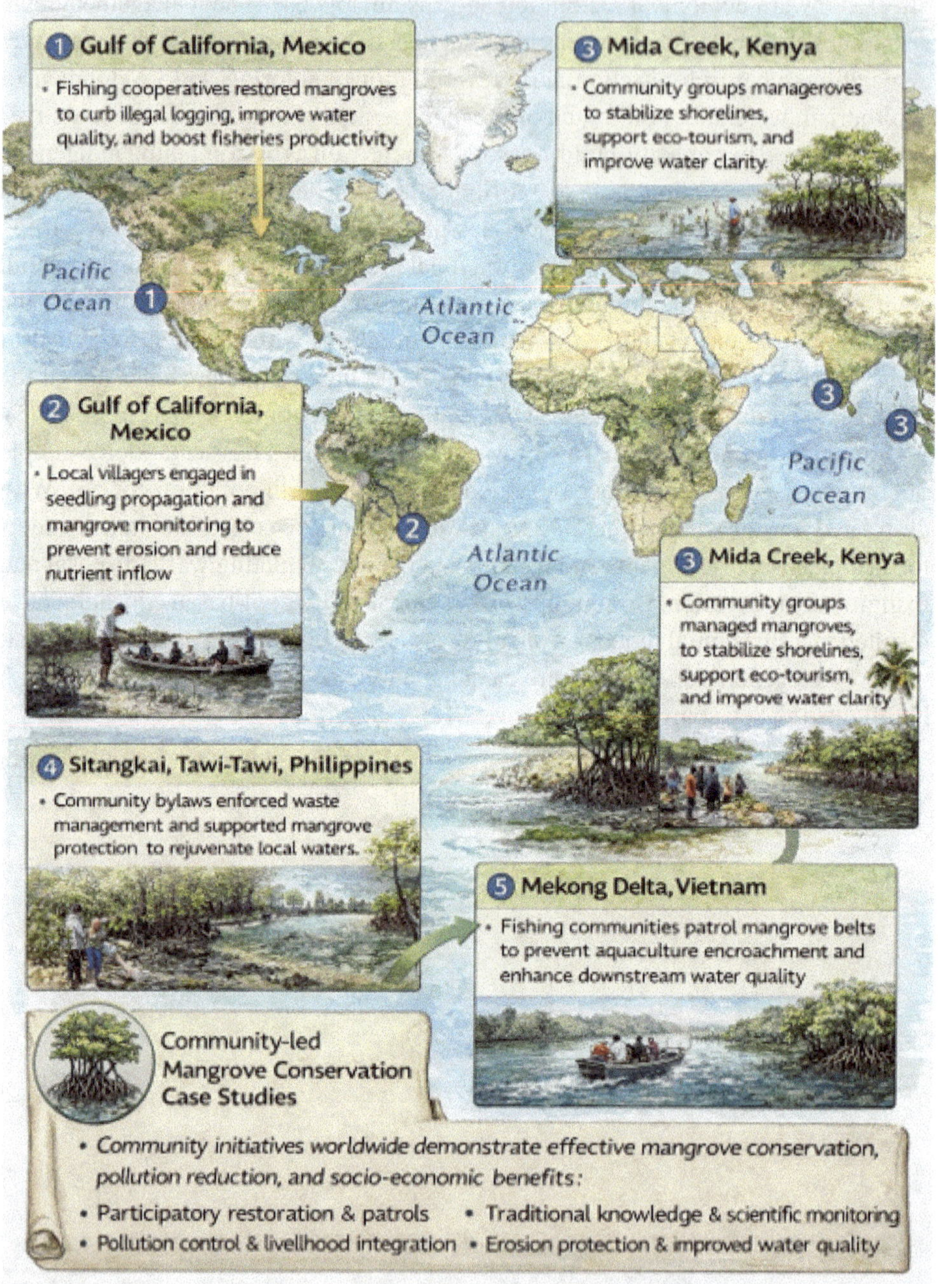

The second example is referring to the Mekong Delta in Vietnam, where fishing communities have used participatory governance structures to manage the growth of aquaculture and the preservation of mangroves. These groups were able to preserve vegetated buffers along the waterways by creating community patrols and ordinances against unsustainable mangrove area conversion. Investigations within the area have demonstrated that the protected mangrove belts are associated with reduced levels of suspended solids and contaminants in downstream, which has highlighted the pollution-reducing nature of undisturbed coastal forests. As shown by the activities of Beach Management Units in the Pangani River Estuary of Tanzania, as well as in the Mikoko Pamoja project in Kenya, community surveillance and co-management agreements are effective at reducing the illegal logging and marine pollution through participatory monitoring (Janmaimool, 2016; Nyangoko et al., 2022; Synergy Solutions for a World in Crisis: Tackling Climate and SDG Action Together, 2023). An example is the Coastal Zone Conservation and Development Program in the Tanga region of Tanzania, where the local people were empowered to manage local natural resources, such as the patrols, capacity building, and the promotion of other livelihoods like beekeeping and seaweed farming (Nyangoko et al., 2022).

These case studies collectively highlight that community-led mangrove conservation can be more than a localized environmental effort; it can serve as an effective mechanism for addressing broader coastal pollution challenges. The success of these initiatives often rests on strong local institutions, meaningful stakeholder participation, integration of traditional knowledge with scientific monitoring, and supportive policy environments. As such, they offer valuable lessons for scaling up community-based conservation and incorporating mangroves into comprehensive coastal management strategies. Furthermore, the expansion of Payment for Ecosystem Services models, such as the Vanga Blue Forest project in Kenya which protects 460 hectares of mangroves and supports nearly 9,000 people through the sale of carbon credits, suggests that financial incentives can sustain community engagement while simultaneously sequestering carbon and filtering pollutants (Synergy Solutions for a World in Crisis: Tackling Climate and SDG Action Together, 2023).

5. ANALYTICAL FRAMEWORK LINKING COMMUNITY-BASED MANGROVE CONSERVATION, POLLUTION MITIGATION, AND ECOLOGICAL OUTCOMES

In order to appreciate the potential of community-based mangrove conservation in coastal pollution reduction, it is necessary to set up a precise analytical framework, connecting with the management action to the ecological process and environmental outcomes. Community-based conservation is not only a governance mechanism, it's also a driver for ecological restoration and pollution reduction.

The framework proposed starts with community participation and collaborative governance. There is participatory conservation planning, restoration work, monitoring and environmental law enforcement by local communities, government agencies, non-governmental groups and other stakeholders. These activities help to minimize pressures like illegal logging, unsustainable expansion of aquaculture, waste disposal, mangrove conversion, etc.

The management interventions have direct impacts on the restoration and protection of mangrove systems. The ecological role of mangrove forests is enhanced by the improvement of the vegetative cover, increase in vegetation density, development of the root system and strengthening of the integrity of the ecosystem. Pollution mitigation processes, such as sediment retention, absorption of nutrients, sequestration of heavy metals, microbial degradation of organic pollutants and the trapping of floating debris and microplastics, are then undertaken by mangrove systems with a greater effectiveness.

Improvements in these pollution abatement processes can cause measurable ecological impacts. A decrease in nutrient loads reduces the potential for eutrophication and harmful algal blooms. Increased sediment retention reduce turbidity and keep adjacent coral reefs and seagrass systems safe. Contaminants are sequestered, which helps to improve water quality and to minimize the ecological risk to marine organisms. At the same time, the preserved mangrove areas serve as nurseries for fish and other aquatic life, enhancing the biodiversity and ecological resilience.

In addition to ecological improvements, these enhancements also help produce positive socio-economic results. Local livelihoods are supported by improving fisheries productivity, providing ecotourism opportunities, promoting blue carbon projects and coastal protection projects, all of which increases community engagement in conservation projects. This feedback loop helps to keep communities engaged and leads to sustainability of conservation programs.

This framework showcases the need for community-based mangrove conservation to go beyond a social or governance approach. Instead, it is a holistic environmental management plan in which the stewardship of local people has real impact on ecological processes that reduce pollution and enhance the health of the coastal environment.

6. PRACTICAL MANAGERIAL SIGNIFICANCE AND IMPLEMENTATION FRAMEWORK

- However, community-based mangrove conservation has faced several challenges that have hitherto hindered the success of such initiatives and their sustainability in the long term. Insecurity of land tenure, uncertainty regarding rights to use resources and the tensions within conservation toward short-term economic imperatives -- for instance, popular coastal development and aquaculture expansion or logging -- constitute major hurdles. Lack of available funding, insufficient technical knowledge and poor participation—especially by women and disadvantaged groups — can undermine fair governance and community motivation. Environmental stressors like climate change, RSL rise, and extreme weather events exacerbate these challenges by heightening vulnerability of coastal habitats such as mangroves and increasing uncertainties in restoration efforts.
- Problems in the workplace are viewed as the likelihood of an adverse event.In the workplace, problems are considered as the probability of an undesirable situation.
- The outcome of this review has various implications for coastal resource managers, environmental agencies, NGO's and local governments involved in mangrove conservation and pollution control. Creating an effective community-based mangrove conservation program is a complex process, which will involve a structured decision-making process that takes into account conservation needs, community engagement, financial sustainability, and adaptive management.
- A four stage management system is suggested. First, managers need to carry out baseline assessments to determine the areas of pollution, degraded mangrove areas, community reliance on mangrove resources, and current governance structures. Evidence-based prioritization of conservation interventions through the integration of remote sensing and field surveys and local ecological knowledge.
- Second, stakeholder participation and co-planning should be set up using participatory governance processes with local communities, public officials, academics, and private sector actors. To enhance transparency and long-term commitment, decision makers need to have a clear understanding of the roles, management responsibilities, and benefit-sharing arrangements of stakeholders.
- Third, there is a need for strategies in implementing restoration, pollution-source reduction and sustainable livelihood development that are done in a community way. The practical solutions involve mangrove replanting,

community monitoring programmes, waste management, cultivation of mangrove-friendly aquaculture, ecotourism development, and environmental education campaign. Further, participation and program sustainability can be enhanced through financial incentives like Payment for Ecosystem Services (PES), blue carbon projects, and conservation grants.

- Fourth, managers should implement adaptive monitoring and evaluation systems, based on measurable indicators like mangrove survival, water quality parameters, concentrations of pollutants, recovery of biodiversity, level of community participation, and livelihood outcomes. Periodic monitoring allows managers to detect potential problems at an early stage, assess the effectiveness of interventions and adapt management techniques to address those needs.
- On a strategic level, good implementation will depend on the cooperation of various sectoral authorities (environment, fisheries, agriculture, tourism and local development). Mangrove protection in the community can provide the greatest benefits in terms of reduction of pollution, whilst also being used to provide benefits for climate adaptation, conservation and sustainable economic development as part of a coastal zone management plan. This framework serves as a practical means of turning conservation goals into management strategies that can be adopted in a variety of coastal areas.

7. FUTURE RESEARCH AND PRACTICE

Future work should be aimed at improving the scientific and institutional basis for CBCM in mangroves. Research needs include quantify of pollution reduction benefits, evaluation of longer-term ecosystem resilience to disturbances, and creating integrative monitoring approaches that blend community-based observations with remote sensing and water quality data. In practice, scale-up of what works will involve investment in capacity building, digital tools for participatory monitoring work and cross-sectoral partnerships that connect conservation to sustainable economic opportunities. The integration of community-based mangrove management with climate finance, ecosystem service valuation, and adaptive governance approaches is important in order to optimize their contribution to coastal pollution control and sustainable coastal development.

8. CONCLUSION

Mangroves form a perfect natural protective measures against pollution of the coasts because of its unique physical structures and biogeochemical processes, which can intercept, transform, and concentrate pollutants along the land-sea boundary. The current review demonstrates that, under the condition of mangrove conservation led by the local communities, such ecosystem services can be quantitatively enhanced in terms of the long-term care effects, adaptive management and reduction of the anthropogenic pressure. The measures taken at a communal level assist in the survival and functionality of mangroves and are also milestones towards fighting pollution at its origin by increasing environmental awareness instilling long term behaviour into the coastal areas. The Asian, African and Latin American case studies have revealed that community-based mangrove conservation can reduce the sediment loads, increase the enrichment nutrients input and storage of solid waste and at the same time allow the local development through sustainable use of resources and social integration. However, on-the-ground efforts are confined by frightening forces such as insecure land tenure, inadequate funds, lack of institutional backing as well as increasing risks posed by climate change. Without the facilitating policies and long-term funding systems, it is unclear how the scale up and maintenance of community conservation efforts will be achieved. Solutions based on nature like mangroves should be utilized to reduce coastal pollution by incorporating integrated strategies. These include Portuguese formal acknowledgement of community management rights, inclusion of mangrove conservation into the national pollution control and climate change adaptation policies, capacity building and participatory (independent monitoring). Future research needs to also seek to measure the impacts of pollution mitigation of these projects, enhance community-based monitoring in a more technological-driven methodology, and develop more interdisciplinary collaboration between scientists, policy-makers, and coastal communities. CBMC can become transformational by incorporating ecological conservation in social empowerment and policy support to acquire sustainable tidal wetland ecosystems with social resilience.

FUNDING

This work was supported by the Deanship of Scientific Research, the Vice Presidency for Graduate Studies and Scientific Research, King Faisal University, Saudi Arabia, under Grant No. KFU260473

REFERENCES

Al Amin, S. F., Pangestu, A. Y., Dharma, Y., Sari, N. A., Maulidia, O., Octaviani, E. A., & Salimah, W. (2025). Mangroves as Carbon Sequesterers: Diversity and Carbon Estimation Study in Pantai Mekar Village, Muara Gembong District, Bekasi Regency. *Jurnal Biologi Tropis*, *25*(1), 936–946. DOI: 10.29303/jbt.v25i1.8618

Alsharif, S. M., Ismaeil, M., Saeed, A., & El-Sayed, W. S. (2024). Metagenomic 16S rRNA analysis and predictive functional profiling revealed intrinsic organohalides respiration and bioremediation potential in mangrove sediment. *BMC Microbiology*, *24*(1), 176. DOI: 10.1186/s12866-024-03291-8 PMID: 38778276

Anu, K., Sneha, V. K., & Busheera, P. (2024). Mangroves in environmental engineering: Harnessing the multifunctional potential of Nature's coastal architects for sustainable ecosystem management. *Results in Engineering*, *21*, 101765. DOI: 10.1016/j.rineng.2024.101765

Arfan, A., Sanusi, W., Rakib, M., Juanda, M. F., & Sukri, I. (2024). Mangrove Ecosystem Management Strategy to Support Sustainable Development Goal 14. *Environmental Research, Engineering and Management*, *80*(1), 64–76. DOI: 10.5755/j01.erem.80.1.33887

Atapaththu, K. S. S., Herath, S. S., Subramanuiam, G., Yapa, Y. M. A. L. W., Gamage, S. M. K. W., Masakorala, K., Jinadasa, B. K. K. K., & Wu, M. (2025). Challenges in coastal ecosystem Sustainability: Drivers of water quality degradation and their ecological impact [Review of *Challenges in coastal ecosystem Sustainability: Drivers of water quality degradation and their ecological impact*]. *Marine Environmental Research, 209*, 107194. Elsevier BV. DOI: 10.1016/j.marenvres.2025.107194

Bajahmoum, E. A., & Almaghamsi, A. (2025). Physicochemical degradation of Avicennia marina mangrove soils in the Red Sea: Implications for coastal ecosystem services. *Frontiers in Soil Science*, *5*, 1621591. Advance online publication. DOI: 10.3389/fsoil.2025.1621591

Bhagarathi, L. K., & DaSilva, P. N. B. (2024). Impacts and implications of anthropogenic activities on mangrove forests: A review [Review of *Impacts and implications of anthropogenic activities on mangrove forests: A review*]. *Magna Scientia Advanced Research and Reviews, 11*(1), 40. GSC Online Press. DOI: 10.30574/msarr.2024.11.1.0074

Bhatti, U. A., Asif, M., Alajmi, M., Awedat, K., Al-Rasheed, A., Madaminov, B., Makhmudov, S., & Shojonov, M. (2026). Spatiotemporal evolution and uncertainty analysis of mangrove ecosystem carbon storage based on multi-source remote sensing and Monte Carlo error propagation: A case study of the mangrove reserves in Hainan Province. *Geocarto International*, *41*(1), 2657626. Advance online publication. DOI: 10.1080/10106049.2026.2657626

Borges, R., Ferreira, A. C., & de Lacerda, L. D. (2017). Systematic Planning and Ecosystem-Based Management as Strategies to Reconcile Mangrove Conservation with Resource Use. *Frontiers in Marine Science*, *4*, 353. Advance online publication. DOI: 10.3389/fmars.2017.00353

Camacho, L. D., Gevaña, D. T., Sabino, L. L., Ruzol, C., Garcia, J. E., Camacho, A. C. D., Oo, T. N., Maung, A. C., Saxena, K. G., Liang, L., Yiu, E., & Takeuchi, K. (2020). Sustainable mangrove rehabilitation: Lessons and insights from community-based management in the Philippines and Myanmar. *APN Science Bulletin*, *10*(1), 18–25. DOI: 10.30852/sb.2020.983

Chen, B. (2022). Current status and trends of research on microplastic fugacity characteristics and pollution levels in mangrove wetlands. *Frontiers in Environmental Science*, *10*, 1070120. Advance online publication. DOI: 10.3389/fenvs.2022.1070120

Cziesielski, M. J., Duarte, C. M., Aalismail, N., Al-Hafedh, Y. S., Antón, A., Baalkhuyur, F., Baker, A. C., Balke, T., Baums, I. B., Berumen, M. L., Chalastani, V. I., Cornwell, B., Daffonchio, D., Diele, K., Farooq, E., Gattuso, J., He, S., Lovelock, C. E., Mcleod, E., & Aranda, M. (2021). Investing in Blue Natural Capital to Secure a Future for the Red Sea Ecosystems. *Frontiers in Marine Science*, *7*, 603722. Advance online publication. DOI: 10.3389/fmars.2020.603722

da Costa, A. B., & Ribeiro, L. P. F. (2017). Mangroves of Maputo, Mozambique: From Threatened to Thriving? *The Plan Journal*, *2*(2). Advance online publication. DOI: 10.15274/tpj.2017.02.02.21

de Souza, L. M. B., Zreik, I. V. R., Salem-Sermanet, A., Seghouani, N., & Pourchier, L. (2024). A Deep Learning-Based Approach for Mangrove Monitoring. *arXiv (Cornell University)*. https://doi.org//arxiv.2410.05443DOI: 10.48550

Earp, H. S., Prinz, N., Cziesielski, M. J., & Andskog, M. A. (2018). *For a World Without Boundaries: Connectivity Between Marine Tropical Ecosystems in Times of Change* (p. 125). DOI: 10.1007/978-3-319-93284-2_9

Elegbede, I. O., Fakoya, K. A., Adewolu, M. A., Jolaosho, T. L., Adebayo, J. A., Oshodi, E., Hungevu, R. F., Oladosu, A. O., & Abikoye, O. (2023). Understanding the social–ecological systems of non-state seafood sustainability scheme in the blue economy. *Environment, Development and Sustainability*, *27*(2), 2721–2752. Advance online publication. DOI: 10.1007/s10668-023-04004-3

Environment and livelihoods in tropical coastal zones: managing agriculture-fishery-aquaculture conflicts. (2006). In *CABI eBooks*. DOI: 10.1079/9781845931070.0000

Evans, L., Buchan, P. M., Fortnam, M., Honig, M., & Heaps, L. (2023). Putting coastal communities at the center of a sustainable blue economy: A review of risks, opportunities, and strategies [Review of *Putting coastal communities at the center of a sustainable blue economy: A review of risks, opportunities, and strategies*]. *Frontiers in Political Science, 4*. Frontiers Media. DOI: 10.3389/fpos.2022.1032204

Fatimatuzzahroh, F., Hadi, S. P., & Purnaweni, H. (2018). Mangrove Cultivation Community Based in Karangsong, Indramayu, West Java. *IOP Conference Series. Earth and Environmental Science*, *145*, 12139. DOI: 10.1088/1755-1315/145/1/012139

Friess, D. A., Aung, T., Huxham, M., Lovelock, C. E., Mukherjee, N., & Sasmito, S. D. (2019). SDG 14: Life below Water – Impacts on Mangroves. In *Cambridge University Press eBooks* (p. 445). Cambridge University Press. DOI: 10.1017/9781108765015.016

Friess, D. A., Yando, E. S., Alemu, J. B., Wong, L.-W., Soto, S. D., & Bhatia, N. (2020). Ecosystem services and disservices of mangrove forests and salt marshes. In *CRC Press eBooks* (p. 107). Informa., DOI: 10.1201/9780429351495-3

Gnansounou, C. S., Sagoe, A. A., Mattah, P. A. D., Salako, K. V., Aheto, D. W., & Kakaï, R. G. (2021). Is the Co-management Approach Effective for Mangrove Conservation in West Africa? *Research Square (Research Square)*. DOI: 10.21203/rs.3.rs-1085890/v1

Hagger, V., Worthington, T. A., Lovelock, C. E., Adame, M. F., Amano, T., Brown, B. M., Friess, D. A., Landis, E., Mumby, P. J., Morrison, T. H., O'Brien, K. R., Wilson, K. A., Zganjar, C., & Saunders, M. I. (2022). Drivers of global mangrove loss and gain in social-ecological systems. *Nature Communications*, *13*(1), 6373. DOI: 10.1038/s41467-022-33962-x PMID: 36289201

Hai, N. T., Dell, B., Phuong, V. T., & Harper, R. J.NGUYEN. (2020). Towards a more robust approach for the restoration of mangroves in Vietnam. *Annals of Forest Science, 77*(1), 18. Advance online publication. DOI: 10.1007/s13595-020-0921-0

Hasim, H. (2021). Mangrove Ecosystem, Seagrass, Coral Reef: Its Role in Self-Purification and Carrying Capacity in Coastal Areas. *International Journal Papier Advance and Scientific Review, 2*(1), 37–49. DOI: 10.47667/ijpasr.v2i1.93

Hewindati, Y. T., & Kusumaningrum, E. N. (2024). Traditional Mangrove Management in Langensari Village, Subang, West Java: A Study on the Practice of Integrating Mangroves into Polyculture Ponds. *E3S Web of Conferences, 483*, 1018. DOI: 10.1051/e3sconf/202448301018

Janmaimool, P. (2016). *The Establishment of a Community-based Mangrove Forest Management Plan: Lessons Learned from Mangrove Forest Conservation in the Nernkhor Sub-district*. Deleted Journal., DOI: 10.35762/AER.2016.38.3.6

Katsanevakis, S., Borja, Á., Gissi, E., Ferse, S. C. A., & Teixeira, H. (2025). Editorial: Understanding the response of ecosystems to increasing human pressures and climate change – management options. *Frontiers in Marine Science, 12*, 1688993. Advance online publication. DOI: 10.3389/fmars.2025.1688993

Khan, W. R., Latiff, A., & Ibrahim, F. (2025). Biodiversity, Health, and Ecosystem Services of Mangroves. *Forests, 16*(8), 1278. DOI: 10.3390/f16081278

Lasaiba, M. A. (2023). Peningkatan Partisipasi Masyarakat melalui Penanaman Mangrove dalam Rehabilitasi Pesisir. *Jurnal ABDINUS Jurnal Pengabdian Nusantara, 7*(3), 623–633. DOI: 10.29407/ja.v7i3.19962

Lawasi, Moh. A., Kenda, N., Yusnikusumah, T. R., Pratama, B. B., Pratiwi, D., Septina, A. D., & Asrawijaya, E. (2025). FOREST-BASED ECOTOURISM IN INDONESIA: A COMPREHENSIVE REVIEW OF POLICY CHALLENGES, DIVERSE PRACTICES, STAKEHOLDER ENGAGEMENT, CONSERVATION EFFORTS, AND SOCIOECONOMIC ASPECTS [Review of *FOREST-BASED ECOTOURISM IN INDONESIA: A COMPREHENSIVE REVIEW OF POLICY CHALLENGES, DIVERSE PRACTICES, STAKEHOLDER ENGAGEMENT, CONSERVATION EFFORTS, AND SOCIOECONOMIC ASPECTS*]. *GeoJournal of Tourism and Geosites, 60*, 1041. Editura Universitatii din Oradea. DOI: 10.30892/gtg.602spl02-1478

Li, R., Yan, C., Wu, Q., Cheng, X., Wang, H., Yang, Z., Liang, Z.-B., Tian, D., Yuan, N., Hou, L., & Gao, L. (2025). New insights into sink-source dynamics of mangrove for microplastics: Quantitative evidences from field observation and resuspension simulation. *Journal of Hazardous Materials, 140590*, 140590. Advance online publication. DOI: 10.1016/j.jhazmat.2025.140590 PMID: 41313852

Macamo, C., da Costa, F. I., Bandeira, S., Adams, J. B., & Balidy, H. (2024). Mangrove community-based management in Eastern Africa: Experiences from rural Mozambique. *Frontiers in Marine Science, 11*, 1337678. Advance online publication. DOI: 10.3389/fmars.2024.1337678

Macera, L., Andrieu, J., Crook, O.-J., Muthusankar, G., & Ben, P. D. (2024). RESTORATION OF MANGROVE: EVALUATING ECOLOGICAL, SOCIAL, AND ECONOMIC INTEGRATION FOR PROJECT SUCCESS – A CASE STUDY IN THE PHILIPPINES. *Environment & Ecosystem Science*, *8*(2), 117–126. DOI: 10.26480/ees.02.2024.117.126

Mahato, P. S. (2024). A Study of Morphological Changes in the Coastal Areas and Offshore Islands of Sudarban Coastline Using Remote Sensing. In *IntechOpen eBooks*. IntechOpen., DOI: 10.5772/intechopen.112243

Maiti, S. K., & Chowdhury, A. (2013). Effects of Anthropogenic Pollution on Mangrove Biodiversity: A Review [Review of *Effects of Anthropogenic Pollution on Mangrove Biodiversity: A Review*]. *Journal of Environmental Protection, 4*(12), 1428. Scientific Research Publishing. DOI: 10.4236/jep.2013.412163

Marquez, G. P. B., & Olavides, R. D. (2024). Integrating science-based and local ecological knowledge: A case study of mangrove restoration and rehabilitation projects in the Philippines. *Disasters*, *48*(S1), e12630. Advance online publication. DOI: 10.1111/disa.12630 PMID: 38840490

Nugraheni, I. K., Zamani, N. P., & Cordova, M. R. (2024). Abundance of Microplastics in Mangrove Sediments on Pari Island, Jakarta Bay, Indonesia. *Research Square (Research Square)*. DOI: 10.21203/rs.3.rs-4403456/v1

Nyangoko, B. P., Shalli, M. S., Mangora, M. M., Gullström, M., & Berg, H. (2022). Socioeconomic determinants of mangrove exploitation and management in the Pangani River Estuary, Tanzania. *Ecology and Society*, *27*(2), art32. Advance online publication. DOI: 10.5751/ES-13227-270232

Pham, T. T., Vu, T. P., Pham, D. C., Dao, L. H. T., Nguyen, V. T., Hoang, N. V. H., ... & Nguyen, D. T. (2019). Opportunities and challenges for mangrove management in Vietnam: Lessons learned from Thanh Hoa, Thai Binh and Quang Ninh provinces.

Que, B. J., Silalahi, P. Y., & Lekatompessy, J. C. (2024). Restoring Maluku's Coast: Action-Driven Beach Cleanups, Mangrove Planting, and Coral Renewal in Ambon by the Faculty of Medicine, University of Pattimura. In *IntechOpen eBooks*. IntechOpen. DOI: 10.5772/intechopen.1007779

Rahmadi, M. T., Yuniastuti, E., Suciani, A., Harefa, M. S., Persada, A. Y., & Tuhono, E. (2023). Threats to Mangrove Ecosystems and Their Impact on Coastal Biodiversity: A Study on Mangrove Management in Langsa City. *Indonesian Journal of Earth Sciences*, *3*(2), A627. Advance online publication. DOI: 10.52562/injoes.2023.627

Roy, S., & Trivedi, S. (2023). Geospatial Assessment of Long-Term Changes (1937–2019) in Mangrove Vegetation and Shoreline Dynamics of Godavari Estuary, East Coast of India. *Journal of the Indian Society of Remote Sensing, 51*(6), 1309. *s41467-023-44037-w.pdf.* (n.d.).DOI: 10.1007/s12524-023-01698-w

Sabai, D. (2023). The Role of Integrated Coastal Management Approach in the Protection of Coastal and Marine Resources in the Eastern Coast of Tanzania. *Transylvanian Review of Systematical and Ecological Research*, *25*(2), 77–86. DOI: 10.2478/trser-2023-0014

Sathiyamoorthy, S., & Sakurai, T. (2024). Effectiveness of community participation in Mangrove restoration: The evidence from northern Sri Lanka. *Environmental Economics and Policy Studies*, *26*(4), 759–779. DOI: 10.1007/s10018-024-00397-1

Singh, R., Zarapkar, A., Samnani, A., Rathore, S. S., & Joshi, M. (2025). Mangrove Conservation - A Survey Based Study on the Awareness and Attitude of General Population of Mumbai. *International Journal of Scientific Research in Science and Technology*, *12*(5), 309–319. DOI: 10.32628/IJSRST25126237

Stadtländer, C. T. K.-H. (2025). Marine pollution: Monitoring, management and mitigation. *The International Journal of Environmental Studies*, *1*(5-6), 2147–2151. Advance online publication. DOI: 10.1080/00207233.2025.2546747

Sukarmen, A., Mubarak, M., Chairilsyah, D., Yoswaty, D., & Hamidy, R. (2023). Community-Based Mangrove Protection to Mitigate Climate Change: A Socio-Ecological Approach. *International Journal of Sustainable Development and Planning*, *18*(8), 2473–2480. DOI: 10.18280/ijsdp.180818

Trégarot, E., D'Olivo, J. P., Botelho, A. Z., Cabrito, A., Cardoso, G. O., Casal, G., Cornet, C. C., Cragg, S. M., Degia, A. K., Fredriksen, S., Furlan, E., Heiss, G. A., Kersting, D. K., Maréchal, J., Meesters, E. H., O'Leary, B. C., Pérez, G., Seijo-Núñez, C., Simide, R., … Juan, S. de. (2023). Effects of climate change on marine coastal ecosystems – A review to guide research and management [Review of *Effects of climate change on marine coastal ecosystems – A review to guide research and management*]. *Biological Conservation, 289*, 110394. Elsevier BV. DOI: 10.1016/j.biocon.2023.110394

United Nations. (2023). Synergy solutions for a World in crisis: tackling climate and SDG action together.

Wang, Y., & Gu, J. (2021). Ecological responses, adaptation and mechanisms of mangrove wetland ecosystem to global climate change and anthropogenic activities. *International Biodeterioration & Biodegradation*, *162*, 105248. DOI: 10.1016/j.ibiod.2021.105248

Yulianto, F., Basyuni, M., Mardiatno, D., Ardiati, A. S., Kusuma, I. K. T. W., Purnomo, M. B., Kongko, W., Wardani, K. S., Fachrudin, I., Baba, S., & Miyagi, T. (2025). Long-term monitoring of mangrove restoration and carbon dynamics in Tanjung Benoa, Bali, Indonesia: An assessment approach using combination remote sensing and field surveys. *Research Square (Research Square)*. DOI: 10.21203/rs.3.rs-6779800/v1

Yulianto, H., Ihsan, Y. N., Sumiarsa, D., Ansari, A., & Hendarmawan, H. (2024). Assessing the sustainability of the blue swimming crab (Portunus pelagicus) on the Eastern Coast of Lampung: A holistic approach to conservation and resource stewardship. *Frontiers in Marine Science*, *11*, 1304838. Advance online publication. DOI: 10.3389/fmars.2024.1304838

Zhang, P., Zhao, W., Zhang, J., Gao, Y., Wang, S., & Jian, Q. (2024). Human activities altered the enrichment patterns of microplastics in mangrove blue carbon ecosystem in the semi-enclosed Zhanjiang Bay, China. *Frontiers in Marine Science*, *11*, 1362170. Advance online publication. DOI: 10.3389/fmars.2024.1362170

Zhao, L., Chen, D., Tan, H., Wang, Y., Qiao, Y., & Xing, Y. (2025). The response of mangrove degradation to mineral composition. *Frontiers in Marine Science*, *12*, 1626088. Advance online publication. DOI: 10.3389/fmars.2025.1626088

Chapter 5
Corporate Social Responsibility as a Tool for Strengthening Blue Carbon and Mangrove Conservation in Malaysia

Sazlina Salleh
http://orcid.org/0000-0002-8789-3387
Universiti Sains Malaysia, Malaysia

Nur Ain Amani Abdul Mubin
Universiti Sains Malaysia, Malaysia

Dina Miza Suhaimi
Universiti Sains Malaysia, Malaysia

Norhanis Mohammad Razalli
http://orcid.org/0000-0001-7752-699X
Universiti Sains Malaysia, Malaysia

Mahadi Mohammad
Universiti Sains Malaysia, Malaysia

ABSTRACT

Mangrove ecosystems are increasingly recognised as nature-based solutions for climate mitigation, coastal protection, and biodiversity conservation. In Malaysia, where approximately 600,000 hectares of mangroves contribute substantially to

DOI: 10.4018/979-8-3373-9320-9.ch005

national carbon storage, Corporate Social Responsibility (CSR) has emerged as a visible actor in restoration and blue carbon initiatives. However, a key governance question remains: are these CSR-driven efforts supported by strong scientific frameworks, measurable carbon accounting, and long-term monitoring systems, or are they treated as one CSR programme among many? This chapter examines how CSR can be structured to enhance measurable conservation outcomes rather than remain limited to short-term or symbolic activities. Using Malaysia as a case, the chapter demonstrates how CSR can contribute to advancing blue carbon conservation. The findings provide practical guidance for designing corporate models that prioritise ecological integrity, measurable outcomes, and long-term sustainability.

1. INTRODUCTION

Mangrove forests are vital to carbon neutrality and climate resilience, sequestering five times more carbon than terrestrial forests and providing a natural defence against the intensifying storm surges and rising sea levels caused by climate change (Alongi, 2008; Donato et al., 2011). They also provide key ecosystem services, including coastal protection, nursery habitats for fisheries, water filtration, and resources supporting coastal livelihoods (Islam et al., 2024). In Malaysia, these functions are especially significant given the country's extensive coastline of approximately 4800 km and the central role of coastal and marine environments in its ecological, economic, and social development. Coastal zones support fisheries, aquaculture, and ecotourism, while also serving as natural buffers and important nursery habitats. Where these areas are characterised by mangrove forests, their contribution to climate mitigation, shoreline stabilisation, erosion control, flood reduction, and protection against coastal hazards becomes even more pronounced (Ward et al., 2016). Malaysia, therefore, has globally important blue carbon ecosystems, including mangroves and seagrasses, with significant potential for conservation, restoration, and rehabilitation.

Mangrove conservation in Malaysia is driven by both local priorities and international commitments that recognise the role of coastal ecosystems in climate mitigation and biodiversity conservation. Global frameworks such as the Paris Agreement, the Kunming–Montreal Global Biodiversity Framework, and the Sustainable Development Goals (SDGs) have increased attention on nature-based solutions and blue carbon ecosystems. These commitments are reflected in national policies, including the National Policy on Biological Diversity 2022–2030 and the National Climate Change Policy, which promote ecosystem conservation as part of climate adaptation and mitigation strategies. Within this evolving policy landscape, Corporate Social Responsibility (CSR) is emerging as a practical mechanism for private-sector participation in mangrove conservation and blue carbon initiatives. Beyond symbolic contributions, CSR can mobilise resources, strengthen partnerships, support restoration, and enhance community and scientific engagement. This chapter examines how CSR can strengthen mangrove and blue carbon conservation in Malaysia and how better coordination among science, policy, and corporate actors can improve conservation effectiveness, accountability, and long-term sustainability.

2. MANGROVE ECOSYSTEMS AND BLUE CARBON POTENTIAL IN MALAYSIA

2.1 Distribution and Ecological Characteristics of Malaysian Mangroves

Malaysia's mangroves cover approximately 0.5 million hectares and include 39 true mangrove species. These forests are structured into distinct zones including *Avicennia-Sonneratia*, *Bruguiera-Rhizophora*, and back mangroves shaped by tidal and soil conditions (Jusoff, 2013). They are concentrated in East Malaysia, while Peninsular mangroves occur mainly along the sheltered west coast (Figure 1, Table 1). The floristic composition is heavily dominated by Rhizophoraceae, with *Rhizophora apiculata* and *Rhizophora mucronata* as the most abundant species. These wetlands are highly productive carbon sinks, storing an average of 500.8 Mg C ha^{-1} across aboveground and belowground biomass, and they provide critical environmental services such as coastal protection against storms and erosion, nutrient cycling, and vital nursery grounds for commercial fisheries.

Figure 1. Distribution of mangrove forests in Malaysia: (a) Peninsular Malaysia, with mangroves concentrated along the west coast; and (b) East Malaysia (Sabah and Sarawak), showing extensive coastal and estuarine mangrove coverage. Red polygons indicate mapped mangrove extents. Base map: Google Earth

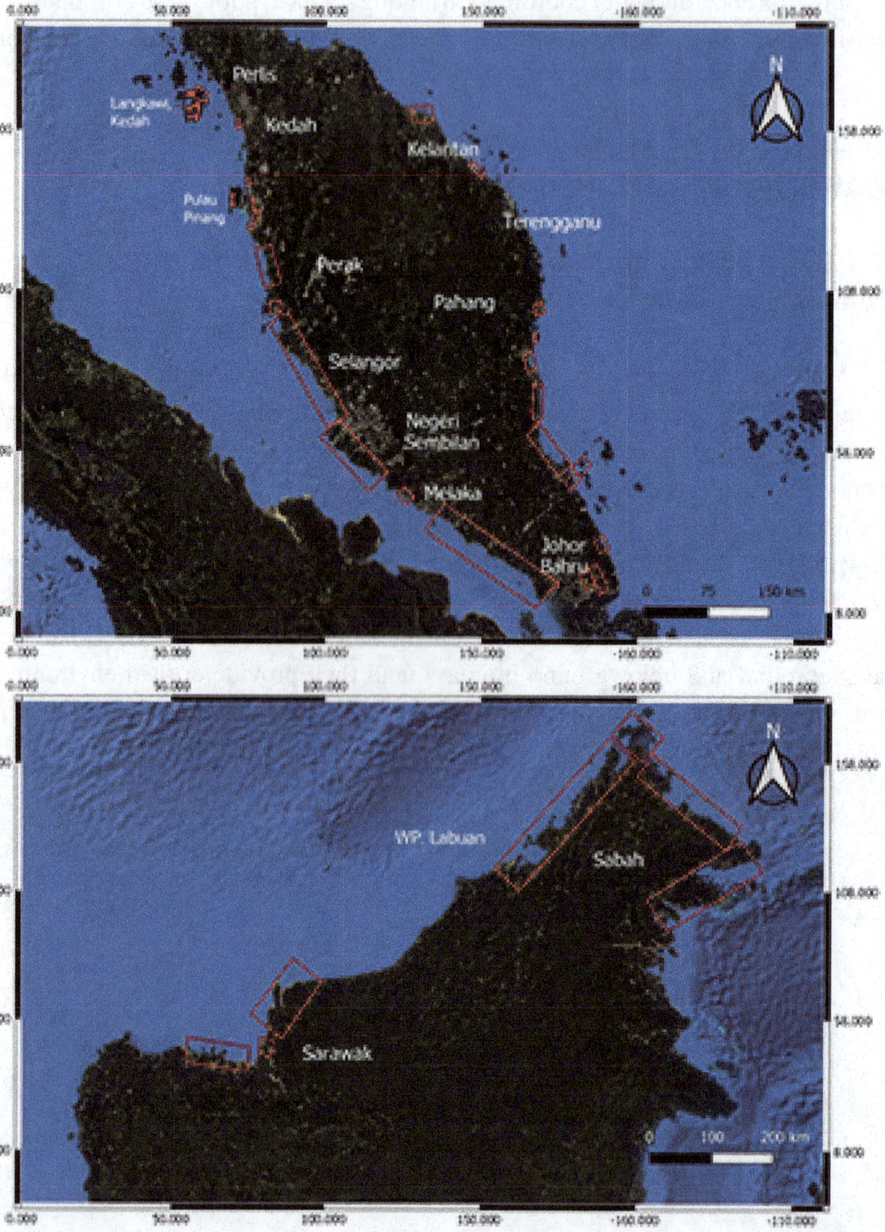

Table 1. Mangrove area and ecological characteristics by region in Malaysia

Region	Mangrove Area (2023) & National Share	Ecological Interpretation	Dominant Species
Peninsular Malaysia	**115,317 ha** 19.7% of national total	>90% concentrated on the sheltered west coast (Strait of Malacca), with limited fragmented patches on the exposed northeast monsoon coast. Largest single reserves: Matang MFR (Perak, ~41,000 ha), Klang (Selangor, ~14,417 ha), and Sungai Pulai Forest Reserve (Johor, ~7,890 ha)	*Rhizophora apiculata (Bakau Minyak) Rhizophora mucronata (Bakau Kurap) Ceriops tagal Avicennia* spp. *(Api-api)*
Sabah	**343,470 ha** 58.6% of national total	Forms the core of Malaysia's mangrove estate and blue carbon resources, concentrated along the east coast (Semporna, Tawau, Lahad Datu, Kinabatangan, Sandakan) and north coast (Pitas, Kota Marudu, Kudat), with smaller stands on the west coast.	*Rhizophora apiculata* (Bakau Minyak) *Rhizophora mucronata* (Bakau Kurap) *Bruguiera gymnorrhiza* (Tumu merah) *Nypa fruticans* (Nipah) *Avicennia marina / A. alba*
Sarawak	**127,416 ha** 21.7% of national total	Major estuarine and deltaic mangrove systems of high ecological and carbon value. The Rajang–Belawai–Paloh complex covers 93,852 ha (58% of Sarawak's mangroves), making it the largest mangrove complex in Malaysia. Other key sites include the Trusan–Lawas estuary and the 6,500 ha Kuching Wetlands National Park (6,500 ha; Ramsar 2005)	*Rhizophora apiculata* (Bakau Minyak) *Rhizophora mucronata* (Bakau Kurap) *Xylocarpus granatum Bruguiera gymnorrhiza / B. cylindrica Nypa fruticans* (Nipah)

Sources: (Azman et al., 2021; Goessens et al., 2014; Hatta et al., 2022; Islam et al., 2024, 2026; Omar, 2024; Suhaili et al., 2024)

The Matang Mangrove Forest Reserve (MMFR) in Peninsular Malaysia is recognised as the world's longest continuously managed mangrove forest and a potential blue carbon REDD+ pilot site (Ammar et al., 2014). It serves as both an ecological asset and a socio-economic resource, supporting timber production, fisheries, ecotourism, and local livelihoods. However, MMFR also reflects broader national trends of mangrove degradation, with Malaysia estimated to have lost approximately 30% of its mangrove extent since the 1970s (Goessens et al., 2014). Although managed under a formal Working Plan since 1902, recent assessments suggest that long-term resilience may be constrained by the dominance of *Rhizophora apiculata*, which accounts for over 95% of tree density. This structural imbalance has prompted recommendations for earlier thinning interventions to maintain productivity and ecological function. Historical analyses further indicate that MMFR lost 8,017.3 ha between 1993 and 2011, largely due to conversion for aquaculture and agriculture.

Recent estimates place MMFR's carbon stock at 107.34 Tg C, approximately five times higher than earlier estimates, highlighting its significance as a national blue carbon asset (Wolswijk et al., 2026). While silvicultural activities may result in carbon losses, protected areas continue to retain substantial sediment carbon pools exceeding 3,000 Mg C ha^{-1}. Recognising this potential, Malaysia has intensified its commitment to blue carbon conservation and mangrove restoration as part of its climate mitigation strategy. Corporate Social Responsibility (CSR) has emerged as an important mechanism for supporting these efforts by mobilising resources, strengthening community participation, and promoting local stewardship. Through partnerships between corporations, communities, and conservation agencies, CSR can help bridge the gap between national climate goals and the local governance needed to protect Malaysia's carbon-rich mangrove ecosystems.

2.2 Mangrove Conservation and Restoration Efforts

Malaysia's mangrove conservation framework is primarily based on the gazettement of Permanent Reserved Forests under the National Forestry Act 1984. While national coordination is provided by the Forestry Department of Peninsular Malaysia (FDPM) and its counterparts in Sabah and Sarawak, jurisdiction remains under state authority. Although this framework has supported mangrove management in Peninsular Malaysia for more than a century, current conservation priorities increasingly focus on East Malaysia. Sabah, which contains 58.6% of the nation's mangrove cover, represents Malaysia's most important blue carbon stronghold. Recent studies identified Kota Marudu as having the country's highest mangrove carbon stocks, reaching 1,010.65 Mg C ha^{-1} (Suhaili et al., 2024). Sarawak contributes a further 21.7% (127,416 ha) through extensive estuarine and deltaic systems, particularly the Rajang–Belawai–Paloh complex, the largest mangrove landscape in Malaysia at 93,852 ha and an estimated 38.9 million Mg of stored carbon. In contrast, Peninsular Malaysia accounts for 19.7% of national mangrove cover (115,317 ha) but hosts some of the country's most intensively studied and managed mangrove ecosystems. Nevertheless, the loss of 34,413 ha between 2017 and 2023 highlights the need to strengthen protection and restoration efforts (Omar, 2024).

Malaysia benefits from a strong ecological and institutional foundation for mangrove conservation, exemplified by the Matang Mangrove Forest Reserve (MMFR), where conservation, sustainable use, and economic benefits coexist under structured management. However, mangrove ecosystems continue to face pressures from land conversion, urban expansion, aquaculture, and competing coastal development priorities (Faridah-Hanum et al., 2014). Conservation is further constrained by sectoral planning, limited public participation, and the tendency to undervalue mangroves relative to alternative land uses (Goh, 2016). At the regional level, overlapping institutional mandates and competing policy objectives continue to weaken conservation effectiveness (Friess et al., 2016). Public awareness also remains uneven. Although respondents expressed a willingness to pay MYR 18.25–18.31 for mangrove conservation, studies indicate limited understanding of the broader ecological and cultural value of mangrove ecosystems (Kamaludin et al., 2022).

To address these challenges, the Forest Plantation and Forest Protection Division of FDPM promotes science-based restoration through gene banks, Seed Production Areas (SPAs), nursery management systems, and technical guidance available via the Official Mangrove Portal. These initiatives enhance genetic diversity, improve restoration success, and support long-term ecological integrity. By translating scientific knowledge into practical restoration guidelines, they also help align CSR-supported conservation programmes with ecological suitability, monitoring requirements, and sustainable stewardship rather than short-term planting targets.

2.3 Blue Carbon Potential of Mangrove Ecosystems

Carbon in mangrove ecosystems is stored in three main pools: aboveground biomass, belowground biomass, and sediment organic carbon. Aboveground components such as stems, branches, and leaves store carbon through photosynthesis, while roots contribute to both carbon storage and sediment stabilisation. The largest proportion of carbon is typically stored in sediments. In Malaysia, sediment carbon often dominates total ecosystem carbon stocks; for example, in Sabah mangroves, soil carbon accounts for approximately 69% of total storage, with overall stocks around 450 Mg C ha^{-1} (Hatta et al., 2024). This reflects the capacity of mangrove environments to accumulate and retain organic matter over time, resulting in substantial subsurface carbon reservoirs.

Mangroves sequester carbon through a combination of biological productivity and sedimentary processes. Atmospheric carbon dioxide is fixed through photosynthesis and incorporated into plant biomass, with some transferred to soils through litterfall, root turnover, and microbial decomposition. Root systems enhance sediment trapping by reducing water flow and promoting fine particle deposition, which facilitates organic matter burial. The waterlogged, anoxic conditions of mangrove soils further slow decomposition, allowing carbon to accumulate and persist over long periods (Choudhary et al., 2024). Mangrove sediments may also contain allochthonous carbon transported from surrounding terrestrial and marine environments.

Mangroves are among the most carbon-rich ecosystems globally, with average total carbon stocks ranging from 800–1000 Mg C ha^{-1} when both biomass and soils are considered (Lee et al., 2025). Studies in Malaysia demonstrate substantial variation depending on environmental conditions and forest structure. For example, mangrove forests in Kota Marudu, Sabah, have been reported to store more than 1,000 Mg C ha^{-1}, highlighting the importance of relatively intact systems for long-term carbon storage (Suhaili et al., 2024). At the national scale, Malaysia's approximately 600,000 ha of mangrove forests represent a significant blue carbon reservoir, storing carbon equivalent to hundreds of millions of tonnes of carbon dioxide.

The blue carbon potential of mangroves is influenced by sediment characteristics, hydrological conditions, salinity, pH, and nutrient availability, all of which affect carbon accumulation and preservation (Sahari et al., 2024). Although sediment carbon can remain stable for centuries under low-oxygen conditions, disturbances that expose sediments to oxygen may accelerate decomposition and release stored carbon, making these ecosystems vulnerable to degradation (Alongi, 2012). Quantifying blue carbon requires integrated field and analytical approaches. Aboveground biomass is commonly estimated using species-specific allometric equations, while sediment carbon is assessed through coring and laboratory analyses. Recent advances in remote sensing and unmanned aerial vehicles (UAVs) have further improved the ability to map mangrove extent and monitor carbon stocks in complex coastal environments (Abu Bakar et al., 2024).

3. CORPORATE SOCIAL RESPONSIBILITY AND ENVIRONMENTAL GOVERNANCE

3.1 Evolution of Corporate Social Responsibility (CSR)

In early writings on Corporate Social Responsibility (CSR), it was more often referred to as social responsibility (SR) than as CSR. The publication of Bowen's 1953 book Social Responsibilities of the Businessman is regarded as a classic and foundational book of this subject (Carroll, 1999). He set forth an initial definition of the SR of businessmen as follows:

> *"What responsibilities to society may businessmen reasonably be expected to assume? It refers to the obligations of businessmen to pursue those policies, to make those decisions, or to follow those lines of action which are desirable in terms of the objectives and values of our society" (Bowen, 2015)*

In contrast to Bowen, Friedman argued that the only social responsibility of business is to increase its profits (Ferrero et al., 2014):

"What does it mean to say that business has responsibilities? Only people can have responsibilities. The SR of business is to increase its profits" (Friedman, 1970)

He argued that only people can have responsibilities, not businesses. The people who are hired by business owners have a responsibility primarily to their employers, to meet their desires, which in most cases equates to profits. Friedman also argued that an individual can have perceived responsibilities in areas away from the business, but says of this:

"If we wish, we can refer to some of these responsibilities as 'social responsibilities.' But in these respects, he is acting as a principal not an agent; he is spending his own money or time or energy, not the money of his employers or the time and energy he has contracted to devote to their purposes. If these are 'social responsibilities,' they are the social responsibilities of the individual, not the business" (Friedman, 1970)

Nonetheless, Friedman's "profits-only" view of CSR was not widely accepted (Davis, 2017). Later in 1984, Freeman popularised the stakeholder concept and emphasized the needs to satisfy not only the shareholders, but multiple stakeholders who can affect the corporation's outcomes. People create organisations to leverage their collective resources in pursuit of common goals. As organisations pursue these goals, they interact with others inside a larger context called society. Society therefore exists as a mix of these different organisational forms. Each of these groups, therefore, can claim a stake in the organisation's operations. (Freeman, 2010) defined these groups as organisations' stakeholders. The wide definition of stakeholder includes any group or individual who can affect or is affected by the corporation (Freeman & Reed, 1983). Freeman's original stakeholder framework comprises seven stakeholders: shareholders (or financiers), suppliers, civil society, employees, customers, competitors and the government. In 2007, Freeman extended the stakeholder framework and introduced a two-tier stakeholder group: (1) tier one – the primary or definitional groups and (2) tier two – the secondary or instrumental groups. The primary or definitional stakeholders (Freeman, 2010) consist of the owners (or financiers), employees, customers, suppliers and local communities. These primary stakeholders are essential to business survival and growth. Secondary stakeholders include competitors, corporate critics, environmentalists, activists, the media, governments, and special-interest groups (SIGs). Although less directly involved in business operations, they can significantly influence primary stakeholder relationships and organisational performance. Loss of support from these groups may expose businesses to operational, reputational, and strategic risks..

Further expansion of CSR was later undertaken when Carroll introduced a four-part of CSR definition in 1979. The original definition of Carroll's CSR definition was originally stated as: "CSR encompasses the economic, legal, ethical, and discretionary (philanthropic) expectations that society has of organisations at a given point in time" (Carroll, 1999). In 1991, Carroll extracted the four parts of the CSR definition and recast them into a CSR pyramid. The purpose of the pyramid was to highlight the definitional aspect of CSR and to illustrate the building-block nature of the four-part framework. Carroll's evolution of his original four-part construct of CSR (1979) and then his pyramidal depiction of CSR (1991) is said to be the most cited, most influential and most relevant and it has proliferated in the literature since the 1950s (Baden, 2016; Wahab et al., 2007).

3.2 The State of CSR in Malaysia

There are numerous CSR initiatives in Malaysia. For instance, in 1999, the Malaysian Institute of Corporate Governance introduced the Malaysian Code on Corporate Governance (MCCG), which identifies a framework for best practices in corporate governance (Wahab et al., 2007). Later in 2004, the National Integrity Plan (NIP) was introduced by Malaysian Institute of Integrity (IIM). The focus of NIP was to strengthen ethics and integrity for all members of the community (Baqutayan & Mayati, 2021). In 2005, the Putrajaya Committee on GLC High Performance (PCG) launched ten initiatives to improve the performance of government-linked companies (GLCs) in Malaysia. Since then, PCG has launched ten initiatives, two of which include the Green Book (Enhance Board Effectiveness) and the Silver Book (Achieving Value through Social Responsibility). In addition, the Bursa Malaysia launched its CSR framework for the public-listed companies (PLCs) 2006. The Framework is a set of voluntary reporting guidelines that outline four CSR initiative areas, covering the environment, community, workplace and marketplace (Lu & Castka, 2009).

3.3 CSR and Environmental, Social, and Governance (ESG) Framework in Malaysia

The first report on sustainable development in Malaysia was introduced in 1987 (Mohammad & Wasiuzzaman, 2021). The effort is further strengthened in 2006 when Bursa Malaysia introduced the CSR framework which is mandatory for the public listed companies (PLCs) in Malaysia. Bursa Malaysia defined CSR as a transparent and open business practice based on ethical values and respect for the employees, community, shareholders, other stakeholders and the environment. It is also recommended that the corporate sustainability strategies should be directed at ESG practices. The government of Malaysia announced the Malaysian code on corporate governance (MCCG) 2012 that emphasized the need for transparency in financial and non-financial disclosures for the shareholders and other potential investors. Predominantly, in non-financial reporting, the MCCG 2012 focused on environmental, social and governance (ESG) reporting and improving its related corporate strategies (Zahid, 2020).

Bursa Malaysia evaluates firms' ESG transparency and performance across 14 categories using the FTSE Russell ESG Rating model (Sulaiman et al., 2024). While FTSE Russell aligns with international ESG standards in its thematic coverage and materiality approach, it functions primarily as a benchmarking and evaluation tool rather than a disclosure framework. To promote sustainable investment and corporate accountability, Bursa Malaysia introduced the FTSE4Good Bursa Malaysia (F4GBM) Index in 2014 with the aims to: (i) support investors in making ESG investments in Malaysian listed companies; (ii) increase the profile and exposure of companies with leading ESG practices; (iii) encourage best-practice ESG disclosure; and (iv) support the transition to a lower-carbon and more sustainable economy. In 2021, Bursa Malaysia and FTSE Russell launched the FTSE4Good Bursa Malaysia Shariah (F4GBMS) Index to meet growing demand for ESG and Shariah-compliant investment solutions. The introduction of both indices has contributed to increased ESG reporting and disclosure among Malaysian listed companies.

Figure 2. F4GBM and F4GBMS models (Bursa Malaysia, 2018)

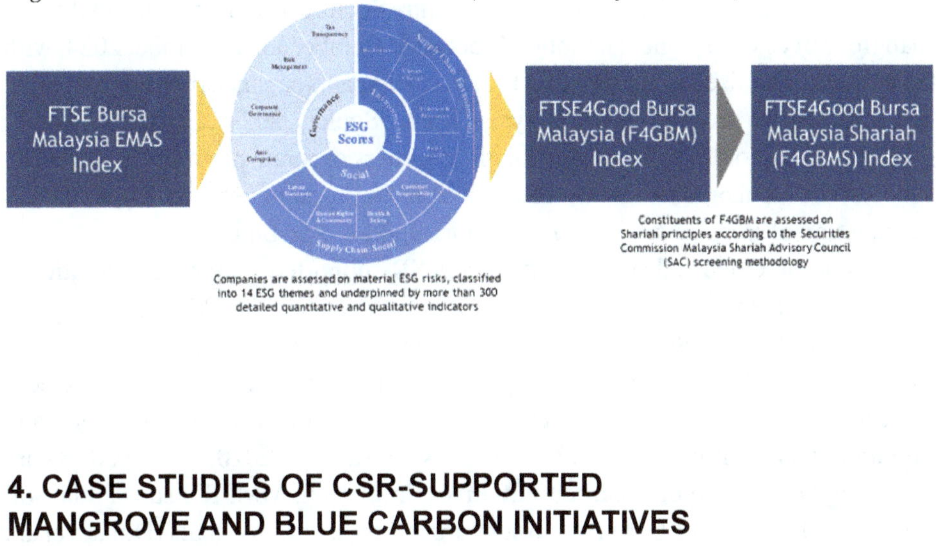

4. CASE STUDIES OF CSR-SUPPORTED MANGROVE AND BLUE CARBON INITIATIVES

4.1 Corporate Participation in Mangrove Conservation

The emergence of Corporate Social Responsibility (CSR) as a mechanism for mangrove conservation in Malaysia reflects a broader shift in which corporations are increasingly recognised not only as funding sources but also as strategic partners in climate governance. Over the past two decades, Government-Linked Companies

(GLCs), private institutions, NGOs, and state agencies have implemented a wide range of mangrove conservation initiatives, from volunteer planting activities to long-term, science-based restoration programmes (Table 2).

This approach is reflected in the Low Carbon Society Blueprint for Iskandar Malaysia 2030, which identifies mangrove rehabilitation under Action 04: Natural Environment and Habitats. Rather than focusing solely on planting, the blueprint promotes a structured restoration process involving habitat assessment, site prioritisation, species selection with agencies such as FRIM, community participation, and long-term monitoring. It also emphasises public–private partnerships and ESG-driven initiatives as key drivers of restoration success. Examples include mangrove conservation programmes led by Petronas Refinery and Petrochemical Corp., Pengerang Terminals in the Sungai Johor estuary, and Port of Tanjung Pelepas in Tanjong Adang (Low Carbon Society Blueprint for Iskandar Malaysia 2030 Climate Action Plan, 2022). The blueprint further notes that such collaboration is already visible in practice, citing examples such as the mangrove conservation programme launched by Petronas Refinery and Petrochemical Corp. and Pengerang Terminals in the Sungai Johor estuary, as well as Port of Tanjung Pelepas' replanting programme in Tanjong Adang.

At the operational level, the Malaysian Nature Society (MNS) reported that 5,293 mangrove trees were planted nationwide between October and December 2024, with 2,223 seedlings raised in nurseries and 3,576 participants engaged through education and conservation programmes (Malaysian Nature Society, 2024). In Kuala Selangor Nature Park, restoration activities brought together corporate partners, universities, schools, and local organisations through awareness programmes, nursery work, and the planting of *Rhizophora apiculata* and *Rizophora mucronata*.

Beyond NGO-led and private initiatives, CSR-supported mangrove restoration is increasingly integrated into state-managed forest systems. In the Matang Mangrove Forest Reserve (MMFR), the Perak State Forestry Department has partnered with PETRONAS and YTL Cement to support restoration and strengthen degraded ecological buffers threatened by aquaculture expansion, industrial encroachment, and pollution (Goessens et al., 2014; Martínez-Espinosa et al., 2020). These efforts are increasingly linked to blue carbon initiatives, leveraging MMFR's substantial soil carbon stocks of up to 1,074 t ha^{-1} to support carbon-neutrality goals (Donato et al., 2011). Collectively, these examples show that mangrove restoration in Malaysia is evolving beyond ecological rehabilitation into a collaborative governance approach that integrates conservation, livelihoods, local institutions, and multi-stakeholder partnerships.

Mangrove and blue carbon initiatives in Malaysia range from small planting campaigns to large-scale conservation and research programmes (Table 2). A key strength is collaboration among universities, research institutes, NGOs, and technical partners, which integrates scientific knowledge into CSR implementation. This is important because restoration success depends on appropriate species selection, site suitability, and long-term survival rather than planting numbers alone. Science-based approaches improve ecological suitability, monitoring, and verification, increasing the likelihood that CSR investments deliver meaningful blue carbon, biodiversity, and coastal protection benefits. When effectively implemented, these initiatives can make a more credible contribution to Malaysia's climate and nature-based solutions targets.

Table 2. Selected mangrove and blue carbon initiatives in Malaysia, highlighting programme scale, partnership arrangements, and reported aims and impacts

Lead Organisation	**Programme**	**Scale / Area**	**Stakeholders**	**Aims and Impact**
Yayasan PETRONAS Sector: Oil and Gas	PETRONAS–Mercedes F1 Blue Carbon Research	100,000 seedlings (~10–20 ha) Sungai Santi Forest Reserve, Johor 5 Year Program	Universiti Putra Malaysia and University of São Paul	Aims: Focuses on researching, conserving, and restoring mangrove ecosystems. Impact: Combines scientific research with nature-based solutions to enhance biodiversity and protect coastal communities, complementing PETRONAS' efforts to achieve net-zero carbon emissions by 2050
Malaysia Aviation Group Sector: Aviation	Malaysia Aviation Group (MAG) CSR	220 seedlings (~0.02 ha) Tanjung Rhu, Carey Island	Local communities, NGOs; Selangor	Aims: Mangrove restoration, community engagement, Impact: support for Mah Meri cultural heritage. Link: (malaysiaairlines.com)
Chery Corporate Malaysia Sdn. Bhd Sector: Automotive	Cherish the Nature Partnership	130 km North-central Selangor Coast, 6 hectares of degraded mangrove area targeted for restoration 2 years program with post planting monitoring	Local communities, academic institution IUCN, Malaysian Nature Society, OMODA and JAECOO Malaysia	Aims: Restore degraded mangroves along the North-Central Selangor Coast, strengthen conservation awareness, and support community-led conservation using a science-based approach. Impact: Improving sapling survival and ecological recovery through science-based site selection, community nurseries, local planting, and post planting monitoring, while also supporting shoreline protection, biodiversity, and livelihoods. Link: IUCN

Lead Organisation	Programme	Scale / Area	Stakeholders	Aims and Impact
Ecosecurities / Blue Natural Capital	Rajang-Belawai Delta Conservation	77,600 ha mangrove landscape with an initial preparatory phase leading to a multi-year blue carbon project.	Blue Natural Capital / Blue Carbon Accelerator Fund, South East Carbon, Sarawak Forest Department, local stakeholders, NGOs, and community groups.	Aim: Restore and protect the mangroves, assess its blue carbon potential, and develop technical, social, and financial pathways for a longer-term conservation initiative. Impact: Baseline for a future CCB and VCS blue carbon project, attract private investment, strengthen community resilience, reduce illegal mangrove harvesting, expand protected areas, and support local livelihoods through carbon revenue sharing and nature-based enterprises. Link: EcoSecurities
CIMB Foundation	CIMB–UKM Mangrove Conservation	7,000 seedlings (~7 ha) Kuala Selangor National Park (KSNP) 2 years program	Local communities, LESTARI-UKM, MyMangrove	Aims: Restore mangrove canopy gaps at KSNP through habitat restoration, awareness, and research on carbon stock, sequestration, and financing. Impact: Supports biodiversity, pond restoration, coastal protection, and livelihood opportunities for five fishing communities; expected to benefit about 40% of Kuala Selangor's population Link: CIMB Foundation
Yayasan Sime Darby	Yayasan Sime Darby Mangrove Conservation	Scale: RM4.5 million; six sites in Johor, Kedah, Perak, and Selangor; target 50,000 trees and 25 ha. Long term program	Global Environment Centre, state forestry departments, local authorities, and community groups.	Aim: Conserve and restore mangroves while strengthening coastal community livelihoods and resilience. Impact: Community-based restoration with training, school awareness programmes, mangrove-based livelihood products, and local patrollers for monitoring. Link: YSB and GEC

While the Malaysian examples illustrate the growing role of CSR in supporting mangrove conservation and blue carbon initiatives, similar approaches have been implemented internationally under diverse governance and socio-economic contexts. In Thailand, initiatives such as the PTTEP Mangrove Reforestation Project and the Brother Thailand Mangrove Forest Restoration Project demonstrate how long-term corporate commitment, community participation, and technological innovation can

continued on following page

Table 2. Continued

enhance restoration effectiveness and ecosystem resilience (Howard et al., 2023; Satyanarayana et al., 2021). n Indonesia, Pertamina-supported programmes link mangrove conservation with biodiversity protection, livelihoods, and ecotourism through collaborative governance (Christian et al., 2021; Gevaña et al., 2018). In contrast, Australian blue carbon projects operate within more mature regulatory frameworks that integrate restoration with carbon accounting, carbon markets, and rigorous monitoring, reporting, and verification (MRV) systems (Lovelock & Duarte, 2019; Macreadie et al., 2019a). Despite differing contexts, these initiatives share common success factors, including long-term funding, stakeholder collaboration, scientific monitoring, and community participation, offering valuable lessons for strengthening blue carbon initiatives in Malaysia.

4.2 Community and Multi-Stakeholder Partnerships

Effective mangrove conservation in Malaysia increasingly depends on multi-stakeholder partnerships involving corporate funders, scientific institutions, government agencies, NGOs, and local communities. This reflects recognition that the ecological complexity of mangrove systems and the socioeconomic drivers of their degradation cannot be addressed by any single sector (Inaku et al., 2025; Kamaludin et al., 2022; Satyanarayana et al., 2021).

Evidence from corporate-led initiatives shows that the most sustained and ecologically effective programmes integrate community participation and scientific oversight as core components of governance, rather than as supplementary elements. In Malaysia, mangrove-related CSR must recognise that Indigenous and marginalised coastal communities often have long-standing cultural, livelihood, and territorial ties to these ecosystems that predate formal conservation frameworks. In areas inhabited by Orang Asli (indigenous) groups, such as the Mah Meri along the Selangor coast, standard corporate engagement approaches may not be sufficient, as mangrove restoration also intersects with customary practices, cultural heritage, and local ecological stewardship (Zuhairi et al., 2020). Treating these issues as equivalent to general livelihood support or planting participation risks reducing community inclusion to a symbolic exercise. Meaningful engagement, therefore, requires prior and informed consent, culturally appropriate processes, and benefit-sharing arrangements that strengthen existing community institutions and knowledge systems.

Science and community partnerships are key determinants of ecological credibility and carbon market eligibility in Malaysian mangrove CSR initiatives. Collaboration with research institutions provides baseline carbon stock assessments, biodiversity monitoring, and sequestration data (Ammar et al., 2014), which are often required for certification under standards such as the Verra Verified Carbon Standard and the Gold Standard. In this context, The Tri Helix framework, involving government, academia, and the private sector, offers a structured approach to scaling these partnerships beyond individual corporate efforts.

Community and multi-stakeholder engagement in Malaysia is thus transitional, shifting from a philanthropic model towards co-governance, where communities, scientists, corporations, and state agencies share responsibility for ecological outcomes. While current initiatives (Table 2) demonstrate this potential, significant gaps remain. Addressing these requires improved programme design alongside reforms in benefit-sharing, MRV systems, and corporate disclosure standards. Such changes are necessary to transform isolated multi-stakeholder arrangements into a coherent national blue carbon governance system aligned with Malaysia's commitments under the National Policy on Climate Change 2.0 and the National Policy on Biological Diversity 2022 to 2030.

4.3 Measuring CSR Outcomes in Mangrove Conservation and Blue Carbon Initiatives and Multi-Stakeholder Partnerships

While CSR-supported mangrove conservation initiatives are increasingly recognised for their environmental and social contributions, evaluating their effectiveness requires measurable indicators that extend beyond the number of trees planted. Quantitative assessments of restoration outcomes, carbon sequestration potential, and community benefits provide a stronger basis for evaluating the contribution of CSR programmes to climate mitigation, biodiversity conservation, and sustainable development. Table 3 summarises selected quantitative indicators reported across CSR-supported mangrove conservation initiatives discussed in this chapter.

Table 3. Quantitative indicators of CSR impacts on mangrove restoration, carbon sequestration, and community livelihoods reported across selected case studies

Initiative	Restoration Indicator	Carbon Indicator	Community indicator
PETRONAS–Mercedes F1	100,000 seedlings planted	Blue carbon monitoring programme established	University and stakeholder collaboration
CIMB–UKM	7,000 seedlings planted (~7 ha)	Carbon stock and sequestration assessments conducted	Benefits to local fishing communities
Yayasan Sime Darby	50,000 trees across 25 ha	Potential carbon storage enhancement	Community participation and awareness programmes
Brother Thailand	>15 years restoration	>40,000 kg CO_2 sequestered annually	Improved fisheries income and community involvement
PTTEP Thailand	Thousands of hectares restored	Carbon sequestration monitored using AI and remote sensing	Community guardianship and livelihood support

The reported indicators demonstrate that CSR-supported initiatives can generate measurable ecological, climate, and socio-economic outcomes. Restoration achievements range from thousands of seedlings planted to large-scale rehabilitation programmes covering multiple hectares of mangrove habitat. Several initiatives have also incorporated blue carbon assessments through carbon stock estimation, carbon sequestration monitoring, and long-term ecological research. Community benefits are evident through livelihood enhancement, local participation in restoration activities, ecotourism development, and support for coastal fisheries.

However, the case studies also reveal inconsistencies in reporting practices. While restoration inputs such as the number of seedlings planted are commonly reported, fewer initiatives provide data on survival rates, biodiversity recovery, long-term carbon sequestration, or livelihood improvements. This highlights the need for more standardised monitoring, reporting, and verification (MRV) approaches to improve transparency and enable meaningful comparisons across projects.

5. MALAYSIA'S CLIMATE AND BIODIVERSITY POLICY LANDSCAPE

5.1 Policy Alignment with Blue Carbon Initiatives

Malaysia's climate and biodiversity policy framework has increasingly recognised the role of blue carbon ecosystems in climate mitigation and biodiversity conservation, creating a supportive environment for CSR-driven initiatives. Mangroves, seagrasses, and tidal marshes are now more closely linked to national commitments under international climate and biodiversity frameworks, including the United Nations Framework Convention on Climate Change and the Convention on Biological Diversity (Lee et al., 2025). National policies promote carbon sinks, nature-based solutions, multi-stakeholder collaboration, and greater corporate accountability through ESG-related disclosures. However, implementation remains limited. Many corporate initiatives continue to be evaluated using input-based indicators, such as the number of trees planted, rather than outcome-based measures including survival rates, carbon sequestration, and ecosystem recovery. This limits the ability to assess the effectiveness, credibility, and long-term contribution of CSR-supported blue carbon initiatives to national climate and biodiversity goals (Bastin et al., 2019; Macreadie et al., 2019).

Emerging developments in Malaysia's carbon market framework signal a shift toward more structured, incentive-based approaches. The introduction of a national carbon market and exploration of blue carbon credit mechanisms create opportunities for CSR to evolve from voluntary activities into formal climate mitigation contributions (Montenegro et al., 2025; Perera et al., 2025). However, key enabling elements remain limited, particularly standardised monitoring, reporting, and verification (MRV) systems and clearly defined carbon rights. Existing projects face challenges including methodological inconsistencies, transparency issues, and high implementation costs (Karpowicz et al., 2024) while unclear carbon ownership hinders equitable benefit-sharing and governance (Afuwape, 2025; World Bank, 2023). Although policy instruments provide entry points for CSR engagement (Table 4), gaps persist in governance coordination, measurement frameworks, and integration of corporate contributions into national climate accounting. At the state level, alignment between policy initiatives and CSR activities remains inconsistent (Becker et al., 2025; Gevaña et al., 2018; Karpowicz et al., 2024).

Table 4. Key Malaysian policies and their relevance to blue carbon initiatives

Policy Instrument	Year	Key Blue Carbon Provision	CSR Relevance	Key Gap
National Policy on Climate Change 2.0 (NPCC 2.0) Link: NPCC 2.0	2024	Increase carbon sink capacity; scale up private sector involvement through carbon pricing; NDC Action Plan & LT-LEDS	Direct mandate for GLC investment in NbS; supports blue carbon credit market development	No binding MRV standard; private sector contributions not yet formally attributed to NDC
National Energy Transition Roadmap (NETR) Link: NETR	2023	GLCs designated as national champions; RM2bn seed funding for private investment	Reinforces GLC mandate for nature-based solutions alongside energy transition	NETR focus is energy, not blue carbon specifically
Green Technology Master Plan (GTMP) Link: GTMP	2017–2030	Whole-of-nation approach; GTFS and GITA fiscal incentives	Reduces financial barriers to corporate investment in mangrove rehabilitation	Incentive uptake for blue carbon projects not yet documented
National Policy on Biological Diversity 2022–2030 (NPBD) Link: NPBD	2022	Mangroves as critical ecosystems; multi-stakeholder conservation; 30x30 alignment; KMGBF	Policy basis for corporate biodiversity claims; targets for ESG disclosure	No binding private sector restoration quota
Sabah Mangrove Action Plan Link: SMAP	2024–2033	Comprehensive 10-year framework for Sabah's 378,038 ha; most advanced state-level blue carbon policy	Framework for aligning Sabah CSR projects with state priorities; 59% of Malaysia's mangroves	No CSR projects currently explicitly reference this plan
Bursa Malaysia Sustainability Reporting Framework (NSRF) Link: NSRF	2024	Phased mandatory TCFD-aligned disclosure (2025–2027); IFRS S1 & S2 alignment	Creates accountability pressure for disclosed conservation metrics to be verifiable	Input metrics (trees planted) currently sufficient for compliance; outcome metrics not yet mandated

5.2 Policies and Frameworks Related to Blue Carbon and Mangrove Conservation

Malaysia's mangrove and blue carbon governance is supported by national programmes, state-level initiatives, and broader environmental policies. While these frameworks demonstrate strong commitment to mangrove conservation, integration with blue carbon remains limited. At the national level, conservation has largely focused on restoration programmes such as the Mangrove and Suitable Species Planting Programme, which has planted millions of trees (Omar, 2024). These efforts have raised awareness and rehabilitated degraded areas but remain centred on planting rather than long-term ecological outcomes. Critical indicators such as survival rates, carbon storage, and ecosystem recovery are not consistently monitored, limiting assessment of their climate mitigation contributions (Donato et al., 2011; Macreadie et al., 2019).

Broader sustainability and environmental policies provide indirect support for mangrove conservation by promoting sustainable development and private-sector participation, including CSR initiatives. However, blue carbon remains poorly defined and not fully integrated into these frameworks, limiting its role in national climate strategies (Economic Planning Unit, 2021; Ministry of Natural Resources and Environment Malaysia, 2016; Zen et al., 2023). At the state level, mangrove management falls under land and forestry jurisdiction, leading to regional variation in implementation. While some states have established protected areas and management plans, overall coordination remains uneven. As a result, many CSR initiatives are not well aligned with state frameworks, resulting in fragmented implementation and missed opportunities for integrated management (Friess et al., 2016; Lee et al., 2025).

A major challenge across policy levels is the lack of clear carbon ownership and governance arrangements. Uncertainty over ownership and benefit rights for mangrove carbon limits blue carbon project development and participation in carbon markets, and is linked to overlapping agency responsibilities and weak federal–state coordination (Donato et al., 2011; Friess et al., 2016; Lee et al., 2025). Another key gap is the absence of standardised monitoring, reporting, and verification (MRV) systems. Without consistent methods to quantify carbon stocks and ecological outcomes, conservation efforts cannot be effectively evaluated or linked to national climate targets, reducing the credibility of CSR initiatives and their integration into carbon markets (Donato et al., 2011; Kauffman & Donato, 2012; Pendleton et al., 2012).

5.3 Policy Directions and Way Forward

Malaysia's transition from a supportive policy environment to an operational blue carbon governance system requires CSR to be more closely integrated into climate mitigation strategies rather than treated as a voluntary contribution. As private sector participation becomes increasingly important for achieving national climate targets, a National Blue Carbon Policy, or a dedicated instrument under NPCC 2.0, could establish minimum Monitoring, Reporting, and Verification (MRV) requirements for all projects, including CSR-led initiatives. Robust MRV systems are essential for credible carbon claims, market participation, and the development of a National Blue Carbon Registry to improve transparency and prevent double counting (Howard et al., 2017; World Bank Group, 2021). Clear carbon rights are equally important for investor confidence and private sector participation (World Bank, 2023).

Stronger federal–state coordination is needed to address jurisdictional fragmentation, particularly where CSR-led restoration occurs on state-controlled mangrove land. Without better alignment, restoration remains vulnerable to insecure tenure, conflicting land uses, and uneven governance. Clear zoning and tenure arrangements are therefore essential to ensure CSR investments deliver durable ecological and carbon benefits (Friess et al., 2016; McLeod et al., 2011). Carbon market access must likewise become more workable for small and medium-scale projects, since many CSR initiatives face high transaction and verification costs. Aggregation mechanisms and simplified methodologies could allow smaller projects to pool verified outcomes and participate more effectively in voluntary carbon markets (Locatelli et al., 2015).

Corporate disclosure should also move beyond symbolic reporting towards measurable outcomes. Reporting frameworks should require site-specific evidence on carbon sequestration, ecosystem survival, and community co-benefits, thereby strengthening accountability and aligning CSR with scientifically credible mitigation outcomes (Bursa Malaysia, 2023). CSR-linked blue carbon outcomes should also be reflected in Malaysia's Nationally Determined Contributions and greenhouse gas inventories so that corporate initiatives are better aligned with national climate commitments (Griscom et al., 2017). Taken together, these reforms would reposition CSR as a more structured, accountable, and policy-aligned component of Malaysia's blue carbon governance system.

6 COORDINATING SCIENCE, POLICY, AND CSR

6.1 Science-Based Alignment of Corporate Conservation Initiatives

The long-term success of mangrove restoration and blue carbon initiatives depends on effective coordination among science, policy, and corporate social responsibility (CSR). Scientific research provides the basis for site selection, species suitability, monitoring, and carbon assessment, while policy frameworks guide implementation, ensure regulatory alignment, and support continuity. CSR contributes by mobilising corporate resources for restoration, conservation, and long-term stewardship. However, effective CSR-driven restoration cannot rely on isolated planting efforts alone. Sustainable outcomes, particularly for long-term carbon sequestration, require collaborative partnerships among government agencies, NGOs, researchers, local communities, and corporate actors.

This coordination is particularly important when mangrove restoration is framed as a nature-based solution (NbS) rather than a tree-planting exercise. In the context of blue carbon, NbS focuses on restoring ecosystem functions that support climate mitigation, coastal protection, biodiversity conservation, and livelihood resilience (Beh, 2022; Gijsman et al., 2021). This framing shifts CSR away from short-term, symbolic greening towards ecologically grounded, socially relevant interventions aligned with policy goals. It also reduces the risk of greenwashing by recognising that tree planting alone does not guarantee environmental benefits, especially without ecological suitability and long-term accountability (Neimark, 2018).

The Building with Nature (BwN) Sungai Haji Dorani Mangrove Project illustrates the importance of integrating science, policy, and CSR in restoration efforts. Rather than relying solely on planting, the project used Bamboo Wave Screens to reduce erosion, promote sediment accretion, and support natural mangrove regeneration (Hashim et al., 2010). In contrast, restoration efforts in Sungai Acheh, Penang, where approximately 400,000 mangroves were planted over two decades with CSR support, faced constraints including limited technical expertise, language barriers, and restricted access to finance (Beh, 2022). These examples demonstrate that corporate support and community participation alone are insufficient without scientific guidance, institutional capacity, and sustained funding. They also highlight the value of mangroves in supporting fisheries, coastal livelihoods, and carbon sequestration. Ultimately, effective CSR-linked restoration depends on strong coordination among science, policy, and communities, supported by robust monitoring systems to verify blue carbon outcomes and enhance ecological resilience (Beh, 2022).

6.2 Strengthening Scientific Monitoring and Data Systems

Strong scientific monitoring is essential for ensuring the credibility and effectiveness of mangrove conservation and blue carbon initiatives. In complex mixed-species mangrove forests, site- and species-specific allometric equations provide more accurate biomass estimates than generic models, which often fail to capture local structural variation (Khan et al., 2025). Consequently, reporting frameworks such as REDD+ should move beyond traditional zonation approaches and incorporate forest structural characteristics to improve carbon accounting and alignment with international blue carbon standards.

Science-based restoration requires corporate conservation programmes to extend beyond symbolic tree planting and adopt robust ecological practices throughout the restoration cycle. This begins with pre-implementation hydrological and ecological assessments to determine site suitability and reduce restoration failure. Species selection should follow natural zonation patterns and local provenance to enhance ecological fit and genetic resilience, while restoration planning should incorporate Traditional Ecological Knowledge from local communities (Firdaus et al., 2019; Rodríguez-Rodríguez et al., 2021).

Monitoring should continue for three to five years, tracking key indicators such as sapling survival, biomass growth, and soil carbon stocks using standardised methods developed by institutions such as the Forest Research Institute Malaysia (FRIM) (Omar, 2024; Suhaili et al., 2024). Alignment with international standards, including Global Mangrove Alliance guidelines, further strengthens project credibility (Howard et al., 2023). o enhance ESG performance and reduce greenwashing risks, restoration outcomes should undergo independent third-party verification consistent with Bursa Carbon Exchange (BCX) requirements(Bursa Malaysia, 2023). Although such approaches require greater investment and collaboration among corporate, academic, government, and NGO stakeholders, they improve long-term ecosystem sustainability and support the recognition of mangrove restoration as a credible blue carbon asset contributing to national Net Zero targets (Al-Mutairi, 2025).

6.3 Proposed CSR Blue Carbon Credibility Framework

Drawing from the chapter's findings, a CSR Blue Carbon Credibility Framework is proposed to guide the evaluation of corporate conservation initiatives. The framework recognises that CSR activities, including funding, planting programmes, volunteer efforts, and ESG commitments, should not automatically be regarded as blue carbon contributions. Instead, credibility depends on whether these commitments generate policy-aligned, scientifically robust, verifiable, and socially beneficial conservation outcomes. As shown in Figure 3, the framework comprises five dimensions: CSR commitment, policy alignment, scientific credibility, blue carbon verification, and shared conservation value. Together, these dimensions distinguish symbolic planting activities from initiatives that support measurable mangrove recovery, carbon protection, biodiversity conservation, coastal resilience, and climate governance. By positioning CSR as a governance mechanism rather than an outcome, the framework provides a practical tool to strengthen accountability, transparency, and long-term effectiveness of CSR-supported conservation initiatives.

Figure 3. CSR blue carbon credibility framework for evaluating CSR-supported mangrove conservation initiatives through five dimensions: CSR commitment, policy alignment, scientific credibility, blue carbon verification, and shared conservation value

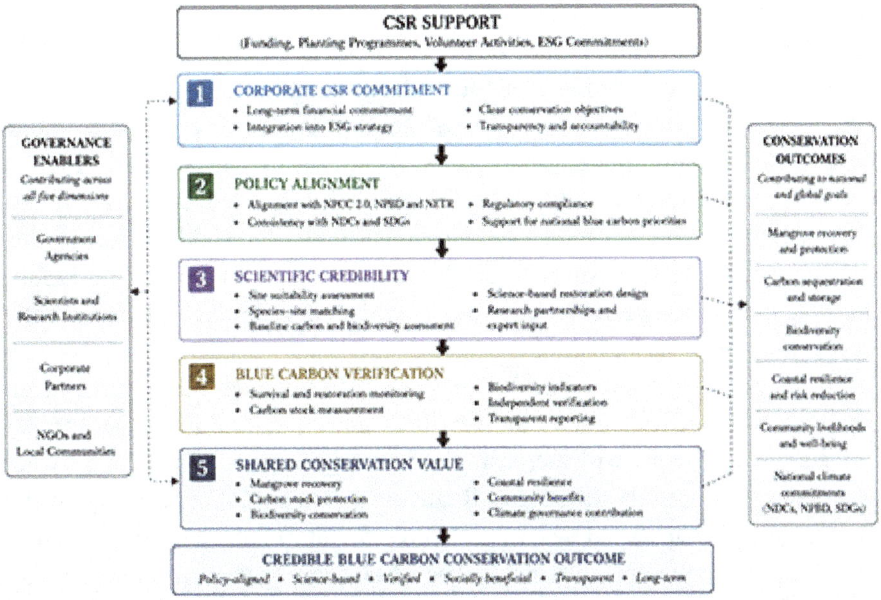

Despite the growing role of CSR in mangrove conservation and blue carbon initiatives, several challenges remain. Funding continuity is a key concern, as short corporate budgeting cycles often do not align with the long timeframes required for ecosystem recovery and carbon sequestration. Governance fragmentation among government agencies, corporate actors, and civil society organisations can also create overlapping responsibilities and inconsistent implementation. In addition, conservation objectives may conflict with coastal development, aquaculture expansion, and other land-use priorities. Monitoring and verification remain uneven, with many initiatives focusing on inputs such as trees planted rather than ecological outcomes including survival rates, biodiversity recovery, and carbon sequestration (Howard et al., 2017; Macreadie et al., 2019). Addressing these challenges is essential to improve the effectiveness and credibility of CSR-supported blue carbon initiatives.

7 CONCLUSIONS

This chapter examined how Corporate Social Responsibility (CSR) can serve as a strategic mechanism for advancing blue carbon conservation in Malaysia rather than remaining a symbolic sustainability commitment. Evidence from the case studies demonstrates that CSR contributes most effectively to mangrove conservation when supported by credible scientific frameworks, long-term partnerships with universities and local communities, alignment with national climate and biodiversity policies, and measurable ecological and socio-economic outcomes.

The findings show that the distinction between symbolic and substantive CSR depends less on tree planting itself than on programme design. Initiatives such as the CIMB–UKM programme, PETRONAS–Mercedes F1 research initiative, Rajang–Belawai Delta project, and Yayasan Sime Darby restoration programme illustrate a more credible model of CSR grounded in science-based restoration, multi-stakeholder collaboration, community participation, and long-term monitoring. Successful initiatives share three characteristics: robust ecological baselines, independent monitoring and verification, and community co-management that supports long-term stewardship.

As Malaysia advances toward its Net Zero 2050 target and commitments under the Kunming–Montreal Global Biodiversity Framework, its approximately 600,000 ha of mangrove forests will increasingly be recognised as climate infrastructure rather than solely conservation areas. Realising this potential requires scientifically accountable CSR supported by corporate investment, scientific rigour, community stewardship, and coherent governance. The Malaysian experience demonstrates how private-sector engagement can strengthen blue carbon conservation and offers lessons for other coastal economies seeking to integrate CSR into national climate strategies.

Future research should focus on developing standardised frameworks that integrate CSR reporting, blue carbon accounting, and ecosystem-based climate policies. Greater collaboration among governments, corporations, communities, academic institutions, and international organisations, together with harmonised MRV systems and carbon credit mechanisms, would enhance the credibility, scalability, and long-term sustainability of CSR-supported blue carbon initiatives.

ACKNOWLEDGEMENT

The authors gratefully acknowledge the financial support provided by Universiti Sains Malaysia RUI NICHE GRANT under Grant No. [R502-KR-ARU001-0000002305K134], which made this work possible.

REFERENCES

Abu Bakar, N. A., Wan Mohd Jaafar, W. S., Abdul Maulud, K. N., Muhmad Kamarulzaman, A. M., Saad, S. N. M., & Mohan, M. (2024). Monitoring mangrove-based blue carbon ecosystems using UAVs: A review. [). Taylor and Francis Ltd.]. *Geocarto International, 39*(1), 2405123. DOI: 10.1080/10106049.2024.2405123

Afuwape, K. (2025). Legal and regulatory frameworks for blue carbon credit markets. *Journal of Foreign Legislation and Comparative Law*, *21*(3), 88. DOI: 10.61205/S199132220034876-9

Al-Mutairi, K. A. (2025). Literature Analysis and Synthesis of the Red Sea Mangrove Ecosystem: Decades of Human Impacts and Knowledge Gaps. *Applied Ecology and Environmental Research*, *23*(4), 7031–7067. DOI: 10.15666/aeer/2304_70317067

Alongi, D. M. (2008). Mangrove forests: Resilience, protection from tsunamis, and responses to global climate change. *Estuarine, Coastal and Shelf Science*, *76*(1), 1–13. DOI: 10.1016/j.ecss.2007.08.024

Alongi, D. M. (2012). Carbon sequestration in mangrove forests. *Carbon Management*, *3*(3), 313–322. DOI: 10.4155/cmt.12.20

Ammar, A. A., Dargusch, P., & Shamsudin, I. (2014). Can the Matang Mangrove Forest Reserve provide perfect teething ground for a blue carbon based REDD+ pilot project? *Journal of Tropical Forest Science*, *26*(3), 371–381.

Azman, M. S., Sharma, S., Shaharudin, M. A. M., Hamzah, M. L., Adibah, S. N., Zakaria, R. M., & MacKenzie, R. A. (2021). Stand structure, biomass and dynamics of naturally regenerated and restored mangroves in Malaysia. *Forest Ecology and Management, 482*(December 2020), 118852.

Baden, D. (2016). A reconstruction of Carroll's pyramid of corporate social responsibility for the 21st century. *International Journal of Corporate Social Responsibility 2016 1:1, 1*(1), 8-.

Baqutayan, S. M. S., & Mayati, S. H. B. (2021). Insight of Recent Strategies and Initiatives in Managing Integrity in Malaysia. *International Journal of Academic Research in Economics and Management Sciences*, *10*(3). Advance online publication. DOI: 10.6007/IJAREMS/v10-i3/10824

Bastin, J. F., Finegold, Y., Garcia, C., Mollicone, D., Rezende, M., Routh, D., Zohner, C. M., & Crowther, T. W. (2019). The global tree restoration potential. *Science*, *364*(6448), 76–79. DOI: 10.1126/science.aax0848 PMID: 31273120

Becker, A., Bercovici, S. K., Choo, J., Fakhrurrozi, F., Fong, A., Fowell, S. E., Hossain, E., Hussein, M. A. S., Jamilah, M., Mujahid, A., Müller, M., Ooi, J. L. S., Quiros, T. E. A. L., Richard, F., Saleh, E., Chee, S. Y., Then, A. Y. H., Ticman, K., Wee, J. L. S., & Evans, C. (2025). Blue carbon management integrating socioeconomic and environmental interconnectivity in Southeast Asia: An urgent climate priority. *Frontiers in Marine Science*, *12*, 1642387. DOI: 10.3389/fmars.2025.1642387

Beh, L. Y. (2022). *Malaysia's mangrove-planting fishermen stumble at finance hurdle*.

Bowen, H. R. (2015). *Social-Responsibilities-Of-The-Businessman*. Harper.

Bursa Malaysia. (2018). *FTSE4Good Bursa Malaysia (F4GBM)*. Index.

Bursa Malaysia. (2023). *Sustainability Reporting Guide*.

Carroll, A. B. (1999). *Corporate Social Responsibility Evolution of a Definitional Construct*.

Choudhary, B., Dhar, V., & Pawase, A. S. (2024). Blue carbon and the role of mangroves in carbon sequestration: Its mechanisms, estimation, human impacts and conservation strategies for economic incentives. *Journal of Sea Research*, *199*, 199. DOI: 10.1016/j.seares.2024.102504

Christian, Y., Budiman, M. K., Purwanto, W., & Damar, A. (2021). Supporting community-based mangrove forest management as Essential Ecosystem Area in Sungai Pakning, Riau. *IOP Conference Series. Earth and Environmental Science*, *744*(1), 012007. DOI: 10.1088/1755-1315/744/1/012007

Davis, K. (2017). The Case for and Against Business Assumption of Social Responsibilities. *Academy of Management Journal*, *16*(2), 312–322. DOI: 10.2307/255331

Donato, D. C., Kauffman, J. B., Murdiyarso, D., Kurnianto, S., Stidham, M., & Kanninen, M. (2011). Mangroves among the most carbon-rich forests in the tropics. *Nature Geoscience*, *4*(5), 293–297. DOI: 10.1038/ngeo1123

Faridah-Hanum, I., Latiff, A., Hakeem, K. R., & Ozturk, M. (2014). Mangrove ecosystems of Asia: Status, challenges and management strategies. In *Mangrove Ecosystems of Asia: Status, Challenges and Management Strategies*. Springer New York.

Ferrero, I., Michael Hoffman, W., & Mcnulty, R. E. (2014). Must Milton Friedman embrace stakeholder theory? *Business and Society Review, 119*(1), 37–59. DOI: 10.1111/basr.12024

Firdaus, L. N., Ma'ruf, R., Elmustian, E., Suarman, S., & Melay, R. (2019). Traditional Ecological Knowledge on Mangrove Ecosystem Utilization: Learning from Orang Suku Laut Kongky Strait, Lingga. *Journal of Educational Sciences, 3*(3), 328. DOI: 10.31258/jes.3.3.p.328-339

Freeman, E. (2010). *Strategic management: A stakeholder approach*. Cambridge University Press. DOI: 10.1017/CBO9781139192675

Freeman, E., & Reed, D. (1983). Stockholders and Stakeholders: A New Perspective on Corporate Governance. *California Management Review, 25*(3), 88–106. DOI: 10.2307/41165018

Friedman, M. (1970). The Social Responsibility of Business Is to Increase Its Profits. In *Corporate Ethics and Corporate Governance* (Vol. 32). The New York Times Magazines.

Friess, D. A., Thompson, B. S., Brown, B., Amir, A. A., Cameron, C., Koldewey, H. J., Sasmito, S. D., & Sidik, F. (2016). Policy challenges and approaches for the conservation of mangrove forests in Southeast Asia. *Conservation Biology : The Journal of the Society for Conservation Biology, 30*(5), 933–949. DOI: 10.1111/cobi.12784 PMID: 27341487

Gevaña, D. T., Camacho, L. D., & Pulhin, J. M. (2018). Conserving mangroves for their blue carbon: Insights and prospects for community-based mangrove management in southeast Asia. *Coastal Research Library, 25*, 579–588. DOI: 10.1007/978-3-319-73016-5_26

Gijsman, R., Horstman, E. M., van der Wal, D., Friess, D. A., Swales, A., & Wijnberg, K. M. (2021). Nature-Based Engineering: A Review on Reducing Coastal Flood Risk With Mangroves. *Frontiers in Marine Science, 8*, 702412. DOI: 10.3389/fmars.2021.702412

Goessens, A., Satyanarayana, B., Van Der Stocken, T., Zuniga, M. Q., Mohd-Lokman, H., Sulong, I., & Dahdouh-Guebas, F. (2014). Is Matang Mangrove Forest in Malaysia sustainably rejuvenating after more than a century of conservation and harvesting management? *PLoS One*, *9*(8), e105069. DOI: 10.1371/journal.pone.0105069 PMID: 25144689

Goh, H. C. (2016). *Assessing mangrove conservation efforts in Iskandar Malaysia. in Jeffrey Cruikshank.*

Griscom, B. W., Adams, J., Ellis, P. W., Houghton, R. A., Lomax, G., Miteva, D. A., Schlesinger, W. H., Shoch, D., Siikamäki, J. V., Smith, P., Woodbury, P., Zganjar, C., Blackman, A., Campari, J., Conant, R. T., Delgado, C., Elias, P., Gopalakrishna, T., Hamsik, M. R., & Fargione, J. (2017). Natural climate solutions. *Proceedings of the National Academy of Sciences of the United States of America*, *114*(44), 11645–11650. DOI: 10.1073/pnas.1710465114 PMID: 29078344

Hashim, R., Kamali, B., Tamin, N. M., & Zakaria, R. (2010). An integrated approach to coastal rehabilitation: Mangrove restoration in Sungai Haji Dorani, Malaysia. *Estuarine, Coastal and Shelf Science*, *86*(1), 118–124. DOI: 10.1016/j.ecss.2009.10.021

Hatta, S. M., Salleh, E., Suhaili, N. S., & Besar, N. A. (2022). Estimation of carbon pool at mangrove forest of Kudat, Sabah, Malaysia. *Biodiversitas (Surakarta)*, *23*(9), 4601–4608. DOI: 10.13057/biodiv/d230927

Hatta, S. M., Suhaili, N. S., Salleh, E., Awang, N., & 1#, B. (2024). Carbon stock estimation in mangrove forest at Pitas, Sabah, Malaysia. In *2024. Transactions on Science and Technology* (Vol. 11, Number 1). https://tost.unise.org/

Howard, J., Lovelock, C. E., Beeston, M., Cameron, C., Sippo, J., Hagger, V., van Bijsterveldt, C., van Eijk, P., & Tonneijck, F. H. (2023). *Best Practice Guidelines for Mangrove Restoration*. www.mangrovealliance.org

Howard, J., Sutton-Grier, A., Herr, D., Kleypas, J., Landis, E., Mcleod, E., Pidgeon, E., & Simpson, S. (2017). Clarifying the role of coastal and marine systems in climate mitigation. *Frontiers in Ecology and the Environment*, *15*(1), 42–50. DOI: 10.1002/fee.1451

Inaku, D. F., Adrianto, L., Kurnia, R., Kurniawan, F., & Kusumo, S. (2025). Social-Ecological System Perspective to Identify Specific Factors in Building Mangrove Resilience on a Small Island. *BIO Web of Conferences, 185.*

Islam, M. A., Billah, M. M., Idris, M. H., Bhuiyan, M. K. A., & Kamal, A. H. M. (2024). Mangroves of Malaysia: A comprehensive review on ecosystem functions, services, restorations, and potential threats of climate change. []. Springer International Publishing.]. *Hydrobiologia*, *851*(8), 1841–1871. DOI: 10.1007/s10750-023-05431-z

Islam, M. A., Kamal, A. H. M., Idris, M. H., Billah, M. M., Rahman, A. A., Sukeri, M. S. B. M., & Bhuiyan, M. K. A. (2026). Mangrove Ecosystem Carbon Stocks: Insights from Setiu Lagoon, Malaysia. *Estuaries and Coasts*, *49*(4), 87. DOI: 10.1007/s12237-026-01707-4

Jusoff, K. (2013). Malaysian mangrove forests and their significance to the coastal marine environment. *Polish Journal of Environmental Studies*, *22*(4), 979–1005.

Kamaludin, M., Azlina, A. A., Wan Ibrahim, W. N., Mat Alipiah, R., Saputra, J., Abdullah, M. M., Anang, Z., & Omar, C. M. (2022). Effectiveness of a conservation education program among school students on the importance of mangrove ecosystems in Setiu Wetlands, Malaysia. *Applied Environmental Education and Communication*, *21*(1), 23–41. DOI: 10.1080/1533015X.2021.1936298

Karpowicz, D. A., Mohan, M., Watt, M. S., Montenegro, J. F., King, S. A. L., Selvam, P. P., Nithyanandan, M., Robyn, B., Ali, T., Abdullah, M. M., Doaemo, W., & Ewane, E. B. (2024). Mangrove-Based Carbon Market Projects: 15 Considerations for Engaging and Supporting Local Communities. *Diversity*, *16*(9), 574. DOI: 10.3390/d16090574

Kauffman, J. B., & Donato, D. C. (2012). *Protocols for the measurement, monitoring and reporting of structure, biomass and carbon stocks in mangrove forests*. Center for International Forestry Research.

Khan, W. R., Giani, M., Bevilacqua, S., Anees, S. A., Mehmood, K., Nazre, M., Bin Abdul Haddy, A. A., Bin Abang Median, A. N., Bujang, J., Mohamad-Ismail, F.-N., Mohamed, J., Samdin, Z., Abiri, R., Tuan-Ibrahim, T.-M., Mohammad, L.-S., Naji, H.-R., Akram, S., Abdul-Hamid, H., & Dube, T. (2025). Derivation of allometric equations and carbon content estimation in mangrove forests of Malaysia. *Environmental and Sustainability Indicators*, *26*, 100618. DOI: 10.1016/j.indic.2025.100618

Lee, S. L., Chee, S. Y., Huxham, M., Jamilah, M., Choo, J., Kaur, C. R., Amir, A. A., Ooi, J. L. S., Rozaimi, M., Omar, H., Sharma, S., Moritz, M., & Then, A. Y. H. (2025). Blue carbon ecosystems in Malaysia – status, threats, and the way forward for research and policy. *Journal of Environment & Development*, *34*(1), 225–265. DOI: 10.1177/10704965241284366

Locatelli, B., Pavageau, C., Pramova, E., & Di Gregorio, M. (2015). Integrating climate change mitigation and adaptation in agriculture and forestry: Opportunities and trade-offs. *Wiley Interdisciplinary Reviews: Climate Change*, *6*(6), 585–598. DOI: 10.1002/wcc.357

Lovelock, C. E., & Duarte, C. M. (2019). Dimensions of Blue Carbon and emerging perspectives. *Biology Letters*, *15*(3), 20180781. DOI: 10.1098/rsbl.2018.0781 PMID: 30836882

Low Carbon Society Blueprint for Iskandar Malaysia 2030 Climate Action Plan. (2022). www.iskandarmalaysia.com.my

Lu, J. Y., & Castka, P. (2009). Corporate social responsibility in Malaysia -experts' views and perspectives. *Corporate Social Responsibility and Environmental Management*, *16*(3), 146–154. DOI: 10.1002/csr.184

Macreadie, P. I., Anton, A., Raven, J. A., Beaumont, N., Connolly, R. M., Friess, D. A., Kelleway, J. J., Kennedy, H., Kuwae, T., Lavery, P. S., Lovelock, C. E., Smale, D. A., Apostolaki, E. T., Atwood, T. B., Baldock, J., Bianchi, T. S., Chmura, G. L., Eyre, B. D., Fourqurean, J. W., & Duarte, C. M. (2019). The future of Blue Carbon science. *Nature Communications*, *10*(1).

Malaysian Nature Society. (2024). *Quarterly Report*.

Martínez-Espinosa, C., Wolfs, P., Vande Velde, K., Satyanarayana, B., Dahdouh-Guebas, F., & Hugé, J. (2020). Call for a collaborative management at Matang Mangrove Forest Reserve, Malaysia: An assessment from local stakeholders' view point. *Forest Ecology and Management*, *458*, 458. DOI: 10.1016/j.foreco.2019.117741

McLeod, E., Chmura, G. L., Bouillon, S., Salm, R., Björk, M., Duarte, C. M., Lovelock, C. E., Schlesinger, W. H., & Silliman, B. R. (2011). A blueprint for blue carbon: Toward an improved understanding of the role of vegetated coastal habitats in sequestering CO2. *Frontiers in Ecology and the Environment*, *9*(10), 552–560. DOI: 10.1890/110004

Ministry of Natural Resources and Environment Malaysia. (2016). *National Policy on Climate Change*.

Mohammad, W. M. W., & Wasiuzzaman, S. (2021). Environmental, Social and Governance (ESG) disclosure, competitive advantage and performance of firms in Malaysia. *Cleaner Environmental Systems*, *2*, 2. DOI: 10.1016/j.cesys.2021.100015

Montenegro, J. F., Mohan, M., Ewane, E. B., Friess, D. A., Selvam, P. P., Dutta Roy, A., Althauser, A., Tidwell, J., Gamboa-Cutz, J., Shoot, C., Watt, M. S., Macreadie, P. I., Karpowicz, D. A., Perera, N., Atwood, T., Chatting, M. E., Burt, J. A., Udagedara, S., Hendy, I., & Alongi, D. M. (2025). Mangrove-based carbon market projects: Current trends and future perspectives. *Forest Policy and Economics, 181*, 181. DOI: 10.1016/j.forpol.2025.103658

Omar, H. (2024). Status of Mangrove in Malaysia. In *Status of Mangroves in Malaysia*. Forest Research Institute Malaysia.

Pendleton, L., Donato, D. C., Murray, B. C., Crooks, S., Jenkins, W. A., Sifleet, S., Craft, C., Fourqurean, J. W., Kauffman, J. B., Marbà, N., Megonigal, P., Pidgeon, E., Herr, D., Gordon, D., & Baldera, A. (2012). Estimating global "Blue Carbon" emissions from conversion and degradation of vegetated coastal ecosystems. *PLoS One*, *7*(9), e43542. DOI: 10.1371/journal.pone.0043542 PMID: 22962585

Perera, N. S., Costa, M. D. P., Macreadie, P. I., & Wartman, M. (2025). Trends in market-based Blue Carbon Projects. *Sustainable Development (Bradford)*, *33*(3), 3377–3402. DOI: 10.1002/sd.3293

Rodríguez-Rodríguez, J. A., Mancera-Pineda, J. E., & Tavera, H. (2021). Mangrove restoration in Colombia: Trends and lessons learned. *Forest Ecology and Management*, *496*, 496. DOI: 10.1016/j.foreco.2021.119414

Sahari, M. S. I., Mohd Razali, N. A., Redzuan, N. S., Shah, A. M., Awang, N. A., Lee, L. H., Juahir, H., & Muhammad Nor, S. M. (2024). Carbon stock variability of Setiu Lagoon mangroves and its relation to the environmental parameters. *Global Ecology and Conservation*, *53*, 53. DOI: 10.1016/j.gecco.2024.e02994

Satyanarayana, B., Quispe-Zuniga, M. R., Hugé, J., Sulong, I., Mohd-Lokman, H., & Dahdouh-Guebas, F. (2021). Mangroves Fueling Livelihoods: A Socio-Economic Stakeholder Analysis of the Charcoal and Pole Production Systems in the World's Longest Managed Mangrove Forest. *Frontiers in Ecology and Evolution*, *9*(October), 1–13. DOI: 10.3389/fevo.2021.621721

Suhaili, N. S., Hatta, S. M., Salleh, E., & Besar, N. A. (2024). Estimating the total carbon stock in the mangrove forest of Kota Marudu, Sabah, Malaysia. *Journal of Sustainability Science and Management*, *19*(7), 203–215. DOI: 10.46754/jssm.2024.07.012

Suhaili, N. S., Mhd Hatta, S., Salleh, E., & Awang Besar, N.N. A. (2024). Estimating the Total Carbon Stock in the Mangrove Forest of Kota Marudu, Sabah, Malaysia. *Journal of Sustainability Science and Management*, *19*(7), 203–215. DOI: 10.46754/jssm.2024.07.012

Sulaiman, N. N., Omar, A. J., Pakir, F., & Mohamad, M. (2024). Evaluating Environmental, Social and Governance (ESG) Practice Among Malaysian Public Listed Construction Companies Using FTSE Russell Rating Model. *International Journal of Sustainable Construction Engineering Technology*, *15*(1), 25–40. DOI: 10.30880/ijscet.2024.15.01.003

Wahab, E. A. A., How, J. C. Y., & Verhoeven, P. (2007). The Impact of the Malaysian Code on Corporate Governance: Compliance, Institutional Investors and Stock Performance. *Journal of Contemporary Accounting & Economics*, *3*(2), 106–129. DOI: 10.1016/S1815-5669(10)70025-4

Ward, R. D., Friess, D. A., Day, R. H., & Mackenzie, R. A. (2016). Impacts of climate change on mangrove ecosystems: A region by region overview. *Ecosystem Health and Sustainability*, *2*(4), e01211. DOI: 10.1002/ehs2.1211

Wolswijk, G., Satyanarayana, B., Abd Rahim, N. H., Che Abdullah, C. M. K. A., Ali, A. N., Wolswijk, L., Hami Hamzah, M. K., Cannicci, S., & Dahdouh-Guebas, F. (2026). Deep Coring Shows That Mangrove Sediments in Matang (Malaysia) Store up to Five Times More Carbon Than Previously Estimated. *Global Change Biology*, *32*(3), 1–11. DOI: 10.1111/gcb.70773 PMID: 41837329

World Bank. (2023). Unlocking blue carbon development: investment readiness framework for governments. *Unlocking Blue Carbon Development: Investment Readiness Framework for Governments.*

World Bank Group. (2021). *Taskforce on Scaling Voluntary Carbon Markets, Final Report | Public Private Partnership.* https://ppp.worldbank.org/library/taskforce-scaling-voluntary-carbon-markets-final-report

Zahid, M. (2020). *ESG in Focus: The Malaysian Evidence.* https://www.researchgate.net/publication/341479990_ESG_in_Focus_The_Malaysian_Evidence

Zen, I. S., Yusop, R., & Mahat, R. (2023). *Navigating Institutional Fragmentation: Mainstreaming Climate Change Policy in Malaysia's Multi-Level Governance System.*

Zuhairi, M. H., Rosnon, M. R., & Shaari, J. (2020). A Systematic Review on the Mah Meri People in Malaysia. *International Journal of Academic Research in Business & Social Sciences*, *10*(16). Advance online publication. DOI: 10.6007/IJARBSS/v10-i16/8293

Chapter 6
Bridging the Gap Toward a Collaborative Governance Framework for Sustainable Coastal Management in Semarang Indonesia

Kismartini Kismartini
http://orcid.org/0000-0001-7253-0957
Diponegoro University, Indonesia

Asih Widi Lestari
Diponegoro University, Indonesia

Nilam Adini Rakhma
http://orcid.org/0009-0003-9436-0030
Diponegoro University, Indonesia

Tri Yuniningsih
Diponegoro University, Indonesia

Alfiyah Alfiyah
Diponegoro University, Indonesia

DOI: 10.4018/979-8-3373-9320-9.ch006

ABSTRACT

Coastal management in Semarang is shaped by land subsidence, tidal flooding, ecological loss, and socio-economic vulnerability. These challenges cannot be solved through fragmented interventions because they also reflect policy incoherence, weak coordination, and limited stakeholder integration. Based on an integrative review of journals, reports, and case studies, this article develops a collaborative governance framework for Semarang. The findings show that coastal vulnerability is mainly a governance problem. Although government, communities, academia, NGOs, and private actors are involved, collaboration remains fragmented, unequal, and weakly institutionalized. Key barriers include overlapping authority, power imbalances, poor communication, and inefficient resource use. The article proposes five dimensions: shared problem framing, multi-actor forums, integrated knowledge platforms, hybrid resource mobilization, and adaptive, accountable implementation. Sustainable governance requires a shift from ad hoc coordination toward an integrated, participatory, and resilient system.

INTRODUCTION

Sustainable coastal management has been a strategic concern in most of the coastal cities because of the significance of these coastal areas to the ecology, society and economy. Not only does the coastal areas protect marine ecosystems, but also the country economical centre of Indonesia, homes and crucial urbanization. Therefore, the governance of these coastal areas must be combined and engage various sectors, other stakeholders, their interests, and resources. In this case, collaborative governance is vital in the process of coming up with adaptive and sustainable solutions.

Sustainable development has become one of the major paradigms in the governance of natural resources such as the coastal and marine regions. This paradigm underlines that there has to be some sort of balance between the economic, social and environmental interests of any development process and that use of natural resources is not only geared towards short term profitability but also has to be sustainable to the generations to come (Kismartini, 2025). Recent studies on the environmental governance of Semarang also suggest that issues of sustainability including waste management are to be tackled via institutional change, policy coherence and cross-sector collaboration rather than sector technical responses per se (Lituhayu et al., 2026) .

The concept of collateral governance as defined by Ansell and Gash (2008), refers to the collective decision-making process involving cross-sector and multilevel refers to the joint decision making process that involves cross-sector and multi-level government actors by participation, dialogue and shared responsibility to the accomplishment of the public goals that cannot be accomplished by the action of one single actor. This approach is focused on decentralization of power and responsibility, formation of both formal and informal alliances that enable the stakeholders to form a shared understanding of the problems, come up with joint solutions, and establish more flexible and sustainable policies (Nabatchi & Emerson, 2015). This concept can be highly utilized in the management of coastlines since the problem of the coastline is never a one-actor (representing various sectors) with cross-cutting interests e.g. government, private sector, civil society and academia. This implies that effective management of the coastal areas does not just require the government policies but rather it requires the involvement of the local communities and the private sector in decision-making (Kismartini, 2025).

Semarang is an Indonesian city located on the coast with major problems of coastal management, especially the constant tidal floods, the sinking of the land, erosion of the coastline, and increased socio-economic fragility of the low-lying locations. Abidin et al., (2013) indicate that the land subsidence in Semarang can go up to 6–7 cm/year with higher rates in a number of coastal areas. Also, Husnayaen et al. (2018) demonstrate that enhanced land subsidence significantly increases coastal vulnerability, demonstrate that the rise in land subsidence is a contributing factor to the fact that the coast becomes more susceptible, jeopardizing the livelihoods, settlements, and ecosystems in coastal areas. These conditions suggest that the issues of Semarang coast cannot be viewed as purely technical, or environmental, but as governance issues, with disjointed policy response and inefficient cross-sectoral coordination. At this point, collaborative governance is going to be a pertinent analytical and policy prism to the weaknesses of sectoral strategies. The results of climate-change governance in Demak also demonstrate that local leaders play a significant role in the process of transforming climate risks into community-based adaptation, coordination, and collective action (Hidayati et al., 2026) .

The presence of cross-sector actors can be seen in the execution of collaborative governance in the coastal management of Semarang in terms of involvement of government, the local communities, the private sector, academia and the media. A recent study by Kismartini et al. (2023) revealed that engagement of the pentahelix actors government at all levels, the private sector, academia, civil society, and media can bolster the capacity of coastal communities and enhance the sustainable coastal management practices by providing more inclusion in the decision-making process. This model of pentahelix is used in Semarang to make decisions, making sure that every actor plays his or her part according to his/her ability and role. Consequently, the policies that are generated are more acceptable, sustainable and universally acceptable.

Although the current studies on collaborative governance in coastal management are still in the developing phases, a significant portion of the studies is still scattered. A number of studies have only examined a single facet of collaboration including the collaboration process, dynamics among actors, or the roles of particular actors in collaboration. According to Ansell and Gash (2008), collaborative governance involves a formal set up whereby public institutions take the initiative but it is not bound to the state. Emerson et al. (2012) also emphasized that this framework also includes other forms of cross-border governance, such as partnerships between the public, the private sector, the public and the civil society or co-management regimes. This framework plays a crucial role in the coastal management of Semarang to address the coastal resources as a unit and to make sure that communities that may not have been included in the past in the policy decision making will also be involved.

One of the key characteristics of collaborative governance is the creation of the public value in the form of a sustainable management of the coastal areas. The elements of the public value in Semarang are the value of the environment, the wellbeing of the coastal communities and the preservation of social and cultural values. Design and development of inclusive and sustainable solutions which involve a number of stakeholders in the management of coastal ecosystems can create the public values. In sustainable resource management, Emersom and Nabatchi (2015) argue, the success of collaborative governance is realized when the collaboration process note that the key to the effectiveness of collaborative governance lies in ensuring that the collaboration process will be useful to the more broad community. Hopefully, with a collaborative governance scheme in the coastal management, the policies that are more integrated and flexible to accommodate the environmental and social challenges of the coastal communities in Semarang will be developed.

Though the collaborative governance framework is more comprehensive in regard to sustainable coastal management, it has many challenges during implementation. Various studies have indicated that the conflicting interests of the actors, lack of role and responsibility, and communication barriers between actors are some of the barriers that often hinder collaborative coastal management. Wever et al. (2012) also noted that participation and decentralization are important components of coastal management but inadequate collaboration due to institutional weaknesses, conflicts among actors, and the limited ability of local governments to coordinate various parties are more common. Thus, a coastal management framework that would take into account not only the stakeholders engagement but also would make the relations between the actors effective and lead to long-term policies is needed. The systematic review research on the waste-management policy in developing countries also reveals that the collapse of the environmental policy is more prevalent when there is weak institutional capacity, stakeholder coordination, and consistency in the implementation (Yudartha et al., 2026)

The purpose of this article is to provide a literature synthesis of collaborative governance and sustainable coastal management so as to develop a context-specific framework of Semarang, Indonesia. Specifically, the study will set out to: (1) identify the main governance issues that underlie the coastal vulnerability in Semarang; (2) explore the roles of key stakeholders, interaction and constraints as discussed in the literature; and (3) establish a participatory, collaborative and dynamic governance structure to coastal management. It is a conceptual and policy-based, rather than an evaluative, contribution based on these purposes. Neither is it an assessment of the performance of the programs in primary field work, but a syncretic conglomeration of the fragmented conclusions of other research in explaining how collaborative governance can be functional in bridging the gap between environmental risk, institutional fragmentation and sustainable coastal policy. In this way, the article contributes to the development of the field of public administration, linking collaborative governance to the perspective of the sustainability of coastal areas of a fast-urbanizing environmentally vulnerable coastal city.

LITERATURE REVIEW

1. Theories on Governance in Coastal Management

The concept of governance has evolved relative to the traditional concept of government where the state was considered as the key player to a more participatory concept of how people carry out their activities involving various stakeholders, including governments, businesses, civil society and local communities. This transformation is particularly relevant to the field of coastal management where numerous interests are at play such as environmental conservation, economic practices and social demands of the communities residing along the coast. Therefore, one actor or sector cannot be used to take effective action on coastal management but a collaboration between all parties should be used.

Rhodes (1996) notes that the concept of governance suggests the networks of interdependent institutions whereby the parties involved in it work together towards achieving the realization of the common good. That is, within the framework of coastal management, the government is not to regulate but mediate and coordinate divergent- interest actors. Stoker (1998) also notes that collective action is made possible by governance. In this regard, sustainable coastal management should include the participation of various stakeholders, including the government, society, the private sector and the civil society organizations. Collaboration is significant since it will reduce the disputes within the industry and come up with more holistic solutions that can be agreeable to both parties. Similar evidence in firebreak policy demonstrates that the involvement of local communities is crucial in preventing environmental risks, as community involvement enhances early intervention, local adherence, and policy acceptance (Awaluddin et al., 2026)

One of the most applicable governance processes that involve a group of actors is collaborative governance as defined by Ansell and Gash (2008) as a form of governance where the non-governmental and governmental stakeholders are included in joint forums to provide a decision that is reached through consensus. This collaboration in coastal management helps policies to address the ecological requirements but consider the social and economic requirements of the coastal communities. Emerson,, Nabatchi, and Balogh (2012) expand on this to establish an elaborate model of collaborative governance that includes the context, internal processes, collective behavior and its impact on sustainability. Collaboration in coastal management under this framework is considered to be a process but existing under a larger system of interconnections where the sustainability of coastal management is based on how the system responds to changes.

2. Collaborative Models in Coastal Management

The use of collaborative models in coastal management has been used in different countries as a means to tackle the complicated nature of coastal areas that are usually characterized by different interests and actors within limited resources. Collaborative based coastal management is based on the requirement to involve all the stakeholders in the planning and implementation. Pomeroy and Douvere (2008) opined that in the marine spatial planning scenario, collaborative coastal management approach would allow various parties to participate in the decision-making process to ensure such decisions are not solely governmental but also community-driven and the private sector. They underline the fact that the more people are involved in the decision-making process, the more legitimacy and acceptance the policies made.

In addition, the co-management and community-based management models have also been tested and found to be effective in managing the coasts. Such models put the local communities at the hub of the management of coastal resources, supported by governments and the business world. Damastuti et al. (2023) show that the effectiveness of coastal management in the example with community-based mangrove management in Central Java, Indonesia, is closely linked with effective collaborations between the government, the private sector, and local communities. Such a joint venture that has proven to be successful in Bedono has served to reduce the erosion along the coastline and also the mangrove forests have been more efficient in conserving the environment besides enhancing the economic well-being of the locals. The reason was that it was a long-term plan that embraced conservation and economic local level development. Recent international surveys of research on mangrove management also highlight that what makes coastal areas sustainable is an integrated ecological restoration, stakeholder involvement, and institutional framework which links conservation with community livelihoods (Purnaweni et al., 2026)

The applicability of collaborative models to the coastal management is particularly relevant to Indonesia which is an archipelagic nation with diverse coastline which is subjected to various challenges. The collaborative model of coastal management will allow the policy to become closer to the local environment, in addition to the strengthening of the local capacity of the institution. Thus, the collaborative strategy is vital in Indonesian coastal management where different actors whose interests are interdependent should collaborate.

3. The Connection Between Collaboration and Sustainability in Coastal Management

The collaboration between different actors is very crucial in ensuring sustainability in coastal management. Ansell and Gash (2008) argue that successful collaboration in coastal management requires dialogue and believe that effective collaboration in the management of the coastal areas must be based on dialogue and agreement, which is achieved through the process of negotiation and information exchange between the parties involved. The dialogue assists the stakeholders to share their opinions and also to be aware of their interests and consensus can make the decisions acceptable by all the parties and can be carried out in the long term.

Furthermore, Emerson et al. (2012) add that the ability of the collaborative system to learn and adjust to the changes, which can be, in fact, ecological or social, is also a determinant of the sustainability of the management of the coastal area. In the case of Semarang and other cities along the Indonesian coastline, the local government, coastal communities and the private sector will be in a position to address the problems of flooding, erosion and climate change in a better manner through cooperation. Based on this, the constructive engagement and involvement of all the stakeholders in the decision-making processes play a critical role in offering sustainability in the long term of managing the coasts.The example of Semarang where inorganic waste and greenhouse gas emissions are assessed based on scenarios indicates that sustainability transitions should be realized by applying circular-economy strategies with the assistance of data, combining policies, and enacting them jointly (Junaidi et al., 2026).

Collaborative coastal management is also an opportunity to enhance local capacity to cope with current challenges. Pomeroy and Douvere (2008) add that the successful integration of the local communities and other stake holders will help in understanding better the coastal ecosystem and the potential conflict in the use of space, which can help in finding the appropriate solution to the issue. By having discussions and ensuring that all parties are involved in the decision making process, more opportunities arise that the resulting policies will be more favorable to the local needs and be accepted by the community hence improving on long term sustainability.

RESEARCH METHODOLOGY

This study uses a qualitative integrative literature review to explore how the concept of collaborative governance has been talked about in terms of sustainable coastal management in Semarang, Indonesia. The reason behind this methodology is that the article is focused on synthesizing scattered research, establishing common governance issues, and elaborating on a conceptual framework, instead of conducting the primary field research to test a hypothesis (Munn et al. (2018)

(et al., Munn 2018; Jesson et al., 2011). The review therefore can be considered as an analytical methodology to unite the empirical and conceptual knowledge of past studies in a more encompassing way of governing.

The reviewed sources in this paper are all secondary sources of information including peer-reviewed journal articles, policy reports, government reports and case based studies, related to coastal governance, land subsidence, tidal flooding, mangrove management, and community participation in Semarang and other coastal settings like Indonesia. The selection criteria of the sources were on three aspects: substantive relevance to coastal management, analytical relevance to collaborative governance and contextual relevance to Semarang or other coastal governance concerns that are closely related.The literature on the disaster innovation and smart technologies also adds to the selection of sustainability and environmental governance literature, pointing to the fact that the technological support and the institutional preparedness and the adaptive governance capacity become more and more the basis of the implementation of the green policies (Maylasari & Al-Akbar, 2026).

The analysis was conducted through thematic coding and qualitative synthesis. The reviewed sources were organized into several analytical dimensions: (1) the main coastal governance problems in Semarang; (2) the actors involved and their respective roles; (3) the forms of interaction, participation, and coordination reported in the literature; (4) institutional and socio-economic constraints affecting collaboration; and (5) the policy implications emerging from previous studies. These themes were then interpreted using collaborative governance concepts drawn primarily from Ansell and Gash (2008) and Emerson et al. (2012)

Based on this, this paper is not a formal actor-network analysis using primary mapping or network measures. Instead actor relations are defined by published literature and argued as a literature based synthesis of governance. The analysis findings are a suggested collaborative governance framework outlining how policy integration, stakeholder inclusion, communication, institutional coordination, and adaptive implementation can be aligned to enable more sustainable coastal management in Semarang.

RESULTS AND DISCUSSION

Main Governance Challenges Underlying Coastal Vulnerability in Semarang

The literature reviewed demonstrates that sustainable coastal management in Semarang is not to be viewed as a strictly technical problem related to mitigating the impacts of tidal flooding, shoreline protection, or ecosystem restoration only. Instead, it is a multidimensional issue of governance that is created by the interplay of ecological degradation, urban sprawl, land subsidence, fragmented policy responses, and uncoordinated institutional responses. The overlapping pressures of tidal inundation, abrasion, mangrove loss, land-use conversion, and groundwater extraction have a long history in the coastal area in Semarang leading to increased environmental vulnerability and social risk. In this regard, Abidin et al. (2013) determine that land subsidence in Semarang has led to greater inundation, infrastructure demolition and seawater intrusion. Husnayaen et al. (2018), support this diagnosis and demonstrate that augmented subsidence makes coastal susceptibility significantly rise. On the side of governance, Huda et al. (2024) also say that policy incoherency, responding to policy, and stakeholder misalignment are also a limitation in the management of such interrelated problems. Combined, these researches suggest that the initial problem that has to be considered is the governance nature of the very concept of coastal vulnerability: the key deficiency of Semarang is not simply the lack of technical intervention, but rather a lack of a comprehensive governance methodology that would help coordinate the policies, actors, and the implementation mechanisms.

1. Stakeholders, Interactions, and Institutional Arrangements in Collaborative Coastal Governance

Reviewed literature verifies that collaborative governance offers a suitable analytical approach to comprehend the complexity of governance in coastal Semarang. In the perspective of Ansell and Gash (2008) collaborative governance may be described as a form of governance that directly engages non-state actors in a shared decision making process by an agency or agencies. This expression is analytically significant, in the sense that it goes beyond the constricted definition of consultation, but rather focuses on the direct, inclusion of non-state actors into the policy-deliberation and decision-making process. Likewise, Emerson et al. (2012) also present collaborative governance as a conceptual map that is broadly based on which the cross-boundary governing systems can be viewed. Together, these formulations bring out the fact that collaboration is not just an administrative methodology, but a governance that is geared towards managing issues that can be handled by no particular institution.

This theoretical premise particularly fits the context of coastal governance whereby there is spatial interdependence of environmental processes and an institutional dispersion. Water moves without any reference to administrative jurisdictions as it is seen by Handayani et al. (2020) This quote explains one of the main peculiarities of the coastal governance in Semarang: ecological processes, such as flooding, runoff, sediment movement, and intrusion of the tidal processes are not circumscribed by the administrative boundaries. Consequently, spatial planning, groundwater, marine management, provision of infrastructures, settlement control, and ecosystem rehabilitation policy areas cannot be treated as a separate area. These results hence indicate that cooperation in the coastal governance of Semarang is not a choice of the institutional set up, but an imperative to overcome the fragmentation of sectors and to attain policy coherence within a socio-ecologically dynamic setting.

2. Core Dimensions of Collaborative Coastal Governance

The inclusion of the stakeholders is the first core dimension brought out by the synthesis. It is evident in the literature that the effective coastal management of Semarang engages a broad array of actors, including municipal and provincial governments, central state agencies, local communities, fishers, pond farmers, conservation groups, universities, non-governmental organizations, and actors in the private-sector. The operations of existing programs in Semarang indicate that these players are not merely peripheral players in the process, but significant players in planning, implementation and environmental stewardship. In the example of Kurniawan et al. (2022), the actors of the mangrove rehabilitation work based on their roles and duties, but the community is the most important actor in the program. The fact that it tends to re-position the local communities as passive recipients of the policy to its key agents is particularly notable. The application of collaborative governance in Semarang therefore should start with institutional acknowledgement of the understanding of community, lived experience, and first-hand reliance on the sustainability of the coastal setting. Studies on women participation in the peripheral space in rural Indonesia have also shown that substantial participation is possible through informal networks, routine activities and communal based roles that are generally disregarded by formal systems of governance (Alfirdaus et al., 2026)

The second dimension that has been established in the literature is the existence of communicative linkages and networks of actors. Working together involves more than having different stakeholders, it is a matter of how well they interact. The reviewed studies indicate that formal and informal relationships among community organizations, local governments, academics, NGOs and in some cases corporate social responsibility projects affect the management of coastal problems in Semarang. Martuti et al. (2020) report local involvement in coastal governance stretches to the preparation, implementation, and monitoring of the program, which suggests that the communities may be involved in various phases of the policy cycle. Good communication and synergy among the stakeholders is also emphasized in the same study. This kind of results depicts that communication cannot be reduced to a one-way flow of information, or socialization of the policies. Rather it ought to be a dialogue mechanism, a mutual learning mechanism, a negotiation mechanism, a coordination mechanism, and a continuous adjusting mechanism of collective action. In practical terms, this would mean that the effectiveness of the collaborative governance in Semarang will be based on the institutionalization of the existing networks into sustained sites of engagement rather than ad hoc or project based relationships.

The third dimension is associated with the inclusive decision-making. A major lesson of the collaborative governance literature is inclusion must be substantive and not procedural. Ansell and Gash (2008) are clear to remind us that collaborative governance cannot be a simple consultative one. This is a particularly relevant suggestion in the context of the coastal environment in Semarang where not all actors engage in governance processes at the same level of authority, resources, and bargaining power. Relegated to policy making, the experience of communities living in risky coastal areas can be of great value which is usually marginalised. The analyzed articles, therefore, indicate that in Semarang, a collaborative model will only be feasible when the local actors do not have a mere presence, but actual power to shape the agenda-setting, priority choice, and implementation decision. This meaning aligns with that of Martuti et al. (2020), who support a government as an enabler and the community as actors. In other words the role of the state must be re-orientated on facilitative coordination and not unilateral control and communities must be regarded as co-producers of the governance outcomes.

3. Institutional and Socio-Economic Constraints Shaping Collaboration

The synthesis also reveals that social, structural and economic conditions also play an important role in defining the quality of collaboration. Social pillars such as trust and communication are discovered to be fundamental. Marfai et al. (2008) note that the coastal community has been experiencing problems of tidal flooding which means that vulnerability is not an abstract category of policy, but a reality that is known and experienced by coastal Semarang. On the whole, collective action can be evoked by a shared experience related to tidal flooding and environmental devastation, however, the literature can show that shared risk does not always result into trust. Rather, it has to be developed by the constant contact, openness, responsiveness, and the public presentation of intermediate results.The synthesis thus indicates that the social basis of cooperation in Semarang is conditional on the capacity of the stakeholders to build mutual trust, acknowledge the legitimacy of local knowledge and sustain the communication throughout the period. The community empowerment research also shows that there is a potential to enhance the local collective capacity by physical activity programs, social solidarity, and good governance, which suggests that resilience is not only defined by the infrastructure but by social solidarity as well (Dharmawan et al., 2026).

The success of collaboration institutionally is anchored on the institutional environment in which they are incorporated. The reviewed literature shows that there is still a problem of policy fragmentation and overlapping authority. Huda et al. (2024) identify "overlapping public policies" and weak coordination as major obstacles in the governance of land subsidence and related coastal issues. Handayani et al. (2020), observe that hydrological and flood-affected zones are often in the category of more than one responsible party, thus creating disjointed responsibility and a lack of policy coordination. This type of fragmentation undermines the potential of effective collaboration, since it spreads responsibility and is an invitation to cross-sectoral solutions to cross-sectoral issues. The results, in turn, imply that Semarang needs an institutional tool that would help to connect spatial planning, environmental regulation, marine affairs, public works, disaster risk reduction, and local economic policy in the framework of a unified governance. Lacking this institutional bridging, the collaboration will probably be still rhetoric or episodic instead of systemic.

Economic factors also impact collaboration. When short-term projects form the basis of environmental governance arrangements or token participation that lacks support by adequate resources, the arrangements are difficult to sustain. The results indicate that in Semarang, collaboration has been enabled through mixed types of resource mobilization that comprise of public agencies, universities, NGOs and CSR initiatives in a number of instances. Kurniawan et al. (2022) show the way the external assistance strengthened the conservation of the mangroves, but Martuti et al. (2020) show that sustainable management of the coastal areas in Tugurejo involves mangroves planting in pond areas, silvofishery, and ecotourism development. Through these plans, it can be seen that ecological restoration will be maintained more when it is connected to the local livelihood opportunities. Damastuti et al. (2023), support this argument by finding that community-based mangrove management in Central Java is more effective when implemented in long-term and combined strategies with complementary forms of coastal protection in existence. Thus, the findings show that the poorly related collaboration will not become better until a linkage of conservation goals to viable economic incentives and resilience of communities. Studies of green tourism products through collaborative governance have shown that the innovations that are sustainability-driven can be realized when the resources available in the community, the economic potential in the locality, and the collaboration of institutions are linked in a common development agenda (Sastrawan et al., 2026).

4. Persistent Governance Constraints in Semarang's Coastal Management

Although collaborative governance has the normative appeal, it is evident that the studies reviewed indicate that its implementation is limited by various long-term challenges. The former relates to conflict of interest in sectors. Conflicting interests exist in terms of settlement expansion, industrial development, aquaculture, infrastructure development, and environmental conservation in the coastal area of Semarang. In the absence of a shared system of governance, all the sectors tend to do what they consider to be logical hence creating policy disjuncture. The second issue is the asymmetrical power relations. The process of collaboration does not start at the same level and there would be actors with greater access to financial resources, technical expertise, institutional power and political power. The substantive exclusion may also be inferred to be used in such cases in conjunction with the formal inclusion. Ansell and Gash (2008) warn that collaborative arrangements are prone when the involved parties (stakeholders) are unequal in power and ability to enter the process. This disproportion may pose a danger to marginalizing communities that bear the brunt of direct effects of coastal degradation in Semarang.

The third problem is the coordination problem and making joint decisions. As the literature reveals, the local initiatives in Semarang have demonstrated local promise in its coastal initiatives, yet these initiatives are project-related and very fragmented. The miscoordination between the stakeholders is one of the long-standing obstacles to governance that is specifically mentioned by Huda et al. (2024). The results thus suggest that the issue is not the lack of actors, but lack of a coordinating mechanism that is institutionalized in order to coordinate the goals, schedules, data infrastructure, financial strategies, and performance management. Such collaboration is more likely to be a temporary assemblage than a stable governance regime in such situations.

Proposed Five-Dimensional Collaborative Governance Framework

Based on the literature synthesis, this article proposes that a governance gap in Semarang can be closed with the help of a collaborative framework that is established with the help of five interdependent dimensions.

1. First, the framing of a common problem should take place. Coastal flooding, erosion and habitat destruction should not be considered a natural or technical process, but an outcome of land-use decisions, groundwater withdrawal, fragmented and piecemeal policies, and unequal development.
2. Second, there is the necessity to institutionalize involvement of the government agencies, community groups, academics, NGOs and the actors of the private-sector in a formal and long-term multi-actor forum.
3. Third, collaboration must be facilitated by a communication and knowledge platform by integrating scientific knowledge, policy knowledge and local experience knowledge.
4. Fourth, governance should be supported by the hybrid resource mobilization relying on the common budgets, academic support, community programs and individual donations.
5. Fifth, implementation should be responsive and accountable in which stakeholders are able to monitor the progress, learn as outcomes unfold and make joint updates to interventions.

Collectively, the review indicates that the concept of collaborative governance in Semarang cannot be interpreted as a mere administrative approach towards engaging stakeholders. More fundamentally it should be considered as an institutional foundation towards overcoming the disjuncture that exists between the discontinuous sectoral approach to policy and wholeness of the needs of sustainable coastal management. Only under this frame, coastal management in Semarang will be able to go beyond the short-term interventions and a more holistic, inclusive and sustainable system of collective action.

To harmonize the empirical synthesis, the title and the objectives of this article, Table 1 presents the five dimensions of the heart of the proposed collaborative governance model of sustainable coastal management in Semarang. Rather than summarizing findings, the table summarizes the literature on the subject reviewed into the key dimensions of governance required to respond to coastal vulnerability, elucidate the role and interaction of stakeholders, and assist in having a more integrated and adaptive policy response.

Table 1. Proposed five-dimensional collaborative governance framework for sustainable coastal management in Semarang

Dimension	Core Problem Addressed	Collaborative Mechanism	Expected Governance Outcome
Shared Problem Framing	Semarang's coastal problems are often treated as separate technical issues even though subsidence, tidal flooding, ecological degradation, land-use change, and social vulnerability are interlinked.	Stakeholders build a shared understanding that coastal risk is a multidimensional governance problem rather than a purely technical issue.	An integrated policy agenda for cross-sectoral coastal management.
Institutionalized Multi-Actor Forum	Coastal governance remains fragmented across agencies, sectors, and levels of government, while collaboration is often ad hoc and project-based.	A formal and sustained forum brings together local government, provincial and national agencies, communities, NGOs, academia, and private-sector actors in a continuous decision-making arena.	More stable coordination, clearer distribution of roles, and stronger collective ownership of coastal policy.
Integrated Communication and Knowledge Platform	Scientific data, policy information, and local experiential knowledge are often disconnected, limiting evidence-based and socially grounded decision-making.	Collaboration is supported by a communication platform that integrates technical expertise, policy information, and local knowledge through dialogue, information exchange, and mutual learning.	Better-informed policy design, stronger trust among actors, and improved capacity for joint problem-solving.
Hybrid Resource Mobilization	Coastal programs are difficult to sustain when they rely only on sectoral budgets or short-term interventions.	Resource support is mobilized through a combination of public funding, academic assistance, community initiatives, NGO support, and private-sector contributions.	More sustainable implementation, stronger program continuity, and better alignment between environmental protection and local livelihoods.
Adaptive and Accountable Implementation	Coastal interventions often remain rigid, fragmented, and weakly monitored, reducing their long-term effectiveness.	Implementation is designed as an adaptive process with joint monitoring, evaluation, learning, and periodic policy adjustment among stakeholders.	More resilient, accountable, and responsive coastal governance capable of adjusting to changing environmental and social conditions.

Source: Researcher's synthesis based on the reviewed literature

According to Table 1, the proposed framework will be structured around five dimensions that are interdependent and which directly address the objectives of this study. The framework starts with common problem framing to reframe coastal vulnerability as a governance challenge, as opposed to a technical challenge. It then points out institutionalization of multi-actor forum, integration of communication and knowledge systems, mobilization of hybrid resources and adaptive and accountable implementation. Blends of each of the five dimensions provide a viable collaborative governance design to explain how sustainable management of the coastal area in Semarang can overcome the sectoral intervention fragmentation and move towards more integrated, participatory, and resilience policy taking.

DISCUSSION AND POLICY IMPLICATIONS: BRIDGING GOVERNANCE GAPS THROUGH THE PROPOSED FRAMEWORK

According to a review of literature in this article, it is not the presence of the environmental risk alone that is the main problem in the management of the coasts in Semarang, but the persistence of the gap in governance between the risk knowledge and policy design capacity, and the realities that the coastal communities experienced. In the last decade, Semarang has offered a number of flood-controlling projects, climate-adaptive projects and multi-stakeholder agreements. New studies however show these initiatives have not always been incorporated, fair Putri, and sustainable model of coastal governance. Putri et al. (2025), for example, identify a major adaptation barrier in the form of "lack of participation and concern from local communities," while Putri (2025) finds that flood and tidal-flood infrastructure development in Semarang is perceived as "largely unsuccessful" by affected residents. At the planning level, Pratama et al. (2026) also indicate that land subsidence is not officially declared in spatial planning documents as a disaster-prone area. The findings of these studies contribute to the initial objective of the article since it shows that the exposure to the hazard is not only what predisposes the coastal susceptibility of Semarang, but also the long-standing absence of ties between scientific knowledge, institutional design and implementation practice.

That is an even more serious gap between that and actual subsidence data when that gap is read in conjunction with actual subsidence data. The literature available leaves one figure a year of the whole city, but rates by the year of observation and method of measurement. Sarah (2022), synthesizing leveling, GPS, InSAR, and microgravity evidence, reports subsidence rates in Semarang of 0–15.81 cm/year for 1996–2000, 0–12.07 cm/year for 1999–2003, 0–8 cm/year for 2002–2006, 0–8 to 0–12 cm/year for 2007–2008, 0–19 cm/year for 2008–2011, and 0–12.7 cm/year for 2017–2019. Earlier geodetic work by Abidin et al. (2010) also found that subsidence in Semarang could reach roughly 15 cm/year in coastal areas. More recent time-series InSAR analysis by Aditiya and Ito (2023) reports a peak of 100120 mm/year (or approximately 1012 cm/year) is now in the north and east of Semarang and Demak and forecasts that perhaps 20% of the area of Semarang is already affected.

Similar extreme subsidence is being reported by Pratama et al. (2026) at the coastal area of Semarang with a rate of approximately 10-13 cm/year. These figures explain that tidal flooding in Semarang cannot be seen as a pure marine threat; it cannot be separated of the long-term, anthropogenic, and spatially unbalanced land subsidence.

These empirical conditions can be viewed in the five-dimensional framework that is proposed in this paper, which means that a set of governance inadequacies are combined that can be employed to elucidate why the situation with coastal management in Semarang remains fragmented. The latter is a problem-framing deficit, in which the coastal flooding continues to be too often considered a technical or marine risk, rather than a governance issue that is linked to subsidence, land use, groundwater abstraction and uneven development. The second is a lack of participation and coordination whereby the concerned actors do exist but are not related to each other in the sense of stable and equal collaborations. The third is a lack of knowledge-integration whereby science, planning tools and local experience is not necessarily translated into common policy priorities. The fourth is shortage of resources and implementation where interventions are yet to be consistent in terms of sustainability, credibility and social acceptance. The fifth is the absence of accountability and deficit in adaptation, in which monitoring, learning and adjustment of policy is not yet institutionalized. These gaps directly explain the second objective of the article which is to explain how the role of stakeholders, interaction and constraints delimit the scope of current collaboration.

These failings of governance are subsequently explained to the stakeholder functions, institutional constraints and the policy implications therefore bearing out the third aim of the article which is to propose a more integrated and responsive system of collaborative governance to the coastal management of Semarang.

According Kismartini et al. (2023) explain The merit of the collaborative model of Semarang is that the city does not start with an institutional vacuum. The Tapak Mangrove Forest stakeholder research reveals that the provincial environmental and forestry agency is a policy maker and facilitator, city agencies are facilitators, village-level institutions are coordinators, academic actors are facilitators and accelerators and local community groups are coordinators and implementers. This separation of responsibilities is significant in that it demonstrates that there was an early system of governing. Likewise, the research on climate adaptation in Semarang suggests that transdisciplinary platforms at the city level should be considered an important learning and knowledge sharing process. That is, the city already has the fundamental constituents of collaboration: the presence of numerous actors, awareness of the issue, and institutional points of entry. The weakness lies in the fact that these ingredients are still not ready to be matured into a rational and sustainable system of governance that will be able to balance ecological protection, engineering interventions, and social justice.

This institutional incompleteness is evident in Tugurejo. Utami and Supratiwi (2026) is an area of collaboration practice that is more of an instructive coordination than a partnering. The importance of that formulation is that it demonstrates that most local governance structures are often failing: they can be seated in the same room, but agenda-setting authority, technical skill and procedural domination are concentrated in the hands of the government. The study proposes a more institutionalized design, redistribution of power and knowledge, and greater grassroots regeneration of leadership to guarantee long-term independence of programs. Governance-wise this implies that it is not just a lack of collaboration that is the problem of Semarang, but under-institutionalized collaboration. There is cooperation which is activity, but not yet well so as a common governing structure.

A comparable gap appears in the assessment of flood and tidal-flood infrastructure. Putri (2025) the total score of performance is 82.19, which makes the program less successful in the eyes of the community. Effectiveness (25.92), responsiveness (19.60), alignment with planning (9.56), and community involvement (12.43) were rated above the average and the latter one was in the unsuccessful category. In the same study, large percentages of the respondents had indicated that the infrastructure was not in harmony with their needs or their preferences and that most of them had never been involved in the planning processes such as deliberation forums. These results are analytically important as they change the focus of the question of whether there is infrastructure to the question of whether it has popular legitimacy and delivers tangible benefits in everyday life. A governance system, which puts completion rates in the spotlight and not perceived usefulness, may generate physical outputs and fail to generate a sense of trust, ownership, and long-term resilience.

The planning gap is equally critical. Despite decades of scientific evidence on land subsidence, Pratama et al. (2026) demonstrate that the issue did not become the mainstream in the formal system of spatial and development planning in Semarang. Their article points out that the subsidence in the coastal Semarang has been continuing at a rate of about 10-13 cm/ year and the risk has not been well integrated into the planning nomenclature and policy agenda. That is no small technical scam. A planning regime: a planning regime based on the assumption that inundation is an episodic, but not structural, concern will always be reactive when the city is sinking at rates that have reached up to 19 cm/year. The challenge here is not about collecting more data, however, but ensuring that the science of subsidence can change zoning, groundwater regulation, urban growth and development, the site of infrastructures and long-term priorities of investments. It is only in the sense that it can be applied to transform shared knowledge into binding spatial decisions that collaborative governance can make any sense.

The social dimension of coastal resilience reinforces this conclusion. In Genuk, Kiswoyo et al. (2025) find out that Genuk does not have the best social capital in terms of trust, norms, and networks. They also note that both the government and the private sector have failed to do their best to control floods. This is important, as the resilience of a subsiding city cannot be diminished to sea walls, pumps, or raised roads. What the inhabitants need is effective communication, trust among individuals, enforceable environmental laws and mechanisms of mutual support that can aid long-term adaptation. Without the strong social capital, technically qualified projects may fail to be mobilized to coordinate the collective maintenance, citizen control, and the neighborhood-level disaster preparedness. The problem of governance the coastal region in Semarang, in this respect, has something to do with the infrastructures, and, at the same time, it is institutional and relational.

A further weakness lies in the unresolved relation between ecological systems and engineering solutions. Harahap et al. (2024) make a case that the development of sea-walls in Tambak Lorok and Genuk has both positive and negative effects, such as the change in the environment in the neighboring coastal regions. Although the protection structure can reduce the exposure in some areas, the same activity may alter the currents, reduce the mangroves plants, and interfere with the ecological processes of the mangroves. Belland et al. (2025) go one step further by condemning stories that only show that Semarang is a city being lost to the sea. The fact that they discuss amphibious histories of the coast suggests that purity-oriented projects that seek to separate land and water may be devastating to coastal relationships and ways of life, and may eventually produce coastal forms of what they term coastal dispossession. It is a central argument in a Scopus-oriented discussion section: the question of how to guard the territory is not only how to manage the dynamic coastal settings, but how to control the socio-ecological interactions, which were used to sustain them traditionally.

All these findings do not suggest that Semarang lacks the spirit of collaboration but rather suggest that the existing forms of collaboration have not been institutionalized to facilitate them to systematically approach coastal vulnerability. This is where the proposed framework is suggested to be. The framework offers a more coherent platform of connecting the environmental risk, stakeholder engagement, and sustainable policy formulation through its emphasis on collective problem framing, institutionalized multi-actor coordination, systematized communication and knowledge sharing, hybrid resource mobilization and adaptive accountability. In this respect, the framework is correlated to the title and the purpose of the article as it goes on to describe the governance issues, to talk about the collaborative arrangements and lastly proposes an inclusive model of sustainable coastal management in Semarang.

Policy implications are thus fairly clear. Firstly, Semarang needs to institutionalize land subsidence as one of the main risks along its spatial plan, detailed zoning plan, medium-term development plan, and climate adaptation agenda. The fact that severe subsidence rates are still present with the highest 100-120 mm/year being recorded by Aditiya and Ito (2023) and the 10-13 cm/year being recorded by Aditiya and Ito (2023) and the approximately 10–13 cm/year rate highlighted by Pratama et al. (2026). Second, the city requires a permanent cross-level collaborative platform involving municipal and provincial authorities, port and industrial actors, coastal communities, and universities. Aditiya and Ito (2023) clearly indicate that governments must balance regulations, develop more elaborated spatial planning maps and develop mitigation and adaptation facilities. Third, the role of the people has to be transformed into co-production with the neighbourhood based coastal forums, participation in the flood and erosion mapping and utilization of the public monitoring systems to affect the budget and project design. These recommendations are consistent with the emphasis of the suggested framework of common framing, institutionalized cooperation, and substantive inclusion.

Fourth, Semarang ought to possess a mixed coastal governance plan. The flooding will be further worsened by the chronic subsidence, so the hard infrastructure will not be sufficient. The hybrid governance would be the combination of sea walls, pumps, polders, enhancement of drainage, restoration of mangroves, sediment control, ground water control and livelihood security of fishing and pond dwelling residents. This is more reflective of the empirical reality of the Semarang coast, where the structural protection, the ecological and social survival are interrelated. It is also more immediately responsive to the alarms raised by recent studies: that infrastructure is capable of creating an ecological trade-off, and that only land-dominant ideas of coastal order can disenfranchise existing amphibious livelihoods. Lastly, joint monitoring, public accountability and learning of the policies periodically should support implementation to make coastal governance adaptive instead of being reactive.

Semarang will be facing a complex and compounding crisis in the future: climate change, rising tidal flooding, ongoing subsidence, industrial pressures, and varying adaptive ability across societies. But the city has its resources as well: previous experience of adaptation, active communities along the coast, support of academia, and proven but less than perfect multi-actor collaboration. The definition of bridging the gap in Semarang is, thus, wider than the bridging of government and society. It also suggests the relation of science to planning, infrastructure to ecology and technical adaptation to social justice. This way, the proposed system of collaborative governance presents a conceptually consistent means of achieving a more integrated and sustainable form of coastal management in Semarang as can be seen in the title thereof, and in the research objectives of this article.

CONCLUSION

The management of coastal sustainability in Semarang needs to move to a more integrated, participatory and adaptive style of governance, rather than piece meal and sector interventions. The coastal area of Semarang is experiencing a set of environmental problems such as land subsidence, tidal flooding, environmental degradation, and socio-economic vulnerability, but not as discrete environmental problems, but as complex governance problems that need to be addressed at institutional, sector, and stakeholder levels. In this case, the sustainability of coastal management does not solely depend on technical interventions or infrastructural provision, but on how the governance actors can create shared problem conceptualizations, priorities in policy, and continue to act as a collective in the long-term.

The five-dimensional collaborative governance framework proposed offers a context-sensitive framework of dealing with these challenges. The framework offers a rational avenue to deepen the integration of policy and improve the long-term performance of the coastal management in Semarang by emphasizing on the commonality of problem framing, institutionalized multi-actor cooperation, integrated communication and knowledge exchange, hybrid resource mobilization, and adaptive and accountable implementation. Lastly, sustainability of the governance of the coasts in Semarang will be determined by how the state and non-state actors can overcome ad hoc coordination and establish a more durable regime of governance, which will unify the environmental protection, spatial planning, infrastructural development, and community resilience under a single joint policy framework.

REFERENCES

Abidin, H. Z., Andreas, H., Gumilar, I., Sidiq, T. P., & Fukuda, Y. (2013). Land subsidence in coastal city of Semarang (Indonesia): Characteristics, impacts and causes. *Geomatics, Natural Hazards & Risk*, *4*(3), 226–240. DOI: 10.1080/19475705.2012.692336

Abidin, H. Z., Andreas, H., Gumilar, I., Sidiq, T. P., Gamal, M., & Murdohardono, D. (2010). Studying Land Subsidence in Semarang (Indonesia) using Geodetic Methods. *FIG Congress*, 11–16.

Aditiya, A., & Ito, T. (2023). Present-day land subsidence over Semarang revealed by time series InSAR new small baseline subset technique. *International Journal of Applied Earth Observation and Geoinformation*, *125*(August), 103579. DOI: 10.1016/j.jag.2023.103579

Alfirdaus, L. K., Edi, A. C., Hanani, R., & Permana, Y. S. (2026). Beyond formal spaces: women's participation in tuberculosis control in rural Indonesia. *Development in Practice*. 10.1080/09614524.2026.2649799

Ansell, C., & Gash, A. (2008). Collaborative Governance in Theory and Practice. *Journal of Public Administration: Research and Theory*, *18*(4), 543–571. DOI: 10.1093/jopart/mum032

Awaluddin, M., Suwitri, S., & Santoso, R. S. (2026). Firebreak policy: Supporting local communities in forest fire prevention. *Multidisciplinary Science Journal, 8*(5). 10.31893/multiscience.2026339

Belland, M., Kausan, B. Y., Kooy, M., & Zwarteveen, M. (2025). Seeing like a pond: Amphibious stories of coastal subsidence in Central Java. *Geoforum, 161*(October 2024), 104248. DOI: 10.1016/j.geoforum.2025.104248

Damastuti, E., van Wesenbeeck, B. K., Leemans, R., de Groot, R. S., & Silvius, M. J. (2023). Effectiveness of community-based mangrove management for coastal protection: A case study from Central Java, Indonesia. *Ocean and Coastal Management, 238*, 106498. DOI: 10.1016/j.ocecoaman.2023.106498

Dharmawan, E. W., & Warsono, H. Kismartini, & Rahman, A. Z. (2026). Village Community Empowerment through Sparko Exercise: Integrating Physical Fitness, Social Solidarity, and Good Governance. *MSW Management*. 10.7492/3yd4d419

Emerson, K., Nabatchi, T., & Balogh, S. (2012). An Integrative Framework for Collaborative Governance. *Journal of Public Administration: Research and Theory*, *22*(1), 1–29. DOI: 10.1093/jopart/mur011

Handayani, W., Chigbu, U. E., Rudiarto, I., & Putri, I. H. S. (2020). Urbanization and Increasing Flood Risk in the Northern Coast of Central Java—Indonesia: An Assessment towards Better Land Use Policy and Flood Management. *Land (Basel)*, *9*(10), 343. DOI: 10.3390/land9100343

Harahap, M., Sudarno, S., & Muhamad, F. (2024). The design study for the development of sea wall construction on mangrove ecosystems : A case study of Tambak Lorok and Genuk district. *Journal of Bioresources and Environmental Sciences*, *3*(3), 121–127. DOI: 10.61435/jbes.2024.19245

Hidayati, I., Prabawani, B., Rahmadhony, A., Adis Nur Hayati, J., & Boddy, J. (2026). The critical role of local leaders in managing the impacts of climate change in Demak, Indonesia. *In Springer Climate*. 10.1007/978-3-032-15334-0_8

Huda, M. N., Nirmala, R. J., & Yusuf, I. M. (2024). Policy Incoherence and Reactive Approaches: Barriers to Effective Management of Land Subsidence in Semarang City, Indonesia. *JURNAL ILMU SOSIAL*, *23*(1), 20–36. DOI: 10.14710/jis.23.1.2024.20-36

Husnayaen, R., Rimba, A. B., Osawa, T., Parwata, I. N. S., As-syakur, A. R., Kasim, F., & Astarini, I. A. (2018). Physical assessment of coastal vulnerability under enhanced land subsidence in Semarang, Indonesia, using multi-sensor satellite data. *Advances in Space Research*, *61*(8), 2159–2179. DOI: 10.1016/j.asr.2018.01.026

Jesson, J. K., Matheson, L., & Lacey, F. M. (2011). *Doing Your Literature Review. Traditional and Systematic Techniques*. SAGE Publications Ltd.

Junaidi, M. A., Purnaweni, H., Maryono, M., & Nasarani, G. (2026). Scenario-based assessment of inorganic waste and greenhouse gas (GHG) emissions in Semarang City: Towards circular economy transition. *IOP Conference Series. Earth and Environmental Science*, *1586*(1), 012035. DOI: 10.1088/1755-1315/1586/1/012035

Kismartini, K., Roziqin, A., & Authori, N. (2023). A stakeholder analysis for sustainable development of Maritime Village in Semarang coastal community, Indonesia. *Public Administration and Policy*, *26*(3), 321–334. DOI: 10.1108/PAP-10-2022-0119

Kismartini. (2025). *Model Kolaborasi Pentahelix dalam Mewujudkan Kebijakan Pesisir Berkelanjutan*. Undip Press.

Kismartini, D., Wijaya, I. K., & L. S. (2023). Collaborative governance for sustainable coastal management in Semarang: A case study of the pentahelix model. *International Journal of Coastal Management, 41*(2), 12–27. 10.1080/09640284.2023.2034735

Kiswoyo, G. P., Suwitri, S., & Yuniningsih, T. (2025). Social Capital in Dealing with Flood Disasters in Genuk Sub-District. *Semarang City. PERSPEKTIF*, *14*(1), 122–132. DOI: 10.31289/perspektif.v14i1.13327

Kurniawan, R. R., Suharini, E., & Banowati, E. (2022). Mangrove Conservation Group Management in Semarang City, Central Java, Indonesia. *Proceedings of the 6th International Conference on Science, Education and Technology (ISET 2020), 574*(Iset 2020), 458–462. DOI: 10.2991/assehr.k.211125.085

Lituhayu, D., Santoso, R. S., Imawan, S. A., & Bernike, D. Bellastuti, & Saputra, Y. A. (2026). Sustainable Governance and Institutional Reform of Waste Management in Semarang City. *P Conference Series: Earth and Environmental Science, 1610*(012001). DOI: 10.1088/1755-1315/1610/1/012001

Marfai, M. A., King, L., Sartohadi, J., Sudrajat, S., Budiani, S. R., & Yulianto, F. (2008). The impact of tidal flooding on a coastal community in Semarang, Indonesia. *The Environmentalist, 28*(3), 237–248. DOI: 10.1007/s10669-007-9134-4

Martuti, N. K. T., Pribadi, R., Sidiq, W. A. B. N., & Mutiatari, D. P. (2020). Community-Based Integrated Coastal Management Strategy in Tugurejo Subdistrict, Semarang. *Proceedings of the International Conference on Research and Academic Community Services (ICRACOS 2019)*. DOI: 10.2991/icracos-19.2020.15

Maylasari, H., & Al-Akbar, N. (2026). Disaster innovation for environmental governance: Smart technologies for green policy implementation. *IOP Conference Series. Earth and Environmental Science, 1610*(1), 012003. DOI: 10.1088/1755-1315/1610/1/012003

Munn, Z., Peters, M. D. J., Stern, C., Tufanaru, C., McArthur, A., & Aromataris, E. (2018). Systematic review or scoping review? Guidance for authors when choosing between a systematic or scoping review approach. *BMC Medical Research Methodology, 18*(1), 143. DOI: 10.1186/s12874-018-0611-x PMID: 30453902

Nabatchi, T., & Emerson, K. (2015). Collaborative Governance Regimes. *Public Administration Review, 75*(6), 1–11. DOI: 10.1353/book44406

Pomeroy, R., & Douvere, F. (2008). The engagement of stakeholders in the marine spatial planning process. *Marine Policy, 32*(5), 816–822. DOI: 10.1016/j.marpol.2008.03.017

Pratama, F. W. N., Kusmayanti, J. D., & Marpaung, D. A. A. (2026). Assessing the Mainstreaming of Land Subsidence Risk in Spatial and Development Planning: Evidence from Coastal Semarang, Indonesia. *The Indonesian Journal of Planning and Development, 10*(2), 85–95. DOI: 10.14710/ijpd.10.2.85-95

Purnaweni, H., & Hadi, S. P. Soeprobowati, Retnaningsih, T., Roziqin, Ali, Nilam, & Rakhma, A. (2026). What are the Main Concerns for Global Mangrove Management Research? Systematic Literature Review and Bibliometric Analysis. *Research in Ecology, 8*(2). 10.30564/re.v8i2.12149

Putri, I. H. S., Handayani, W., Mulia, F. A., Ichihara, J., Anggraeni, M., Setiadi, R., & Sudarno, S. (2025). Examination of barriers to climate change adaptation: Experience from Semarang city, –Indonesia. *Urban Governance*, *5*(3), 314–330. DOI: 10.1016/j.ugj.2025.08.005

Putri, S. P. (2025). Assessing The Succes of Flood and Tidal Flood Control Infrastructure Program in Semarang. *Ganaya : Jurnal Ilmu Sosial Dan Humaniora*, *8*(3), 130–141. DOI: 10.37329/ganaya.v8i3.4412

Rhodes, R. A. W. (1996). The New Governance: Governing without Government. *Political Studies*, *44*(4), 652–667. DOI: 10.1111/j.1467-9248.1996.tb01747.x

Sarah, D. (2022). Land subsidence hazard in Indonesia: Present research and challenges ahead. *Riset Geologi Dan Pertambangan*, *32*(2), 83–100. DOI: 10.14203/risetgeotam2022.v32.1195

Sastrawan, B., Suwitri, S., Aisyah, S., & RumantaMaman. (2026). Development of functional aromatherapy candles from banana peel extract: A techno-economic study for green tourism products based on collaborative governance. *Indonesian Journal of Science and Technology, 11*(2), 439–448. 10.17509/ijost.v11i2.89775

Stoker, G. (1998). Governance as theory: Five propositions. *International Social Science Journal*, *50*(155), 17–28. DOI: 10.1111/1468-2451.00106

Utami, R., & Supratiwi, S. (2026). Top-down collaborative governance: Dinamika kolaborasi multi aktor dalam pengelolaan Program Kampung Iklim di Kelurahan Tugurejo Kota Semarang. *Journal of Politic and Government Studies*, *15*(2), 504–5016.

Wever, L., Glaser, M., Gorris, P., & Ferrol-Schulte, D. (2012). Decentralization and participation in integrated coastal management: Policy lessons from Brazil and Indonesia. *Ocean and Coastal Management*, *66*, 63–72. DOI: 10.1016/j.ocecoaman.2012.05.001

Yudartha, P. D., Yuwono, T., & Suwitri, S. (2026). Waste management in developing countries: A systematic literature review of policy implementation. *Frontiers in Environmental Science.*

Chapter 7
Reframing Mangrove Governance Under Forest Area Enforcement:
Policy Responses to Mangrove Degradation

Doni Sugianto Sihotang
http://orcid.org/0009-0008-5222-4475
Universitas Negeri Jakarta, Indonesia

Zakirul Hasri Ramadhan
Universitas Sumatera Utara, Indonesia

Usep Suhud
http://orcid.org/0000-0002-1728-0737
Universitas Negeri Jakarta, Indonesia

Agung Wahyu Handaru
Universitas Negeri Jakarta, Indonesia

Ruth Stany Melisa
Universitas Paramadina, Indonesia

ABSTRACT

This chapter delves into the restructuring of mangrove governance in Indonesia through a focus on the enforcement of forest areas as a response to the rapid degradation of mangroves. Specifically centered on Langkat Regency, the chapter scrutinizes how state-led enforcement measures interact with coastal political-economic dynamics, institutional fragmentation, and community-based resource utilization.

DOI: 10.4018/979-8-3373-9320-9.ch007

Drawing from political ecology and coastal forest governance perspectives, the chapter demonstrates that enforcement-driven governance produces varying outcomes: it bolsters territorial control and conservation authority while also engendering socio-ecological tensions and governance trade-offs. The analysis highlights the constraints of coercive approaches and advocates for hybrid governance frameworks that integrate enforcement with community participation, institutional coherence, and adaptive management. By offering pragmatic insights, the chapter contributes to the advancement of more just and sustainable mangrove governance strategies.

INTRODUCTION

Mangrove ecosystems are widely recognized as critical coastal forest systems that deliver a distinctive combination of ecological resilience, climate regulation, and socio-economic benefits. Located at the interface between terrestrial and marine environments, mangroves provide essential ecosystem services, including shoreline stabilization, storm surge attenuation, nursery habitats for fisheries, and long-term carbon sequestration. Recent scholarship consistently positions mangroves as key nature-based solutions for climate change mitigation and adaptation, particularly due to their disproportionately high blue carbon storage relative to terrestrial forests (Kairo et al., 2021; Lovelock et al., 2024). This growing recognition has elevated mangroves within global climate and biodiversity agendas. However, a closer synthesis of the literature reveals an emerging tension between ecological valuation and socio-political realities. While global studies emphasize the protective and carbon sequestration functions of mangroves (Hülsen et al., 2023), parallel research highlights their increasing vulnerability to climate-induced stressors such as sea-level rise, altered sedimentation, and extreme weather events (Hülsen et al., 2025). This dual narrative suggests that mangroves are simultaneously framed as both resilient natural infrastructure and fragile ecosystems under threat, raising questions about the adequacy of current conservation paradigms.

Indonesia occupies a pivotal position in this discourse as the country with the largest extent of mangrove forests globally. Indonesian mangroves contribute substantially to global carbon stocks, coastal protection, and fisheries productivity (Lokollo et al., 2024). At the same time, they sustain local livelihoods, particularly for small-scale fishers and coastal communities who depend on mangrove-related resources (Arkham et al., 2024). Empirical findings from coastal regions, including North Sumatra, demonstrate that mangroves function as socio-economic safety nets, providing food, fuelwood, and income diversification during periods of environmental or economic uncertainty. These insights reinforce the notion that mangroves operate as socio-ecological systems embedded within local cultural and economic practices. Despite extensive recognition of their ecological and socio-economic importance, previous studies often treat mangrove ecosystems in functional or sectoral terms, emphasizing either environmental services or livelihood contributions without fully integrating these dimensions. As a result, existing literature frequently overlooks how these dual roles interact within governance contexts, particularly in regions characterized by competing development priorities. This gap limits understanding of how mangroves can be managed as socio-ecological infrastructure that simultaneously supports climate objectives and social sustainability (Sreelekshmi & Nandan, 2025; Al-Asif, 2025).

Accordingly, the global and national significance of mangrove ecosystems should be understood not only through their ecological functions but also through the inherent tensions between conservation imperatives, climate resilience, and human dependency. Recognizing and addressing these tensions is essential for advancing more integrated and context-sensitive approaches to mangrove governance.

Indonesia's mangrove ecosystems continue to experience persistent degradation despite substantial policy attention and large-scale rehabilitation commitments. Over the past decade, Indonesia has simultaneously positioned itself as a global leader in mangrove restoration while remaining a hotspot for mangrove loss driven by extractive economic activities. Existing studies consistently identify aquaculture expansion, oil palm plantations, coastal infrastructure development, and energy projects as primary drivers of mangrove conversion, particularly in regions such as Sumatra and Kalimantan (Bhowmik et al., 2022; Arifanti et al., 2025; Rahim et al., 2026). These findings highlight the structural pressures exerted by national development priorities on coastal ecosystems. Beyond land-use change, a growing body of literature emphasizes governance failures as a central factor underpinning mangrove degradation. Fragmented authority, overlapping regulations, and conflicting sectoral priorities have historically characterized mangrove management in Indonesia. Mangroves are governed through intersecting legal frameworks, including forestry laws, coastal management policies, and spatial planning systems, resulting in institutional ambiguity and weak enforcement capacity (Gerona-Daga et al., 2022; Utami et al., 2024). This fragmentation enables large-scale concessions to persist while diffusing accountability across multiple agencies (Utami et al., 2024).

Decentralization has further complicated governance dynamics. Although intended to enhance local autonomy, decentralization has often incentivized short-term revenue generation through land conversion, particularly in resource-constrained coastal regions (Arifanti et al., 2022). Empirical studies demonstrate that local governments frequently lack the technical capacity and political leverage to regulate extractive investments effectively, leading to selective enforcement and insufficient monitoring (Friess et al., 2021). Consequently, governance arrangements tend to prioritize economic development over long-term ecological sustainability.

Field-based evidence from coastal North Sumatra illustrates how these governance challenges translate into socio-ecological impacts. Ethnographic findings reveal a gradual shift from mangrove-based livelihoods toward plantation labor and extractive industries, reducing access to traditional fishing and foraging areas. These transitions reflect broader patterns of marginalization, particularly among small-scale fishers whose livelihoods are closely tied to mangrove ecosystems. Moreover, top-down conservation and restoration initiatives have often produced exclusionary outcomes. Government-led rehabilitation programs frequently overlook customary land tenure systems and local ecological knowledge, generating resistance and undermining long-term effectiveness (Charles et al., 2020). This suggests that mangrove degradation persists not merely due to insufficient policy intervention but due to unresolved conflicts between conservation objectives and extractive development interests (Sreelekshmi & Nandan, 2025).

However, existing studies have not sufficiently addressed how these governance failures intersect with emerging enforcement-based approaches and their implications for power relations, access, and livelihood transformation. This creates a critical gap in understanding how contemporary policy shifts, particularly those emphasizing territorial control and enforcement, reshape mangrove governance outcomes. Addressing this gap is essential for advancing more coherent and socially inclusive strategies to manage mangrove ecosystems in Indonesia.

This chapter seeks to advance understanding of mangrove governance in Indonesia by critically examining the recent policy shift toward forest area enforcement as a primary strategy to address mangrove degradation. While global scholarship increasingly recognizes mangroves as essential ecosystems and nature-based solutions, limited attention has been directed toward how enforcement-oriented governance reshapes power relations, resource access, and livelihood dynamics in coastal forest landscapes. In this context, the chapter aims to reposition mangrove governance not merely as a technical or ecological issue but as a political and institutional process embedded in broader dynamics of state territorialization and socio-ecological transformation (Friess et al., 2021). Three interrelated objectives guide the chapter. First, it analyzes how Indonesia's enforcement-driven approach, particularly the reclassification of mangroves as state forest areas, reconfigures authority across forestry, coastal, and spatial planning institutions, contributing to persistent policy incoherence in Southeast Asia (Gerona-Daga et al., 2022). Second, it evaluates the socio-ecological consequences of enforcement-based governance, focusing on access restrictions, exclusion, and livelihood transitions among coastal communities (Charles et al., 2020). Third, it situates Indonesia's approach within broader debates on adaptive, multi-level, and hybrid governance models for mangrove management (Lovelock et al., 2024).

To support this analysis, the chapter adopts a political ecology framework. Political ecology is an analytical approach that examines how power relations, institutional arrangements, and economic interests shape access to natural resources and produce uneven environmental outcomes. Rather than viewing environmental degradation as purely biophysical, this perspective highlights how policies, markets, and governance structures determine who benefits from and who is excluded from resource use. In the context of mangroves, political ecology is particularly relevant for understanding how enforcement policies may simultaneously promote conservation while restricting traditional resource access.

A concrete example of this dynamic can be observed in Langkat Regency, where the reclassification of mangrove areas into state forest zones has altered community access to fishing and gathering areas. Practices that were historically considered customary have increasingly been framed as illegal under enforcement regimes, illustrating how governance interventions reshape local livelihoods and social relations. Methodologically, this study employs an ethnographic approach, based on two months of fieldwork conducted in Desa Bubun, Langkat Regency, from February to April 2025. Data were collected through participant observation, informal interviews, and field notes documenting daily interactions between communities and mangrove environments. This qualitative approach enables a grounded understanding of how national policies are experienced, negotiated, and contested at the local level. By combining policy analysis with ethnographic insights, the chapter adopts a multi-scalar perspective that connects national governance reforms with local socio-ecological realities. Rather than offering a comprehensive national assessment, the focus on Langkat provides a critical entry point for examining broader challenges in mangrove governance and contributes to comparative discussions on enforcement-driven conservation and socially inclusive governance frameworks (Sreelekshmi & Nandan, 2025).

BACKGROUND

1. Conceptualizing Mangrove Governance

Mangrove governance can be understood in operational terms as the set of rules, institutions, and practices that determine how mangrove resources are accessed, used, and managed by different actors. In coastal areas such as Langkat Regency, this governance is not abstract but directly observable in everyday interactions. For instance, local communities in Desa Bubun rely on mangrove areas for fishing, shellfish gathering, and small-scale resource extraction, while simultaneously facing increasing restrictions due to land reclassification and plantation expansion.

These empirical conditions illustrate that mangrove governance is fundamentally about negotiating access, authority, and sustainability in contested coastal spaces. From a theoretical perspective, mangrove governance is increasingly recognized as a complex socio-ecological process involving multiple stakeholders, overlapping institutions, and competing interests across the land–sea interface (Purnaweni et al., 2025). Mangroves are no longer viewed solely as ecological assets but also as common-pool resources whose sustainability depends on interactions among state regulation, market forces, and community-based management (Friess et al., 2021). Their common-pool characteristics, difficulty of exclusion, and high competition among users make them particularly vulnerable to overexploitation and institutional conflict.

The concept draws on several theoretical traditions, including commons governance, adaptive management, and social-ecological systems theory. Governance outcomes are shaped not only by formal regulations but also by informal norms, historical practices, and power relations embedded within communities and institutions. In Indonesia, this complexity is intensified by the coexistence of customary systems and formal state governance, often leading to overlapping claims and contested authority.

In coastal forest contexts, mangrove governance becomes even more complicated due to jurisdictional ambiguity between terrestrial forestry and marine and coastal management regimes. Treating mangroves exclusively as forests or as coastal resources oversimplifies their hybrid ecological nature and contributes to fragmented policy responses (Friess et al., 2021). As a result, the concept of coastal forest governance has emerged to emphasize integrated approaches that align land-use planning, fisheries management, and conservation strategies.

Political ecology perspectives further deepen this understanding by highlighting how power asymmetries and access inequalities shape governance processes. Interventions such as protected areas or restoration programs often produce uneven outcomes, benefiting state or corporate actors while marginalizing small-scale users (Bennett et al., 2025). This dynamic is evident in Indonesia, where governance reforms can simultaneously promote conservation objectives and restrict traditional livelihoods. An integrated framework that combines commons theory, political ecology, and coastal forest governance is therefore essential. Such a framework enables a more grounded understanding of how policies interact with local practices and why governance interventions may generate unintended socio-ecological consequences (May 2022; Huber et al., 2022).

2. Forest Area Enforcement and State Territorialization

Forest area enforcement refers operationally to a set of state-led actions aimed at asserting legal control over land through boundary demarcation, zoning, monitoring, and sanctioning. In the context of mangroves, this process is not merely administrative. Still, it materializes through concrete interventions such as mapping forest boundaries, installing physical markers, revoking land permits, and conducting joint enforcement operations involving forestry authorities, police, and sometimes military actors. In coastal areas like Langkat Regency, these mechanisms are evident in the reclassification of previously accessible mangrove zones as strictly regulated state forest areas, accompanied by increased surveillance and restrictions on community activities. Conceptually, forest area enforcement is closely linked to the process of state territorialization. Territorialization can be understood as the way the state defines, controls, and governs space by transforming landscapes into legally recognized territories. In environmental governance, this involves reclassifying land, formalizing boundaries, and enforcing rules that determine who can access and use natural resources. Studies indicate that such processes are often justified through narratives of environmental crisis, illegality, and the need for ecological protection (Gruby et al., 2013). However, territorialization is not a neutral or purely technical process. Political ecology perspectives emphasize that it reconfigures power relations by expanding state authority while potentially limiting the role of local and customary governance systems (Nuttall, 2023). In mangrove ecosystems, which straddle land and sea boundaries, this process becomes particularly complex, as reclassification into forest areas may override existing coastal or customary management arrangements.

In Indonesia, forest area enforcement has intensified through national initiatives aimed at addressing deforestation, illegal land use, and carbon emissions. This includes the expansion of forest boundaries into coastal zones, thereby bringing mangrove ecosystems under stricter state control due to their strategic importance for blue carbon and climate mitigation (Kairo et al., 2022). Practically, this enforcement often involves revisiting land status, canceling overlapping permits, and restricting access to areas deemed part of the forest estate.

Evidence from Langkat Regency illustrates both the intended and unintended consequences of these interventions. The enforcement of forest boundaries has limited community access to mangrove areas traditionally used for fishing, gathering, and small-scale resource use. Activities previously considered customary are increasingly subject to regulation or prohibition, creating uncertainty and tension among local users. While such measures may help prevent large-scale corporate encroachment, they can also disrupt long-standing livelihood systems and generate perceptions of exclusion. Comparative research suggests that enforcement-driven territorialization is most effective when embedded within broader governance frameworks that recognize local rights and enable negotiated forms of authority (Schafft, 2021). Without such integration, enforcement risks prioritizing spatial control over social legitimacy, reinforcing conflicts rather than resolving them. In mangrove governance, this underscores the need to situate forest area enforcement within more inclusive and adaptive governance arrangements that balance ecological protection with community access (Mbatha, 2022).

3. Political Ecology and Coastal Resource Management

Political ecology can be operationally defined as an approach that examines the relationship between power, resource access, and environmental change. In this perspective, environmental issues are not viewed solely as ecological problems but as outcomes of political, economic, and institutional processes that determine who controls resources and who is excluded from them. In coastal settings, including mangrove ecosystems, political ecology helps explain how governance interventions shape unequal access and produce differentiated impacts across social groups. In practical terms, this dynamic is clearly observable in coastal communities such as those in Langkat Regency. Small-scale fishers and mangrove-dependent households have historically relied on open access to coastal forests for fishing, shellfish collection, and fuelwood gathering. However, as enforcement-based governance expands, these practices are increasingly regulated or restricted under formal legal frameworks. Activities once considered customary are at times redefined as illegal, illustrating how shifts in governance can reconfigure access and transform everyday livelihood practices.

Theoretically, political ecology emphasizes that coastal resource management is embedded within competing sectoral interests, including fisheries, forestry, conservation, and industrial development. Each sector is supported by distinct institutional mandates and forms of authority, often leading to overlapping claims and governance conflicts (Bennett et al., 2025). In mangrove landscapes, this results in contested spaces where conservation policies, extractive activities, and local livelihood systems intersect. A central concept within political ecology is access and exclusion. Rather than focusing only on formal ownership, this approach highlights how access to resources is mediated by power relations, economic capacity, social networks, and state regulation (Ribot & Peluso, 2020). In mangrove governance, enforcement-driven conservation can create exclusionary effects by restricting local practices while allowing large-scale actors to operate under formal permits (Mbatha, 2022). This asymmetry often reinforces existing inequalities and undermines local stewardship systems.

Political ecology also critically examines dominant environmental narratives such as crisis discourse and restoration imperatives. While these narratives can mobilize policy attention and funding, they may oversimplify complex socio-ecological realities and legitimize top-down interventions that overlook local knowledge (Sundnes, 2021). In Langkat, for example, state-led conservation initiatives have reframed traditional mangrove use as environmentally harmful, contributing to tensions between communities and authorities.

Recent scholarship further links political ecology with discussions on environmental justice and participatory governance. Sustainable coastal resource management requires addressing historical inequalities, recognizing diverse knowledge systems, and ensuring fair distribution of benefits and burdens (Berkes et al., 2025). This implies moving beyond purely technical solutions toward governance approaches that explicitly consider power dynamics and livelihood security (Morrissey, 2023). By integrating power, access, and ecological change, political ecology provides a more accessible and grounded framework for understanding why mangrove degradation persists despite policy interventions and why enforcement-based approaches can generate both environmental benefits and new forms of vulnerability (Bennett et al., 2025).

MAIN FOCUS OF THE CHAPTER

Issues, Controversies, Problems

1. Mangrove Degradation in Langkat Regency

Situated along the eastern coast of North Sumatra, Langkat Regency represents a critical case of mangrove degradation shaped by the interaction of agricultural expansion, coastal infrastructure development, and extractive industries. Over the past two decades, large-scale conversion of mangrove areas into oil palm plantations has significantly transformed coastal landscapes, hydrological systems, and local livelihoods. Spatial analyses indicate that Sumatra has experienced rapid mangrove loss driven primarily by plantation expansion and associated infrastructure such as road networks (Eddy et al., 2021; Stewart, 2021). While these patterns are observable across Indonesia, Langkat provides a grounded illustration of how these dynamics unfold at the local level.

Figure 1. Community gathering in Bubun Village, Langkat Regency

Field observations from Desa Bubun reveal that mangrove degradation has occurred incrementally rather than through abrupt deforestation. Since the early 2000s, mangrove forests, nipa palm ecosystems, and coastal wetlands have been progressively replaced by oil palm monocultures, followed by energy-related land reclamation and offshore exploration activities. This gradual transformation has reduced fisheries productivity and weakened coastal protection functions. Similar patterns have been identified in other regions, such as the Mahakam Delta in East Kalimantan, where the expansion of aquaculture, particularly shrimp ponds, has driven extensive mangrove conversion, leading to long-term ecological degradation and declining fishery yields (Bhowmik et al., 2022). This comparison underscores that different economic drivers, plantations in Langkat and aquaculture in Mahakam, produce comparable socio-ecological outcomes.

Figure 2. Conversion of mangrove forest into oil palm plantation in Bubun Village, Langkat Regency

Beyond plantation expansion, coastal infrastructure and energy development have further intensified environmental change in Langkat. Activities such as dredging, land reclamation, and seismic surveys disrupt sediment dynamics and degrade nearshore habitats. Local testimonies indicate that areas previously used for shellfish gathering and as fallback resources during periods of low fishing have become inaccessible or ecologically impaired. Comparable infrastructure-driven pressures are also observed in northern Java, where coastal development has fragmented mangrove ecosystems and increased vulnerability to coastal flooding, reinforcing broader national patterns of degradation. These environmental shifts have led to significant transitions in livelihoods. As traditional activities such as fishing, crab harvesting, and fuelwood collection decline, many residents have shifted toward plantation labor or informal employment in extractive sectors. Similar transitions have been documented in mangrove-dependent communities across Indonesia and South Asia, where environmental degradation increases dependence on volatile commodity markets (Islam et al., 2021). Importantly, mangrove degradation in Langkat reflects broader political-economic forces rather than isolated local decisions. National priorities promoting agribusiness and energy development, combined with governance fragmentation and tenure ambiguity, shape these outcomes. Such structural drivers often normalize gradual environmental decline, with social consequences becoming visible only after ecological thresholds are crossed (Davies et al., 2022). Thus, while Langkat serves as a specific case, it reflects wider regional trajectories of mangrove degradation and governance challenges (Gajre et al., 2025; Utami et al., 2024).

2. Policy Instruments and the Role of Satgas PKH

In response to escalating forest and mangrove degradation, the Indonesian government has implemented enforcement-focused policies, leading to the establishment of the *Satuan Tugas Penertiban Kawasan Hutan* (Satgas PKH). Satgas PKH represents a substantial shift in environmental governance, emphasizing legal enforcement, territorial control, and the reclamation of state forest areas from what is identified as illegal occupation. Initially targeted at combating deforestation in terrestrial forests, Satgas PKH has broadened its scope to encompass coastal regions, including mangrove ecosystems, now recognized as part of the national forest estate. Policy instruments employed by Satgas PKH comprise the redefinition of forest areas, revocation of permits, land reclamation, and coordinated operations with forestry authorities, law enforcement agencies, and the military. Recent policy analyses indicate that these measures are structured to bolster compliance and underscore state authority, particularly in regions plagued by land disputes and governance deficiencies (Lo, 2024). Concerning mangroves, these interventions aim to prevent their conversion for plantations, aquaculture, and infrastructure development by upholding forestry regulations (Astuti & McGregor, 2022). In Langkat Regency, the enforcement of Satgas PKH has resulted in the reclassification of mangrove areas previously utilized for fishing, gathering, or informal agroforestry into tightly regulated forest zones. Findings from Desa Bubun demonstrate that this reclassification has limited community access to mangrove areas, leading to heightened surveillance and uncertainty regarding permissible activities. While these actions have deterred significant land conversions, they have also engendered new socio-ecological conflicts, particularly for households dependent on mangrove resources.

Figure 3. Degraded coastal terrain in Bubun Village, Langkat Regency

Current scholarly discourse increasingly perceives enforcement-based strategies as possessing both positive and negative ramifications. On one hand, they can effectively dissuade influential extractive entities and tackle persistent enforcement challenges (Gronow et al., 2021). Conversely, in the absence of robust social safeguards and community involvement, these strategies may criminalize local practices and undermine local stewardship (Myers et al., 2021).

Research in Southeast Asia indicates that enforcement strategies are most effective when integrated into comprehensive governance frameworks that acknowledge local rights, incorporate conflict-resolution mechanisms, and allow for adaptive management (Mbatha, 2022). Without such integration, policies like Satgas PKH may secure short-term control while perpetuating enduring governance deficiencies. This chapter perceives Satgas PKH not merely as an enforcement body but as a pivotal arena where conflicting ideologies of mangrove governance, territorial control, ecological rehabilitation, and social justice are deliberated. A thorough comprehension of its role is essential in evaluating whether forest enforcement can transition from a coercive tool to a component of more inclusive and sustainable mangrove governance.

3. Socio-Ecological Tensions and Governance Trade-offs

The enforcement-oriented reorganization of mangrove governance in Indonesia has led to notable socio-ecological tensions in coastal areas, including Langkat Regency. These tensions arise from the conflict between the goals of ecological preservation and the realities experienced by communities whose existence, identities, and means of survival are intricately linked to mangrove ecosystems. Recent scholarly findings indicate that governance interventions require a delicate balance between conservation efficacy, social fairness, and economic progress, rather than universally favorable outcomes (Bennett et al., 2025; Buntaine et al., 2024). In Langkat, the emphasis on enforcement of mangrove conservation has restricted access to zones traditionally used for fishing, gathering shellfish, collecting fuelwood, and sustaining livelihoods during periods of marine instability. Fieldwork in Desa Bubun demonstrates that these limitations disrupt daily routines, particularly for small-scale fishers and households lacking alternative sources of income. While these measures may deter large-scale degradation, they also jeopardize local adaptive practices crucial for the preservation of mangroves (Friess et al., 2021).

Figure 4. Community protest banner in a mangrove–oil palm interface area, Langkat Regency, North Sumatra

A protest banner exhibited by local inhabitants in a coastal plantation setting depicts the tangible socio-ecological repercussions of infrastructural neglect and environmental change. The Indonesian inscription states: "Beginilah nasib jalan kami: Musim panas kami makan debu, musim hujan kami berkubang," which translates to: "This is the condition of our road: In the dry season we consume dust; in the rainy season we wallow in mud." The banner conveys community concerns about deteriorating road access in regions affected by plantation expansion and altered hydrology. Figure 5 illustrates the convergence of environmental degradation, disparities in infrastructure provision, and local livelihoods' susceptibility. This illustrates how enforcement-driven forest governance and extractive land-use practices create socio-ecological conflicts, especially in marginal coastal areas where state territorial consolidation does not guarantee enhanced public services or equal access to resources. These tensions underscore a broader governance dilemma concerning ecological well-being and societal acceptance. Scholarship indicates that conservation initiatives lacking community support often encounter compliance challenges and may result in covert resistance, illicit behavior, or confrontations (Mbatha, 2022). In Langkat, reports of heightened supervision and uncertainty about permitted activities suggest that enforcement has altered the community's perception of government authorities, shifting trust dynamics from a remote overseer to an intrusive presence.

Unequal power dynamics exacerbate socio-ecological trade-offs. Studies in political ecology demonstrate that enforcement measures disproportionately affect disadvantaged groups while shielding powerful extractive interests (Morrissey, 2023). In Langkat, despite plantation expansion and energy ventures historically causing mangrove degradation, the repercussions of enforcement actions are chiefly felt at the local community level, reinforcing notions of partial justice. Nonetheless, governance trade-offs are not insurmountable. Recent research highlights the potential of hybrid governance frameworks that combine state enforcement with community participation, acknowledgment of customary practices, and adaptive co-management (Gerona-Daga et al., 2022). These strategies seek to transform trade-offs into negotiated resolutions where ecological preservation and livelihood security can coalesce.

Hence, socio-ecological tensions in Langkat are interpreted as reflections of unresolved governance choices rather than policy inadequacies. Resolving these trade-offs is essential for transitioning from coercive conservation to mangrove governance systems that are both environmentally sustainable and socially just (Berkes et al., 2025).

SOLUTIONS AND RECOMMENDATIONS

1. Rebalancing Enforcement and Community-Based Governance

The growing reliance on enforcement-based mangrove governance in Indonesia, particularly through forest area enforcement and initiatives such as Satgas PKH, necessitates a strategic rebalancing toward more operational and context-sensitive community-based governance. While enforcement has proven effective in curbing large-scale illegal conversion, its long-term success depends on integrating state authority with local stewardship systems rather than displacing them (Myers et al., 2021). In practice, this requires clearly defined co-management arrangements that specify roles, responsibilities, and decision-making processes among stakeholders. A practical co-management model can be conceptualized through a division of roles between the state and local communities. The state, represented by forestry agencies and local governments, is responsible for the legal designation of mangrove areas, boundary enforcement, and oversight of large-scale commercial activities. In contrast, local communities are engaged as active partners in monitoring, small-scale resource management, and restoration activities. In Langkat Regency, particularly in Desa Bubun, such an approach could involve community-based monitoring groups that collaborate with forestry officers to report illegal clearing while maintaining access for traditional practices such as small-scale fishing and silvofisheries.

Empirical evidence from other regions demonstrates the feasibility of such arrangements. In the Mahakam Delta, East Kalimantan, co-management initiatives involving local fishers, government agencies, and NGOs have been implemented to restore mangroves while sustaining aquaculture livelihoods. These initiatives allocate zones for conservation, community use, and rehabilitation, with local groups participating in planting, monitoring, and enforcement support. This model has shown that combining formal regulation with local knowledge improves compliance and ecological outcomes (Friess et al., 2021; Kaliyamurthy & Mao, 2025). Similarly, in Vietnam's Mekong Delta, community forestry schemes have granted local user groups conditional tenure rights over mangrove areas in exchange for maintaining forest cover. Under this system, households are permitted to engage in sustainable harvesting while adhering to conservation rules that are jointly monitored by local authorities and community organizations. Such arrangements illustrate how tenure clarity and shared authority can reduce conflict and enhance long-term sustainability (Pham et al., 2023).

To operationalize this approach in Indonesia, several mechanisms are required. First, legal recognition of customary access through social forestry or coastal tenure reforms is essential to reduce ambiguity and conflict (Lo, 2024). Second, participatory platforms must be established to facilitate negotiation between state agencies, local governments, and communities, ensuring that enforcement measures are socially acceptable (Charles et al., 2020). Third, adaptive management systems should be introduced, allowing policies to evolve based on ecological monitoring and community feedback (Etemadi et al., 2021). Rebalancing enforcement and community-based governance does not weaken state authority but recalibrates it toward collaborative governance. By combining regulatory control with local participation, such hybrid models can better address the dual challenge of preventing large-scale degradation while sustaining community livelihoods (Haque et al., 2022; Eddy et al., 2021; Nunoo et al., 2022).

2. Institutional Coordination and Policy Coherence

Institutional fragmentation remains a central challenge in mangrove governance in Indonesia, where overlapping mandates across forestry, marine and fisheries, spatial planning, agrarian affairs, and regional governments produce regulatory inconsistencies and enforcement gaps. Addressing this issue requires moving beyond conceptual calls for coordination toward concrete institutional mechanisms that enable policy coherence across sectors and administrative levels. Without such operational integration, enforcement-based approaches risk functioning in isolation and perpetuating governance inefficiencies. One key mechanism is the establishment and strengthening of inter-ministerial coordination bodies with clear mandates and decision-making authority. In Indonesia, existing platforms such as the National Mangrove Rehabilitation Program (BRGM – Badan Restorasi Gambut dan Mangrove) provide a partial foundation for inter-agency coordination. However, their effectiveness depends on formalizing roles in aligning forestry regulations, coastal zoning, and spatial planning frameworks. A more robust model would involve a permanent inter-agency task force that integrates ministries responsible for environment and forestry, marine affairs and fisheries, agrarian affairs, and regional planning, with clearly defined responsibilities for mangrove governance.

In addition to institutional alignment, the development of integrated data-sharing systems is critical. Currently, discrepancies in spatial data across agencies contribute to conflicting land classifications and overlapping permits. Establishing a unified geospatial database for mangrove areas, linked to Indonesia's One Map Policy (Kebijakan Satu Peta), would enable consistent decision-making and reduce regulatory ambiguity. Such a system should include real-time updates on land status, concession boundaries, and ecological conditions, accessible to both national and subnational actors. This approach aligns with broader governance recommendations emphasizing transparency and information integration as prerequisites for effective environmental management (Friess et al., 2022). Concrete examples from Southeast Asia illustrate the potential of coordinated governance. In the Mekong Delta, Vietnam has implemented integrated coastal zone management (ICZM) frameworks that synchronize mangrove conservation with aquaculture planning and climate adaptation strategies. These frameworks involve cross-sectoral coordination between ministries and local authorities, supported by shared data systems and joint monitoring mechanisms. Evidence suggests that such integration reduces policy conflicts and enhances ecological outcomes while maintaining economic productivity (Pham et al., 2023). Vertical coordination between national and local governments is equally important. In Indonesia, decentralization grants significant authority to local governments, yet coordination with central agencies often remains weak. Strengthening this relationship requires clear implementation guidelines, fiscal incentives for conservation, and capacity-building programs to enhance local governance capabilities. Without such support, local governments may continue prioritizing short-term economic gains over sustainable mangrove management (Arifanti et al., 2022).

Ultimately, achieving institutional coordination is not merely a technical adjustment but a governance reform that restructures authority, accountability, and information flows. Sustained political commitment, legal harmonization, and institutional learning are essential to ensure that mangrove governance operates as an integrated system rather than fragmented sectors (Utami et al., 2024).

3. Adaptive and Participatory Mangrove Management Models

Adaptive and participatory mangrove management models require clear operationalization to move beyond normative recommendations. In practice, these models can be implemented through a structured, step-by-step framework consisting of three main stages: stakeholder mapping, participatory planning, and integrated monitoring systems. This structured approach ensures that governance is not only inclusive but also responsive to ecological and social dynamics.

3.1. Stakeholder Mapping

The first step involves systematically identifying and categorizing all relevant stakeholders, including government agencies, local communities, private sector actors, and non-governmental organizations. In Langkat Regency, this includes small-scale fishers, plantation operators, village authorities, and forestry institutions. This stage aims to clarify roles, interests, and power relations among actors, which is critical for preventing exclusion and conflict. Research indicates that governance systems that recognize diverse stakeholders are more likely to achieve equitable and sustainable outcomes (Bennett et al., 2025). In addition, mapping enables policymakers to identify marginalized groups whose access to mangrove resources may be disproportionately affected by enforcement policies.

3.2. Participatory Planning

The second step focuses on collaborative decision-making processes where stakeholders jointly define management objectives, zoning arrangements, and resource-use rules. This includes determining which areas are designated for conservation, sustainable use, or restoration. In mangrove contexts, participatory planning integrates scientific knowledge with local ecological understanding, ensuring that policies are grounded in both ecological realities and livelihood needs. Empirical evidence from Southeast Asia shows that participatory approaches improve compliance and reduce governance conflicts by aligning conservation goals with community interests (Kaliyamurthy & Mao, 2025; Gerona-Daga et al., 2022). In Langkat, this could involve formalizing community access to specific zones while maintaining protection in ecologically sensitive areas.

3.3. Monitoring and Adaptive Management Systems

The third step involves establishing continuous monitoring mechanisms supported by both community participation and institutional oversight. Community-based monitoring groups can collect data on mangrove conditions, illegal activities, and resource use, which are then integrated into broader governance systems. These data inform adaptive management, allowing policies to be periodically revised based on ecological indicators and social feedback. This iterative process is essential in dynamic coastal environments where static policies often fail to address changing conditions (Charles et al., 2020; Fischer et al., 2025). Comparative experiences reinforce the effectiveness of this framework. In the Mekong Delta, Vietnam, participatory mangrove management has successfully combined community monitoring with state regulation, improving both forest cover and livelihoods. Similarly, community-based restoration initiatives in Java demonstrate that local involvement enhances ecological recovery and strengthens the legitimacy of governance (Friess et al., 2021). Collectively, these steps illustrate that adaptive and participatory models provide a practical pathway for achieving sustainable and socially inclusive mangrove governance (Lynch et al., 2022).

FUTURE RESEARCH DIRECTIONS

1. Evaluating Long-Term Impacts of Enforcement-Based Governance

The growing reliance on enforcement-based governance, particularly in forest areas and the deployment of Satgas PKH, underscores the necessity for a systematic evaluation of its enduring socio-ecological impacts on mangrove systems. While immediate benefits such as reduced land conversion and bolstered state authority are recognized, current research lacks evaluation of whether these advantages culminate in sustained ecological recuperation and socially acceptable governance in the long run. Subsequent studies should pivot towards longitudinal analyses that account for ecological patterns, shifts in livelihoods, and the legitimacy of governance (Myers et al., 2021). A notable research void exists concerning the ecological efficacy of enforcement-based governance in mangrove restoration. Studies suggest that halting illicit activities may not automatically translate into ecosystem rejuvenation, particularly in landscapes altered by plantations, hydrological modifications, and infrastructure (Friess et al., 2022). Continuous monitoring is imperative to determine whether enforced protection enhances mangrove structure, biodiversity, and ecosystem services, such as fisheries support and coastal protection.

Equally significant are the social repercussions of persistent enforcement. Political ecology scholarship posits that the prolonged exclusion of local users could jeopardize livelihoods, heighten vulnerability, and undermine traditional stewardship practices that are pivotal to mangrove resilience (Mbatha, 2022). Ethnographic evidence from Langkat Regency reveals how enforcement modifies daily access, work methodologies, and community-state relations, often gradually and subtly. Longitudinal investigations should trace the impact of enforcement on household finances, migration patterns, and livelihood decisions across successive generations.

Another critical area of inquiry involves scrutinizing the legitimacy and compliance of governance over time. Findings indicate that enforcement mechanisms lacking social approval may encounter diminishing compliance, informal resistance, or policy reversals (Bennett et al., 2025). Longitudinal research can help determine whether the initial backing for enforcement endures, diminishes, or evolves into negotiated governance arrangements. Moreover, forthcoming research should adopt comparative and mixed-method approaches, integrating remote sensing, ecological indicators, and qualitative fieldwork. Comparative analysis across Indonesian coastal regions and Southeast Asia can help researchers identify contextual factors, such as land tenure clarity, community structure, and institutional synchronization, that influence enforcement outcomes (Gajre et al., 2025).

Evaluating the enduring ramifications of enforcement-based governance is pivotal in discerning whether it serves as a transitional instrument towards sustainable mangrove management or leads to unintended social ramifications. Addressing this query will significantly enrich both mangrove policy formulation and broader dialogues on state-driven environmental governance (Weldon & Hoffman, 2024).

2. Comparative Studies Across Coastal Regions in Asia

Comparative research in coastal regions of Asia plays a vital role in deepening understanding of mangrove governance across diverse political, ecological, and socio-economic contexts. Indonesia stands out as a notable example of enforcement-based mangrove governance; however, understanding its dynamics requires broader contextualization. Cross-regional comparisons offer scholars the opportunity to discern patterns, disparities, and insights that can be extrapolated across different governance systems and their long-term implications for mangrove outcomes (Friess et al., 2021; Yando et al., 2021). Asian coastal areas, encompassing Southeast Asia, South Asia, and East Asia, exhibit a wide array of mangrove tenure systems, institutional frameworks, and developmental stresses. Research has highlighted that nations like Vietnam, Thailand, and the Philippines tend to gravitate towards co-management and community forestry models, while Indonesia and Myanmar emphasize state-driven enforcement and territorial dominance (Kaliyamurthy & Mao, 2025; Barbanera et al., 2022). Examining these contrasting methodologies helps clarify how enforcement-centric governance can either bolster or impede local stewardship efforts.

Comparative research is crucial for assessing the alignment of governance models with socio-ecological settings. Studies suggest that factors such as tenure clarity, decentralization structures, and community mobilization significantly influence mangrove governance outcomes (Pham et al., 2023; Aljahdali et al., 2021). For example, the Langkat case, characterized by conflicting claims, plantation expansions, and enforcement-oriented reclassifications, serves as a valuable reference point for regions with established customary tenure or co-management institutions. Future comparative inquiries should delve into how regional political economies shape mangrove governance trajectories. While export-driven aquaculture and plantation economies pose analogous challenges across Asian coastal zones, governance responses vary with governmental capacity and development priorities (Islam et al., 2021; Romero-Uribe et al., 2022). Comparative assessments can elucidate why certain nations effectively balance mangrove conservation with economic expansion, whereas others grapple with persistent trade-offs.

Methodologically, enriching comparative mangrove research through mixed-method and multi-scalar designs, integrating remote sensing, policy scrutiny, and ethnographic fieldwork, can help establish links between governance reforms at varying scales and livelihood and ecological consequences (Gajre et al., 2025; Lee et al., 2021).

In summary, comprehensive comparative studies across Asian coastal regions can bolster the development of theories on mangrove governance. By juxtaposing Indonesia's enforcement-centered strategy with alternative approaches, forthcoming research can contribute to developing more nuanced, adaptable, and region-specific governance frameworks to address shared challenges in Asia's mangrove ecosystems (Weldon & Hoffman, 2024; Cardenas et al., 2022).

3. Emerging Governance Paradigms for Mangrove Management

As pressures on mangrove ecosystems intensify due to climate change, coastal development, and global commodity expansion, future research needs to focus on engaging with emerging governance paradigms that transcend traditional state-centric or enforcement-driven approaches. Recent studies highlight a shift towards more adaptive, multi-level, and justice-oriented governance frameworks, recognizing that mangrove sustainability presents intertwined governance and ecological challenges (Lovelock et al., 2024; Purnaweni et al., 2026; Savari et al., 2022).

A notable emerging paradigm is multi-level governance, emphasizing coordination across local, national, and transnational levels. Global climate finance, blue carbon markets, and international biodiversity commitments are increasingly influencing mangrove governance, necessitating the harmonization of global standards with local circumstances (Kairo et al., 2021; Soeprobowati et al., 2024). Subsequent research should explore how these global mechanisms interact with national enforcement systems and the implications for local governance autonomy.

Another emergent paradigm centers on co-management and hybrid governance, integrating state authority with community-based institutions, non-governmental organizations, and the private sector. Studies suggest that hybrid arrangements can enhance ecological outcomes and governance legitimacy through authentic power-sharing and clear mandates (Gerona-Daga et al., 2022; Cahyaningsih et al., 2022). In Indonesia, initial evidence indicates that social forestry and participatory restoration initiatives offer alternative avenues to strictly coercive governance, though their long-term efficacy necessitates further investigation.

A third progressively influential paradigm is rights-based and justice-oriented mangrove governance. Scholars in political ecology and environmental justice argue that forthcoming governance models must directly confront historical dispossession, inequitable access, and distributive impacts of conservation (Bennett et al., 2025; Dayal et al., 2022). Findings from Langkat Regency illustrate how enforcement-driven mangrove protection can inadvertently marginalize coastal communities in the absence of justice considerations, underscoring the necessity for governance paradigms that incorporate social safeguards and respect customary practices.

Technological advancements are also reshaping mangrove governance. The incorporation of remote sensing, participatory mapping, and digital monitoring platforms enhances transparency and accountability but raises concerns about data ownership, surveillance, and exclusion (Friess et al., 2021). Subsequent research should critically assess the integration of these tools into governance systems and whose interests they ultimately serve.

Lastly, emerging paradigms increasingly position mangrove management as a component of transformative sustainability pathways, linking conservation with broader socio-economic transitions in coastal regions (Fischer et al., 2025). Advancing this agenda necessitates interdisciplinary research that integrates ecology, political economy, and institutional analysis.

In conclusion, future research on mangrove governance should pivot from assessing individual policies to comprehending how emerging paradigms redefine power, participation, and resilience. This investigation is essential for cultivating governance systems capable of safeguarding mangroves amidst rapid environmental and social transformations (Weldon & Hoffman, 2024).

CONCLUSION

This chapter examines the reframing of mangrove governance in Indonesia through the lens of forest area enforcement, situating it within broader debates on political ecology, state territorialization, and coastal resource management. The analysis demonstrates that mangrove degradation is not solely an ecological phenomenon but is deeply rooted in governance fragmentation, extractive political-economic dynamics, and contested authority across coastal regions. While enforcement-based approaches, particularly through mechanisms such as Satgas PKH, have strengthened state control and reduced certain forms of large-scale illegal conversion, they have also generated complex socio-ecological consequences. Empirical findings from Langkat Regency highlight how enforcement-driven governance reshapes access to mangrove resources, alters livelihood systems, and produces new forms of exclusion. Restrictions on traditional practices such as fishing, gathering, and small-scale resource use illustrate how conservation policies, when implemented without adequate social consideration, can disrupt local adaptive strategies. These findings reinforce broader arguments in political ecology that environmental governance interventions often produce uneven outcomes, benefiting some actors while marginalizing others. At the same time, the chapter underscores that governance challenges in mangrove ecosystems are not merely the result of insufficient policy but arise from conflicting objectives between conservation and development. Institutional fragmentation across forestry, coastal, and spatial planning sectors further exacerbates these tensions, limiting the effectiveness of both enforcement and restoration efforts. As such, mangrove governance in Indonesia is characterized by a persistent tension between ecological protection, economic development, and social equity. The findings also suggest that enforcement should not be viewed as a standalone solution but as one component within a broader governance framework. Hybrid approaches that integrate enforcement with community-based management, institutional coordination, and adaptive governance mechanisms offer more promising pathways for achieving sustainable outcomes. These approaches emphasize the importance of aligning state authority with local knowledge, improving policy coherence, and fostering participatory decision-making processes. This chapter contributes by providing a multi-scalar, empirically grounded analysis that connects national enforcement policies to local socio-ecological realities, thereby bridging the gap between policy discourse and lived experience in mangrove governance. Furthermore, it advances the literature by demonstrating how enforcement-based approaches function as dual instruments, simultaneously generating conservation gains and new vulnerabilities, and by proposing more operational, hybrid governance pathways that reconcile ecological objectives with social justice considerations. This study highlights that the future of mangrove governance lies in moving beyond enforcement-centric models toward more integrated, adaptive, and inclusive frameworks that address the complex interactions between environmental change and human livelihoods (Sreelekshmi & Nandan, 2025).

ACKNOWLEDGMENT

We thanked the Indonesian Ministry of Higher Education, Science, and Technology (Kemendiktisaintek) for the funding of this manuscript making through the PMDSU (Pendidikan Magister menuju Doktor untuk Sarjana Unggul) program (Number: 0419/C3/DT.05.00/2025 and 083/C3/ DT.05.00/PL/2025).

REFERENCES

Al-Asif, A. (2025). Protecting mangrove habitats contributes to achieving global climate goals. *Journal of Aquatic Research and Sustainability*, *2*(3), 1–3. DOI: 10.69517/jars.2025.02.03.0001

Aljahdali, M. O., Munawar, S., & Khan, W. R. (2021). Monitoring mangrove forest degradation and regeneration: Landsat time series analysis of moisture and vegetation indices at Rabigh Lagoon, Red Sea. *Forests*, *12*(1), 52. DOI: 10.3390/f12010052

Arifanti, V. B., Basyuni, M., Suharti, S., Slamet, B., Karlina, E., Sidik, F., Helbert, H., Yeny, I., Yulianti, M., Marwayana, O. N., Macklin, P. A., Rahmania, R., Suyadi, S., Wahyuni, T., Halwany, W., Rahmila, Y. I., Faubiany, V., Mubaraq, A., Aznawi, A. A., & Ali, H. M. (2025). Assessing the environmental and socioeconomic impacts of mangrove loss in Indonesia: A synthesis for science-based policy. *Forest Science and Technology*, *21*(4), 430–446. DOI: 10.1080/21580103.2025.2536595

Arifanti, V. B., Kauffman, J. B., Subarno, , Ilman, M., Tosiani, A., & Novita, N. (2022). Contributions of mangrove conservation and restoration to climate change mitigation in Indonesia. *Global Change Biology*, *28*(15), 4523–4538. DOI: 10.1111/gcb.16216 PMID: 35470521

Arkham, M. N., Riadi, S., Wahyudin, Y., & Krisnafi, Y. (2024). Economic value of mangrove ecosystem services in the coastal area of Bintan Island, Indonesia. *Wetlands Ecology and Management*, *32*(5), 727–740. DOI: 10.1007/s11273-023-09955-y

Barbanera, A., Markesteijn, L., Kairo, J., Juma, G. A., Karythis, S., & Skov, M. W. (2022). Functional responses of mangrove fauna to forest degradation. *Marine and Freshwater Research*, *73*(6), 762–773. DOI: 10.1071/MF21257

Bennett, N. J., Relano, V., Roumbedakis, K., Blythe, J., Andrachuk, M., Claudet, J., Dawson, N., Gill, D., Lazzari, N., Mahajan, S. L., Muhl, E.-K., Riechers, M., Strand, M., & Villasante, S. (2025). Ocean equity: From assessment to action to improve social equity in ocean governance. *Frontiers in Marine Science*, *12*, 1473382. DOI: 10.3389/fmars.2025.1473382

Berkes, F., & Franz, N. (2025). *Governing for transformation towards sustainable small-scale fisheries*. Food & Agriculture Org.

Bhowmik, A. K., Padmanaban, R., Cabral, P., & Romeiras, M. M. (2022). Global mangrove deforestation and its interacting social-ecological drivers: A systematic review and synthesis. *Sustainability (Basel)*, *14*(8), 4433. DOI: 10.3390/su14084433

Buntaine, M. T., Greenstone, M., He, G., Liu, M., Wang, S., & Zhang, B. (2024). Does the squeaky wheel get more grease? The direct and indirect effects of citizen participation on environmental governance in China. *The American Economic Review*, *114*(3), 815–850. DOI: 10.1257/aer.20221215

Cahyaningsih, A. P., Deanova, A. K., Pristiawati, C. M., Ulumuddin, Y. I., Kusumaningrum, L., & Setyawan, A. D. (2022). Causes and impacts of anthropogenic activities on mangrove deforestation and degradation in Indonesia. *International Journal of Bonorowo Wetlands*, *12*(1). Advance online publication. DOI: 10.13057/bonorowo/w120102

Cardenas, S. M., Cohen, M. C., Ruiz, D. P., Souza, A. V., Gomez-Neita, J. S., Pessenda, L. C., & Culligan, N. (2022). Death and regeneration of an Amazonian Mangrove Forest by Anthropic and natural forces. *Remote Sensing (Basel)*, *14*(24), 6197. DOI: 10.3390/rs14246197

Charles, A., Loucks, L., Berkes, F., & Armitage, D. (2020). Community science: A typology and its implications for governance of social-ecological systems. *Environmental Science & Policy*, *106*, 77–86. DOI: 10.1016/j.envsci.2020.01.019

Davies, T. (2022). Slow violence and toxic geographies:'Out of sight'to whom?. *Environment and Planning C: Politics and space, 40*(2), 409-427.

Dayal, S. R., Ravuiwasa, K., & Bhat, J. A. (2022). Deforestation and Degradation in the Mangrove Ecosystem: Implication on Environment and Livelihoods. In *Land Degradation Neutrality: Achieving SDG 15 by Forest Management* (pp. 99–116). Springer Nature Singapore. DOI: 10.1007/978-981-19-5478-8_6

Eddy, S., Milantara, N., & Basyuni, M. (2021). Carbon emissions as impact of mangrove degradation: A case study on the Air Telang Protected Forest, South Sumatra, Indonesia (2000-2020). *Biodiversitas (Surakarta)*, *22*(4). Advance online publication. DOI: 10.13057/biodiv/d220464

Etemadi, H., Smoak, J. M., & Abbasi, E. (2021). Spatiotemporal pattern of degradation in arid mangrove forests of the Northern Persian Gulf. *Oceanologia*, *63*(1), 99–114. DOI: 10.1016/j.oceano.2020.10.003

Fischer, J., Farny, S., Pacheco-Romero, M., & Folke, C. (2025). Resilience and regeneration for a world in crisis. *Ambio*, •••, 1–11. PMID: 41291163

Friess, D. A., Dahdouh-Guebas, F., Satyanarayana, B., Cannicci, S., Chua, S. C., Duke, N., Feller, I. C., Jaafar, Z., Jayatissa, L. P., Koedam, N., Krauss, K. W., Lee, S. Y., Mancera Pineda, J. E., Ruwa, R., Vannini, M., & Yando, E. S. (2021). Scientific contributions of the Mangrove Macrobenthos and Management (MMM) conference series, 2000–2019. *Estuarine, Coastal and Shelf Science*, *248*, 106742. DOI: 10.1016/j.ecss.2020.106742

Gajre, R. B., Barchiesi, S., Friess, D. A., Lang, D., Lee, Y. K., Lim, H. K., Nasry, M., Ochavo, K. G. C., S-H Peh, K., Pina-Covarrubias, E., Rege, A., Yong, D. L., Zeng, Y., & Tang, H. (2025). Bridging conservation and policy: Evaluating national targets to reduce mangrove loss under the Kunming–Montreal biodiversity framework. *Environmental Research Letters*, *20*(11), 114044. DOI: 10.1088/1748-9326/ae1150

Gerona-Daga, M. E. B., & Salmo, S. G.III. (2022). A systematic review of mangrove restoration studies in Southeast Asia: Challenges and opportunities for the United Nation's Decade on Ecosystem Restoration. *Frontiers in Marine Science*, *9*, 987737. DOI: 10.3389/fmars.2022.987737

Gronow, A., Brockhaus, M., Di Gregorio, M., Karimo, A., & Ylä-Anttila, T. (2021). Policy learning as complex contagion: How social networks shape organizational beliefs in forest-based climate change mitigation. *Policy Sciences*, *54*(3), 529–556. DOI: 10.1007/s11077-021-09418-2

Gruby, R. L., & Campbell, L. M. (2013). Scalar politics and the region: Strategies for transcending Pacific Island smallness on a global environmental governance stage. *Environment & Planning A*, *45*(9), 2046–2063. DOI: 10.1068/a45420

Haque, C. E., Berkes, F., Fernández-Llamazares, Á., Ross, H., Chapin, F. S.III, Doberstein, B., Reed, M. G., Agrawal, N., Nayak, P. K., Etkin, D., Doré, M., & Hutton, D. (2022). Social learning for enhancing social-ecological resilience to disaster-shocks: A policy Delphi approach. *Disaster Prevention and Management*, *31*(4), 335–348. DOI: 10.1108/DPM-03-2021-0079

Hein, J., Del Cairo, C., Gallego, D. O., Gutiérrez, T. V., Velez, J. S., & De Francisco, J. C. R. (2020). A political ecology of green territorialization: Frontier expansion and conservation in the Colombian Amazon. *DIE ERDE–Journal of the Geographical Society of Berlin*, *151*(1), 37–57.

Huber, R. A., Greussing, E., & Eberl, J. M. (2022). From populism to climate scepticism: The role of institutional trust and attitudes towards science. *Environmental Politics*, *31*(7), 1115–1138. DOI: 10.1080/09644016.2021.1978200

Hülsen, S., Dee, L. E., Kropf, C. M., Meiler, S., & Bresch, D. N. (2025). Mangroves and their services are at risk from tropical cyclones and sea level rise under climate change. *Communications Earth & Environment*, *6*(1), 262. DOI: 10.1038/s43247-025-02242-z

Hülsen, S., McDonald, R. I., Chaplin-Kramer, R., Bresch, D. N., Sharp, R., Worthington, T., & Kropf, C. M. (2023). Global protection from tropical cyclones by coastal ecosystems, past, present, and under climate change. *Environmental Research Letters*, *18*(12), 124023. DOI: 10.1088/1748-9326/ad00cd

Islam, M. M., Rahman, M. A., Khan, M. S., Mondal, G., & Khan, M. I. (2021). Transformational adaptations to climatic hazards: Insights from mangroves-based coastal fisheries dependent communities of Bangladesh. *Marine Policy*, *128*, 104475. DOI: 10.1016/j.marpol.2021.104475

Kairo, J., & Langat, J. (2021). Ocean climate solutions: Blue carbon now incorporated in the updated Kenya' s nationally determined contributions to Paris Agreement. *Kenya Aquatica Journal*, *6*(1), 52–58.

Kaliyamurthy, K., & Mao, A. A. (2025). Mangroves of India: Diversity, Conservation and Restoration Prospects. In *Ecological Restoration in India: Perspective and Practices* (pp. 111–135). Springer Nature Singapore. DOI: 10.1007/978-981-96-8943-9_6

Lee, C. K., Duncan, C., Nicholson, E., Fatoyinbo, T. E., Lagomasino, D., Thomas, N., Worthington, T. A., & Murray, N. J. (2021). Mapping the extent of mangrove ecosystem degradation by integrating an ecological conceptual model with satellite data. *Remote Sensing (Basel)*, *13*(11), 2047. DOI: 10.3390/rs13112047

Lo, K. (2024). The politics of just transition: Authoritarian environmentalism and implementation flexibility in forest conservation. *Political Geography*, *109*, 103066. DOI: 10.1016/j.polgeo.2024.103066

Lokollo, F. F., Wawo, M., Ceanturi, A., Lewerissa, Y. A., Hulopi, M., Handayani, L. D., & Wardiatno, Y. (2024). Blue Carbon Potential of Mangrove Ecosystems and Its Management to Promote Climate Change Mitigation in Indonesia. *Jurnal Ilmu Kehutanan*, *18*(2), 208–218. DOI: 10.22146/jik.v18i2.11447

Lovelock, C. E., Bennion, V., de Oliveira, M., Hagger, V., Hill, J. W., Kwan, V., Pearse, A. L., Rossini, R. A., & Twomey, A. J. (2024). Mangrove ecology guiding the use of mangroves as nature-based solutions. *Journal of Ecology*, *112*(11), 2510–2521. DOI: 10.1111/1365-2745.14383

Lynch, A. J., Thompson, L. M., Morton, J. M., Beever, E. A., Clifford, M., Limpinsel, D., Magill, R. T., Magness, D. R., Melvin, T. A., Newman, R. A., Porath, M. T., Rahel, F. J., Reynolds, J. H., Schuurman, G. W., Sethi, S. A., & Wilkening, J. L. (2022). RAD adaptive management for transforming ecosystems. *Bioscience*, *72*(1), 45–56. DOI: 10.1093/biosci/biab091

May, C. K. (2022). Complex adaptive governance systems: A framework to understand institutions, organizations, and people in socio-ecological systems. *Socio-Ecological Practice Research*, *4*(1), 39–54. DOI: 10.1007/s42532-021-00101-7 PMID: 35071989

Mbatha, P. (2022). Unravelling the perpetuated marginalization of customary livelihoods on the coast by plural and multi-level conservation governance systems. *Marine Policy*, *143*, 105143. DOI: 10.1016/j.marpol.2022.105143

Morrissey, J. (2023). Coastal communities, blue economy and the climate crisis: Framing just disruptions. *The Geographical Journal*, *189*(2), 283–299. DOI: 10.1111/geoj.12419

Myers, R., Luttrell, C., Harjanthi, R., Fisher, M. R., Menton, M., Läderach, P. R., & Wollenberg, E. K. (2021). Climate change mitigation in forests: Conflict, peacebuilding, and lessons for climate security-Position Paper.

Nunoo, F. K. E., & Agyekumhene, A. (2022). Nunoo, Francis KE, and Andrews Agyekumhene. "Mangrove degradation and management practices along the coast of Ghana.". *Agricultural Sciences*, *13*(10), 1057–1079. DOI: 10.4236/as.2022.1310065

Nuttall, M. (2023). *The Shaping of Greenland's Resource Spaces: Environment, Territory, Geo-security*. Routledge. DOI: 10.4324/9781003175421

Palathoti, S. (2025). Comprehensive review of three decades of developments on conservation and restoration of mangroves: A publications data-driven approach. *Multidiscip. Rev.*, *8*(5), 2025156. DOI: 10.31893/multirev.2025156

Pham, T. T., Nguyen, T. T. A., Nguyen, T. V. A., Tran, N. M. H., Nguyen, T. S., Nguyen, D. Y. K., & Tang, T. K. H. (2023). *Forest stakeholders and forestry-based mitigation options: Contributions to low-emission food systems in the Mekong Delta*. CIFOR.

Purnaweni, H., Hadi, S. P., Soeprobowati, T. R., Roziqin, A., & Rakhma, N. A. (2026). What are the Main Concerns for Global Mangrove Management Research? Systematic Literature Review and Bibliometric Analysis. *Restoration Ecology*, *8*(2), 46–66. DOI: 10.30564/re.v8i2.12149

Purnaweni, H., Soeprobowati, T. R., Chatterjee, U., Kiswoyo, G. P., Rakhma, N. A., & Manik, J. S. T. (2025). The Role of Stakeholders in Community Based Mangrove Management in Tapak Area of Semarang City. In *E3S Web of Conferences* (Vol. 650, p. 02047). EDP Sciences. DOI: 10.1051/e3sconf/202565002047

Rahim, A., Soeprobowati, T. R., Helmi, M., & Kruszewski, Ł. (2026). Spatiotemporal analysis of mangrove cover change and future projections in West Sumatra and Northern Java. *Landscape and Ecological Engineering*. Advance online publication. DOI: 10.1007/s11355-026-00714-8

Romero-Uribe, H. M., López-Portillo, J., Reverchon, F., & Hernandez, M. E. (2022). Effect of degradation of a black mangrove forest on seasonal greenhouse gas emissions. *Environmental Science and Pollution Research International, 29*(8), 11951–11965. DOI: 10.1007/s11356-021-16597-1 PMID: 34558043

Samanta, S., Hazra, S., Mondal, P. P., Chanda, A., Giri, S., French, J. R., & Nicholls, R. J. (2021). Assessment and attribution of mangrove Forest changes in the Indian Sundarbans from 2000 to 2020. *Remote Sensing (Basel), 13*(24), 4957. DOI: 10.3390/rs13244957

Savari, M., Damaneh, H. E., & Damaneh, H. E. (2022). Factors involved in the degradation of mangrove forests in Iran: A mixed study for the management of this ecosystem. *Journal for Nature Conservation, 66*, 126153. DOI: 10.1016/j.jnc.2022.126153

Schafft, K. A. (2021). Rurality and Crises of Democracy: What Can Rural Sociology Offer the Present Moment? *Rural Sociology, 86*(3), 393–418. DOI: 10.1111/ruso.12408

Soeprobowati, T. R., Sularto, R. B., Hadiyanto, H., Puryono, S., Rahim, A., Jumari, J., & Gell, P. (2024). The carbon stock potential of the restored mangrove ecosystem of Pasarbanggi, Rembang, Central Java. *Marine Environmental Research, 193*, 106257. DOI: 10.1016/j.marenvres.2023.106257 PMID: 38086137

Sreelekshmi, S., & Nandan, S. B. (2025). Sustainability challenges of mangrove ecosystems in the Anthropocene: Current perspectives and prospects. *Habitable Planet [this journal has moved to new platform at:* https://www. sciltp. com/journals/hp*], 1*(1&2), 129-144.

Stewart, H. A. (2021). *The role of foundation species in shaping the biodiversity of mangrove islands*. McGill University.

Sundnes, F., Johnsen, K. I., & Sundsvold, B. (2025). Mosaics of nature, meaning, and politics: Conservation of coastal heathlands in Western Norway. *Environment and Planning. E, Nature and Space*, *8*(4), 25148486251342987. DOI: 10.1177/25148486251342987

Utami, W., Sugiyanto, C., & Rahardjo, N. (2024). Mangrove area degradation and management strategies in Indonesia: A review. *Journal of Degraded and Mining Lands Management*, *11*(3), 6037–6047. DOI: 10.15243/jdmlm.2024.113.6037

Weldon, I., & Hoffman, S. J. (2024). "Fit for Purpose?" Assessing the Ecological Fit of the Social Institutions that Globally Govern Antimicrobial Resistance. *Perspectives on Politics*, *22*(4), 1157–1178. DOI: 10.1017/S1537592723002906

Yando, E. S., Sloey, T. M., Dahdouh-Guebas, F., Rogers, K., Abuchahla, G. M., Cannicci, S., Canty, S. W. J., Jennerjahn, T. C., Ogurcak, D. E., Adams, J. B., Connolly, R. M., Diele, K., Lee, S. Y., Rowntree, J. K., Sharma, S., Cavanaugh, K. C., Cormier, N., Feller, I. C., Fratini, S., & Friess, D. A. (2021). Conceptualizing ecosystem degradation using mangrove forests as a model system. *Biological Conservation*, *263*, 109355. DOI: 10.1016/j.biocon.2021.109355

Chapter 8
Collaborative Governance for Mangrove Sustainability:
Lessons From Global Policy and Local Practice

Rifka Resi Pawestri
http://orcid.org/0009-0002-2810-2001
Diponegoro University, Indonesia

Teguh Yuwono
http://orcid.org/0009-0009-8561-2949
Diponegoro University, Indonesia

Nunik Retno Herawati
Diponegoro University, Indonesia

Bangkit Aditya Wiryawan
http://orcid.org/0000-0001-7840-6914
Diponegoro University, Indonesia

ABSTRACT

Mangrove ecosystems are coastal resources whose management is influenced by the dynamics of environmental governance. This chapter discusses mangrove management from an environmental governance perspective, emphasizing the shift from a state-centered approach to multi-actor governance. The discussion focuses on the power relations between actors, including central and local governments, the private sector, local communities, non-governmental organizations, and academics. Using

DOI: 10.4018/979-8-3373-9320-9.ch008

a collaborative governance framework, this book chapter examines the implementation of co-management as a model of mangrove governance and the challenges in implementing it, such as institutional fragmentation, weak policy enforcement, and limited meaningful participation from the community. It shows that there is a gap between policy and practice, which often becomes a major obstacle to mangrove sustainability. Therefore, studies on the importance of inclusive and adaptive collaborative governance are a prerequisite for sustainable mangrove management.

I. INTRODUCTION

As one of the ecosystems in coastal areas, mangroves play a crucial role by providing significant ecological, economic, and social benefits, including maintaining environmental balance while supporting the social and economic well-being of local communities. In maintaining environmental balance while supporting the social and economic well-being of local communities, they provide coastal protection, carbon sequestration, and support for biodiversity (Hartuti Purnaweni et al., 2026). This role is reflected in the various ecosystem services provided, as mangrove ecosystems offer a wide range of important benefits for ecological stability and human well-being, and also serve as a natural resource that contributes to the national economy (Marlianingrum et al., 2019). Mangroves are located in the transitional zone between land and sea, giving them complex ecological functions and making them interconnected with other coastal ecosystems, such as seagrass beds and coral reef ecosystems. The presence of mangrove ecosystems makes a significant contribution not only to the conservation of the coastal environment but also to the well-being of local communities.

From an ecological perspective, mangrove ecosystems play a role in protecting coastal areas from various environmental threats through their function as global carbon sinks, the preservation of biodiversity, and the enhancement of coastal resilience. Mangroves act as carbon sinks that help reduce CO_2 concentrations in the atmosphere, provide habitats for a variety of flora and fauna, and support other ecological processes (Das et al., 2026). Mangroves have complex root systems that act as sediment traps, thereby helping to maintain coastal stability and reduce erosion.

In addition to their ecological functions, mangroves also play a role in mitigating climate change; mangrove ecosystems are classified as blue carbon ecosystems due to their ability to absorb and store large amounts of carbon (Singh et al., 2025). This ability enables mangroves to play a role in reducing greenhouse gas emissions and contributes to strategies for adapting to global climate change. This aligns with the ecological role of mangroves in maintaining ecosystem balance and preserving biodiversity. Although they cover only a small portion of the world's coastal areas, mangroves possess an enormous carbon storage capacity, and their carbon absorption rate is even higher than that of mature tropical forests. Additionally, mangroves provide habitat for thousands of species and support critical ecological processes that play a role in maintaining the sustainability of coastal ecosystems.

From an economic perspective, mangrove ecosystems serve as a source of livelihood for the well-being of coastal communities through various resource utilization activities. Coastal communities often utilize mangrove areas for capture fisheries, aquaculture, the collection of forest products, and the development of ecotourism activities. This is supported by research estimating the economic value of mangrove ecosystem services in the city of Dumai at 110.95 billion rupiah per year, with provisioning services such as fisheries and firewood contributing significantly to community income (Insusanty et al., 2025). This demonstrates that mangroves have tangible economic value that plays a role in sustaining the livelihoods of coastal communities while also supporting sustainable development.

Meanwhile, from a social perspective, mangroves are closely intertwined with the lives of coastal communities. The management of mangrove ecosystems often involves various social practices, cultural values, and local wisdom that have evolved within these communities. Various coastal communities have traditional rules, norms, or practices for utilizing and conserving coastal resources sustainably. The presence of mangroves also plays a role in strengthening social cohesion through collective activities such as rehabilitation, conservation, or other community-based management efforts. The significant role of mangroves in ecological, economic, and social aspects demonstrates that this ecosystem plays a strategic role in supporting sustainable development. This complexity means that mangrove management cannot be carried out through a sectoral approach alone, but also requires governance capable of accommodating various interests and involving many actors. In line with this, the complex management of mangroves—both at the local and global levels—is becoming increasingly challenging as pressure on these ecosystems grows. At the local level, common challenges include land conversion for fish ponds, residential development, industrial expansion, and infrastructure construction in coastal areas, as well as resource exploitation and weak coordination among stakeholders. At the global level, the challenges relate to climate change, rising carbon emissions, the loss of biodiversity, and the demands of achieving the sustainable development agenda. These conditions indicate that managing mangroves cannot be done merely through sectoral or piecemeal approaches; rather, it requires an approach capable of integrating the various interests of stakeholders and policy scales. Consequently, a collaborative governance approach is becoming increasingly relevant for achieving mangrove sustainability through collaboration and cooperation among government, communities, the private sector, academia, and even non-governmental organizations.

In addition, a study of the oil spill on Bintan Island shows that Indonesia, as an oil-producing country, is highly vulnerable to the risk of marine pollution (Purnaweni et al., 2022). Although the issue of oil spills has been extensively discussed in various studies, research on policy management and governance strategies for their handling remains relatively limited (Wiryawan et al., 2026). The findings of these studies indicate that local governments tend to be slow in responding to oil spills that recur annually, while the policy responses implemented remain repressive and conventional. This situation indicates that managing coastal environmental issues requires more than just remedial actions after damage occurs; it also demands a more preventive, coordinated approach that involves multiple stakeholders (Wiryawan et al., 2025). In the context of collaborative mangrove governance, these findings reinforce the argument that the sustainability of coastal ecosystems requires a multi-stakeholder governance model involving the government, local communities, the private sector, academia, and other relevant institutions (Lituhayu et al., 2026). For example, Dinegoro University's current involvement in supporting the net-zero emissions target through mangrove conservation, biodiversity protection, research enhancement, and community service demonstrates that universities play a strategic role in this multi-stakeholder governance system (Ambariyanto et al., 2023). Higher education institutions serve not only as providers of scientific knowledge but also as key actors capable of bridging the needs of policy, conservation practices, and the empowerment of coastal communities. Through multi-stakeholder governance, environmental policies can be directed not only to address damage but also to prevent ecological risks, strengthen institutional coordination, and safeguard the sustainability of coastal resources and the livelihoods of communities dependent on these ecosystems.

Changes in the dynamics of the global environment, the increasing complexity of socio-ecological issues, and growing demands for more sustainable natural resource management are driving a paradigm shift in environmental governance. The approach, which was initially dominated by the "government" model, is gradually shifting toward the "governance" paradigm. This shift in approach not only reflects a change in terminology but also signifies a fundamental transformation in how resources are managed. In the government paradigm, the state is positioned as the sole actor with dominant authority in formulating policies, establishing regulations, and controlling the implementation process of natural resource management. This paradigm is hierarchical, top-down, and positions the government as the primary decision-maker, while other actors, such as the public, are positioned merely as recipients of policy.

Initially, this approach was viewed as effective in addressing various public issues because the state possessed formal legitimacy and strong regulatory instruments. However, over time, the government paradigm has faced various limitations, particularly when confronted with environmental issues that are multidimensional, cross-sectoral, and involve numerous actors with diverse interests. Modern environmental issues, such as climate change, environmental degradation, and the overexploitation of natural resources, cannot be understood as purely technical problems. These issues are also influenced by the interplay of interconnected economic, social, political, institutional, and cultural factors. Research on mangrove degradation in the coastal areas of Brebes Regency has shown that the decline in mangrove coverage is influenced by a combination of ecological and anthropogenic factors, such as coastal erosion, vegetation die-off, logging, and weak protection of coastal areas (Faperi & Hendrarto, 2015). This indicates that the issue of mangrove degradation cannot be understood merely as an environmental problem, but also as a governance issue. Given these conditions, it is impossible for the state to act alone in effectively addressing the complexity of environmental challenges.

Criticism of this government-centric approach has subsequently spurred the emergence of a new paradigm—the governance paradigm—which emphasizes the importance of involving various actors in the processes of resource management and public decision-making. The governance paradigm arises from the recognition that the complexity of environmental issues cannot be resolved solely through state control mechanisms; rather, it requires more horizontal, participatory, and collaborative patterns of interaction. In this context, the state retains its role as a regulator and facilitator but is no longer the sole center of power. Responsibility for environmental management is viewed as a process involving a network of actors comprising the government, civil society, local communities, the private sector, non-governmental organizations, research institutions, academics, and even international organizations.

The governance paradigm emphasizes the importance of negotiation, coordination, and the establishment of cooperation mechanisms among actors in achieving more effective management objectives. The decision-making process is not merely one-directional but is built through various interactions and knowledge exchanges among different parties with varying resources, capacities, and interests. Governance is understood not merely as a formal institutional structure, but also encompasses social relationships, policy networks, norms, and coordination mechanisms that govern interactions among actors. In the context of environmental management policy, the governance approach is considered better equipped to accommodate uncertainty, manage conflicts of interest, and build policy legitimacy through the involvement of those directly affected.

The paradigm shift from "government" to "governance" has significant implications for the management of coastal resources. Coastal areas are regions with complex socio-ecological characteristics, as they serve as a space for interaction between terrestrial and marine ecosystems, while also serving as a hub for various human activities. These areas are the sites of various economic activities, such as fisheries, transportation, industry, tourism, aquaculture, and infrastructure development. At the same time, coastal areas also serve vital ecological functions as habitats for biodiversity, natural barriers against erosion, carbon sinks, and supports for other ecological systems. This diversity of functions makes coastal areas a meeting ground for various interests that often compete with one another. In practice, the sectoral management of coastal resources often leads to policy fragmentation and overlapping jurisdictions among agencies. Various institutions often operate based on their respective sectoral mandates without adequate coordination mechanisms. Consequently, the resulting policies tend to be fragmented and fail to adequately address the interconnections among ecological, social, and economic aspects. Furthermore, local communities that directly depend on coastal resources often have limited involvement in decision-making processes. This situation has the potential to spark conflicts

In this context, the evolution of governance paradigms has led to the emergence of multi-stakeholder governance as an increasingly common approach to coastal resource management. Multi-stakeholder governance has emerged in response to the need for a management system capable of integrating various actors, interests, and policy scales within a single collaborative framework. This approach involves the government, local communities, businesses, academics, civil society organizations, and international institutions in the process of planning, implementing, and evaluating policies.

Multi-stakeholder governance is based on the assumption that each actor possesses different resources, knowledge, capacities, and perspectives when addressing an issue. Local communities, for example, possess experience-based knowledge regarding environmental characteristics and traditional management practices. The government holds political legitimacy and regulatory authority. Meanwhile, the private sector possesses financial resources and investment capacity, while academics provide scientific support and research-based knowledge. The integration of these various resources is seen as capable of producing a more inclusive and adaptive decision-making process.

In mangrove management, multi-stakeholder governance is becoming increasingly important given that the complexity of the issues faced relates not only to ecological aspects but also to social, economic, institutional, and political dimensions. Mangroves are not only viewed as conservation ecosystems but also as a source of livelihood for coastal communities, providers of environmental services, and a component of global climate change mitigation strategies. Therefore, the sustainability of mangroves requires governance mechanisms capable of linking local interests with global agendas, as well as building synergy between formal policies and management practices at the community level.

II. METHOD

This chapter employs a qualitative approach using a literature review to analyze the development of collaborative governance in mangrove management and to identify lessons learned from global policies and their implementation at the local level. The data collected consists of secondary data sourced from various scientific articles and publications obtained through the Scopus database. The literature search was conducted using keywords such as mangrove governance, collaborative governance, multi-actor governance, community-based mangrove management, and sustainable mangrove management. The selected literature focused on studies discussing the roles of actors, collaboration mechanisms, challenges in mangrove governance, and practices in managing mangrove ecosystems across various regions. The analysis was conducted through content analysis by identifying, grouping, and synthesizing key findings into several themes: the transformation of governance from state-centered governance toward multi-actor governance; the role of stakeholders; collaborative governance; community-based management; gender and inclusivity; and challenges in mangrove governance. This chapter also employs a comparative approach to compare mangrove management experiences across various global and local cases to identify factors that support or hinder the success of collaborative governance. The results of the analysis are then presented narratively

III. DISCUSSION

The Dynamics of Mangrove Governance: From State-Centered to Multi-Actor Governance

Mangrove ecosystem management has undergone significant developments in tandem with shifts in social, political, and environmental dynamics at both local and global levels. From a traditional perspective, natural resource management was initially based on a state-centered approach—a governance model that positions the state as the sole actor in formulating policies, making decisions, and implementing resource management programs. The state was viewed as an actor possessing the legitimacy, authority, and adequate policy instruments to regulate resource utilization while safeguarding the public interest. This approach developed within a hierarchical, top-down governance framework, where the government played a dominant role while other actors were positioned as recipients of policy.

However, as environmental issues have grown increasingly complex, state-centered approaches have begun to face numerous challenges, such as ecosystem degradation, climate change, conflicts over resource use, development pressures in coastal areas, and rising economic needs of communities. These challenges demonstrate that resource management requires not only technical and administrative involvement but is also influenced by the interplay of various social, economic, and political interests. These increasingly complex issues mean that the state does not always have sufficient capacity to manage resources effectively when acting as a single, dominant actor.

In the context of mangrove ecosystem management, the limitations of a state-centered approach are becoming increasingly apparent because mangrove ecosystems possess complex and multidimensional characteristics. Mangroves function as ecological systems that support biodiversity and provide protection for coastal areas, and they also serve economic and social functions directly linked to people's livelihoods. Various groups are closely tied to these mangrove ecosystems, such as coastal communities, fishermen, business operators, environmental organizations, academics, and government agencies. Given these diverse interests, a management framework is needed that is far more open and capable of integrating the interests of various actors into a single governance mechanism. This has driven the development of a governance paradigm that emphasizes the involvement of all non-state actors in the resource management process. Based on this approach, mangrove ecosystem management is not merely understood as an exclusive authority held by the government, but also as a collective process involving the interaction of various actors in cooperation, coordination, and the sharing of responsibilities. This shift in approach marks the emergence of multi-stakeholder governance, where non-state actors have a greater role in the decision-making process and policy implementation.

Research conducted on mangrove management in the Payum Beach area of Merauke indicates that collaboration between the government and the Marind community has led to a shift in mangrove governance dynamics from a state-centered approach toward a model involving various actors, known as multi-actor governance (Sajriawati et al., 2021). From a state-centered perspective, the government is positioned as the sole actor with full control over the planning and decision-making processes involved in implementing resource management policies. The research findings indicate that the success of mangrove management is not determined solely by government intervention but is also influenced by non-state actors, particularly local communities, who are both users and managers of these resources.

Research conducted among the Marind tribe shows that local communities are no longer viewed as mere objects of policy but rather as actors who are recognized as possessing the capabilities and knowledge to play an active role in maintaining the sustainability of mangrove ecosystems. Before the government implemented a mangrove management program, local communities had already taken the initiative to plant mangroves on their own as an effort to prevent sand flooding and coastal erosion. This situation demonstrates that non-state actors have a concern for the sustainability of resources that can be directly utilized. The government's involvement through incentive policies does not replace the role of the community but rather strengthens their capabilities and encourages broader participation. The increase in community participation, which reached 906%, and the 969% increase in the number of mangrove seedlings demonstrate that collaboration and cooperation between the state and non-state actors can yield more significant impacts compared to an approach that relies solely on the government. This underscores that mangrove governance can no longer be understood as a hierarchical process unilaterally controlled by the state, but has evolved into a process involving interaction, negotiation, and the division of roles among actors. In the context of multi-actor governance, local communities are regarded as strategic partners who make significant contributions

Stakeholder Relationships in Mangrove Governance

Mangrove ecosystem governance involves complex interactions among various stakeholders with differing capabilities, interests, authorities, and objectives. These stakeholders include government agencies, local communities, nongovernmental organizations (NGOs), the private sector, academic institutions, and international organizations that are directly or indirectly involved in the management of mangrove ecosystems. The involvement of these actors demonstrates that mangrove management is no longer viewed as the sole responsibility of the government, but has evolved into a collective process requiring multi-stakeholder interaction and cooperation to ensure the sustainability of the ecosystem.

Relationships among actors in mangrove governance are shaped by various factors, such as the distribution of authority, power dynamics, levels of participation, institutional capacity, and coordination mechanisms that develop within a governance system. Each actor possesses different resources and levels of influence in the decision-making process. For example, the government possesses formal legitimacy and regulatory authority in formulating policies and overseeing resource utilization. On the other hand, local communities also possess ecological knowledge based on experience and direct connection to the mangrove ecosystem, as their livelihoods often depend on coastal resources. NGOs play a role in advocacy, supporting communities, and facilitating collaboration, while academics provide scientific support through research and the development of evidence-based knowledge. Meanwhile, the private sector plays an increasingly important role through investment, corporate social responsibility programs, and various market-based conservation initiatives.

The Role of the Central Government

The central government holds a strategic position in mangrove governance, as it serves as the national policy-maker and sets the regulatory framework that forms the basis for management at the local level. The central government's role includes drafting regulations, setting national policy directions, providing funding, and integrating mangrove management into the national development agenda and global commitments, such as climate change mitigation, biodiversity conservation, and the achievement of the Sustainable Development Goals (SDGs). The central government also plays a coordinating role in aligning cross-sectoral policies to ensure that mangrove management is not fragmented.

However, in practice, the involvement of numerous ministries and agencies often leads to overlapping jurisdictions. Mangrove management may involve institutions responsible for forestry, marine affairs, the environment, development planning, and even agrarian affairs. This situation indicates that the central government serves not only as the primary decision-maker but also as a coordinator in fostering policy integration.

The Role of Local Governments

Local governments hold a crucial position as actors who interact directly with local conditions and the implementation of policies on the ground (Yuwono, 2025). In the context of decentralization, local governments have the authority to translate national policies into programs tailored to regional needs. This role includes formulating local policies, overseeing the use of coastal areas, implementing mangrove rehabilitation programs, empowering communities, and facilitating collaboration among local stakeholders.

The proximity of local governments to the community gives them a strategic role as a bridge between national interests and local needs. However, institutional capacity, resource constraints, and differing regional development priorities often affect the effectiveness of mangrove management policy implementation.

The Role of Local Communities

Local communities are the actors with the closest ties to mangrove ecosystems, as their livelihoods often depend on these resources. They utilize mangroves for various economic activities, such as fishing, aquaculture, the collection of non-timber forest products, and eco-tourism. Additionally, these communities possess local knowledge that has been developed through long-standing experience interacting with the environment. In the modern governance paradigm, communities are no longer positioned as objects of policy but as active actors involved in the management process (Farhan et al., 2025). Community engagement is considered crucial as it enhances policy legitimacy, strengthens social oversight, and fosters a sense of ownership toward conservation programs. Nevertheless, the position of communities within governance relationships is often still constrained by limited access to information, resources, and decision-making processes.

The Role of NGOs (Non-Governmental Organizations)

Non-governmental organizations (NGOs) play a vital role in strengthening mangrove governance through advocacy, community support, capacity building, and monitoring policy implementation. NGOs often act as facilitators bridging communication between communities and the government and helping to encourage the participation of underrepresented groups

The Role of Academics

Academics serve as producers of knowledge and providers of scientific support in mangrove governance. Through research, policy analysis, and innovation development, academics generate evidence-based knowledge that can be used to support decision-making processes. Academics also play a role in knowledge transfer through education, training, and community service activities.

In mangrove governance, the contribution of academics is crucial for bridging the gap between policy needs and scientific data. Scientific knowledge can be used to understand ecosystem conditions, evaluate policy effectiveness, and formulate more adaptive management strategies. However, a common challenge is the gap between research findings and policy implementation on the ground.

The Role of the Private Sector

The private sector is becoming an increasingly influential actor in mangrove governance, particularly through investment, economic activities, and corporate social responsibility (CSR) programs. Private sector involvement can provide financial and technological support for mangrove conservation and rehabilitation efforts. Additionally, companies can serve as partners in the development of environment-based economic initiatives and incentive schemes such as Payment for Ecosystem Services (PES).

However, the private sector can also exert pressure on mangrove ecosystems through industrial activities, infrastructure development, or land-use conversion. Therefore, private sector involvement requires adequate oversight and regulatory mechanisms to ensure that economic interests are balanced with environmental sustainability goals (Kismartini et al., 2025). The diversity of actors involved in mangrove governance makes this ecosystem a complex arena of social and political interaction. The relationships formed are not always equal, as they are influenced by differences in capacity, access to resources, institutional positions, and the power to influence policy processes. Several case studies indicate that certain actors may exert a more dominant influence compared to others, thereby creating power imbalances. This imbalance affects the processes of participation, representation of interests, and the direction of decision-making in mangrove management. In addition to being influenced by power dynamics, relationships among actors are also shaped by the level of collaboration established among stakeholders. Effective collaboration requires trust, effective and intensive communication, clear role division, and a shared understanding of common goals. The governance framework plays a role in shaping the patterns of relationships among actors. Institutional structures, regulatory systems, and policies

Dynamics of Key Actors

1. Power Imbalances

In mangrove governance, state actors at the national level often play an overly dominant role in decision-making processes by leveraging various instruments of power. This situation is reflected in mangrove ecosystem governance practices in the Sundarbans region of Bangladesh, where the state exerts strong and dominant influence in determining policy directions for resource management (Khan et al., 2024). Although non-governmental organizations (NGOs) have functioned as bridging actors—facilitating interaction, communication, and coordination among various actors—the relationships that develop between these actors do not occur on an equal footing. The involvement of NGOs in governance structures is often still constrained by limited influence in the formulation or decision-making processes regarding strategic policies. The existence of power imbalances—where state actors at the national level hold the greatest authority in determining policy direction, while NGOs function as intermediaries, and local stakeholders or communities still have relatively little influence over strategic decisions—demonstrates that the transition from state-centered governance to multi-actor governance does not automatically eliminate state dominance.

This situation indicates that efforts to involve multiple actors in environmental governance do not always result in inclusive and balanced relationships. In some contexts, unequal distribution of power can actually reinforce the dominance of certain actors and create imbalances in the decision-making process. Similar findings are evident in the context of environmental governance in China, where a governance system that tends toward authoritarianism actually exacerbates power imbalances among actors. Policies formulated by the central government alone often fail to fully accommodate the interests of local communities, particularly when such policies are perceived as threatening their livelihoods (Wang et al., 2024). Consequently, this has led local communities to mount various forms of resistance against the implemented policies. The power imbalance among these coalitions of actors can ultimately trigger conflict and intensify social resistance. This demonstrates that the primary challenge in transforming environmental governance is not merely expanding the involvement of non-state actors, but also creating a governance mechanism capable of balancing the distribution of power, strengthening participation, and ensuring that every actor has sufficient space to contribute to the decision-making process.

2. Collaborative Governance

Collaborative governance is viewed as a strategic approach to achieving sustainable mangrove ecosystem management. This approach has gained traction as a response to the complexity of environmental issues that cannot be resolved through sector-specific mechanisms. Collaborative governance emphasizes the importance of involving various stakeholders in the process of planning, decision-making, implementation, and policy evaluation. This approach is based on the assumption that each stakeholder possesses different capabilities, resources, knowledge, and interests; thus, the involvement of multiple parties can create a management mechanism that is far more adaptive and sustainable.

In Indonesia, the implementation of collaborative governance has demonstrated the important role played by various stakeholders in supporting the sustainability of mangrove ecosystem management. Research conducted in Tangerang Regency, Banten, shows that the role of stakeholders is one of the key factors determining the success of mangrove management (Mahardika et al., 2022). The findings of this study indicate that the central government and local governments are key actors in the development and implementation of mangrove governance. Both actors have the capacity to build partnerships with various other stakeholders, such as local communities, the private sector, universities, nongovernmental organizations, research institutions, and educational institutions.

The study conducted in Tangerang Regency shows that strategies for managing mangroves can be implemented by strengthening the roles of each stakeholder, specifically by encouraging the central and local governments to implement programs that support the achievement of strategic objectives, particularly through a collaborative management approach. A collaborative approach is considered capable of producing more effective management because it involves various parties with a high level of collaboration. Communities as resource users, NGOs, universities, local governments, and research institutions demonstrate strong collaborative capacity in supporting the implementation of strategies for managing mangrove ecosystems. Findings from this study indicate that the success of mangrove management does not solely depend on the state's ability to regulate and control resources, but is also influenced by the quality of relationships among actors within the established governance system. The high level of collaboration indicates the presence of knowledge exchange, shared responsibility, and strengthened coordination among stakeholders. This reflects a shift from state-centered governance toward multi-actor governance. Within this paradigm, the sustainability of mangrove management is increasingly determined by the ability of various actors to build synergies, strengthen collaborative networks, and create inclusive and adaptive governance mechanisms.

3. Community-Based Management

Local communities, as the primary actors in the resource management process, must be involved through local communities as the centers of decision-making and management implementation (Warsono et al., 2025). Unlike state-centered approaches, community-based management provides a platform for communities to manage resources based on their needs, experiences, local knowledge, and social mechanisms that have developed at the community level. Research findings on community-based mangrove management in four coastal areas of Central Java reinforce the view that local communities, as the primary actors in the resource management process, must be actively involved in environmental management decision-making. This research demonstrates that the success of mangrove management also heavily depends on community participation. The village of Surodadi has shown better outcomes in resource utilization and improvements in community livelihoods due to more effective local community involvement (Damastuti & de Groot, 2017).

The significant relevance of a community-based approach can also be seen in a study on the governance of sustainable tourism development in Glawan Village, Semarang Regency (Rahman et al., 2025). Although the subject of this study does not focus on mangrove ecosystems, this research offers conceptual relevance to discussions on collaborative mangrove governance, particularly in examining how local leadership, community participation, and institutional innovation can serve as key factors in ensuring environmental sustainability. In the context of Glawan Village, the village head serves as a policy entrepreneur capable of integrating tourism initiatives into village planning documents, mobilizing community groups, and fostering coordination with local government regarding spatial management. This aligns with the principles of collaborative mangrove governance, which require synergy among government, local communities, environmental groups, and other stakeholders to preserve the ecological functions of coastal areas (Inayah Hidayati et al., n.d.).

This interconnection is also evident in the management of land-use changes. In Glawan Village, the conversion of agricultural land into homestays and tourism facilities is not carried out freely, but is regulated through participatory zoning and village regulations to protect agricultural and ecological areas. This is relevant to mangrove governance, as one of the main challenges to the sustainability of mangrove ecosystems is the pressure of land conversion for aquaculture ponds, settlements, industry, and coastal infrastructure. Thus, the experience from Glawan Village demonstrates that local regulations, village deliberations, and community engagement can serve as important tools to control spatial changes so as not to undermine ecological functions.

In addition to the case in Indonesia, research conducted in Ecuador also shows that active community involvement can foster the collective capacity to resolve conflicts independently, without relying entirely on intervention by external actors (Chávez-Páez & Beitl, 2025). This demonstrates that local communities have the capacity to establish rules, manage diverse interests, and create governance mechanisms that support the sustainability of natural resources. Communities are no longer viewed merely as resource users, but also as decision-makers and key actors in environmental governance.

In practice, the involvement of local communities in mangrove ecosystem management does not always proceed without obstacles. Differences in the level of investment among actors—whether in terms of resources, time, knowledge, or institutional capacity—can create an imbalance of roles and influence in the management process. Additionally, conflicting interests and political fragmentation within the community can trigger conflicts that may hinder the effectiveness of governance. From a multi-actor governance perspective, it is evident that local community involvement can strengthen the legitimacy, participation, and sustainability of mangrove management. However, the success of a community-based approach can be influenced by the quality of social relationships, institutional capacity, a balanced distribution of roles, and the ability of actors to establish mechanisms for collaboration and conflict resolution (Awaluddin et al., 2026). Thus, community-based mangrove ecosystem management not only emphasizes the importance of community participation but also highlights the need for governance that can effectively manage conflicting interests and disparities among stakeholders.

4. Gender and Inclusivity

The dimensions of gender and inclusivity are crucial aspects in discussions of mangrove governance because they influence patterns of participation, the distribution of roles, and access to decision-making processes (Pawestri et al., 2025). In various contexts of mangrove management in Indonesia, women have made significant contributions to conservation activities, the development of coastal resource-based economies, and community capacity building. However, in practice, women's involvement in decision-making is often still limited. Research on managing mangrove ecotourism in Siak, Riau, indicates that Malay women play a vital role as environmental stewards, economic actors, and custodians of cultural values in the management of mangrove ecosystems (Nisaa et al., 2025). This role is manifested through active involvement in mangrove conservation activities, the development of ecotourism enterprises, and environmental education. Research findings indicate that women not only participate as implementers of activities at the community level but also possess the capacity to fulfill leadership and ecological management functions. This study also shows that women in Siak possess adaptive capabilities in balancing domestic responsibilities with public roles and environmental management.

Although women's involvement in management practices is active, their representation in formal decision-making processes remains relatively limited. This indicates a gap between the level of participation and the level of influence within governance structures. These findings reinforce the argument that environmental governance issues are not merely about the involvement of actors, but also concern the distribution of access, power relations, and equal opportunities in decision-making processes. When compared to several cases in Vietnam and Bangladesh, mangrove governance in Siak demonstrates more inclusive gender dynamics and a stronger level of women's integration in ecological and economic decision-making. This situation indicates that more inclusive governance can create broader spaces for participation for groups that were previously underrepresented.

Challenges in Collaborative Mangrove Management

Mangroves are a coastal ecosystem that plays a role in maintaining environmental balance, supporting the economic sustainability of coastal communities, and strengthening regional resilience to climate change. Additionally, mangroves serve as natural coastal barriers against erosion, storms, and tidal waves, while also functioning as vital habitats for diverse marine species. Mangroves hold significant economic and social value, contributing to the ecotourism sector, fisheries, and local livelihoods, as well as acting as major carbon sinks and reservoirs. Nevertheless, the sustainability of mangrove ecosystems faces various serious threats due to land-use conversion, overexploitation of resources, coastal development, environmental pollution, and weak integrated resource management. This situation indicates that mangrove management cannot be carried out in isolation by a single actor or institution, but rather requires a collaborative governance approach involving various stakeholders. Collaborative governance of mangrove management is essential because coastal issues are not limited to ecological aspects alone but also encompass social, economic, cultural, and political dimensions. In practice, mangrove management involves various stakeholders—including the central and local governments, local communities, nongovernmental organizations, academics, the private sector, and indigenous communities—each of whom has distinct interests and roles regarding the utilization of mangrove resources. A collaborative approach is seen as capable of strengthening cross-sectoral coordination, increasing community participation, and creating a balance between conservation goals and the economic development needs of coastal communities. Through collaboration, each actor can share responsibilities, resources, knowledge, and capabilities in maintaining the sustainability of mangrove ecosystems.

The implementation of collaborative governance in mangrove ecosystem management still faces a range of complex challenges. Conflicting interests among stakeholders often lead to conflicts in the decision-making process, particularly when short-term economic interests clash with long-term environmental conservation goals. Furthermore, weak institutional coordination, policy overlaps, power imbalances, limited capacity among local communities, and the lack of integration of local knowledge into formal policies are major obstacles to achieving effective collaborative governance.

In many studies, collaboration is merely administrative and a formality, without a fair distribution of roles or benefits among all stakeholders; as a result, various mangrove rehabilitation and conservation programs often do not function optimally and are difficult to sustain in the long term. The phenomena of conflicting interests and policy fragmentation in mangrove governance increasingly highlight the gap between policies designed in theory and their implementation on the ground. Mangrove ecosystem governance fundamentally involves various sectors and institutions with differing interests, priorities, and agendas; thus, mangrove management is not the sole responsibility of a single agency but encompasses the central government, local governments, and local communities. The complexity of these actors often results in formulated policies failing to operate in an integrated manner; instead, they tend to become fragmented due to weak coordination and differing development orientations across sectors. In many cases, environmental conservation interests must contend with demands for economic development, industrial expansion, or coastal infrastructure development, leading to suboptimal implementation of mangrove conservation policies.

Research in Panama shows that the existence of various wetland and mangrove ecosystem protection policies does not automatically guarantee successful conservation if the institutional structure is still less than optimal and the coordination between actors is not yet effective (Chamberland-Fontaine et al., 2022). This reflects a gap between formal written policies and their implementation in practice. Therefore, successful mangrove governance depends not only on the number of regulations established but also requires policy integration, alignment of goals between actors, strong institutional capacity, and consistent, sustainable policy implementation. Without effective cross-sectoral collaboration, mangrove ecosystem management will tend to be fragmented, overlapping, and less able to ensure ecosystem sustainability or the well-being of coastal communities.

Another challenge faced in collaborative mangrove management is the weak enforcement of environmental laws and policies. In Nigeria, there is an environmental policy, the Environmental Impact Assessment Act, but this regulation lacks specific technical provisions and strong enforcement mechanisms. As a result, despite the existence of environmental protection policies, mangrove ecosystems continue to be damaged due to oil spills, deforestation, and environmental pollution. This situation shows that formal policies alone are insufficient to guarantee the sustainability of mangrove ecosystems if they are not supported by an effective monitoring system, strict sanctions, and adequate institutional capacity. Weak policy implementation causes existing regulations to tend to be merely administrative documents without being able to prevent real environmental damage.

Furthermore, the challenges of collaborative mangrove governance also relate to the unclear division of roles and responsibilities between levels of government. A case in Japan demonstrates the ambiguity in the distribution of authority between institutions, leading to inconsistent policy implementation (Chiba T, 2026). In an environmental governance system, every level of government, nationally, regionally, and locally, should have clear duties and authorities in implementing environmental policies. However, the division of responsibilities is not clearly defined, resulting in policy implementation often experiencing confusion, overlapping authority, and shifting responsibilities between institutions. As a result, local governments must adopt ad hoc or reactive solutions to emerging problems rather than implementing more strategic and sustainable management. This results in ineffective coastal ecosystem management due to weak coordination and a clear policy direction. Therefore, the success of mangrove governance still depends heavily on a clear institutional structure, a clear distribution of authority, and strong cross-sectoral coordination for effective and sustainable implementation of conservation policies. Numerous cases at both the local and global levels demonstrate that collaborative governance of mangrove ecosystems encompasses not only ecological aspects but also institutional, political, coordination between actors, law enforcement, and the distribution of authority. Therefore, strengthening collaborative governance that is more integrative, participatory, and adaptive requires all stakeholders to play a vital role in the coastal resource management process (Ardianto et al., 2025).

The challenges identified in collaborative mangrove management indicate that future governance arrangements need to move beyond conventional forms of stakeholder participation toward more innovative mechanisms that connect ecosystem conservation, community benefits, scientific knowledge, and sustainable financing. Existing co-management arrangements provide an important institutional foundation because they already involve governments, local communities, non-governmental organizations, academics, and private-sector actors. However, these arrangements can be further strengthened by integrating blue carbon initiatives, Payment for Ecosystem Services, and green and blue financing instruments into collaborative mangrove governance.

Blue carbon initiatives can serve as a common environmental agenda that connects the interest and resources of multiple actors within existing co-management arrangements. Central and local governments should establish enabling regulations, integrate blue carbon objectives into coastal development and mangrove management policies, and provide institutional mechanism for cross-sectoral coordination. Local communities should be positioned as co-managers responsible for mangrove restioration, ecosystem protection, and community-based monitoring rather than merely as beneficiaries of conservation programs Their local ecological knowledge can support the identification of ecosystem changes and strengthen the long term sustainability of restoration inititatives. NGOs can act as bridging organization by facilitating communication between communities and government agencies, strengthening community capacity, and ensuring the marginalized groups are represented in decision making processes. Academic and research institutions should provide scientific support through carbon stock assessment, ecosystem mapping, monitoring, reporting, and verification, as well as evaluating the ecological and social impacts of blue carbon program. Meanwhile, private sectoc actors can contribute financial resources, technogoly, and investment to support restoration and conservation activities under clear environmental accountability and monitoring mechanisms. Payment for Ecosystem Services can futher strengthen co-management by creating a direct connection between conservation responsibilities and economic benefits for local communities. Within this mechanism, governments should develop regulatory frameworks and transparent benefit-sharing arrangements that recognize communities as providers of ecosystem services. Local communities can undertake conservation, restoration, and monitoring activities and receive incentives based on agreen ecological performance indicators. NGOs can facilitate negotiations, strengthen community organizations, and monitor the fairness of benefit distribution. Academics can contribute to the development of measurable indicators for ecosystem services and evaluate the effectiveness of incentive schemes. Private-sector actors, particularly those benefiting from coastal ecosystem services or seeking to support environmental commitments, can participate as financing partners. Such arrangements can help address the persistent imbalance between conservation objectives and livelihood needs by ensuring that communities receive tangible benefits from maintaining mangrove ecosystems.

Green and blue bonds offer another potential mechanism for addressing the financial limitations of long-term mangrove conservation. Governments can facilitate the integration of mangrove restoration and coastal resilience programs into eligible green or blue financing portfolios while establishing environmental and social safeguards. Private-sector and financial actos can mobilize investment for large-scale restoration, sustainable coastal enterprises, and ecosystem monitoring technologies. Academics should support the development of science-based project indicators and independent impact assessments, while NGOs can strengthen social accountability and ensure that financing mechanism do not marginalize local communities. Local communities should be directly involved in determining local priorities and implementing funded activities to ensure that environmental financing responds to both ecosystem needs and community livelihoods.

The integration of these innovations requires an adaptive co-management platform that clearly defines actor responsibilities, decision-making mechanism, benefit-sharing arrangements, and accountability systems. Rather that creating entirely new governance structures, blue carbon inititatives, PES, and green and blue financing should be embedded within existing collaborative arrangements. Governments can provide regulatory legitimacy and coordination; communities contribute local knowledge and direct ecosystem management; NGOs facilitate participation and social accountability; academics provide scientific evidence and monitoring systems; and private-sector actos mobilize financial and technological resources. Through this role-based integration, collaborative mangrove governance can evolve from administrative stakeholder participation toward an adaptive governance system that links ecological sustainability, community welfare, and innovative environmental financing.

IV. CONCLUSION

Mangrove ecosystem management has undergone a paradigm shift from state-centered governance toward multi-stakeholder governance, which positions various stakeholders as integral parts of the decision-making process and the management of coastal resources. This shift is driven by the complexity of mangrove functions, which are not only ecologically valuable but also possess social and economic dimensions closely linked to the well-being of coastal communities. This study demonstrates that collaborative governance is crucial in supporting mangrove sustainability through the involvement of the government, local communities, non-governmental organizations, academics, and the private sector. In practice, collaboration in managing mangrove ecosystems still faces various challenges, such as power imbalances among actors, policy fragmentation, weak law enforcement, and unclear delineation of authority. Therefore, strengthening mangrove governance requires a more integrated, inclusive, and adaptive approach through enhanced cross-sectoral coordination, institutional capacity building, and expanded participation for all stakeholders. Thus, collaborative mangrove ecosystem governance serves as a vital foundation for achieving mangrove sustainability and supporting sustainable development in coastal areas.

REFERENCES

Ambariyanto, A., Utama, Y. J., Sugianto, D. N., Ariyanti, D., & Handayani, E. P. (2023). Mangrove Conservation and Biodiversity Protection Strategies in Universitas Diponegoro to Achieve Net Zero Emission. *Journal of Sustainability Perspectives, 3*(2), 202–208. DOI: 10.14710/jsp.2023.20482

Ardianto, H. T., & Martini, R., Wijayanto, & Prameswari, Y. P. (2025). Community-Based Waste Bank: Collaborative Governance in Waste Management in Two Regions. *IOP Conference Series. Earth and Environmental Science, 012016*(1), •••. DOI: 10.1088/1755-1315/1537/1/012016

Awaluddin, M., Suwitri, S., & Santoso, R. S. (2026). Firebreak policy: Supporting local communities in forest fire prevention . *Multidisciplinary Science Journal, 8*(5).

Chamberland-Fontaine, S., Heckadon-Moreno, S., & Hickey, G. M. (2022). Tangled Roots and Murky Waters: Piecing Together Panama's Mangrove Policy Puzzle. *Frontiers in Forests and Global Change, 5*, 818722. Advance online publication. DOI: 10.3389/ffgc.2022.818722

Chávez-Páez, W., & Beitl, C. M. (2025). Who holds the power in the absence of the state? Polycentric governance and community justice in Ecuador's mangrove management. *Environmental Development, 56*, 101263. Advance online publication. DOI: 10.1016/j.envdev.2025.101263

Chiba T. (2026). *What hinders the implementation of coastal litter policies? Multilevel governance and structural gaps in the Osaka Bay area, Japan.*

Damastuti, E., & de Groot, R. (2017). Effectiveness of community-based mangrove management for sustainable resource use and livelihood support: A case study of four villages in Central Java, Indonesia. *Journal of Environmental Management, 203*, 510–521. DOI: 10.1016/j.jenvman.2017.07.025 PMID: 28841518

Das, S., Dutta, S., Roy, J., Roy Choudhury, M., Sen, P., Chanda, A., Dutta, H., Nandi, S., & Ghosh, T. (2026). Mangroves in the Anthropocene: A global synthesis of carbon storage, biodiversity, and coastal resilience under climate and anthropogenic stressors. *The Science of the Total Environment, 1015*, 181405. Advance online publication. DOI: 10.1016/j.scitotenv.2026.181405 PMID: 41558381

Faperi, S., & Hendrarto, I. B. (2015). Management Strategies of Mangrove Degradation in Coastal Areas of Brebes Regency, Central Java, Indonesia. *Journal of Coastal Zone Management, 18*(2). Advance online publication. DOI: 10.4172/2473-3350.1000401

Farhan, E. K., Setiyono, B., Yuwono, T., & Dewi, R. S. (2025). Governance perspective: Foreign investment in decentralization in Indonesia. *Frontiers in Political Science*, *7*, 1562656. Advance online publication. DOI: 10.3389/fpos.2025.1562656

Hidayati, I., Prabawani, B., Rahmadhony, A., Hayati, A. N., & Boddy, J. (n.d.). *The Critical Role of Local Leaders in Managing the Impacts of Climate Change in Demak.* Springer Climate. DOI: 10.1007/978-3-032-15334-0_8

Insusanty, E., Mulyadi, A., Efriyeldi, E., & Warningsih, T. (2025). Valuing mangrove ecosystem services in Dumai, Indonesia, for conservation and policy. *Asian Journal of Forestry*, *9*(2), 428–439. DOI: 10.13057/asianjfor/r090223

Khan, M. F. A., Rahman, M. S., Maryudi, A., Schusser, C., & Giessen, L. (2024). Revealing the Multilevel Actors Power Network in Mangrove Forest Governance-Insights from the Sundarbans, Bangladesh. *Forest and Society*, *8*(2), 372–401. DOI: 10.24259/fs.v8i2.32924

Kismartini, Saputra, Y. A., & Hanani, R. (2025). Sustainable Tourism in Crisis? Environmental Challenges and Economic Potentials in Rembang's Coastal Area. *E3S Web of Conferences, 650*, 02014. DOI: 10.1051/e3sconf/202565002014

Lituhayu, D., Santoso, R. S., Imawan, S. A., Bellastuti, D. B., & Saputra, Y. A. (2026). Sustainable Governance and Institutional Reform of Waste Management in Semarang City. *IOP Conference Series. Earth and Environmental Science*, *1610*(1), 012001. DOI: 10.1088/1755-1315/1610/1/012001

Mahardika, S. M. A. H., Yulianda, F., Adrianto, L., & Sulistiono, S. (2022). Prospective analysis of the role of actors in governing mangrove ecosystem area in Tangerang District, Indonesia. *Biodiversitas (Surakarta)*, *23*(9), 4940–4947. DOI: 10.13057/biodiv/d230964

Marlianingrum, P. R., Kusumastanto, T., Adrianto, L., & Fahrudin, A. (2019). Economic analysis of management option for sustainable mangrove ecosystem in Tangerang District, Banten Province, Indonesia. *IOP Conference Series. Earth and Environmental Science*, *241*(1), 012026. Advance online publication. DOI: 10.1088/1755-1315/241/1/012026

Nisaa, F., Rismawati, R., Azwar, W., & Zabua, B. H. (2025). Women's Environmental Governance Roles in Mangrove Ecotourism. *E3s Web of Conferences, 671*. DOI: 10.1051/e3sconf/202567105009

Pawestri, R. R., Azahra, J. M., & Alfirdaus, L. K. (2025). Gender Equality and Environmental Justice: Assessing Prospects and Challenge in Indonesia. *IOP Conference Series. Earth and Environmental Science*, *1537*(1), 012058. DOI: 10.1088/1755-1315/1537/1/012058

Hartuti Purnaweni, Sudharto Prawata Hadi, Tri Retnaningsih Soeprobowati, Ali Roziqin, & Nilam Adini Rakhma. (2026). What are the Main Concerns for Global Mangrove Management Re-search? Systematic Literature Review and Bibliometric Analysis. *Research in Ecology, 8*(2).

Purnaweni, H., Saputra, J., Roziqin, A., Kismartini, K., Djumiarti, T., & Seitz, T. (2022). Oil Spill Governance: Evidence from Bintan Island, Indonesia. *Sustainability (Basel)*, *14*(3), 1603. Advance online publication. DOI: 10.3390/su14031603

Rahman, A. Z., Pradana, A. E., & Rakhma, N. A. (2025). Sustainable Tourism Development in Glawan Village: Balancing Infrastructure Growth and Environment through Local Governance. *IOP Conference Series. Earth and Environmental Science*, *1537*(1), 012001. Advance online publication. DOI: 10.1088/1755-1315/1537/1/012001

Sajriawati, Amir, A., Hiskya, H. J., Uspayanti, R., & Hidayat, S. H. (2021). Collaboration of Government Policies and the Marind Tribe Communities on Mangrove Resources Management in the Payum Coastal Area of Merauke Regency in Papua. *Review of International Geographical Education Online*, *11*(5), 280–286. DOI: 10.48047/rigeo.11/5/28

Singh, N., Yadav, N., & Srivastava, S. (2025). Nutrient Cycles, Protection and Management of Mangroves Ecosystem: A Sustainable Approach. In *Green Energy and Technology: Part F492*. DOI: 10.1007/978-981-96-3993-9_19

Wang, W., Zhai, D., Li, X., Fang, H., & Yang, Y. (2024). Conflicts in mangrove protected areas through the actor-centred power framework - Insights from China. *Forest Policy and Economics*, *158*, 103122. Advance online publication. DOI: 10.1016/j.forpol.2023.103122

Warsono, H., Muhamad, J. A., Fitrianto, M. R., Yusuf, I. M., Astuti, R. S., & Priyadi, B. P. (2025). Analysis of Village Development Policy and Village Community Empowerment in Indonesia. *IOP Conference Series. Earth and Environmental Science*, *1537*(1), 012032. DOI: 10.1088/1755-1315/1537/1/012032

Wiryawan, B. A., Agustin, E. S. A. S., Manar, D. G., Otchia, C., & Phommachanh, N. (2026). Industrial park policy and regional industrialization in Indonesia. *Applied Economics Letters*, •••, 1–7. DOI: 10.1080/13504851.2026.2658243

Wiryawan, B. A., Kasyfilham, F., Pawestri, R. R., Herawati, N. R., Yuwono, T., & Annef, A. B. (2025). Inclusive Environmental Governance in Indonesia: Empirical Analysis Using Geospatial Forest Fire Data. *IOP Conference Series. Earth and Environmental Science*, *1537*(1), 012015. DOI: 10.1088/1755-1315/1537/1/012015

Yuwono, T. (2025). Examining the Limitations of Translatability in Stunting Policy: An Actor-Network Theory (ANT) Perspective by Bruno Latour. *The International Journal of Interdisciplinary Civic and Political Studies*, *21*(1), 17–43. DOI: 10.18848/2327-0071/CGP/v21i01/17-43

Chapter 9
Strengthening Mangroves Management in Local Areas:
A Community-Based Mangrove Management

Nabil Fariz Nashrullah
http://orcid.org/0009-0008-4078-1856
Diponegoro University, Indonesia

Nilam Adini Rakhma
http://orcid.org/0009-0003-9436-0030
Diponegoro University, Indonesia

ABSTRACT

Mangrove ecosystems are crucial for maintaining the equilibrium of coastal settings. This study examines the enhancement of mangrove management through the implementation of a Community-Based Mangrove Management (CBMM) approach. This approach empowers local communities to oversee the design, execution, and evaluation of the activity. The research employs qualitative approach which data collected through in-depth interviews, field trips, and document analysis from official government report. The locus study is Rembang Regency, Indonesia, particularly in several areas which have mangrove areas. The findings indicate that effective mangrove management is enhanced and sustained when communities are actively engaged, local organizations are strengthened, and collaboration occurs across communities, government, and other entities.The study indicates that community-based mangrove management effectively safeguards the environment, supports local populations, and ensures the long-term resilience of coastal regions.

DOI: 10.4018/979-8-3373-9320-9.ch009

INTRODUCTION

The dynamic ecological interface where terrestrial landscapes meet maritime environments harbors some of the most biologically productive ecosystems on Earth. Among these coastal boundary zones, mangrove forests stand out as critical ecological structures that provide a wide array of vital ecosystem services, maintaining both environmental equilibrium and the socio-economic security of human settlements (Wahyudi et al., 2025). Composed of specialized halophytic woody plants that thrive in the harsh, fluctuating conditions of intertidal zones across tropical and subtropical regions, these unique coastal wetlands serve as indispensable buffers for coastal area. Structurally, the complex, interlaced aerial root systems of various mangrove species, such as Rhizophora and Avicennia, serve as natural wave attenuators. They break the destructive kinetic energy of storm surges, seasonal monsoonal swells, and oceanic tsunamis, while effectively trapping sediments to stabilize shorelines against catastrophic coastal erosion.

Beyond their immediate physical protective functions, mangrove ecosystems are vital biodiversity hotspots. They provide critical breeding grounds, nursery sanctuaries, and feeding habitats for a vast matrix of marine life, including commercially important fish, crustaceans, and mollusks, thereby directly sustaining regional marine food webs. On a global scale, the significance of these coastal wetlands has attracted intense international attention due to their extraordinary capacity to mitigate climate change through carbon sequestration (Hidayati et al., 2026). Often classified under the paradigm of "blue carbon" repositories, the waterlogged, anaerobic soils beneath healthy mangrove canopies are capable of trapping and locking away organic carbon at rate boundaries that significantly outpace terrestrial tropical rainforests, making their preservation a core pillar of international climate resilience frameworks. Transparent accountability mechanisms are equally important to ensure that environmental governance remains responsive to community interests and public oversight (Husna et al., 2023).

Figure 1. Spatial cartographic distribution of mangrove sub-ecosystems in Rembang Regency (Author's illustration based on BPS Rembang (2025), KLHK Indonesia, and Google Earth satellite imagery, supported by (Safitri et al., 2017) and (Rahman Halim et al., 2018))

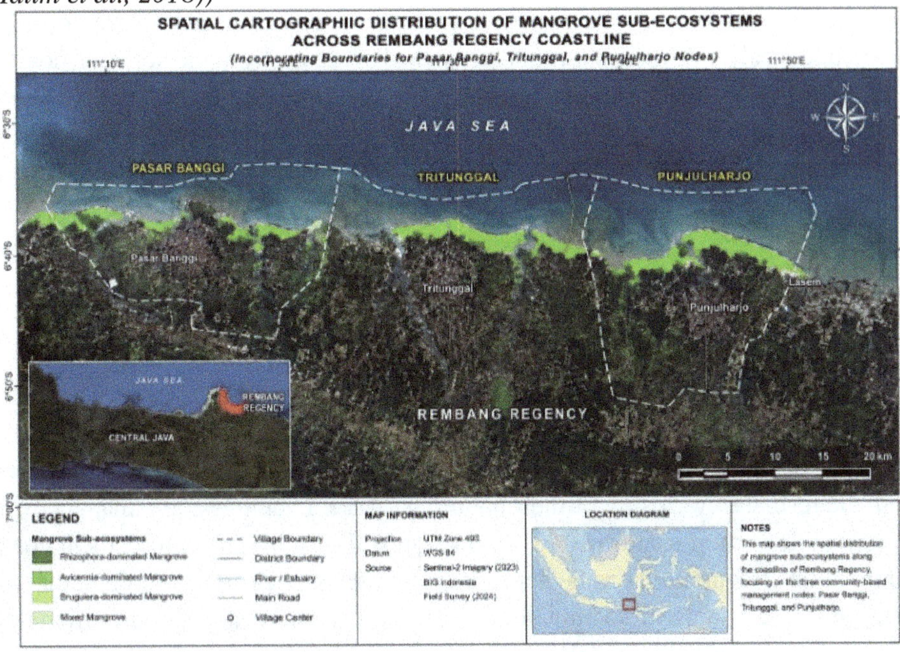

According to empirical data compiled by the Central Bureau of Statistics (BPS) of Rembang Regency, the region spans a strategic coastline of approximately 63.5 kilometers, making it highly vulnerable to climate-driven oceanographic shifts. Prior to systematic community intervention, extensive monitoring indicated that severe coastal abrasion had compromised over 24% of this total shoreline infrastructure, resulting in a net loss of approximately 152 hectares of productive brackish-water aquaculture land and localized residential boundaries. Ongoing environmental deterioration within the Rembang coastal zone has increasingly undermined local economic resilience and the sustainability of community-based development activities (Kismartini et al., 2025). This severe environmental displacement highlights why decentralized coastal resource stabilization has become an immediate ecological necessity for the municipality.

As known that Indonesian archipelago, the structural integrity of mangrove forests is directly intertwined with daily economic survival and community resilience. These ecosystems underpin small-scale artisanal fisheries, provide non-timber forest derivatives, offer raw materials for localized livelihoods, and increasingly serve as focal points for sustainable ecotourism developments that inject vital alternative income into rural coastal households (Prabawani et al., 2025). However, despite the clear ecological and socio-economic importance of these habitats, global mangrove coverage has faced unprecedented degradation over the last half-century. Anthropogenic pressures manifesting as rapid aquaculture expansion, widespread clearing for commercial shrimp ponds, aggressive coastal infrastructure developments, and localized overexploitation Empirical evidence from the northern coastal region of Central Java indicates that declining mangrove conditions, coupled with climate-related stressors, contribute significantly to escalating coastal hazards and environmental risks (Widjonarko et al., 2025), have severely fragmented these fragile habitats.

Historically, the governance of marine and coastal resources within the Indonesian state was defined by highly centralized, rigid, top-down administrative frameworks. Under the bureaucratic structures of previous decades, coastal conservation agendas, land-use zoning laws, and forestry management blueprints were drafted, dictated, and enforced exclusively by central government ministries located far from the actual coastal communities. This centralized approach treated coastal inhabitants merely as passive subjects or economic extractors, completely alienating local populations from the formal stewardship of their natural heritage. Lacking secure management rights, viable economic alternatives, or institutional trust in state enforcement, local communities frequently became active participants in environmental degradation, clearing native mangroves to build traditional brackish-water aquaculture structures or expansion zones for solar salt production. Centralized enforcement mechanisms eventually collapsed under logistically unfeasible monitoring demands across thousands of kilometers of coastline, culminating in widespread open-access exploitation and severe ecosystem degradation (Abdillah et al., 2025). Consequently, modern attempts to establish local mangrove management models frequently encounter complex structural barriers and institutional friction.

A primary challenge in contemporary coastal administration is the systemic deficit in consistent, highly capable local leadership at the grassroots and municipal levels. This leadership gap often manifests as a lack of political will, insufficient technical expertise regarding ecological restoration, and short-term economic prioritization among village administrators who favor immediate financial gains from aquaculture conversions over long-term environmental sustainability (Rahman et al., 2025). This governance deficit is further aggravated by fragmented, overlapping institutional jurisdictions, where regional environmental agencies, marine affairs offices, and forestry departments operate within non-cohesive policy silos, creating regulatory gridlocks that stall conservation initiatives (Afonso et al., 2026).

The relentless advance of coastal erosion and climate-driven sea-level rise actively undermines the natural regenerative capacity of degraded coastlines. This environmental decay makes active restoration attempts technically challenging and financially prohibitive for local village budgets (Hanani et al., 2021). Most critically, previous top-down governance failures have left a legacy of inadequate community involvement in formal planning and decision-making structures. When local resource users are excluded from the early design, continuous execution, and transparent evaluation phases of environmental policies, the public strategies lose their grounding in localized ecological realities, giving rise to persistent resource-use conflicts and a profound lack of local compliance with environmental regulations (Akram et al., 2023).

To counter these systemic failures, contemporary public administration and environmental policy models have shifted toward decentralized paradigms, specifically through the development of the Community-Based Mangrove Management (CBMM) framework. Rooted in the broader theoretical tenets of Common-Pool Resource (CPR) governance, CBMM redefines coastal communities not as destructive threats to environmental integrity, but as the primary, most effective stewards of their local ecological resources. This decentralized approach works by structurally empowering local citizens to actively direct the complete lifecycle of environmental governance, ensuring that conservation laws, zoning restrictions, and replanting schemes are designed and maintained from the bottom up. By centering management rights, enforcement responsibilities, and economic benefits directly within formalized grassroots institutions, CBMM seeks to mend the historical fracture between environmental preservation and community development (Buncag, 2022).

This study investigates the strategic enhancement of mangrove management through the operational lens of the CBMM model within Rembang Regency, located on the vulnerable northern coast of Central Java Province, Indonesia. Rembang Regency presents a compelling empirical environment for this analysis; it possesses an extensive, highly strategic coastline that has historically suffered from extreme ecological pressures, catastrophic abrasion, and massive land conversion for salt pans and aquaculture. In recent years, however, specific coastal nodes within this regency most notably the administrative villages of Pasar Banggi, Tritunggal, and Punjulharjo have emerged as prominent regional benchmarks for community-led environmental restoration, local institutional building, and mangrove-based ecotourism (Purnaweni et al., 2025; Roziqin, 2018).

By evaluating the diverse socio-ecological realities, regulatory dynamics, and multi-stakeholder interactions across these three distinct villages, this paper aims to provide a nuanced evaluation of how localized governance structures can be strengthened. Ultimately, this research seeks to offer meaningful insights and scalable strategies to the global discourse on sustainable coastal zone administration, decentralized environmental policy, and the structural empowerment of frontline communities facing the realities of global climate change. Within Rembang Regency, the historical reliance on nearshore capture fisheries has grown increasingly untenable as localized habitat destruction triggers a steep decline in maximum sustainable yields. This economic contraction forces a critical reassessment of community-led stewardship, transforming environmental conservation from an altruistic luxury into an absolute prerequisite for household survival. By exploring the institutional micro-dynamics that govern how village assemblies allocate common-pool resources, this study uncovers the latent administrative friction that frequently subverts decentralized mandates. Understanding these internal community power structures is vital for designing public policies that possess the necessary social legitimacy to withstand commercial market pressures.

To achieve a comprehensive understanding of these decentralized dynamics, it is necessary to contextualize how regional environmental shifts over the past two decades have altered community perceptions of risk and economic opportunity. In the specific context of Rembang Regency, the rapid expansion of industrial aquaculture and coastal manufacturing zones has fundamentally rewritten the spatial politics of the intertidal fringe. Historically, community participation was viewed by municipal authorities as merely a secondary compliance mechanism a procedural checkbox to satisfy regional development protocols rather than an active administrative asset. This historical oversight resulted in widespread policy failure, as community members, alienated from decision-making, frequently dismantled state-funded breakwaters or cleared protected forestry zones to secure short-term agricultural survival.

2. LITERATURE REVIEW

2.1 The Concept of Community-Based Natural Resource Management (CBNRM) and CBMM

The theoretical foundation of Community-Based Mangrove Management (CBMM) is deeply embedded in the historical evolution of common-pool resource theory, a paradigm shift pioneered by political economist Elinor Ostrom. For decades, traditional environmental governance was dominated by the classic tragedy of the commons framework, which argued that individuals utilizing an open-access, shared natural resource would inevitably deplete it through rational self-interest. Early theoretical assumptions asserted that the only remedies to prevent catastrophic ecological collapse were either total privatization of the resource or heavy-handed, top-down state regulation.

However, subsequent empirical field research dismantled this rigid dichotomy by demonstrating that local communities are fully capable of self-organizing to govern shared common-pool resources sustainably without relying on corporate privatization or centralized state coercion (de Oliveira et al., 2024). Community-Based

Natural Resource Management (CBNRM) emerged directly from this theoretical breakthrough, operating on the foundational premise that coastal populations are not inherently destructive extractors. Instead, because they possess an intimate, multi-generational understanding of their local ecosystems often referred to as traditional ecological knowledge they have a direct, long-term stake in the sustainability of those resources.

When granted secure property, tenure, or management rights, local communities are intrinsically motivated to implement effective monitoring, collective choice arrangements, and conflict-resolution mechanisms to safeguard their long-term survival (Cudiamat & Querijero, 2023). This localized approach structuralizes a sense of shared responsibility, converting resource extractors into active guardians of the coastal fringe.

When applied specifically to coastal wetlands, CBMM translates these broader common-pool resource principles into localized institutional structures governing intertidal mangrove forests. According to contemporary marine policy literature, a successful CBMM framework requires a deliberate departure from paternalistic state structures toward genuine devolution, wherein actual executive decision-making authority is transferred to grassroots institutions (Buncag, 2022). Contemporary mangrove management literature consistently highlights the interdependence of participatory governance, institutional innovation, and ecological sustainability in achieving effective conservation outcomes(Purnaweni et al., 2026). The core architecture of CBMM demands that local communities are not merely utilized as temporary, low-cost volunteer labor for seasonal tree-planting initiatives. Rather, they must be structurally integrated as equal partners or primary directors in designing comprehensive management plans, executing local conservation bylaws, and evaluating socio-ecological outcomes (Buncag et al., 2023).

Empirical research across Southeast Asia underscores that when local resource users are structurally embedded in these institutional processes, the systemic costs of resource monitoring and legal enforcement decrease significantly. This efficiency occurs because community-enforced social norms, cultural taboos, and village regulations carry far greater social legitimacy and compliance within the community than distant, poorly enforced national legislation. This structural integration effectively closes the enforcement gap that often plagues far-removed centralized regimes.

2.2 Global Challenges and Structural Barriers in Local Coastal Governance

Despite the normative appeal of community-centric frameworks, the practical execution of local mangrove management is consistently hindered by intense structural, institutional, and socio-economic barriers. A comprehensive global review highlights a persistent paradox: while mangroves are celebrated internationally for their climate change mitigation and blue carbon potential, localized management attempts frequently fail due to deep-seated structural challenges (Akram et al., 2023). These obstacles include intense, ongoing conflicts over coastal land use, a lack of strict regulatory enforcement by municipal authorities, highly fragmented and non-cohesive legislative frameworks, and a widespread deficit in localized environmental awareness among marginalized populations.

Local authorities often face immense systemic pressure to prioritize short-term economic gains from commercial aquaculture expansion or industrial infrastructure over the long-term, non-market ecological benefits of mangrove conservation. This short-term valuation cycle routinely undermines community efforts to establish permanent ecological reserves in the intertidal zone, as regional economic indicators frequently fail to account for the invaluable protective services and ecosystem functions provided by these marine habitats.

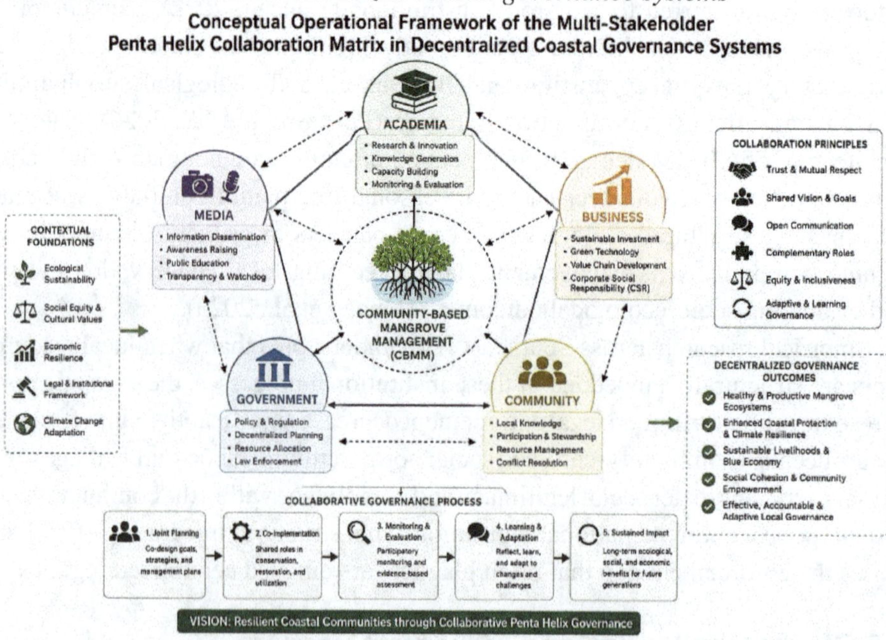

Figure 2. Conceptual operational framework of the multi-stakeholder penta helix collaboration matrix in decentralized coastal governance systems

This institutional friction is further illuminated by contemporary analyses of local government institutions handling vulnerable coastal zones. Empirical findings demonstrate that even when decentralized environmental units are legally established at the district and village levels, their practical efficacy is heavily constrained by severe financial limitations, inadequate technical capacity, and pervasive conflicts of interest among diverse coastal stakeholders (Mwanja et al., 2024). The long-term success of community-driven environmental initiatives largely depends on the preparedness of local stakeholders, active public engagement, and continuous capacity enhancement (Kurniati et al., 2024). The intense economic demand for immediate resource extraction such as clearing mangroves for charcoal production, wood fuel, or illegal agricultural encroachment frequently outpaces the slow, deliberative consensus-building processes of local administrative units.

Furthermore, local governance is severely hindered by overlapping jurisdictional boundaries and a profound lack of clarity regarding which state or local ministries hold ultimate authority over intertidal zones (Tuisuva & Shibata, 2023). This institutional fragmentation creates structural gridlocks, allowing illegal encroachment to go unchecked by exploiting bureaucratic blind spots, while simultaneously stalling community-led conservation initiatives due to regulatory inertia (Afonso et al., 2026). Without a unified legislative framework, localized conservation groups are often left to navigate a chaotic regulatory environment that weakens their statutory authority, decreases public confidence, and allows commercial actors to exploit loopholes at the expense of ecological integrity.

2.3 The Dynamics of Community Participation and Empowerment

The ultimate success of decentralized environmental governance is entirely dependent on the quality, depth, and structural nature of community participation. Public administration scholarship emphasizes that true participation must transcend superficial, tokenistic engagement such as public information meetings, unidirectional consultation, or forced compliance and achieve meaningful, structural empowerment. Structural equation modeling demonstrates that systematic institutional support directly enhances community participation capabilities (Tong et al., 2024).

When local institutions provide communities with structural rights, technical training, and transparent financial mechanisms, the public's psychological and political empowerment increases, leading to proactive, sustained engagement in local resource management (Gunawan et al., 2025). This cyclical empowerment builds a robust democratic arena where local stakeholders actively shape regional conservation goals rather than remaining passive recipients of administrative mandates. Collaborative governance mechanisms provide an effective platform for reconciling environmental conservation priorities with local development aspirations (Sari et al., 2024).

Conversely, when participation processes are poorly designed, exclusionary, or implemented too late in the policy cycle, they can generate severe systemic dysfunction. Critical meta-analyses of public participation warn that shallow or superficial engagement frequently triggers public resistance, institutional distrust, and intense social polarization (Radtke, 2025). If local communities perceive that decision-making processes are dominated by elite capture or that their input is structurally marginalized to fulfill a bureaucratic checklist, environmental initiatives lose all social legitimacy. Meaningful public participation plays a critical role in strengthening environmental stewardship and improving ecological management within environmentally sensitive coastal regions (Anas et al., 2023).

This highlights the necessity of embedding robust, transparent, and equitable procedural justice into the core design of local environmental governance systems to prevent internal alienation and community pushback (Boxshall, 2022). True empowerment requires institutional designs that guarantee equitable access to resources, active representation for marginalized subgroups, and binding decision-making authority over local environmental policies, thereby fostering an enduring sense of collective ownership and securing long-term civic compliance.

2.4 Socio-Ecological Resilience and Sustainable Coastal Development

Sustainable coastal development requires a balanced integration of human livelihood needs and ecological preservation, a concept encapsulated in the study of socio-ecological systems. Theoretical models argue that coastal risk governance must be anchored in the continuous, empirical interaction between humans and nature (Dakey et al., 2022). By utilizing community-based approaches to map how local populations interact with and depend on ecological elements, planners can develop localized resilience indicators that reflect true community preferences and adaptive capacities. This ensures that disaster risk reduction and climate adaptation policies are co-designed by the very populations facing environmental vulnerabilities, creating highly responsive intervention strategies that strengthen both bio-physical defenses and local economic systems.

Policy-driven, large-scale ecological restoration using hybrid Nature-based Solutions frameworks such as combining soft vegetation restoration with community-enforced protective zones can rapidly restore coastal morphology, enhance biodiversity, and maximize protective ecosystem services (Liu et al., 2025). However, a persistent gap remains between the high-level endorsement of blue Nature-based Solutions frameworks in global agendas and their concrete integration into local municipal policies and bylaws (Lise et al., 2025). This policy-to-practice gap often leaves local coastal managers without the legal mechanisms required to penalize landscape modification or enforce habitat restoration requirements.

Bridging this implementation gap requires innovative local administrative approaches, such as the implementation of integrated economic or sustainable blue economy zones. Ecological communication models can bridge multi-stakeholder interests within coastal zones (Nugroho et al., 2026). By applying structured, thematic communication frameworks across public and private sectors, local authorities can foster Penta Helix collaborations uniting government, academia, business, civil society, and local communities to transform coastal resource management into a circular, sustainable blue economy that guarantees both environmental longevity and localized prosperity (Abdillah et al., 2025).(Purnaweni et al., 2025) This modern governance matrix repositions community-managed intertidal zones not merely as isolated conservation enclaves, but as primary drivers of regional socioeconomic transformation and long-term ecological security against the accelerating impacts of climate change.

3 METHODE

3.1 Research Design and Approach

This study implements a qualitative descriptive research design with a multi-case study approach to comprehensively evaluate the operational dynamics of Community-Based Mangrove Management (CBMM) across three specific coastal nodes in Rembang Regency, Indonesia. Qualitative methodology is highly appropriate for this investigation as it allows the researchers to explore complex public administration structures, institutional behaviors, and localized socio-ecological phenomena within their real-world contexts. Rather than reducing administrative dynamics to isolated numerical variables, a qualitative framework captures the nuanced perceptions, structural constraints, and behavioral motivations of various stakeholders involved in coastal governance.

By utilizing a multi-case study design focused on the administrative villages of Pasar Banggi, Tritunggal, and Punjulharjo, this research identifies shared operational patterns, structural commonalities, and unique localized divergence in policy execution across different village administrations (Purnaweni et al., 2025). This comparative approach strengthens the analytical validity of the study, transforming localized empirical data into scalable governance insights that can inform broader environmental policy frameworks.

Incorporating a multi-case approach within this descriptive framework specifically allows for an inductive analytical process that exposes how historical paths of development affect current policy resistance. By evaluating these three distinct spatial typologies under a single comparative matrix, the research identifies how variance in localized institutional design ranging from highly formalized village bylaws to informal cooperative pacts influences the overall adaptive capacity of coastal social-ecological systems (Purnaweni et al., 2025). This empirical depth prevents the study from generating overgeneralized administrative claims, ensuring that the final policy recommendations are closely tailored to the complex, shifting jurisdictional realities of decentralized coastal environments.

3.2 Informant Selection Matrix

To ensure a comprehensive and balanced collection of empirical data, informants were selected using a combination of purposive and snowball sampling techniques. Purposive sampling was initially deployed to systematically identify key institutional actors who hold formal administrative authority, regulatory responsibilities, or direct operational involvement in regional coastal governance. This targeted selection process ensured that the sample included high-level representatives from municipal environmental offices, regional marine affairs departments, and village administrative leaders who dictate local zoning bylaws.

Simultaneously, snowball sampling was utilized to identify and engage grassroots environmental activists, local coastal fishers, salt farmers, and community leaders who might otherwise remain marginalized within formal public policy evaluations. This approach successfully gathered a diverse array of perspectives, balancing top-down administrative viewpoints with the lived, practical experiences of frontline resource users (Abdillah et al., 2025). The final informant matrix consists of key stakeholders categorized into four core groups: state policy makers, village authorities, community task forces, and localized economic actors.

To operationalize this qualitative investigation, the finalized research sample comprised a cohort of exactly 28 key institutional and grassroots informants. This structured group included 4 high-level administrative officers from the municipal environmental and marine affairs agencies, 6 village government executives across the selected nodes, 12 active field members of local community mangrove task forces, and 6 localized economic actors who operate small-scale artisanal fisheries and salt farming pans within the coastal fringes of Pasar Banggi, Tritunggal, and Punjulharjo.

The operationalization of this dual sampling matrix was deliberately structured to prevent institutional elite capture, a common methodological vulnerability in public policy field research. By ensuring that non-state community actors, marginalized small-scale fishers, and seasonal female salt workers were granted equal representation in the interview schedule, the study establishes a democratic data foundation that reflects genuine grassroots consensus (Abdillah et al., 2025). This deliberate multi-tier informant structure ensures that data saturation was achieved across all administrative segments, allowing for a comprehensive cross-examination of how official state environmental directives are interpreted, modified, or resisted at the ground level (Gunawan et al., 2025).

Figure 3. Research methodological flowchart and data triangulation framework (Author's elaboration adapted from Miles, Huberman, and Saldaña's Interactive Model of Qualitative Data Analysis (2014))

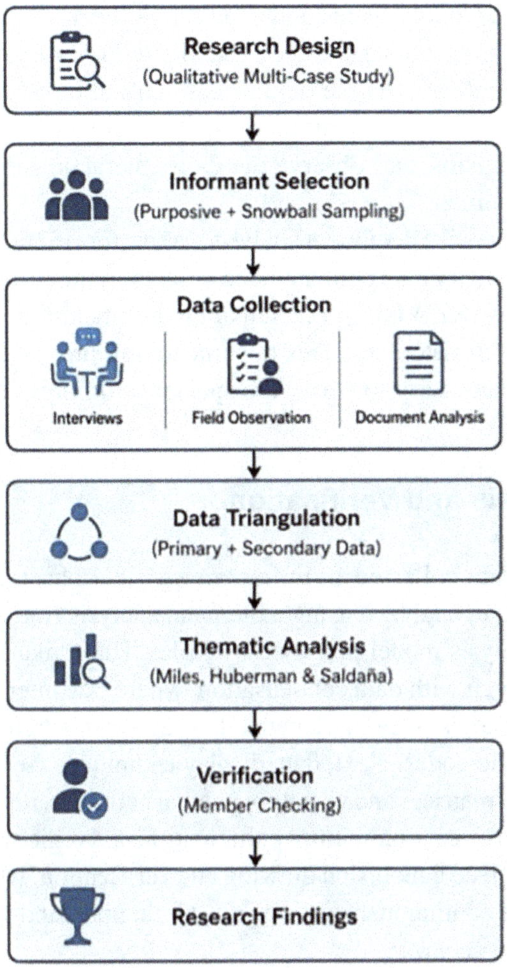

3.3 Data Collection Techniques

Empirical data for this study were gathered through three integrated collection methods designed to ensure maximum depth and context. First, in-depth, semi-structured interviews were conducted with the selected informants. These sessions were guided by flexible open-ended protocols, allowing participants to thoroughly describe systemic challenges, institutional friction, historical top-down policy failures, and the socio-economic impact of ecotourism expansion. Second, systematic field observations were conducted across the intertidal zones, mangrove boardwalks, community nurseries, and rehabilitated coastlines of Pasar Banggi, Tritunggal, and Punjulharjo. These field assessments allowed the research team to directly evaluate the physical condition of the mangrove canopies, examine local enforcement mechanisms, and observe the daily interactions between ecotourists and the local environment.

Third, a comprehensive document analysis was performed on relevant regulatory materials. This secondary data pool included village statutory instruments, regional coastal zoning maps (RZWP3K), municipal environmental progress reports, and localized historical climate data. This documentation provided the baseline contextual framework necessary to verify the spoken testimonies of the interviewees (Afonso et al., 2026).

3.4 Data Analysis and Verification

The empirical data collected from interviews, field observations, and institutional documents were analyzed using a thematic analysis framework based on the interactive data analysis model proposed by Miles, Huberman, and Saldaña. This iterative process began with data condensation, where raw interview transcriptions and extensive field notes were systematically organized, filtered, and categorized into relevant thematic codes. Next, data display techniques were used to construct structured analytical matrices and visual conceptual networks, illustrating the explicit connections between local community participation and regional policy outcomes. The final phase involved conclusion drawing and verification, where emerging analytical patterns were continuously tested against field notes and secondary literature to ensure conceptual accuracy.

To guarantee rigorous academic validity and reliability, the study adhered to strict data triangulation protocols. By cross-referencing verbal testimonies from interviews with direct physical evidence from field observations and written records from municipal administrative archives, the research team minimized subjective bias and elite capture. Furthermore, member-checking procedures were implemented, allowing key informants to review, verify, and validate the accuracy of the transcribed perspectives prior to final synthesis. This measure ensured that the study accurately reflects the socio-ecological realities of the Rembang Regency coastline (Gunawan et al., 2025).

4. RESULTS AND DISCUSSION

4.1 Empirical Findings Across the Three Coastal Nodes (Expanded Analysis)

The operational execution of Community-Based Mangrove Management (CBMM) across Rembang Regency reveals distinct structural variations, driven by the unique socio-ecological histories, institutional leadership, and economic priorities of each administrative village. In Pasar Banggi, the mangrove management paradigm has achieved an advanced stage of institutional maturity, shifting from an early coastal defense strategy to a highly lucrative eco-enterprise. Historically facing severe coastal erosion that threatened critical infrastructure, the village mobilized grassroots task forces to establish extensive Rhizophora and Avicennia barriers. Over two decades, this localized initiative evolved into a premier regional ecotourism destination, featuring expansive boardwalk networks and dedicated research zones.

The administrative success of Pasar Banggi is heavily rooted in its robust local statutory instruments. The village government enacted strict local regulations (Peraturan Desa) that formally designate the mangrove zones as protected common-pool resources, completely banning unauthorized logging and industrial encroachment. Furthermore, the financial model of Pasar Banggi is highly self-sustaining; a substantial percentage of the revenue generated from ecotourism admission tickets is directly channeled back into the community cooperative fund. This fund is transparently managed to finance continuous seedling propagation, maintenance of the physical boardwalks, and localized social welfare programs. This circular economic structure has successfully aligned the financial interests of the local population with long-term ecological preservation (Purnaweni et al., 2025).

Figure 4. Spatial land-use zonation map of the Pasar Banggi mangrove eco-enterprise node

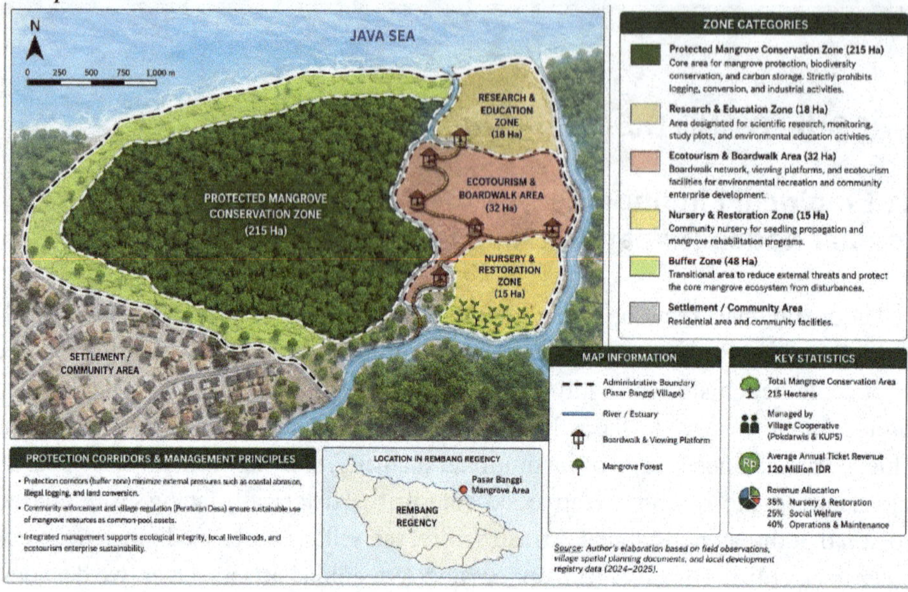

Source: *Author's elaboration based on field observations, village spatial planning documents (RZWP3K), local development registry data, and Community-Based Mangrove Management (CBMM) practices in Pasar Banggi Village, Rembang Regency (2024–2025).*

Statistical records from local development registries show that the mangrove conservation zone in Pasar Banggi spans a highly consolidated area of approximately 215 hectares, predominantly populated by mature Rhizophora apiculata canopies. The self-sustaining financial framework established by the village cooperative manages an average annual ticket revenue of approximately 120 million IDR directly from ecotourism operations. Financial audits indicate that exactly 35% of these gross corporate proceeds are directly reinvested into community nursery infrastructure and continuous seedling propagation, while 25% is allocated to local social welfare funds, creating an enduring baseline for community-based economic empowerment.

To further scrutinize this structural layout, field investigations reveal that the institutional resilience of Pasar Banggi is heavily supported by continuous multi-generational knowledge transfer. The local governance board has successfully integrated environmental conservation modules into the primary school curricula within the village, preventing the loss of traditional ecological values among the younger demographic. This systematic approach effectively buffers the local administrative apparatus against external socio-economic shocks, as the community possesses a unified ideological stance against coastal destruction.

In contrast, Tritunggal presents a different administrative landscape, where community-based management is explicitly intertwined with the urgent demands of coastal climate adaptation and physical resilience. Unlike Pasar Banggi, which benefits from an established tourism market, Tritunggal focuses its institutional energy on combating severe beach abrasion and mitigating the destructive impacts of seasonal monsoonal tidal flooding. Experiences from coastal communities in northern Central Java demonstrate the importance of combining adaptation and mitigation strategies to enhance resilience against interconnected environmental threats(Muktiali et al., 2025). The community governance model here is characterized by highly adaptive, informal civic groups working in close alignment with regional non-governmental organizations and academic institutions.

Without extensive initial state funding, the village utilized traditional ecological knowledge to construct permeable hybrid structures made from local bamboo and netting. These structures break the kinetic energy of oncoming waves while trapping marine sediments to create stable substrates for natural mangrove regeneration. While the commercialization of ecotourism in Tritunggal remains in its developmental infancy, the community's high level of environmental awareness operates as a powerful informal enforcement mechanism. Local fishers and coastal inhabitants actively monitor the intertidal zones, identifying habitat disruption and organizing spontaneous replanting drives to protect their vulnerable residential boundaries (Abdillah et al., 2025).

Meanwhile, Punjulharjo exhibits a transitional governance model that highlights the intense structural friction between rapid industrial aquaculture expansion and community-led conservation frameworks. The coastline of Punjabi has historically been dominated by heavy commercial investment in brackish-water shrimp ponds and intensive solar salt production pans, which led to widespread deforestation of the native mangrove fringe. In response to catastrophic land loss and declining artisanal fish catches caused by this ecological destruction, a dedicated group of grassroots environmental activists initiated a systematic counter-movement to reclaim the intertidal commons.

The CBMM framework in Punjulharjo operates under high-stress conditions, navigating intense resource-use conflicts with powerful local aquaculture elites who possess significant economic influence over village politics. Despite these barriers, the local community task force successfully negotiated with village administrators to secure formal zoning protections for a designated mangrove conservation corridor. This reclaimed zone is strategically positioned to serve as a biological filter, capturing destructive chemical effluents from upstream aquaculture ponds before they discharge into the open sea, thereby protecting the broader marine ecosystem (Afonso et al., 2026).

Figure 5. Comparative structural matrix of devolved governance parameters across Pasar Banggi, Tritunggal, and Punjabi coastal nodes

Source: Author's elaboration based on field interviews, direct observations, document analysis, and cross-case institutional assessment conducted in Pasar Banggi, Tritunggal, and Punjulharjo Villages, Rembang Regency (2024–2025).

To contextualize the operational variances across these three empirical sites, a cross-case institutional analysis reveals that the success of decentralized governance is not uniform. The structural differences can be systematically cross-examined using parameters such as regulatory formalization, primary ecological drivers, and economic distribution mechanisms as detailed below:

Table 1. Economic distribution mechanisms

Coastal Node	Primary Ecological Driver	Regulatory Instrument	Economic Mechanism
Pasar Banggi	Ecotourism Development & Canopy Protection	Formalized Village Bylaws (Peraturan Desa)	Community Cooperative Revenue-Sharing
Tritunggal	Coastline Abrasion & Tidal Flood Mitigation	Informal Civic Accords & NGO Pacts	Voluntary Collective Action Carbon Pooling
Punjulharjo	Aquaculture Effluent Filtering & Land Reclamation	Disputed Zoning Demarcations	Fragmented Private-Grassroots Compensations

The systemic divergence highlighted in the analytical matrix underscores that a one-size-fits-all policy approach is fundamentally unviable for the wider Indonesian archipelago. In Punjulharjo, the persistent threat of elite capture continues to destabilize the statutory authority of the grassroots task force. Unlike Pasar Banggi, where environmental enforcement is backed by the coercive legal power of the village government, the guardians of the Punjulharjo mangrove fringe must rely on social pressure and ecological advocacy to deter illegal land conversion.

Furthermore, field observation notes indicate that the localized boundaries in Punjulharjo remain highly fluid, often shifting based on the political cycles of village head elections, which introduces a high degree of institutional instability into long-term resource preservation planning (Akram et al., 2023).

4.2 Institutional Dynamics and the Penta Helix Collaboration Model

The empirical analysis of the three coastal nodes demonstrates that the long-term survival of decentralized coastal governance depends heavily on the structural transition from isolated community action to a multi-stakeholder partnership. This dynamic is best understood through the theoretical lens of the Penta Helix collaboration model, which integrates the distinct capacities of government agencies, academic institutions, private business entities, civil society organizations, and local communities into a cohesive administrative matrix. In Rembang Regency, the practical implementation of this five-fold integration has altered the traditional top-down administrative hierarchy, creating a more horizontal, collaborative policy arena (Purnaweni et al., 2025).

Government entities, particularly the regional environmental and marine affairs offices, have gradually shifted their operational roles from rigid, centralized enforcers to structural facilitators. Instead of imposing distant bureaucratic decrees, these state agencies now provide essential legal recognition, formalizing village-level conservation regulations into district-level statutory blueprints, which grants local communities the secure tenure rights necessary to exclude external extractors (Boxshall, 2022). This bureaucratic formalization effectively bridges the gap between state laws and local social codes, transforming village guards into legally recognized agents of coastal security.

Academic institutions have also played a crucial role in strengthening the technical capacity of these local governance models. Universities and marine research centers have consistently deployed field researchers to conduct rigorous spatial mapping, carbon stock assessments, and botanical health inventories across Pasar Banggi, Tritunggal, and Punjulharjo. This continuous flow of scientific data provides local community task forces with the empirical evidence needed to design effective replanting patterns, optimize seedling survival rates, and defend their conservation zones against industrial expansion claims (Purnaweni et al., 2025).

Figure 6. Methodological interaction diagram illustrating the flow of academic scientific data and corporate CSR capital into grassroots management systems

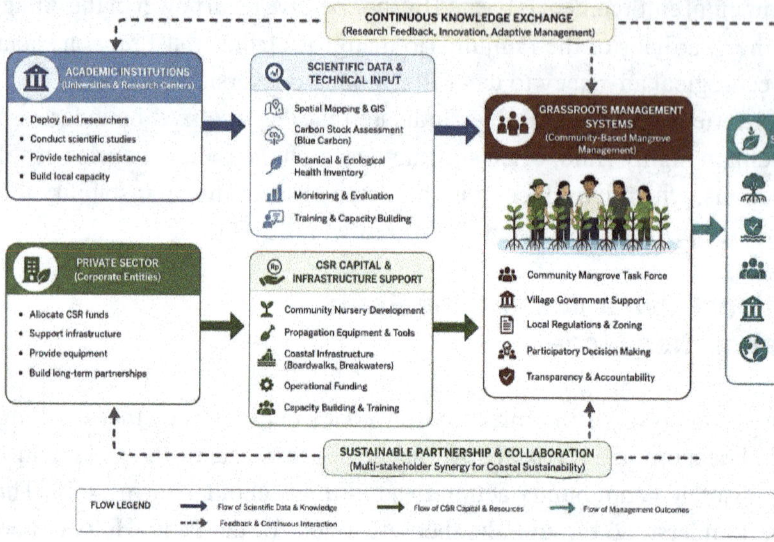

Source: Author's elaboration based on field interviews, stakeholder engagement processes, Corporate Social Responsibility (CSR) program documentation, and academic–community collaboration practices observed in Pasar Banggi, Tritunggal, and Punjulharjo Villages, Rembang Regency (2024–2025).

Furthermore, the integration of private business entities through Corporate Social Responsibility (CSR) programs has injected vital capital into these rural coastal nodes. When corporate funding is channeled transparently into community-managed accounts rather than being spent on short-term, superficial corporate marketing campaigns it provides the infrastructure support needed to build robust community nurseries, purchase high-quality propagation equipment, and construct durable coastal breakwaters. This structural funding allows local village units to undertake capital-intensive ecological restoration projects that would otherwise completely exhaust limited village budgets (Gunawan et al., 2025).

However, the operationalization of the Penta Helix matrix is not without systemic friction. The divergent motivations, operational timelines, and power dynamics of the different stakeholders frequently create administrative gridlocks and platf breakdowns within the coastal zone. Private corporate entities often demand rapid, highly visible aesthetic results from their investments to satisfy short-term corporate reporting cycles, clashing with the slow, non-linear realities of ecological restoration and the deliberative, consensus-building nature of community governance.

Additionally, political scientists warn that multi-stakeholder platforms are consistently vulnerable to elite capture, where wealthy aquaculture interests or powerful municipal bureaucrats manipulate collaborative spaces to advance private agendas(Subektie et al., 2025). effectively marginalizing the voices of traditional fishers and vulnerable coastal women (Radtke, 2025). To prevent these structural imbalances, the governance frameworks in Rembang Regency must establish formal, transparent conflict-resolution platforms and strict procedural rules that guarantee equal voting rights and decision-making weight to grassroots community representatives (Abdillah et al., 2025). Strengthening transparency and accountability within governance systems is essential to maintaining public trust and ensuring responsiveness to community concerns(Purnaweni et al., 2024).

To further unpack these institutional dynamics, the integration of civil society organizations refwithin the Rembang Regency Penta Helix network serves as a critical democratic buffer. These non-governmental actors systematically monitor the deployment of corporate capital and state subsidies, preventing the weaponization of financial incentives by local political elites who seek to consolidate power during village election cycles. Quantitative evaluations of collaborative governance frameworks demonstrate that multi-stakeholder interactions operate most effectively when protected by statutory performance bonds and transparent reporting mechanisms (Boxshall, 2022). In coastal nodes like Pasar Banggi, the institutionalized transparency of the local ecotourism board has effectively neutralized the risk of administrative opacity, transforming the cooperative into a trusted financial intermediary capable of processing external grants without incurring high bureaucratic overheads.

Conversely, the collaborative friction observed in Punjulharjo emphasizes the danger of structural asymmetry within the Penta Helix framework. When private aquaculture enterprises operate with significant regulatory impunity, the communicative arenas established by municipal authorities frequently degenerate into spaces of tokenistic consultation. To remedy this structural imbalance, contemporary public policy models advocate for the implementation of binding co-management pacts that grant grassroots task forces the statutory veto power to halt corporate land-use expansions that violate localized zoning boundaries (Radtke, 2025). By legalizing these defensive veto mechanisms at the district level, the regional government can insulate vulnerable common-pool resources from market-driven degradation, ensuring that multi-stakeholder platforms foster genuine socio-ecological equity rather than facilitating sophisticated forms of environmental greenwashing (Gunawan et al., 2025). This institutional stabilization ultimately changes the role of local fishers from vulnerable, reactive victims of coastal industrial development into legally empowered co-directors of the regional blue economy masterplan, creating an enduring administrative model for peripheral maritime zones across the global South.

4.3 Structural Barriers: Capacity Deficits and Leadership Dynamics

A critical obstacle to the widespread replication of successful CBMM models is the systemic capacity deficit and volatile leadership dynamics observed at the grassroots level. The long-term continuity of community-led environmental governance is highly dependent on the presence of charismatic, technically capable local leaders who can navigate complex state bureaucracies while maintaining deep social trust within their villages. In coastal nodes where such leadership is vibrant, environmental initiatives thrive; these local leaders successfully bridge the communicative gap between rural populations and institutional investors, translating complex scientific data into clear village bylaws and securing external grants to fund local conservation work (Mwanja et al., 2024).

However, this reliance on individualized leadership exposes a profound systemic vulnerability: when a key leader retires, relocates, or enters local politics, the entire institutional architecture of the village mangrove task force frequently collapses due to inadequate institutional succession planning and a lack of younger leadership development (Mwanja et al., 2024). This vulnerability is highly visible in emerging coastal nodes, where the institutional framework is built around a single charismatic figure rather than being embedded into the bureaucratic structure of the village government. Without a systematic protocol to train the next generation of environmental administrators, community-led initiatives remain highly vulnerable to sudden leadership transitions, which can erase decades of conservation progress in a single electoral cycle.

This capacity gap is further exacerbated by the unequal distribution of technical and administrative skills within rural coastal populations. While local fishermen and salt farmers possess deep experiential knowledge of tidal patterns and marine behavior, they often lack formal training in financial management, grant proposal writing, and complex environmental litigation. This technical deficiency leaves local organizations highly dependent on external consultants, NGOs, or academic institutions to handle their administrative affairs, making them vulnerable to exploitation or sudden drops in project funding (Tuisuva & Shibata, 2023).

To overcome this, regional governments must implement long-term training programs focused on administrative capacity-building within local management boards, ensuring that rural communities can independently manage the financial, legal, and operational aspects of coastal zone governance.

Addressing these systemic capacity deficits requires a fundamental shift from ad-hoc technical workshops to the institutional formalization of coastal governance academies at the municipal leve(Lituhayu et al., 2026)l. Financial dependency on volatile provincial subventions or unpredictable corporate donations frequently paralyses village environmental task forces, stalling critical infrastructural maintenance and leaving newly propagated seedling beds highly vulnerable to extreme monsoonal surges. Institutional analyses of decentralized environmental regimes confirm that long-term policy continuity cannot be achieved without the structural integration of independent, community-managed revenue generation streams (Tuisuva & Shibata, 2023). By training grassroots leaders in diversified financial mechanisms, such as blue carbon asset accounting, localized eco-taxation models, and sustainable community-owned aquaculture frameworks, the regional administration can foster a resilient cadre of environmental entrepreneurs who are fully equipped to navigate complex transnational carbon markets.

Furthermore, the deep-seated resistance to institutional succession planning within traditional village hierarchies must be actively dismantled through formalized legal structures. Village governments must enact statutory mandates requiring the inclusion of younger demographics and women's representatives within the executive councils of all mangrove management boards, ensuring that traditional ecological knowledge is systematically preserved and modernized (Mwanja et al., 2024). This continuous administrative evolution effectively insulates the community against the disruptive shocks of individualized leadership transitions, translating temporary personal charisma into permanent, rule-bound institutional memory. When technical litigious skills, complex bookkeeping methodologies, and scientific monitoring techniques are horizontally distributed across multiple community layers, the structural vulnerability of elite capture decreases significantly, providing a robust, democratic foundation for grassroots climate resilience that can independently withstand external economic pressures and shifting municipal political agendas over long-term developmental horizons.

Figure 7. Leadership succession and capacity-building framework for community-based mangrove management

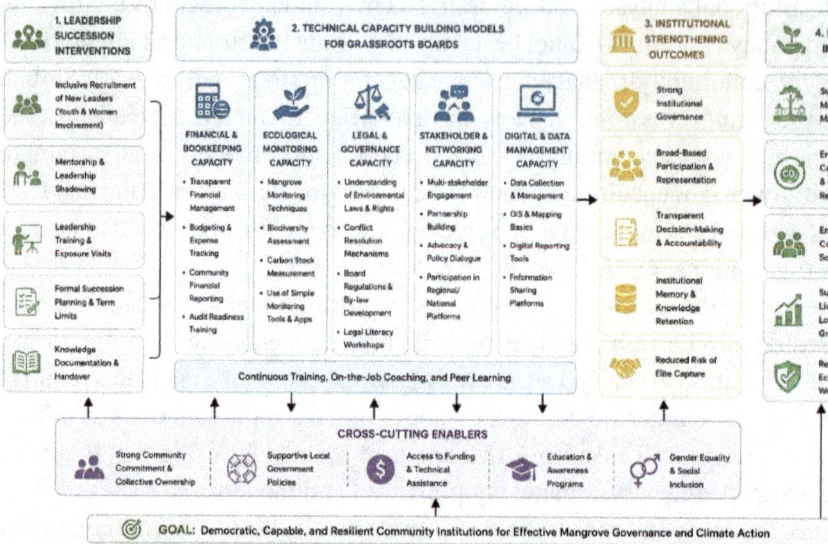

Source: *Developed by the author based on (Mwanja et al., 2024a), (Tong et al., 2024), (Cudiamat & Querijero, 2023), and field findings.*

4.4 Socio-Ecological Impacts and Community Empowerment

Despite the persistent challenges, the implementation of community-driven management across the three coastal nodes has generated profound positive socio-ecological impacts, completely reshaping the coastal landscape and the socioeconomic status of its inhabitants. Ecologically, the systematic replanting and strict protection of mangrove buffers have successfully halted severe shoreline recession, stabilized localized soil structures, and catalyzed a significant recovery in regional marine biodiversity.

Field assessments show a substantial return of native bird species, intertidal crabs, and commercially valuable fish larvae to the restored root networks. This ecological revival has directly revitalized the local artisanal fishing economy, as fishermen report higher daily catches within closer proximity to the shore, reducing their operational fuel costs and increasing their overall household income (Dakey et al., 2022).

From a social perspective, the CBMM framework has acted as a powerful engine for community empowerment, shifting local populations from passive resource users to active policymakers(Warsono et al., 2025). Through participation in village assemblies and conservation groups, marginalized individuals including traditional women's groups and young people have gained a political voice, actively shaping the environmental regulations and economic strategies of their villages(Tong et al., 2024).

This democratic engagement has built a strong sense of collective ownership and civic pride, making communities resilient against external political and economic pressures that seek to exploit their natural heritage. The transformation of these coastal nodes proves that when local communities are trusted with the stewardship of their natural resources, environmental sustainability and human development can advance together in harmony (Cudiamat & Querijero, 2023).

4.5 Policy Alignment and Future Scalability

To guarantee the long-term viability of community-based frameworks, local initiatives must be aligned with broader national and provincial environmental policies. In Indonesia, coastal zone management is governed by a complex web of legislation, including national laws on coastal and small island management (UU WP3K) and provincial spatial plans (RZWP3K). Historically, a lack of harmonization between local village regulations and these overarching state frameworks created significant legal ambiguity, leaving community-managed zones vulnerable to overlapping industrial concessions granted by higher-level ministries (Akram et al., 2023).

Addressing this regulatory mismatch requires local governments to proactively integrate village-level conservation maps and bylaws into official provincial spatial planning registries, giving community-managed zones full legal protection against external industrial development. Greater integration between environmental objectives and regional planning frameworks is increasingly viewed as a fundamental requirement for sustainable development governance(Manik et al., 2025). When these bottom-up zoning rules are successfully embedded within macro-level state legislation, it creates a secure statutory barrier that insulates local ecosystems from shifting political priorities

Figure 8. Structural policy alignment architecture for integrating local village bylaws into national marine spatial planning frameworks

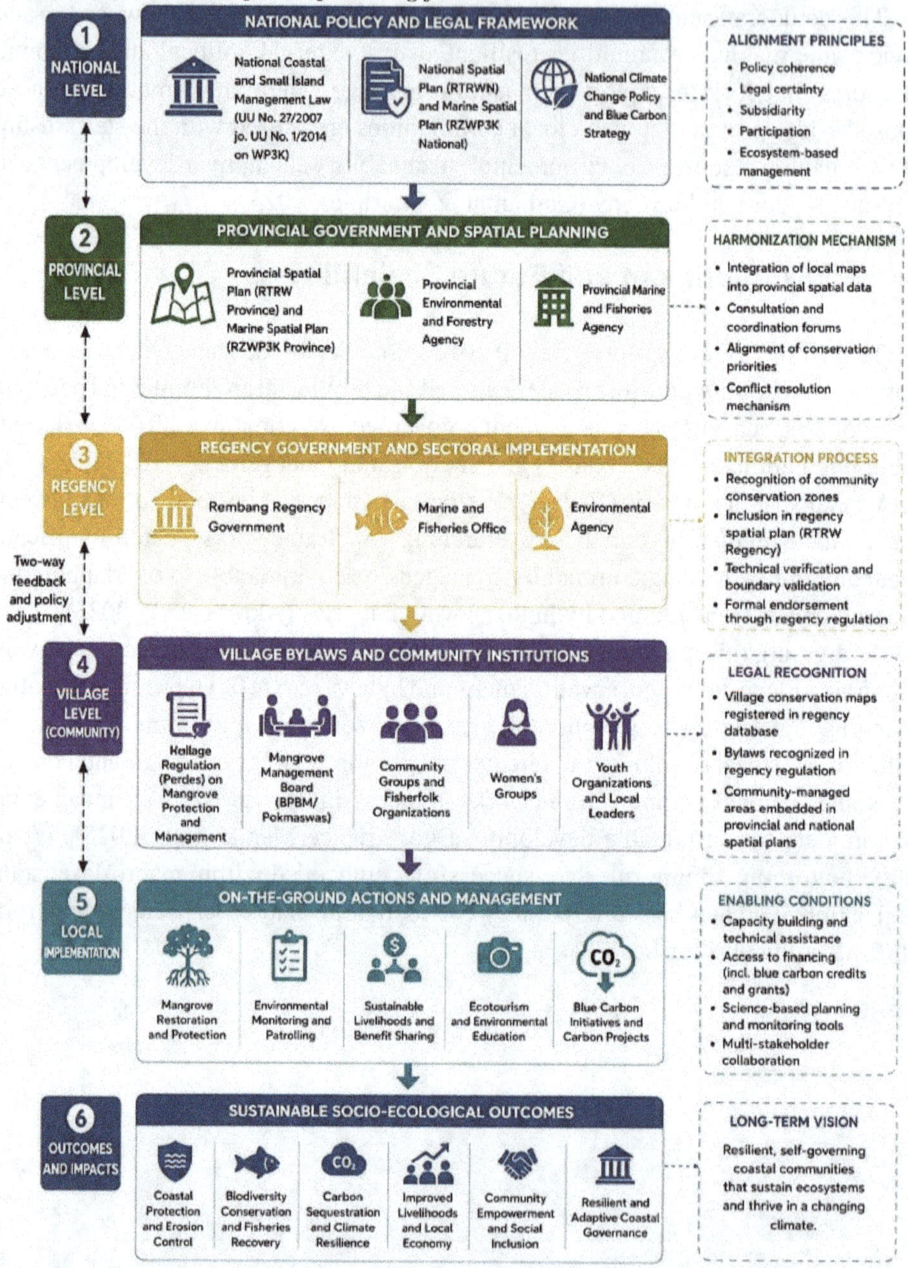

Source: *Developed by the author based on (Akram et al., 2023), (Dakey et al., 2022), (Abdillah et al., 2025), (Lise et al., 2025), (Nugroho et al., 2026), and field findings*

Looking to the future, the scalability of the Rembang Regency governance model depends on creating standardized institutional blueprints that can be adapted to other vulnerable coastlines across the Indonesian archipelago. While the unique socio-ecological features of individual villages prevent a one-size-fits-all approach, the core principles of CBMM such as devolved decision-making, equitable benefit-sharing, and multi-stakeholder collaboration are universally applicable. By capturing these lessons in formal policy toolkits, regional development agencies can help other coastal communities replicate these successes.

This approach can transform vulnerable coastlines into resilient, self-governing socio-ecological systems capable of thriving amid global climate change (Lise et al., 2025)(Nugroho et al., 2026). Furthermore, future expansion strategies must leverage international climate finance networks, such as blue carbon credit markets and global conservation grants, to provide localized task forces with the long-term financial independence required to sustain their monitoring operations without relying solely on volatile municipal budgets (Abdillah et al., 2025)(Purnaweni et al., 2025).

Ultimately, the global scalability of the Rembang Regency decentralization paradigm depends on its capacity to convert localized empirical successes into macro-level statutory revision. The persistent friction between bottom-up community zoning guidelines and top-down national industrial masterplans highlights a critical structural vulnerability that threatens the longevity of all decentralized conservation enclaves in Southeast Asia. To bridge this structural rift, contemporary environmental policy frameworks must pioneer the legal mechanism of "nested spatial planning," wherein village-level statutory instruments are granted automatic, non-discretionary incorporation into provincial maritime zoning registries (Akram et al., 2023). This regulatory integration ensures that external industrial licenses issued by federal ministries are legally invalid if they overlap with conservation boundaries designated by grassroots task forces, providing frontline communities with robust, sovereign defense mechanisms against corporate encroachment.

In conclusion, the strategic evolution of Community-Based Mangrove Management in Pasar Banggi, Tritunggal, and Punjulharjo demonstrates that coastal environmental protection is not an obstacle to rural economic growth, but its primary foundation. By restructuring the institutional matrix through the Penta Helix model, building independent administrative capacities at the village level, and harmonizing localized bylaws with overarching national spatial plans, regional authorities can establish a highly adaptive, self-sustaining ecosystem of climate resilience (Lise et al., 2025). As climate-driven coastline recession accelerates along the vulnerable northern coast of Java, the universal deployment of these formalized, community-centric governance frameworks offers a scalable, empirically proven pathway to protect both global blue carbon repositories and the socio-economic sovereignty of vulnerable maritime populations facing the cascading realities of global climate change (Nugroho et al., 2026).

5. CONCLUSION

The empirical investigation into Community-Based Mangrove Management (CBMM) across the coastal nodes of Pasar Banggi, Tritunggal, and Punjulharjo in Rembang Regency provides profound insights into the shifting paradigm of decentralized environmental governance. This research fundamentally demonstrates that transferring management authority from rigid, centralized state bureaucracies to empowered, grassroots institutions represents a highly robust strategy for mitigating severe coastal degradation, arresting shoreline recession, and fostering the

restoration of vital intertidal marine biodiversity. The varying degrees of institutional maturity evidenced across the three study sites underscore a critical reality: a community's ability to govern common-pool resources is not merely a product of intent, but is deeply contingent upon the precision of local statutory instruments, the absolute transparency of economic benefit-sharing mechanisms, and the sustained adaptive capacity of local leadership figures.

Furthermore, the operationalization of the Penta Helix collaboration model within Rembang Regency reveals that the enduring viability of coastal conservation cannot rely upon isolated, localized volunteerism. Instead, it necessitates a sophisticated, multi-stakeholder architecture where state agencies provide the essential legal security of land tenure, academic institutions supply continuous scientific validation, and private-sector entities offer transparent, long-term infrastructure capital. Despite the presence of systemic friction such as the precarious nature of leadership transitions, technical deficits within marginalized demographics, and the relentless economic pressure exerted by industrial aquaculture expansion the socio-ecological dividends of community-led stewardship are indisputable. Empowering frontline coastal populations both politically and economically not only bolsters physical resilience against the cascading threats of climate change but also catalyzes sustainable, long-term economic development for rural households.

Ultimately, this research concludes that for decentralized environmental frameworks to be effectively scaled and replicated across other vulnerable maritime fringes, regional authorities must proactively harmonize village-level regulations with provincial spatial planning registries. By securing comprehensive legal recognition and institutional backing, these community-centric approaches can effectively transform vulnerable, open-access zones into highly resilient, self-governing socio-ecological systems. Such frameworks successfully reconcile the historical dichotomy between ecological preservation and human prosperity, offering a scalable blueprint for coastal administration throughout the Indonesian archipelago. Moving forward, future investigations should prioritize longitudinal analyses to quantify the complex socioeconomic trade-offs of transitioning from traditional aquaculture to intensive mangrove ecotourism. Moreover, integrating emerging global carbon credit markets into village cooperative systems will remain vital to securing the permanent financial autonomy required to sustain these critical environmental monitoring operations amidst a rapidly warming climate.

REFERENCES

Abdillah, A., Widianingsih, I., Buchari, R. A., Nurasa, H., & Rogayan, D. V. (2025). Community-Based Sustainable Mangrove Management for Environmental Resilience in Indonesia: A Case from Sinjai, North Luwu, and Pangkep Regencies. In *Advances in Global Change Research* (Vol. 80, pp. 283–301). Springer Science and Business Media B.V. DOI: 10.1007/978-3-031-85359-3_12

Afonso, F., Fernandes, M., Austen, M. C., Broszeit, S., & Brito, A. C. (2026). The role of legislation, policy and regulation in estuarine management: From an international to a regional scale. *Ocean and Coastal Management*, *277*, 108184. Advance online publication. DOI: 10.1016/j.ocecoaman.2026.108184

Akram, H., Hussain, S., Mazumdar, P., Chua, K. O., Butt, T. E., & Harikrishna, J. A. (2023). Mangrove Health: A Review of Functions, Threats, and Challenges Associated with Mangrove Management Practices. *Forests*, *14*(9), 1698. Advance online publication. DOI: 10.3390/f14091698

Anas, A. K., Maryono, , & Purnaweni, H. (2023). Community participation in waste management in abrasion areas. *IOP Conference Series. Earth and Environmental Science*, *1268*(1), 012028. DOI: 10.1088/1755-1315/1268/1/012028

Boxshall, A. (2022). Perspectives on building climate resilience via marine and coastal management from the governance frontline in Victoria, Australia. *Ocean and Coastal Management*, *228*, 106291. Advance online publication. DOI: 10.1016/j.ocecoaman.2022.106291

Buncag, M. J. J. (2022). Community-Based Mangrove Forest Management Sustainability Analysis in Tagpait, Aborlan and Bacungan, Puerto Princesa City, Palawan, Philippines. *Environment and Ecology Research*, *10*(3), 325–333. DOI: 10.13189/eer.2022.100301

Buncag, M. J. J., Arzaga, J. S., Tangonan, L. F., de Castro, J. H., Villanueva, M. C. M., Margallo, L., Lactuan, I. R., Docto, S. G., Garcia, A. V., Denosta, P. E. C., & Villaruel, S. A. L. (2023). Sustainability Evaluation of Mangrove Forest Management System of Tagbanua Tribe in Bgy. Manalo, Puerto Princesa City, Palawan, Philippines. *Journal of Environment and Earth Science*, *5*(2), 36–49. DOI: 10.30564/jees.v5i2.5756

Cudiamat, M. A., & Querijero, B. L. (2023). Building sustainability in community-managed mangrove forest using biocultural approach to conservation. In *Plant Diversity in Biocultural Landscapes* (pp. 17–40). Springer Nature., DOI: 10.1007/978-981-19-8649-9_2

Dakey, S., Joshi, S., Sukhwani, V., & Deshkar, S. (2022). A Community-based Approach to Mainstream Human-Nature Interactions into Coastal Risk Governance: A case of Katrenikona, India. *Geographica Pannonica, 26*(1), 65–78. DOI: 10.5937/gp26-35582

de Oliveira, M., Morrison, T., O'Brien, K. R., & Lovelock, C. E. (2024). Governance of coastal wetlands: Beyond the community conservation paradigm. *Ocean and Coastal Management, 255*, 107253. Advance online publication. DOI: 10.1016/j.ocecoaman.2024.107253

Gunawan, H., Basyuni, M., Subarudi,, Suharti, S., Kustanti, A., Wahyuni, T., Arifanti, V. B., Yeny, I., Affandi, O., Sugiarti,, Zuhriana, D., Lastini, T., Herawati, T., Riswati, M. K., & Effendi, R. (2025a). please change to Gunawan, H., Basyuni, M., Subarudi,, Suharti, S., Kustanti, A., Wahyuni, T., Ari- fanti, V. B., Yeny, I., Affandi, O., Sugiarti,, Zuhriana, D., Lastini, T., Herawati, T., Riswati, M. K., & Effendi, R. (2025a). Empowering conservation: The transformative role of mangrove education in Indonesia's climate strategies. *Forest Science and Technology, 21*(4), 374–396. DOI: 10.1080/21580103.2025.2519475

Hanani, R., Rahman, A. Z., & Kristanto, Y. (2021). COVID-19 and Adaptation Strategies in Community-Based Tourism: Insights from Community-Based Tourism Sector in Central Java. In *E3S Web of Conferences* (Vol. 317, p. 01054). EDP Sciences. DOI: 10.1051/e3sconf/202131701054

Hidayati, I., Prabawani, B., Rahmadhony, A., Hayati, A. N., & Boddy, J. (2026). The Critical Role of Local Leaders in Managing the Impacts of Climate Change in Demak, Indonesia. *Springer Climate*, (Part F1651), 119–133. DOI: 10.1007/978-3-032-15334-0_8

Husna, J., Wijayanti, I., Budiastusi Wiratmo, L., Indrahti, S., Naryoso, A., Edzan Che Nasir, N., Patria Ratna, M., Mulyaning Andini, B., & Raihan Putra Ratarno, D. (2023). Mapping the Scientific Literature on Mangrove Conservation in Indonesia: A Bibliometric Analysis to Environmental Research. *E3S Web of Conferences, 448*, 03036. DOI: 10.1051/e3sconf/202344803036

Kismartini, Saputra, Y. A., & Hanani, R. (2025). Sustainable Tourism in Crisis? Environmental Challenges and Economic Potentials in Rembang's Coastal Area. *E3S Web of Conferences, 650*, 02014. DOI: 10.1051/e3sconf/202565002014

Kurniati, R., Handayani, W., Prabawani, B., Nida, R. S., & Savitri, S. F. (2024). COMMUNITY PARTICIPATION, COMMUNITY READINESS, AND PREFERENCES TO PROMOTE EDU-TOURISM IN MANGROVE AREAS OF TELUK AWUR VILLAGE, JEPARA, INDONESIA. *PLANNING MALAYSIA*, *22*(4), 370–384. DOI: 10.21837/pm.v22i33.1555

Lise, F., Gianluca, F., & Pierre, F. (2025). Blue Nature-Based Solutions in marine and coastal EU policies: Challenges, recommendations and policy opportunities throughout the policy cycle. *Marine Policy*, *177*, 106687. Advance online publication. DOI: 10.1016/j.marpol.2025.106687

Lituhayu, D., Santoso, R. S., Imawan, S. A., Bellastuti, D. B., & Saputra, Y. A. (2026). Sustainable Governance and Institutional Reform of Waste Management in Semarang City. *IOP Conference Series. Earth and Environmental Science*, *1610*(1), 012001. DOI: 10.1088/1755-1315/1610/1/012001

Liu., (2025). *Liu 2025 pollution mangrove ecosystem - Google Scholar*. https://scholar.google.com/scholar?q=Liu+2025+pollution+mangrove+ecosystem

Manik, J. S. T., Purnaweni, H., Damayanti, I., & Djumiarti, T. (2025). Integrating policy strategy documents: SDGs and KLHS/SEA, into the Regional Medium-Term Development Plan/RPJMD as a foundation for sustainable regional development. *E3S Web of Conferences*, *650*, 02040. DOI: 10.1051/e3sconf/202565002040

Muktiali, M., Sudharto Hadi, P., & Purnaweni, H. (2025). Multi-hazard systems approach to overcome multi-hazard disaster in the northern coastal region of central jawa province. *E3S Web of Conferences*, *605*, 03035. DOI: 10.1051/e3sconf/202560503035

Mwanja, R. I., Peter, K. H., & Said, M. K. (2024a). The efficacy of management practices in combating mangrove forest degradation: A case study from Pangani District-Tanzania. *East African Journal of Science, Technology and Innovation*, *6*(Special issue 1), 1–17. DOI: 10.37425/17mqsq04

Nugroho, W., Tunggali, A. P. P. W., Isa, J. R. S. P. K. M., Suhendra, E. S., & Nugroho, A. Z. (2026). Ecological Communication Approach in Sustainable Coastal Development: A Case Study of Minapolitan in Indonesia. *Springer Proceedings in Business and Economics*, 379–394. DOI: 10.1007/978-981-95-4017-4_24

Prabawani, B., Nugroho Sugianto, D., Ariyanti, D., Pribadi, R., Wigati, M., & Fisher, M. R. (2025). Turning Tides: Sustainable Community Practices in Restoring Mangrove Ecosystems in Coastal Riau. In *E3S Web of Conferences* (Vol. 673, p. 01004). EDP Sciences. DOI: 10.1051/e3sconf/202567301004

Purnaweni, H., Djumiarti, T., Roziqin, A., & Santoso, B. (2024). How do local government strategies advance social accountability? The challenges from environmental management of Slaughterhouse in Semarang City, Indonesia. *Local Environment, 29*(2), 245–261. DOI: 10.1080/13549839.2023.2282094

Purnaweni, H., Hadi, S. P., Soeprobowati, T. R., Roziqin, A., & Rakhma, N. A. (2026). What are the Main Concerns for Global Mangrove Management Research? Systematic Literature Review and Bibliometric Analysis. *Research in Ecology, 8*(2), 46-66–46–66. DOI: 10.30564/re.v8i2.12149

Purnaweni, H., Soeprobowati, T. R., Chatterjee, U., Kiswoyo, G. P., Rakhma, N. A., & Manik, J. S. T. (2025). The Role of Stakeholders in Community Based Mangrove Management in Tapak Area of Semarang City. *E3S Web of Conferences, 650*. DOI: 10.1051/e3sconf/202565002047

Radtke, J. (2025). Barriers and benefits of public participation in energy transitions: A meta-analysis of empirical evidence from Central Europe. *Renewable & Sustainable Energy Reviews, 221*, 115693. Advance online publication. DOI: 10.1016/j.rser.2025.115693

Rahman, A. Z., Pradana, A. E., & Rakhma, N. A. (2025, September). Sustainable Tourism Development in Glawan Village: Balancing Infrastructure Growth and Environment through Local Governance. In *IOP Conference Series: Earth and Environmental Science* (Vol. 1537, No. 1, p. 012001). IOP Publishing. DOI: 10.1088/1755-1315/1537/1/012001

Rahman Halim, M. A., Ihrami Rahmila, Y., Muhammad, F., & Safitri, Y. (2018). The Effect on Mangrove Density with Sediment Rate in Coastal Pasar Banggi of Rembang Regency. *E3S Web of Conferences, 73*, 04020. DOI: 10.1051/e3sconf/20187304020

Roziqin, A. (2018). Environmental policy of mangroves management in Rembang Regency. In *E3S Web of Conferences* (Vol. 31, p. 09002). EDP Sciences., DOI: 10.1051/e3sconf/20183109002

Safitri, Y., Saputro, S., & Hariadi, H. (2017). HUBUNGAN LAJU SEDIMENTASI TERHADAP KERAPATAN MANGROVE DI PANTAI PASAR BANGGI KABUPATEN REMBANG. *Journal of Oceanography*, *6*(4), 553–563. https://ejournal3.undip.ac.id/index.php/joce/article/view/20402

Sari, N., Baiquni, M., Damanik, J., Tasikmalaya, N., Dewantara, Y. F., Facrureza, D., Hanifah, R. D., Kristanti, D., Purnaweni, H., Dwimawanti, I. H., & Yuwono, T. (2024). Harmonizing tourism and nature conservation: A governance model for sustainable tourism villages in Indonesia. *IOP Conference Series. Earth and Environmental Science, 1414*(1), 012078. DOI: 10.1088/1755-1315/1119/1/012078

Subektie, R. W., Huda, M. N., Safira, C., & Sari, A. (2025). Mapping the Frontiers of Digital Political Activism: Insights from Bibliometric Analysis. *Studies in Media and Communication*, *14*(1), 402–414. DOI: 10.11114/smc.v14i1.8289

Tong, J., Li, Y., & Yang, Y. (2024). System Construction, Tourism Empowerment, and Community Participation: The Sustainable Way of Rural Tourism Development. *Sustainability (Basel)*, *16*(1), 422. Advance online publication. DOI: 10.3390/su16010422

Tuisuva, S., & Shibata, S. (2023). Exploring sustainable mangrove management policy, challenges and structure: A case study of Nadi and Momi districts, Fiji Island. *Journal of Forest Research*, *28*(5), 313–323. DOI: 10.1080/13416979.2023.2220189

Wahyudi, R., Kismartini, Puspita, A. S., & Arief Budihardjo, M. (2025). The Socio-Economic Impacts of Land Subsidence in Coastal Semarang. *E3S Web of Conferences*, *650*, 02043. DOI: 10.1051/e3sconf/202565002043

Warsono, H., Muhamad, J. A., Fitrianto, M. R., Yusuf, I. M., Astuti, R. S., & Priyadi, B. P. (2025). Analysis of Village Development Policy and Village Community Empowerment in Indonesia. *IOP Conference Series. Earth and Environmental Science*, *1537*(1), 012032. DOI: 10.1088/1755-1315/1537/1/012032

Widjonarko, Purnaweni, H., & Maryono. (2025). Spatio-Temporal pattern of coastal flooding due to microclimate change and mangrove deforestation in Semarang and Demak Coastal Area. *E3S Web of Conferences*, *605*, 03012. DOI: 10.1051/e3sconf/202560503012

Chapter 10
Community-Based Mangrove Management to Reduce Poverty in the Coastal Areas of Central Java

Gandisa Pangestuti Kiswoyo
http://orcid.org/0009-0001-0068-3740
Universitas Negeri Yogyakarta, Indonesia

Rizki Hidayati
http://orcid.org/0009-0007-7092-884X
Universitas Negeri Yogyakarta, Indonesia

Antonia Eka Sari Widyastuti
http://orcid.org/0009-0000-0092-6848
Universitas Negeri Yogyakarta, Indonesia

Muhamad Nur Chozin
http://orcid.org/0000-0002-2071-7485
Universitas Negeri Yogyakarta, Indonesia

ABSTRACT

Mangrove ecosystems play a vital role in maintaining the stability of coastal areas, but they currently face serious threats due to human activities and weak institutional systems. This study analyzes the role of Community based Mangrove Management in reducing poverty and ensuring environmental sustainability along the northern coast of Central Java. Using a Systematic Literature Review and semi structured in-

DOI: 10.4018/979-8-3373-9320-9.ch010

terviews, this study explores various management models, ranging from participatory and collaborative models to those based on local knowledge. The study's findings indicate that active community involvement, supported by economic incentives and clear management rights, is crucial for the success of ecosystem restoration while simultaneously improving economic well being through livelihood diversification such as ecotourism and the utilization of mangrove products. This study concludes that strengthening the capacity of local institutions and coordination among stakeholders are key to achieving inclusive and sustainable mangrove management.

1. INTRODUCTION

Mangroves are ecosystems that play a vital role in maintaining the stability of coastal areas. The increasingly alarming global condition of mangroves indicates that the challenges of managing complex natural resources are a serious issue(Purnaweni et al., 2026). Several countries have lost mangrove land cover due to weak institutional systems, a lack of enforcement of environmental regulations, and minimal involvement of local communities in management processes. Top-down management approaches often fail to address local needs and fall short of achieving long-term sustainability (K. Datta et al., 2012). To address these limitations, a community-based approach or mangrove management that emphasizes the importance of local community participation is needed.

Mangroves, which possess high ecological and economic value, are among the world's most threatened ecosystems. Global reports indicate that approximately 30–50% of the world's mangroves have been lost in recent decades due to human activities (FAO, 2023). The primary causes of this loss include the expansion of aquaculture ponds, conversion to agricultural land, and coastal urbanization (Friess et al., 2019). Additionally, weak governance and uncoordinated policies further exacerbate the situation. This indicates that conventional management approaches remain inadequate to address the complexity of mangrove-related issues.

Community involvement often occurs only during specific events or celebrations, serving merely as evidence of documented activities without any long-term sustainability. The community must be present and involved in the entire process from initial planning and implementation to periodic and ongoing evaluation of management. This approach aligns with the concept of participatory development, which emphasizes community empowerment as the primary actors in resource management (Chambers, 1994). Furthermore, community-based management must also optimize local knowledge and strengthen social capital, which are key factors in the success of conservation (Berkes, 2009).

Collaboration between communities and the government is essential for achieving greater socio-ecological impacts in improving mangrove management, particularly in coastal areas (Armitage et al., 2009). However, the success of CBMM implementation is highly dependent on the local context, including policy support, institutional capacity, funding sustainability, and the quality of coordination among stakeholders(Purnaweni et al., 2025). Moreover, community-based mangrove management has demonstrated success in improving community well being while restoring ecosystems, it still faces various challenges, such as conflicts of interest, capacity constraints, and the phenomenon of elite capture in resource management (DasGupta & Shaw, 2013).

Indonesia has the largest mangrove area in the world, accounting for approximately 20% of the global total (FAO, 2023). Mangroves in Indonesia play a vital role in maintaining the stability of coastal ecosystems and supporting the livelihoods of millions of people who depend on these resources. This situation calls for more effective and sustainable management strategies. The Indonesian government has initiated various mangrove rehabilitation programs as part of its national strategy. The involvement of local communities through community empowerment programs that include mangrove rehabilitation activities also helps to raise environmental awareness and build local capacity.

Based on these various dynamics, a systematic study is needed to understand the developments, challenges, and opportunities in community based mangrove management. Through this study, it is hoped that key factors influencing the success of CBMM in addressing poverty, as well as their implications for management policies and practices, can be identified. Thus, this research not only provides an academic contribution but is also relevant for evidence-based policy making. Consequently, sustainable mangrove management becomes an essential component in achieving inclusive and long term development.

2. MATERIALS DAN METHODS

This research uses a qualitative case study approach using the Systematic Literature Review (SLR) method, combined with primary data collection through interviews to analyze the role of Community-Based Mangrove Management (CBMM) in reducing poverty in the northern coastal area of Central Java. This approach was chosen because it provides a comprehensive understanding of the CBMM phenomenon through interviews and various relevant scientific literature and policy documents, enriched by empirical perspectives from local actors (Creswell, 2009). This method focuses on exploring the "how" and "why" aspects, particularly regarding the relationship between community-based mangrove management, public policy, and its impact on the socio-economic conditions of coastal communities. The research area is focused on the northern coastal area of Central Java, specifically areas with mangrove ecosystems and coastal poverty issues, such as Brebes Regency, Demak Regency, Kendal Regency, Rembang Regency, and Semarang City. This selection of areas was based on the numerous studies and reports discussing mangrove management practices and the socio-economic dynamics of coastal communities in the region.

The data sources in this study consist of secondary and primary data. Secondary data were obtained through the Systematic Literature Review (SLR) method, which was carried out systematically through several stages. The first stage was identifying the research question, namely how CBMM contributes to poverty reduction and coastal environmental sustainability. The second stage was a literature search, conducted through various scientific databases such as Google Scholar, Scopus, and ScienceDirect using keywords such as community-based mangrove management, mangrove conservation, coastal poverty, and sustainable coastal governance. The third stage was determining inclusion and exclusion criteria. The selected literature included scientific journal articles, proceedings, and official government reports published within the last 10–15 years and relevant to the research topic, particularly in the context of Indonesia or tropical coastal areas. Literature that was irrelevant or did not meet academic standards was eliminated. The fourth stage was the selection and evaluation of the quality of the literature, assessing the suitability of the topic, methodology, and contribution of the findings to CBMM studies and poverty reduction. The fifth stage was data extraction and synthesis, which was conducted by grouping the findings based on key themes such as community-based mangrove management, economic empowerment, coastal policies, and their impact on community welfare.

In addition to secondary data, this study also utilized primary data obtained through semi-structured interviews. Interviews were conducted with informants selected using purposive sampling techniques, including individuals with knowledge and direct involvement in community-based mangrove management, such as mangrove management groups, coastal communities, and local leaders. The interviews aimed to gather in-depth information about CBMM practices, forms of community participation, perceived economic benefits, and challenges in implementation at the local level. These interviews served to complement and validate the findings from the SLR.

The data analysis technique in this study employed a thematic analysis approach combined with source triangulation (Miles et al., 2019). The collected data, from both the SLR and interviews, were then reduced, classified, and coded to identify patterns, relationships, and key themes. The analysis results were then critically interpreted to gain a comprehensive understanding of CBMM's role in poverty reduction in coastal areas. The triangulation process involved comparing findings from various sources, including scientific literature, policy documents, and field interviews, to ensure data consistency and validity. Thus, this research is expected to be able to produce in-depth analysis and provide a strong basis in formulating recommendations for sustainable community-based mangrove management policies.

3. RESULTS AND DISCUSSION

3.1. Conditions of Coastal Communities in Northern Central Java

The northern region of Central Java Province is a coastal area encompassing the coast from Brebes Regency to Rembang Regency, including coastal cities such as Semarang, Pekalongan, and Tegal, which directly border the Java Sea. The northern coast of Central Java boasts a mangrove forest ecosystem, sandy and muddy beaches (river deltas), and extensive shrimp and fishponds. These ecosystems are what cause coastal communities to rely on the abundant marine resources for their livelihoods (Fama, 2016). Therefore, some coastal communities work as fishermen, fish farmers, and other seafood processors, a tradition passed down from generation to generation (Sunartiningsih & Larasati, 2020). Coastal communities are characterized by a strong sense of community interaction, mutual cooperation, and mutual assistance. Although coastal areas possess abundant natural resources such as crabs, fish, shrimp, and various other marine life, this does not significantly impact the quality of life of coastal communities. They are often categorized as a vulnerable group and have a relatively low level of well-being compared to non-coastal areas. This is due to their high dependence on natural resources and limited access (Sofianto, 2017). Limited access to technology and fragmented governance hinder effective adaptation and resource management(Sutanto et al., 2022).

Economically, coastal communities, such as those in northern Central Java, face social vulnerabilities such as poverty, influenced by group participation and fishing distance. This is due to declining fish catches and non-adaptive behaviors, such as unsustainable fishing practices, which threaten long-term economic stability, resulting in low incomes and dependence on marine resources (Mulyadi, 2024; Prawito & Mulyasari, 2021). Research shows that poverty in coastal areas is not only caused by low incomes, but also by unequal access to resources, dependence on financiers (middlemen), and development policies that do not fully support traditional fishermen (Sofianto, 2017). This situation places coastal communities in a cycle of economic vulnerability that is difficult to break. Furthermore, industrial development in the northern coastal areas.

Furthermore, from an environmental perspective, the northern coastal areas of Central Java are vulnerable to ecological disasters such as significant erosion and flooding due to sea level rise and the conversion of mangrove ecosystems. Rudhi Pribadi, a mangrove expert from the Faculty of Marine Affairs and Fisheries at Diponegoro University, stated that in general, mangroves in the northern coast of Java, particularly in Semarang, have been degraded since the 1920s due to the clearing of shrimp ponds. Sea level rise, land subsidence, and tidal flooding also threaten livelihoods, agriculture, and infrastructure. The Semarang-Demak toll road project threatens the sustainability of 42.7 hectares of mangrove ecosystems, leading to the loss of mangrove forests. Not to mention the development of industrial areas on the coast, reinforced by the issuance of Presidential Regulation (Perpres) Number 60 of 2022, which has been damaged. Even the fate of mangrove forests is currently in jeopardy due to the impact of the construction of a toll road that also serves as a sea wall and industrial area(Turisno et al., 2026). Data from the Central Java Provincial Maritime Affairs and Fisheries Service (DKP), the impact of the Semarang-Demak toll road construction damaged 42.7 hectares of mangrove forests. The mangrove forest was directly affected by the construction of 14.1 hectares, and the indirect impact was 28.5 hectares. In detail, the affected area in Trimulyo includes a direct impact of 12.4 hectares, and an indirect impact of 21.9 hectares. Then in Sayung, Demak Regency, the direct impact was 1.6 hectares and the indirect impact was 6.4 hectares. The Bedono area, Demak Regency, was directly affected by 0.15 hectares, and indirectly by 0.15 hectares. The government's efforts to relocate mangroves for the toll road is not the right step. This was revealed by researchers Teran Yohana Putri and Fitri Liana Dewi in a study published in the 2023 Journal of Education and Counseling (JPDK) (Aninda Putri K et al., 2023). his is a clear example of the government's inconsistency in promoting mangrove conservation. However, they are also destroying it.

The impact of this phenomenon not only causes environmental damage but also directly impacts the socio-economic lives of communities, such as land loss, damaged settlements, and decreased fishing catches (Susanto & Nugrahajati, RR. Susilastuti Dwi Paripurno, 2025). In fact, in several areas, such as the coasts of Demak and Semarang, tidal flooding and abrasion have caused significant changes in the spatial structure of the region, including the submergence of settlements and productive land (Pandu, 2023).

In the face of these pressures, coastal communities have demonstrated a strong capacity for adaptation. Adaptation strategies include landfilling to elevate houses and infrastructure, despite the high costs (Marfai, 2012; Purnomo & Widiyanto, 2023; Solihuddin et al., 2021). Mangrove forest restoration and climate-resilient infrastructure have been proposed as sustainable solutions (Marfai, 2012; Suntoro et al., 2025) Communities along the northern coast of Central Java have diversified their livelihoods to reduce their dependence on the fisheries sector. Some coastal communities have begun adapting by changing occupations, such as shifting to tourism or ecotourism, which promotes economic resilience and empowerment (Prabawani et al., 2025). However, this adaptive capacity is still limited and is highly dependent on government policy support and sustainable community empowerment programs.

3.2. Model Community based Mangrove Management

Community based mangrove management (CBMM) is an approach to coastal resource management that places local communities as the main actors in the entire management process, from planning, implementation, to evaluation. This approach developed in response to the limitations of top-down management models that tend to overlook local contexts and are less capable of ensuring long-term sustainability. CBMM is not only understood as a technical conservation strategy but also as a governance framework that simultaneously integrates social, economic, and ecological dimensions(Irsadi et al., 2020).

CBMM is no longer understood as a simple participatory approach, but rather as part of a social-ecological system (SES) that involves dynamic interactions between humans and the environment (Nijamdeen et al., 2024). Mangroves do not only function as a biophysical ecosystem but also as a source of livelihood and an arena for social interactions influenced by interests, power, and policies (Otieno et al., 2026). The CBMM model develops in various typologies that reflect the diversity of local contexts, institutional capacities, and the dynamics of actor relationships.

The CBMM model can be classified into several main approaches. Each model has different operational characteristics, advantages, and limitations in addressing the challenges of mangrove management in coastal areas. However, in practice these models rarely stand alone and tend to operate by forming an adaptive hybrid model configuration. This hybrid approach allows for the integration of community participation, institutional mechanisms, economic incentives, and ecological conservation efforts within a more comprehensive management framework (Buncag, 2022; Kusumadewi et al., 2024). Here are various models in community-based mangrove management (CBMM).

a. Participatory Model

The participatory model is an approach in CBMM that places the community as the central actor in the entire management process, from the planning stage to evaluation. This approach is rooted in the assumption that community involvement not only enhances policy legitimacy but also strengthens the effectiveness and sustainability of coastal resource management. Participation does not emerge automatically, but is formed thru the interaction of various determinant factors, especially economic incentives, level of knowledge, and perceptions of the benefits of mangroves. Economic incentives have proven to be the most dominant in driving community involvement, even surpassing the influence of other social factors (Jadloc et al., 2025). This indicates that participation is better understood as a rational response to available economic opportunities, rather than merely a reflection of ecological awareness.

A positive perception of the importance of mangroves does not always transform into tangible actions in the form of active involvement, both in the decision-making process and program implementation (Qurniati et al., 2024; Wibisono et al., 2023). This phenomenon shows that cognitive awareness does not automatically result in collective action, but rather requires institutional support that can bridge the gap between knowledge and practice. Participation can develop to the level of delegated power, where the community has relatively wide autonomy in managing resources, although still within the framework of formal regulations (Majesty et al., 2018).

Participation tends to be limited when the community is only involved at the implementation stage without access to planning and decision-making. This condition gives rise to tokenism (Arnstein, 1969) which is participation that is symbolic without a substantive distribution of power. As a result, the community is positioned as implementers or labor providers, without access to strategic information or the authority to determine management direction, so participation does not lead to empowerment, but merely legitimizes top-down interventions. Therefore, the participatory model cannot stand alone and needs to be combined with other approaches such as co-management and institution-based governance to achieve sustainable and equitable mangrove management. Although it serves as the normative foundation of CBMM, this model still has structural limitations without strong institutional support and equitable power distribution.

b. Co-Management Model

The Co-management model or collaborative management emerges as an evolution of the participatory model that emphasizes collaboration between communities, governments, and other actors such as NGOs and the private sector. This approach is based on the principle of sharing authority and responsibility in resource management. In this model, the rights and responsibilities of resource management are formally divided between government authorities and local community groups(Martuti et al., 2020).

This approach is widely practiced in various countries, such as in Zanzibar, Tanzania, where the government grants partial management rights to communities thru Community Forest Management Agreements to reduce pressure on mangrove forests (Mohamed et al., 2023). Similarly, in Lampung, Indonesia, the success of mangrove restoration is significantly linked to the effective implementation of co-management between the Forestry Department, NGOs, and local fishing groups, as reflected in the drastic increase in dense mangrove cover by more than 100% over a decade (Hasani et al., 2023).

However, the effectiveness of co-management model heavily depends on the quality of governance, including transparency, coordination, and clarity of roles among actors (Ahmed et al., 2023). This model is also prone to failure if the division of roles and responsibilities is unclear or if there is a capacity imbalance between the state and the community, which often leads to state capture in the collaboration process (Mohamed et al., 2023). In many cases, weaknesses in coordination and overlapping authorities become major obstacles in the implementation of CBMM (Mohamed et al., 2023).

c. Livelihood-based Model

Livelihood-based models place the improvement of community economic welfare as the main entry point to achieve conservation goals. This approach is based on the premise that communities will be motivated to protect mangroves if they can derive tangible and sustainable economic benefits from the ecosystem. This approach is supported by findings that the success of CBMM is greatly influenced by the presence of direct economic benefits for the community, such as thru silvofishery, ecotourism, and the utilization of mangrove products (Basyuni et al., 2018; Tjahjono et al., 2022). Economic valuation studies also show that mangroves have significant economic value, but this value is often not internalized in policies, leading to unsustainable exploitation (Otieno et al., 2026; Tunyo et al., 2026).

d. Local Wisdom-Based Model

The Local Wisdom-Based Model is an approach that recognizes and integrates traditional knowledge systems, norms, and informal institutions that have long been embedded within communities as the foundation for mangrove management. This approach is grounded in the strength of customary leadership, strong networks of trust, and culturally embedded conflict resolution mechanisms. When communities are able to sustain and revitalize their local institutions, such as customary forest management committees that establish effective rules and enforce socially binding sanctions, the resulting condition of mangrove forests is significantly superior to those managed by the state (Sudtongkong & Webb, 2008).

Social capital which is the core of this model, can even be forged and strengthened by traumatic experiences. In post-conflict Sri Lanka, social cohesion that initially formed as a mechanism to survive amidst the crisis, instead transformed into an extraordinary collective energy, which then became the main foundation for the success of collective actions in restoring the damaged mangrove ecosystem (Sathiyamoorthy & Sakurai, 2024). Although it possesses the strength to create adaptive governance and high legitimacy at the local level, this model contains intrinsic limitations in terms of scalability. This approach is highly tied to specific and unique socio-cultural contexts, making it difficult to replicate on a large scale in other regions with different characteristics. Its resilience continues to be tested by the rapid penetration of modernization, demographic changes due to migration flows, and the pressures of the global market that tend to overlook local values in pursuit of short-term economic efficiency.

e. Ecological and Restoration-Based Model

The Ecological and Restoration-Based Model places the recovery of the structure and function of mangrove ecosystems as the main orientation in management. This model emphasizes integrated efforts in the rehabilitation and protection of the area. Unlike approaches that solely focus on quantitative seedling planting targets, this model is based on the understanding that the success of restoration cannot be reduced to just planting activities. A comprehensive synthesis of various CBMM case studies emphasizes that significant and sustainable ecological achievements are highly determined by a combination of more holistic interventions, including the protection of areas from external pressures, the development of economic alternatives for communities, and the strengthening of local community institutional capacities(Muhammad et al., 2025).

These findings underscore that successful ecological restoration cannot be separated from the dimensions of governance and the economic empowerment of the surrounding communities. The effectiveness of this model depends on its ability to adapt to the local context, as changes in the mangrove ecosystem are greatly influenced by location-specific spatial factors. Population density, distance to shrimp farming areas, and settlement patterns are the main determinants influencing the rate and direction of mangrove cover change (Purwanto et al., 2025). Consequently, the CBMM approach cannot be standardized, but must be contextual and location based.

f. Governance based Model

Local autonomy is often hindered by the phenomenon of recentralization, which occurs when the government formally decentralizes responsibilities but retains control over strategic decisions. In Kenya, the establishment of Community Forest Associations (CFA) faces complex and multi-tiered technical requirements. Communities must develop a management plan that requires multi-level bureaucratic approval, making the process lengthy, costly, and slow. This complexity often becomes an instrument to delay or restrict the transfer of rights to the community, reducing participation to an administrative formality (Kairu et al., 2024).

Moreover, weak coordination and overlapping authorities among institutions are the main obstacles to the implementation of CBMM. In Senegal, decentralization has actually triggered the fragmentation of authority and confusion at the local level due to unclear role distribution (Arumugam et al., 2020). In Zanzibar, the implementation of Community Forest Management Agreements is hindered by the unclear relationship between central and local authorities, as well as conflicts between formal law and customary practices (Mohamed et al., 2023). Meanwhile, in Vietnam, the involvement of multiple ministries and local authorities leads to policy overlaps and a lack of synchronization between central planning and local implementation (Ebeler et al., 2025). These conditions emphasize that without institutional clarity and effective coordination, decentralization has the potential to weaken governance.

The success of interventions generally requires strengthening community institutions as the main pillar. The combination of active restoration, rule enforcement, and institutional capacity development has proven to be the most effective in simultaneously generating ecological and socio-economic impacts (Kusumadewi et al., 2024). Therefore, an effective CBMM governance model demands clarity of roles among actors, strong and transparent coordination, and political commitment to distribute power more equitably to the community. Without fundamental governance reforms, the CBMM initiative risks becoming mired in conflicts of interest, bureaucratic inefficiencies, and structural injustices that ultimately threaten the sustainability of the mangrove ecosystem.

g. Conservation-Based Model

The conservation based model prioritizes ecological goals in mangrove management, with local communities acting as stewards or strategic partners. The targets include the expansion of protected areas, the restoration of critical habitats, and the increase of blue carbon stocks. This approach is generally driven by the climate change mitigation agenda and supported by the Payment for Ecosystem Services mechanism, which provides financial incentives to communities(Rahim et al., 2024). For example, the Socio Manglar program in Ecuador provides incentives to fishermen's associations to patrol and monitor the area, which has proven effective in protecting the forest while also strengthening the community's sense of ownership (Chávez-Páez & Beitl, 2025).

Effective conservation interventions cannot stand alone as exclusive protection efforts; rather, they must be integrated with active restoration and community institutional strengthening. The combination of rule enforcement, science-based restoration, and strengthening local institutions has proven to be the most effective in generating dual impacts, namely the increase in carbon stocks, improvement in land cover, and enhancement of community welfare (Kusumadewi et al., 2024). These findings are consistent with a study in Vietnam that shows conservation-oriented mangrove restoration enhances well-being thru fishery productivity and ecotourism opportunities, although the distribution of benefits is not yet equitable (Ebeler et al., 2025). In Southern Thailand, community-based mangrove management also results in better ecosystem quality, with basal area reaching 78.3 m^2/ha, far exceeding the state forest which is only 15.4 m^2/ha (Sudtongkong & Webb, 2008). This confirms that community-based conservation is not only socially superior but also more ecologically effective.

However, overly rigid conservation approaches that are focused on global targets often overlook the short-term socio-economic needs of local communities. A study in Bali shows that although the community's knowledge and perception of mangroves are high, actual participation remains low due to the lack of significant economic benefits (Utami et al., 2018). When access to resources is restricted without compensation or alternative livelihoods, resistance and social conflict tend to increase. Therefore, the success of conservation models requires a balance between achieving ecological targets and meeting the economic needs of the community. Without that balance, conservation has the potential to trigger injustice and new conflicts in coastal areas.

3.3. Ecological Impact of Community-Based Mangrove Management

Community based Mangrove Management (CBMM) is a coastal management approach that simultaneously impacts the socio-economic and ecological dimensions. In the framework of the social-ecological system (SES), CBMM functions as a mechanism that connects the collective actions of the community with the biophysical dynamics of the ecosystem. Therefore, its success is not only measured by participation or improvements in welfare but also by its ability to restore the structure, function, and resilience of the mangrove ecosystem.

The ecological impact is the result of the convergence of key reinforcing factors, namely local autonomy, effective leadership, economic incentives, and adequate external support. This combination forms a social ecosystem conducive to successful management. In this context, CBMM encompasses two main dimensions, namely ecological achievements and the social-institutional prerequisites that enable the attainment of these achievements.

Empirically, the ecological impact of CBMM is evident in the increase in coverage, density, and structure of mangrove vegetation. In Teluk Pandan, Lampung, the implementation of co-management over a decade increased dense mangrove categories by 109.71% and reduced sparse mangroves by 49.35% (Hasani et al., 2023). This achievement shows that clarity of ownership and local autonomy foster a strong sense of belonging, enabling the community to actively participate as ecosystem guardians with a high level of commitment.

A similar phenomenon is observed in Southern Thailand, particularly in the villages of Tong Tasae and Laem Makham, where the community forms management committees, establishes zoning, and autonomously designs sanction systems. As a result, the forest structure is better preserved with a basal area reaching 78.3 and 22.1 m^2/ha, much higher compared to the surrounding state forests which only have 15.4 and 9.8 m^2/ha (Sudtongkong & Webb, 2008). At the regional scale, public participation in reforestation policies and programs has also been proven to increase the extent of mangroves in Southeast Asia (Koga et al., 2026). This shows that CBMM not only has a local impact but also cumulative implications on a broader landscape.

The success of knowledge integration is highly determined by local leadership and the strength of social capital. Autonomy will not be effective without key figures capable of mobilizing collective action and building consensus at the community level. In Southern Thailand, the community's ability to withstand the pressure of commercial charcoal concessions is closely linked to the role of charismatic leaders operating within a strong context of social capital (Sudtongkong & Webb, 2008).

Social capital does not always develop in a linear manner, rather it is often shaped by experiences of crisis. In Sri Lanka, prolonged exposure to conflict and civil war has, paradoxically, strengthened social cohesion and community solidarity (Sathiyamoorthy & Sakurai, 2024). This cohesion subsequently transforms into collective capacity that supports the success of post-conflict mangrove restoration. As a result, conflict-affected communities in certain cases demonstrate higher restoration outcomes compared to others.

The ecological impacts of CBMM are also reflected in the recovery of essential ecosystem functions, including coastal protection, hydrological regulation, and carbon sequestration (blue carbon)(Ariyanti et al., 2025). Strengthening mangrove cover through community-based initiatives has been shown to enhance coastal protection capacity, particularly in reducing the impacts of tidal surges and tsunamis (Gultom et al., 2021). Community-restored and managed mangrove forests along the Demak coastline, especially in Bedono Village, have significantly improved wave attenuation capacity, where a 200-meter-wide greenbelt has been shown to reduce incoming wave height by more than 50 percent (Damastuti et al., 2023).

The sustainability of ecosystem function recovery heavily depends on the consistency of community participation, which can only be maintained if supported by tangible economic incentives. An approach that relies solely on moral awareness has proven inadequate to encourage long-term engagement. In Davao City, Philippines, community participation is primarily driven by expected economic benefits, such as increased income and job stability (Reciproco et al., 2023). On the other hand, in the Ngurah Rai Forest Park, Bali, high levels of knowledge and perception do not correlate with participation due to the lack of direct economic benefits (Utami et al., 2018). These findings emphasize that conservation policies need to substantively integrate economic dimensions, so that conservation is not only an ecological obligation but also provides financial benefits to the community.

However, these achievements are highly contingent upon the role of communities supported by external actors acting as linking agents. Non-governmental organizations, academics, and donor institutions function as catalysts by providing technical support and facilitating access to policy networks. In Southern Thailand, the Yadfon Foundation plays a critical role in training and advocacy, effectively bridging communities with bureaucratic structures (Sudtongkong & Webb, 2008). A synthesis of multiple studies further indicates that the combination of active restoration, rule enforcement, and the development of community-based business models represent the most effective strategy for achieving ecological and socio-economic outcomes simultaneously. Without the integration of meaningful participation, institutional capacity, science-based design, and external support, CBMM risks producing limited and unsustainable impacts.

3.4. Socio-Economic Impact of Mangrove Management on Poverty

While mangrove restoration is significantly effective in maintaining the stability of coastal areas, many scholars believe that it would also benefit the surrounding community in terms of the socio-economic impact. The restoration strategy brings a new economic opportunity that's not only focus on the fishery activities but also the utilization of the mangrove ecosystem in any form, from edutourism to handcraft based product. This condition results in the coastal community economic development particularly the poverty alleviation.

a. Poverty alleviation through livelihood diversification

Livelihood diversification plays a critical role in improving the community economic resilience. Through diversification, the community had a more access in a multiple stream of livelihood income that could reduce their dependency in a single and conventional income sources. The mangrove management initiatives, such as rehabilitation and conservation, have created opportunities for income diversification. For instance, mangrove-based ecotourism and sustainable resource utilization (e.g. fishing, aquaculture) have provided alternative livelihood for coastal communities, reducing their dependence on traditional, often vulnerable, income sources (Elida et al., 2025). While the traditional income strategy is vulnerable to environment degradation during climate variability, the livelihood diversification offers a better opportunity for coastal communities by integrating the ecological rehabilitation with the economic activities.

In the Surodadi village of Central Java, the extraction of mud crabs from managed mangrove areas provides an annual monetary value of approximately 1,928 USD per household (Damastuti & de Groot, 2017). This demonstrates a significant contribution of alternative income stream to the local community. Furthermore, in Jangkaran Village, the community has successfully developed alternative livelihood such as shrimp pond management and ecotourism activities, which have improved socio-economic resilience through varied income sources. In Ayah village, the potential for income generation is further demonstrated through tourism services, including ticket sales, boat rentals, and the marketing of local food and souvenirs (AFIFAH et al., 2023). Consequently, the transition from mono-sectoral livelihoods to diversified mangrove-based activities provides a sustainable foundation for improving the economic welfare of coastal residents (Liza & Santoso, 2025).

b. Resilience against socio-economic vulnerabilities

Regardless of its natural resource potential, the coastal areas also give a vulnerable impact towards the community through many environmental hazards such as natural disasters, abrasion, and periodic sea tides. In this situation, mangrove ecosystem function as a natural infrastructure that enhances the resilience of coastal communities against socio-economic vulnerabilities triggered by environmental hazards. The mangrove forest physical morphology/structure creates a protective buffer that mitigate the impacts of natural disasters in coastal areas as in tidal flooding or coastal erosion. This protective mechanism protects the vulnerable community against financial asset depletion.

Analysis in Tugu District of Semarang shows that residents in areas with stable mangrove coverage, such as Tugurejo, possess higher resilience scores compared to regions where the vegetation has been lost (Martuti et al., 2022). Align with this finding, scientific modeling of the Semarang and Demak coastline indicates that mangrove deforestation is a major driver increased flood areas, which have expanded by an average of 84 hectares per year (Widjonarko et al., 2025). On the other hand, in Bedono Village, community led efforts have utilized hybrid structures to restore land and protect housing from permanent inundation (Damastuti & de Groot, 2017). These findings evidence that the mangrove management positively supports the long-term socio-economic stability of coastal households by reducing the frequency of coastal disaster impacts.

c. Positive Multiplier Effects

Successful mangrove community-based management not only gives a direct impact towards community income. The initiatives generate positive multiplier effects beyond immediate financial gains to support regional development. The collective action that focus on the restoration of biodiversity and ecosystem services creates a foundation for educational tourism and scientific research, providing long term benefit to the broader social-ecological system. The environmental sustainability act as a living laboratory that not only providing knowledge but also preserving the nature for further generation. The empirical evidence presented by study in Karawang Regency which reveal that rehabilitation programs improve ecosystem quality while simultaneously increasing community participation in environmental maintenance(Kurniati et al., 2024). Programs focused on education ecotourism have successfully fostered stakeholder partnership that support broader sustainable development goals (Rakhmanissazly et al., 2018). Furthermore, the presence of healthy mangrove belts supports the preservation of life support system and provides habitats for aquatic biotas that are economically valuable to local fishers (AFIFAH et al., 2023). In summary, the integration of social economic, and ecological goals through community management ensures a holistic approach to achieve regional sustainability.

Some studies had provide an empirical evidence on significant potential of mangrove management in poverty alleviation across several study locus (Abbas et al., 2025; Gravitiani et al., 2018). However, some institutional technical constrains limit the program full potential as a tool for poverty alleviation. One of the inhibiting factors is the lack of financial and limited technical expertise mastered by the community involved. This prevent locals from scaling the conservation efforts into sustainable business models. Some supporting interventions are required to support the long term program implementation such as, training, resources mobilization, and financial support. Research in Semarang City identifies that despite a high willingness to participate in management, low financial capital and limited technological skills remain primary obstacles for local groups (Martuti et al., 2022). Many coastal households in Northern Java are characterized by high livelihood vulnerability indices due to their limited adaptive capacity and exposure to climate stressors (Gravitiani et al., 2018). Additionally, insufficient market access hinders the ability of communities to maximize the revenue potential of sustainable mangrove products (Abbas et al., 2025). These systemic obstacles indicate that targeted technical and financial support are necessary to ensure that conservation efforts result in meaningful poverty reduction.

3.5. Challenges in Implementing Community-Based Mangrove Management

Community based mangrove management is a human-driven innovation aimed at environmental sustainability, focusing on coastal management through the optimization of mangroves to address coastal erosion. This approach remains a viable long-term strategy. However, such optimization cannot be fully sustained by future generations. Economic priorities often underlie the lack of community participation in the agricultural sector. The community is now beginning to shift its employment focus toward the industrial sector, particularly as more industrial zones are being developed in coastal areas.

The participatory approach positions local communities as key actors in the sustainable management of coastal ecosystems. This approach is considered effective in maintaining a balance between ecological and economic interests, particularly in mitigating coastal erosion, enhancing coastal resilience, and providing alternative livelihoods. However, the implementation of CBMM faces various multidimensional challenges that could potentially hinder the program's long-term sustainability. Some aspects that may pose challenges to the sustainability of community-based mangrove management are as follows.

a. Social economic

Socioeconomic challenges are a crucial factor in determining the level of participation among coastal communities. Structural poverty often forces communities to exploit mangrove resources unsustainably in order to meet short-term needs. This is consistent with the findings of Walters et al. (2008), who state that economic pressures are the primary determinant of mangrove degradation in developing countries. Additionally, social inequality and unequal income distribution further weaken social cohesion in community-based management (D. Datta et al., 2012). On the other hand, time constraints and the younger generation's low interest in the natural resource based sector pose challenges for the regeneration of CBMM practitioners.

A study by (Friess et al., 2019) emphasizes that the lack of youth involvement in mangrove conservation threatens the long term sustainability of communitybased management, particularly in the coastal regions of Southeast Asia.

b. Institution

Weak institutional capacity poses a significant barrier to the implementation of CBMM. The lack of law enforcement against environmental violations such as the conversion of mangroves into fish ponds or industrial zones highlights a gap between policy and on-the-ground implementation. Primavera et al. (2019) emphasize that the success of mangrove rehabilitation heavily depends on strong policy support and consistent monitoring. Furthermore, the criminalization of coastal communities in resource management often arises due to overlapping regulations and unclear management rights. This can erode public trust in the government and weaken collective participation (Armitage et al., 2009).

c. External and environmental

From an external perspective, the conversion of mangrove land into economic zones such as fish ponds, settlements, and tourism areas poses the primary threat to the sustainability of the ecosystem. Friess et al. (2019) found that Southeast Asia is experiencing the fastest rate of mangrove loss in the world due to land-use change. Additionally, environmental pollution, particularly plastic waste, further exacerbates the degradation of mangrove ecosystems. Plastic waste not only disrupts the ecological functions of mangroves but also impacts biodiversity(Halim et al., 2023). Low levels of public education regarding environmental management and coastal disaster mitigation pose a significant challenge. Insufficient environmental literacy leads to management practices that are not adaptive to climate change and disaster risks. According to Alongi (2015), building community capacity through education and training is key to strengthening the resilience of mangrove ecosystems and coastal communities.

3.6. Ideal Model for Community based Mangrove Management to Reduce Poverty

Based on the analysis, it was found that mangrove management can contribute to poverty reduction in coastal areas. Efforts implemented through a composite strategy are considered the most effective in combining cross sectoral policies by integrating ecosystem conservation, economic empowerment of households (fishermen and women's groups), and ecological risk mitigation to achieve long term sustainability.

A composite strategy for community-based mangrove management is an integrated approach that simultaneously combines ecological, economic, and social dimensions. This strategy is considered the most effective because it integrates cross-sectoral policies into a mutually reinforcing framework. This approach emphasizes not only the conservation of mangrove ecosystems but also the economic empowerment of coastal communities, particularly fishermen and women's groups (Alongi, 2020). Thus, mangrove management is no longer sectoral in nature but has become an inclusive sustainable development strategy. Multi sectoral integration can significantly enhance the effectiveness of coastal resource management (Primavera, 2000).

The key pillars of this strategy include cross-sectoral integration, active community participation, and the restoration of ecosystem services. Cross-sectoral integration enables synergies between environmental, economic, and social policies, thereby creating broader and more sustainable impacts (D. Datta et al., 2012; K. Datta et al., 2012). Meanwhile, active community participation is key to success, as it involves local actors in the decision-making process and the implementation of restoration programs. The restoration of mangrove ecosystem services also plays a vital role in enhancing ecological functions such as carbon sequestration, coastal protection, and marine habitat(Pringgenies et al., 2026). Direct community involvement can double the success of restoration efforts compared to top-down approaches (Pelyukh et al., 2021).

The poverty reduction measures in this strategy are implemented through the development of silvofishery and ecotourism, the diversification of mangrove-based products, and the strengthening of the role of non-governmental organizations (NGOs). Silvofishery combines aquaculture with mangrove conservation, thereby providing economic benefits without damaging the environment. Product diversification, such as processed non-timber mangrove products, opens opportunities for home-based industries that can increase the income of coastal communities. Additionally, the role of NGOs is highly strategic in enhancing community capacity, including in digital marketing and technology transfer. This approach has proven effective in boosting local economic resilience while maintaining ecosystem sustainability (Farella et al., 2020; Huxham et al., 2018).

The simulation results show that the composite strategy has a far more significant impact than the zero action scenario. Over a 10 year period, chronic poverty rates can be reduced to near zero, while fisheries production and the success of ecological restoration increase dramatically. Conversely, without an integrated strategy, poverty tends to rise and ecosystem conditions degrade. These findings underscore the importance of a community-based approach to mangrove management as a long-term solution for poverty alleviation. Thus, the composite strategy is not only ecologically relevant but also has tangible socio-economic impacts.

CONCLUSIONS

Community based Mangrove Management (CBMM) on the northern coast of Central Java has proven to be an effective strategy for addressing environmental degradation while reducing poverty. Through active participation in every stage from planning to evaluation, local communities have transitioned from being mere objects to becoming key actors in safeguarding the sustainability of coastal ecosystems. The success of this model heavily depends on tangible economic incentives for residents, such as the development of ecotourism, the utilization of non-timber products, and increased fishing yields. Thus, mangrove restoration not only ecologically restores the landscape but also fosters economic self-reliance among vulnerable communities. This participatory approach has proven capable of delivering better conservation outcomes compared to management carried out entirely by the state.

Although it has had a positive impact, the implementation of CBMM still faces major challenges in the form of weak inter-agency coordination and conflicts of interest in infrastructure development policies. Large-scale projects, such as the construction of a toll road in the Semarang-Demak region, often threaten the sustainability of mangrove ecosystems that have been painstakingly restored by local residents. Therefore, strengthening the capacity of local institutions and integrating traditional knowledge with modern science are key to maintaining the resilience of coastal areas. Consistent government policy support and fair distribution of authority are necessary so that the benefits of management can be felt equally by all segments of society. The sustainability of this program ultimately depends on the synergy between ecological protection, economic well-being, and transparent governance.

REFERENCES

Abbas, M. H. I., Qurrata, V. A., Yusida, E., Priambodo, M. P., Mohd. Shafiai, M. H., Saputra, J., & Cahayati, N. (2025). Sustainable Development in Coastal Regions: Integrating Blue Economy and Community Ecotourism for Poverty Eradication. *Environment and Ecology Research, 13*(2), 198–212. DOI: 10.13189/eer.2025.130202

Afifah, R. N., Putri, A., Hartanti, A. N., Negari, S. I. T., Pratama, M. S. R., Zuaini, P. A. K., Al Madani, A. R., Muryanto, B. S., Muhammad, F., Astikasari, L., Indriyani, S., Kurniawati, I., Sunarto, S., Kusumaningrum, L., Budiharta, S., Flores, A. B., & Setyawan, A. D.AFIFAH. (2023). Ecotourism development as a community-based conservation effort in Ayah Mangrove Forest, Kebumen, Central Java, Indonesia. *Asian Journal of Forestry, 7*(1). Advance online publication. DOI: 10.13057/asianjfor/r070105

Ahmed, J., Kathambi, B., & Kibugi, R. (2023). POLICY PERSPECTIVE ON GOVERNANCE STANDARDS SETTING USING COMMUNITY PARTICIPATION FOR SUSTAINABLE MANGROVE MANAGEMENT IN LAMU KENYA. *International Journal of Conservation Science, 14*(1), 315–326. DOI: 10.36868/IJCS.2023.01.20

Alongi, D. M. (2015). The Impact of Climate Change on Mangrove Forests. *Current Climate Change Reports, 1*(1), 30–39. DOI: 10.1007/s40641-015-0002-x

Alongi, D. M. (2020). Global Significance of Mangrove Blue Carbon in Climate Change Mitigation. *Sci, 2*(3), 67. DOI: 10.3390/sci2030067

Aninda Putri, K., Mulyawan, A., Annizar, B., Santoso, H. C., Arifianto, I., Nashr, J. A., & Yusuf, M. D. (2023). *PD, M. R., Wibisono, P., & A, U. A.* HITAM PEKAT HIDUP WARGA DI JAWA TENGAH Laporann Kolaboratif LBH Semarang Bersama Media Jawa Tengah.

Ariyanti, D., Sugianto, D. N., Purbasari, A., Azizah, N. U., Nurwidiyanto, A. R., & Gao, W. (2025). Activated mangrove biochar for sustainable carbon sequestration. *E3S Web of Conferences, 655*, 01028. DOI: 10.1051/e3sconf/202565501028

Armitage, D. R., Plummer, R., Berkes, F., Arthur, R. I., Charles, A. T., Davidson-Hunt, I. J., Diduck, A. P., Doubleday, N. C., Johnson, D. S., Marschke, M., McConney, P., Pinkerton, E. W., & Wollenberg, E. K. (2009). Adaptive co-management for social–ecological complexity. *Frontiers in Ecology and the Environment, 7*(2), 95–102. DOI: 10.1890/070089

Arnstein, S. R. (1969). A Ladder Of Citizen Participation. *Journal of the American Institute of Planners, 35*(4), 216–224. DOI: 10.1080/01944366908977225

Arumugam, M., Niyomugabo, R., Dahdouh-Guebas, F., & Hugé, J. (2020). The perceptions of stakeholders on current management of mangroves in the Sine-Saloum Delta, Senegal. *Estuarine, Coastal and Shelf Science, 247*. Advance online publication. DOI: 10.1016/j.ecss.2020.106751

Basyuni, M., Yani, P., & Hartini, K. S. (2018). Evaluation of mangrove management through community-based silvofishery in North Sumatra, Indonesia. *IOP Conference Series. Earth and Environmental Science, 122*(1), 012109. Advance online publication. DOI: 10.1088/1755-1315/122/1/012109

Berkes, F. (2009). Evolution of co-management: Role of knowledge generation, bridging organizations and social learning. *Journal of Environmental Management, 90*(5), 1692–1702. DOI: 10.1016/j.jenvman.2008.12.001 PMID: 19110363

Buncag, M. J. J. (2022). Community-Based Mangrove Forest Management Sustainability Analysis in Tagpait, Aborlan and Bacungan, Puerto Princesa City, Palawan, Philippines. *Environment and Ecology Research, 10*(3), 325–333. DOI: 10.13189/eer.2022.100301

Chambers, R. (1994). Participatory rural appraisal (PRA): Challenges, potentials and paradigm. *World Development, 22*(10), 1437–1454. DOI: 10.1016/0305-750X(94)90030-2

Chávez-Páez, W., & Beitl, C. M. (2025). Who holds the power in the absence of the state? Polycentric governance and community justice in Ecuador's mangrove management. *Environmental Development, 56*, 101263. Advance online publication. DOI: 10.1016/j.envdev.2025.101263

Creswell, J. W. (2009). Research Design: Qualitative, Quantitative, and Mixed Methods Approaches. In *SAGE Publications, Inc. All.* DOI: 10.1016/j.math.2010.09.003

Damastuti, E., & de Groot, R. (2017). Effectiveness of community-based mangrove management for sustainable resource use and livelihood support: A case study of four villages in Central Java, Indonesia. *Journal of Environmental Management, 203*, 510–521. DOI: 10.1016/j.jenvman.2017.07.025 PMID: 28841518

DasGupta, R., & Shaw, R. (2013). Cumulative Impacts of Human Interventions and Climate Change on Mangrove Ecosystems of South and Southeast Asia: An Overview. *Journal of Ecosystems, 2013*, 1–15. DOI: 10.1155/2013/379429

Datta, D., Chattopadhyay, R. N., & Guha, P. (2012). Community based mangrove management: A review on status and sustainability. *Journal of Environmental Management, 107*, 84–95. DOI: 10.1016/j.jenvman.2012.04.013 PMID: 22595074

Datta, K., Baisakh, N., Ganguly, M., Krishnan, S., Yamaguchi Shinozaki, K., & Datta, S. K. (2012). Overexpression of *Arabidopsis* and Rice stress genes' inducible transcription factor confers drought and salinity tolerance to rice. *Plant Biotechnology Journal, 10*(5), 579–586. DOI: 10.1111/j.1467-7652.2012.00688.x PMID: 22385556

Ebeler, L., Balke, T., Laurie, E., Le, H., Nguyen, T., & Renaud, F. G. (2025). Local-scale impacts of mangrove restoration and conservation on coastal communities in two Vietnamese deltas: Socio-economic and institutional dynamics. *Ocean and Coastal Management, 269*, 107812. Advance online publication. DOI: 10.1016/j.ocecoaman.2025.107812

Elida, F., Anggoro, S., Putro, S. P., & NR Wardhani, W. (2025). Conceptual framework for assessing the ecotourism carrying capacity of mangrove areas at ujung piring beach, Mlonggo. *E3S Web of Conferences, 605*, 03007. DOI: 10.1051/e3sconf/202560503007

Fama, A. (2016). Komunitas Masyarakat Pesisir Di Tambak Lorok, Semarang. *Sabda, 11*(2), 65–75.

FAO. (2023). The world's mangroves 2000–2020. In *The world's mangroves 2000–2020*. FAO., DOI: 10.4060/cc7044en

Farella, G., Menegon, S., Fadini, A., Depellegrin, D., Manea, E., Perini, L., & Barbanti, A. (2020). Incorporating ecosystem services conservation into a scenario-based MSP framework: An Adriatic case study. *Ocean and Coastal Management, 193*, 105230. DOI: 10.1016/j.ocecoaman.2020.105230

Friess, D. A., Rogers, K., Lovelock, C. E., Krauss, K. W., Hamilton, S. E., Lee, S. Y., Lucas, R., Primavera, J., Rajkaran, A., & Shi, S. (2019). The State of the World's Mangrove Forests: Past, Present, and Future. *Annual Review of Environment and Resources, 44*(1), 89–115. DOI: 10.1146/annurev-environ-101718-033302

Gravitiani, E., Fitriana, S. N., Bernadeta, R., & Pratiwi, S. R. (2018). Climate change impact to socio-economy vulnerability in northern Java coastal area. *IOP Conference Series. Earth and Environmental Science, 200*, 012058. DOI: 10.1088/1755-1315/200/1/012058

Halim, M. A. R., Soeprobowati, T. R., & Hadiyanto, H. (2023). Ecological Assessment of the Role of Mangrove Trees in Carbon Sequestration and Biodiversity In Karimunjawa National Park Indonesia. *Geography, Environment, Sustainability, 16*(3), 32–42. DOI: 10.24057/2071-9388-2022-2565

Hasani, Q., Anisa, A., Damai, A. A., Yuliana, D., Yudha, I. G., & Julian, D. (2023). Changes in density level and mangrove land cover on Teluk Pandan Coast, Lampung, Indonesia, after 10 years of community-based management. *Biodiversitas (Surakarta), 24*(7), 3735–3742. DOI: 10.13057/biodiv/d240710

Huxham, M., Whitlock, D., Githaiga, M., & Dencer-Brown, A. (2018). Carbon in the Coastal Seascape: How Interactions Between Mangrove Forests, Seagrass Meadows and Tidal Marshes Influence Carbon Storage. *Current Forestry Reports, 4*(2), 101–110. DOI: 10.1007/s40725-018-0077-4

Irsadi, A., Anggoro, S., & Soeprobowati, T. R. (2020). Mangrove conservation and its implication on community life of Bedono Village. *Demak, Central Java, 040041*, 040041. Advance online publication. DOI: 10.1063/5.0002627

Jadloc, J. A., Legaspina, T., Gabrinao, W. P., & Gorgonio, C. (2025). Drivers that motivate participation in mangrove forest management among the local people at Barangay Lasang, Davao City, Philippines. *BIO Web of Conferences, 157*. DOI: 10.1051/bioconf/202515704005

Kairu, A., Mbeche, R., Kotut, K., & Kairo, J. (2024). From centralization to decentralization: Evolution of forest policies and their implications on mangrove management in Kenya. *Forest Policy and Economics, 168*, 103290. Advance online publication. DOI: 10.1016/j.forpol.2024.103290

Koga, K., Kayo, C., Quevedo, J. M. D., & Kohsaka, R. (2026). What drives mangrove forest gains? Evidence from six Southeast Asian countries using fixed effects panel data models. *Forest Ecology and Management, 601*, 123330. Advance online publication. DOI: 10.1016/j.foreco.2025.123330

Kurniati, R., Handayani, W., Prabawani, B., Nida, R. S., & Savitri, S. F. (2024). COMMUNITY PARTICIPATION, COMMUNITY READINESS, AND PREFERENCES TO PROMOTE EDU-TOURISM IN MANGROVE AREAS OF TELUK AWUR VILLAGE, JEPARA, INDONESIA. *PLANNING MALAYSIA, 22*. Advance online publication. DOI: 10.21837/pm.v22i33.1555

Kusumadewi, S. D., Purnomo, H., Nadhira, S., & Putrizulfan, R. (2024). Systematic review on the implementation of mangrove community-based restoration in Indonesia and beyond. *IOP Conference Series. Earth and Environmental Science, 1315*(1), 012052. Advance online publication. DOI: 10.1088/1755-1315/1315/1/012052

Liza, N., & Santoso, B. J. (2025). Socio-economic resilience of coastal communities in the mangrove rehabilitation area of the Bogowonto River, Indonesia. *IOP Conference Series. Earth and Environmental Science, 1506*(1), 012014. DOI: 10.1088/1755-1315/1506/1/012014

Majesty, K. I., & Fadmastuti, M. (2018). Degree of community participation in mangrove resources management as livelihood support in West Java, Indonesia. *E3S Web of Conferences, 74*. DOI: 10.1051/e3sconf/20187410005

Marfai, M. A. (2012). Preliminary assessment of coastal erosion and local community adaptation in sayung coastal area, central java -indonesia. *Quaestiones Geographicae, 31*(3), 47–55. DOI: 10.2478/v10117-012-0028-2

Martuti, N. K. T., Pribadi, R., Dewi, N. K., Sidiq, W. A. B. N., & Mutiatari, D. P. (2022). Analysis of environment, socio-economic, and stakeholder partnership for integrated coastal management in Semarang City, Indonesia. *Gestão Costeira Integrada, 22*(1), 11–25. DOI: 10.5894/rgci-n431

Martuti, N. K. T., Pribadi, R., Dewi, N. K., Sidiq, W. A. B. N., & Nugraha, S. B. (2020). The Dynamics of Coastline and Mangrove Ecosystems in Coastal Area of Mangkang Kulon Subdistrict, Semarang. *IOP Conference Series. Earth and Environmental Science, 550*(1), 012011. DOI: 10.1088/1755-1315/550/1/012011

Miles, M. B., Huberman, A. M., & Saldaña, J. (2019). *Qualitative data analysis: A methods sourcebook*. SAGE Publications, Inc.

Mohamed, M. K., Adam, E., & Jackson, C. M. (2023). Policy Review and Regulatory Challenges and Strategies for the Sustainable Mangrove Management in Zanzibar. *Sustainability (Basel), 15*(2), 1557. Advance online publication. DOI: 10.3390/su15021557

Muhammad, F., Marsum, M., Putro, S. P., Hadi, M., Murningsih, M., Hidayat, J. W., Ahmad, W., Nabiha, P. I., & Wardoyo, S. (2025). Vulnerability analysis and sustainable rehabilitation strategies for mangrove ecosystems in Bedono coastal area, Central Java. *BMC Ecology and Evolution, 25*(1), 86. DOI: 10.1186/s12862-025-02406-7 PMID: 40877764

Mulyadi, A. (2024). Adaptation of climate change and environmental pollution in Cirebon's coastal community. In N. null, I. M.H., & G. K. (Eds.), *E3S Web of Conferences* (Vol. 600). EDP Sciences. DOI: 10.1051/e3sconf/202460005005

Nijamdeen, T. W. G. F. M., Peruzzo, S., Kodikara, K. A. S., Ratsimbazafy, H. A., Nijamdeen, T. W. G. F. A., Thahira, T., Sajeevan, T., Kugathasan, D., Hugé, J., & Dahdouh-Guebas, F. (2024). Stakeholder perceptions in mangrove management in the Jaffna Peninsula, Sri Lanka. *Forest Policy and Economics, 164*, 103236. Advance online publication. DOI: 10.1016/j.forpol.2024.103236

Otieno, P., Owuor, M., Pacheco, C. F., Coppo, G. C., Santos, T. M., Ferreira, T. O., & Bernardino, A. F. (2026). Valuing mangrove ecosystem services along the Brazilian Amazon coast. *Ecosystem Services*, *78*, 101831. Advance online publication. DOI: 10.1016/j.ecoser.2026.101831

Pandu, P. (2023). *Adaptasi yang Menjaga Asa Masyarakat Pesisir Utara Jawa.* Kompas.

Pelyukh, O., Lavnyy, V., Paletto, A., & Troxler, D. (2021). Stakeholder analysis in sustainable forest management: An application in the Yavoriv region (Ukraine). *Forest Policy and Economics*, *131*, 102561. DOI: 10.1016/j.forpol.2021.102561

Prabawani, B., Nugroho Sugianto, D., Ariyanti, D., Pribadi, R., Wigati, M., & Fisher, M. R. (2025). Turning Tides: Sustainable Community Practices in Restoring Mangrove Ecosystems in Coastal Riau. *E3S Web of Conferences, 673*, 01004. DOI: 10.1051/e3sconf/202567301004

Prawito, P., & Mulyasari, G. (2021). Comparative social vulnerability of fishermen in the coastal area of bengkulu and central java, Indonesia. *Aquaculture, Aquarium, Conservation & Legislation*, *14*(5), 3045–3054.

Primavera, J. H. (2000). Development and conservation of Philippine mangroves: Institutional issues. *Ecological Economics*, *35*(1), 91–106. DOI: 10.1016/S0921-8009(00)00170-1

Primavera, J. H., Friess, D. A., Van Lavieren, H., & Lee, S. Y. (2019). The Mangrove Ecosystem. In Sheppard, C. (Ed.), *World Seas: An Environmental Evaluation* (2nd ed., pp. 1–34). Academic Press., https://doi.org/https://doi.org/10.1016/B978-0-12-805052-1.00001-2 DOI: 10.1016/B978-0-12-805052-1.00001-2

Pringgenies, D., Setyati, W. A., & Ariyanto, D. (2026). Bacterial presence and nutrient dynamics in mangrove crab mounds: Links to blue carbon revealed by molecular approaches. *Estuarine, Coastal and Shelf Science*, *334*, 109790. DOI: 10.1016/j.ecss.2026.109790

Purnaweni, H., Hadi, S. P., Soeprobowati, T. R., Roziqin, A., & Rakhma, N. A. (2026). *What are the Main Concerns for Global Mangrove Management Research? Systematic Literature Review and Bibliometric Analysis*. Research in Ecology., DOI: 10.30564/re.v8i2.12149

Purnaweni, H., Soeprobowati, T. R., Chatterjee, U., Kiswoyo, G. P., Rakhma, N. A., & Manik, J. S. T. (2025). The Role of Stakeholders in Community Based Mangrove Management in Tapak Area of Semarang City. *E3S Web of Conferences, 650*, 02047. DOI: 10.1051/e3sconf/202565002047

Purnomo, S. N., Widiyanto, W., & Suroso, . (2023). Landfilling as individual adaptation for coastal flooding (Rob) countermeasures in North Central Java. *IOP Conference Series. Earth and Environmental Science, 1173*(1), 012057. Advance online publication. DOI: 10.1088/1755-1315/1173/1/012057

Purwanto, A. D., Wikantika, K., Deliar, A., Darmawan, S., Suprapedi, , Saepuloh, A., Handayani, A. P., Wisha, U. J., Rahmawati, E., Falah, F. N., & Prayudha, B. (2025). Suprapedi, Saepuloh, A., Handayani, A. P., Wisha, U. J., Rahmawati, E., Falah, F. N., & Prayudha, B. (2025). Spatial Heterogeneity of Mangrove Changes and Its Local Driving Factors on the Southern Coast of Java Island, Indonesia. *Scientifica, 2025*(1), 6670733. Advance online publication. DOI: 10.1155/sci5/6670733

Qurniati, R., Duryat, D., Darmawan, A., & Inoue, M. (2024). Ecological Perspective, Perception, and Attitude of Local Communities Toward Managing and Utilizing the Mangrove Ecosystem in Lampung Province, Indonesia. *Small-scale Forestry, 23*(3), 471–491. DOI: 10.1007/s11842-024-09572-4

Rahim, A., Soeprobowati, T. R., Putranto, T. T., Al Falah, M. H., & Gell, P. (2024). Contribution of mangrove forest carbon stocks on climate change mitigation: A case study at Tuntang Estuary, Central Java. *Journal of Coastal Conservation, 28*(4), 65. DOI: 10.1007/s11852-024-01059-w

Rakhmanissazly, A., Permatasari, A. I., & Peranginangin, E. C. (2018). Edco-tourism; A Coastal Management Program to Improve Social Economics. *IOP Conference Series. Earth and Environmental Science, 116*, 012038. DOI: 10.1088/1755-1315/116/1/012038

Reciproco, A., Suarez, J., Padilla, J., Rick Flora, J., & Cañada, B., M. C., R. Velasco, C., & Buhong, N. (. (2023). Awareness and Participation in Mangrove Management of Coastal Communities in Baler, Aurora, Philippines. *Ecology. Environmental Conservation, 29*(02), 928–934. DOI: 10.53550/EEC.2023.v29i02.064

Sathiyamoorthy, S., & Sakurai, T. (2024). Effectiveness of community participation in Mangrove restoration: The evidence from northern Sri Lanka. *Environmental Economics and Policy Studies, 26*(4), 759–779. DOI: 10.1007/s10018-024-00397-1

Sofianto, A. (2017). Prinsip-Prinsip Penanggulangan Kemiskinan Di Wilayah Pesisir Utara Jawa Tengah. *Buletin Ilmiah Marina Sosial Ekonomi Kelautan Dan Perikanan, 2*(2), 81. DOI: 10.15578/marina.v2i2.4679

Solihuddin, T., Husrin, S., Salim, H. L., Kepel, T. L., Mustikasari, E., Heriati, A., Ati, R. N. A., Purbani, D., Mbay, L. O. N., Indriasari, V. Y., & Berliana, B. (2021). Coastal erosion on the north coast of Java: Adaptation strategies and coastal management. In T. J.M.S., U. A., & P. J.A. (Eds.), *IOP Conference Series: Earth and Environmental Science* (Vol. 777, Number 1). IOP Publishing Ltd. DOI: 10.1088/1755-1315/777/1/012035

Sudtongkong, C., & Webb, E. L. (2008). *Outcomes of State-vs*. Community-Based Mangrove Management in Southern Thailand.

Sunartiningsih, A., & Larasati, Z. W. (2020). Inisiatif dari Pesisir Utara Jawa: Merangkul Kearifan Lokal dan Kerjasama Antar Aktor Berkepentingan dalam Pemberdayaan Masyarakat Pesisir. *Gulawentah:Jurnal Studi Sosial*, *5*(1), 28. DOI: 10.25273/gulawentah.v5i1.6617

Suntoro, A., Marbun, A. A. Y., & Sayekti, C. W. (2025). Chapter 10 Climate Change Challenge: Coastal Community Adaptation Towards Achieving Sustainable Community Resilience in Central Java, Indonesia. In *Advances in Global Change Research* (Vol. 80, pp. 235–255). Springer Science and Business Media B.V. DOI: 10.1007/978-3-031-85359-3_10

Susanto, N. A., Nugrahajati, R. R., & Susilastuti Dwi Paripurno, E. T. (2025). Kemampuan Adaptasi Masyarakat Pesisir Menghadapi Dampak Bahaya Abrasi Akibat Banjir Rob Dan Perubahan Iklim (Studi Kasus: Perubahan Desa Bedono Sayung Demak Kurun Waktu Tahun 1990-2020). *Pendas : Jurnal Ilmiah Pendidikan Dasar*, *10*(4), 285–303.

Sutanto, H. A., Susilowati, I., Iskandar, D. D., & Waridin, . (2022). Mitigation and adaptation to climate change through sustainable mangrove management on the coast of Rembang Regency. *IOP Conference Series. Earth and Environmental Science*, *1036*(1), 012014. DOI: 10.1088/1755-1315/1036/1/012014

Tjahjono, A., Adi Intyas, C., & Fattah, M. (2022). MANGROVE MANAGEMENT STRATEGY FOR SUSTAINABLE BUSINESS BASED ON INDONESIAN ECOLOGICAL PRODUCTS. *Geo Journal of Tourism and Geosites*, *43*(3), 1045–1055. DOI: 10.30892/gtg.43325-919

Tunyo, D. A., Vondolia, G. K., & Xuan, B. B. (2026). Valuing mangrove management options in Ghana using discrete choice experiments. *Ocean and Coastal Management*, *276*, 108163. Advance online publication. DOI: 10.1016/j.ocecoaman.2026.108163

Turisno, B. E., Dewi, G. S., & Al Asy'arie, M. A. H. (2026). Designation of Reclaimed Land as Mangrove Conservation Area in Sustainable Coastal Zone Management. *International Journal of Sustainable Development and Planning*, *21*(01), 443–451. DOI: 10.18280/ijsdp.210138

Utami, N. D. N., Susiloningtyas, D., & Handayani, T. (2018). Community Perception and Participation of Mangrove Ecosystem in Ngurah Rai Forest Park Bali, Indonesia. *IOP Conference Series. Earth and Environmental Science*, *145*(1), 012147. Advance online publication. DOI: 10.1088/1755-1315/145/1/012147

Walters, B. B., Rönnbäck, P., Kovacs, J. M., Crona, B., Hussain, S. A., Badola, R., Primavera, J. H., Barbier, E., & Dahdouh-Guebas, F. (2008). Ethnobiology, socio-economics and management of mangrove forests: A review. *Aquatic Botany*, *89*(2), 220–236. DOI: 10.1016/j.aquabot.2008.02.009

Wibisono, G., Sardjono, M. A., Rujehan, , & Suhardiman, A. (2023). Community-based mangrove management in Berau Regency. *IOP Conference Series. Earth and Environmental Science*, *1192*(1), 012019. Advance online publication. DOI: 10.1088/1755-1315/1192/1/012019

Widjonarko, Purnaweni, H., & Maryono. (2025). Spatio-Temporal pattern of coastal flooding due to microclimate change and mangrove deforestation in Semarang and Demak Coastal Area. *E3S Web of Conferences, 605*, 03012. DOI: 10.1051/e3sconf/202560503012

Chapter 11
Gender Mainstreaming in Mangrove Conservation

Aga Natalis
http://orcid.org/0000-0001-7077-9169
Diponegoro University, Indonesia

Ani Purwanti
Diponegoro University, Indonesia

ABSTRACT

Conserving mangroves is essential to protecting coastal ecosystems, mitigating climate change, and improving residents' welfare. This chapter emphasises women's roles in mangrove management via rehabilitation efforts, sustainable ecotourism, and livelihoods derived from mangrove forest resources. Women possess profound local knowledge and significantly contribute to ecological sustainability while bolstering household earnings; however, they frequently encounter structural and epistemological marginalisation and limited access to decision-making processes. Global and national evaluations indicate that effective conservation relies on the synergy of legal policy, acknowledgement of women's knowledge, and their empowerment. Incorporating a gender perspective into mangrove conservation improves the efficacy of ecological management, social sustainability, and equity for coastal communities.

I. INTRODUCTION

Mangrove ecosystems cover more than 137,600 km^2 worldwide, according to the Global Mangrove Watch baseline, with the largest extend found in Indonesia, followed by Brazil and Australia. Despite their significant ecological value for coastal

DOI: 10.4018/979-8-3373-9320-9.ch011

defence, carbon sequestration, and biodiversity preservation, mangroves remain under severe pressure from deforestation and land change. Between 1996 and 2020, mangrove coverage globally declined by approximately 3.4%, indicating ongoing the continued degradation of these critical ecosystems. Particularly alarming is the fact that only 6.9% of global mangrove regions are formally protected, underscoring the urgent need for more targeted, effective, and sustainable conservation initiatives to ensure the long-term preservation of mangrove ecosystems (Bunting et al., 2022).

Mangrove forests have considerable advantages and considerable potential to enhance the economic circumstances of coastal towns. These ecosystems possess substantial biodiversity and generate ecological, economic, and tourism dimensions. Mangrove trees are immensely beneficial and can be employed in several ways to sustain livelihoods through a wide range of uses (Das et al., 2022; Hussain & Badola, 2010; Nuraeni & Kusuma, 2023). As resilient woody plants with dense foliage, mangrives provide various usable components — including roots, bark, trunks, leaves, and flowers — all of which can be used for varipus purposes (Asante et al., 2023). Understanding women's contributions to mangrove forest management therefore requires an examination of gender-based labour divisions, which dictate who is most intimately connected to and actively engaged in the stewardship of coastal ecosystems (Begum et al., 2025; Cormier-Salem, 2017; Purwanti et al., 2025). In this setting, women's entrenched responsibilities for protecting coastal ecosystems have been strengthened through co-management strategies that, in numerous countries, including Indonesia, have demonstrated efficacy in engaging communities in sustainable forest resource management.

Three key concepts underpin this discussion and required clarification. First, co-management refers to institutional arrangements that allocate authority and responsibility for coastal resources management between the government and local communities through clearly separated roles, rights, and obligations. Second, gender mainstreaming is positioned as a strategy for integrating the interests and needs of both women and men into all stages of policymaking, ensuring that equality is embedded within governance rather than treated as peripheral concern. Third, epistemic injustice stems from Fricker's idea, reveals the disadvantages experienced by someone in their capacity as kowners, particularly when when women's credibility is doubted, and their local knowledge is ecluded from, or undervalued in, decision-making processess. Co-management strategies increasingly highlight women's essential contributions to the preservation of coastal ecosystems through their involvement in mangrove management. Begum et al. (2022) assert that co-management, widely adopted in many countries, serves as a participatory framework for forest resource management to enhance conservation outcomes, especially in developing nations overseeing protected areas. In South Asia, co-management is implemented through stakeholder partnerships that establish clearly defined roles, rights, and responsibilities. In Indonesia, co-management has been implemented to rehabilitate mangrove forests and mitigate forest degradation through law enforcement, community incentives, governmental oversight, and conflict resolution mechanisms. similarly, the Vietnamese government has sought to institutionalise co-management by fostering shared commitments and collaborative decision-making with local stakeholders to safeguard forest biodiversity. The participation of local populations, particularly women, in forest co-management is essential for enhancing livelihoods, augmenting access to forest resources, and mitigating inequalities. Furthermore, women's involvement enhances forest governance and promotes sustainable biodiversity conservation by ensuring equitable access to forest resources and meaningful participation in decision-making within protected areas.

Women significantly contribute to ecotourism initiatives that enhance conservation awareness and promote ecosystem sustainability within broader mangrove management activities (Nisaa et al., 2025; Samal & Dash, 2024; Suhartanto et al., 2025). This co-management strategy aligns with sustainable ecotourism techniques implemented by enabling women to participate actively in nursery development, planting, and guiding visitors, thereby strengthening public awareness of conservation. Women participate in seedling production, planting, maintenance, and visitor engagement, exemplifying sustainable ecotourism methods (Begum et al., 2022; Pearson et al., 2019; Purwanti, Mahfud, et al., 2022; Purwanti, Wijaningsih, et al., 2022; Ramoni-Perazzi et al., 2026; Rasquinha, 2024; Senis et al., 2026; Tijjani et al., 2021). These actions not only preserve the ecological integrity of mangrove ecosystems as tourism destination but also facilitate environmental education, enhancing conservation consciousness. Numerous studies further indicate that women exhibit heightened environmental awareness and sensitivity, serving as stewards of local knowledge and practices relevant to coastal resource management. Indigenous women in Papua, for example, manage "women's forests" in accordance with customary law, contributing to both food security and cultural preservation (Senis et al., 2026; Tijjani et al., 2021). In Ecuador, women maintain strong cultural and emotional connections with mangrove ecosystems, highlighting their significance in cultural ecosystem services (Treviño, 2022). Moreover, women participate economically through several activities, such as shellfish harvesting, ecotourism, and alternative revenue generation, which not only benefit household livelihoods but also elevate their social standing. In certain areas, women are increasingly assuming leadership roles in mangrove conservation efforts, as exemplified in the Philippines, where they hold executive-level positions in marine conservation activities (Cañada et al., 2022).

Despite women's substantial contributions to mangrove management, they persistently encounter numerous hurdles that restrict their access to wider decision-making processes. These obstacles are fundamentally based on socio-cultural norms that limit women to household tasks, thereby hindering their involvement in public affairs and environmental governance, such as mangrove governance. Patriarchal frameworks and gender prejudice frequently lead to the underrecognition of women as pivotal contributors in resource management, despite their active participation in conservation, the use of mangrove resources, and ecotourism-related economic endeavours (Begum et al., 2022; Nisaa et al., 2025; Purwanti et al., 2025).

Consequently, women's participation within formal decision-making bodies and institutions is restricted, even though the decisions rendered significantly impact their lives and livelihoods. This scenario is further intensified by the multiple burdens women shoulder as they balance domestic responsibilities with participation in environmental management efforts — often without sufficient support or acknowledgement from policies and institutions that remain predominantly male-dominated (Begum et al., 2025; Dogeje, 2026; Purwanti et al., 2025).

Furthermore, restricted access to resources, training opportunities, infrastructure, and institutional support further limits women's capacity to impact decision-making processes. These challenges are further reinforced by broader economic and political disparities, as changes in policy modifications and resource management frequently disadvantage women, especially those who serve as the primary income earners in their families.

II. MANGROVE CONSERVATION IN GLOBAL AND NATIONAL PERSPECTIVES: SYNERGIES BETWEEN ECOLOGY, LOCAL COMMUNITIES, AND STRENGTHENING LEGAL FRAMEWORKS

Mangrove conservation worldwide has increasingly demonstrated that these coastal ecosystems are indispensable for global climate stability and the livelihoods of coastal communities (Ashrafuzzaman, 2025). The World Wildlife Fund (WWF), through its Mangroves for Community and Climate initiative, funded by the Bezos Earth Fund, aims to conserve coastal ecosystems by enhancing scientific research, management tools, and the capabilities of communities and mangrove managers in Mexico, Madagascar, Colombia, and Fiji. Since 2020, WWF has contributed to the protection, restoration, and governance of about 2.3 million acres of mangroves forests, which collectively sequester over 324 million tonnes of carbon. Beyond these natural advantages, continued mangrove conservation provides coastal protection to about 292,000 individuals annually (World Wildlife Fund, 2026).

This accomplishment is inextricably linked to the vital role of local communities as the principal agents in preserving mangrove ecosystems. In the Tsiribihina Delta in western Madagascar, women serve as primary stewards of mangrove forests. Rasina Jamaly engages in mangrove nurseries that support the restoration of damaged forests, whilst Clémentine Soava relies on honeybees production inhabiting mangrove regions for her livelihood and concurrently promotes local biodiversity. Simultaneously, Voloniaina Nomenjanahary Victorine undertakes regular patrols to deter unlawful logging activities. Their endeavours illustrate that mangrove conservation extends beyond environmental protection by simultaneously generating sustainable economic prospects for women and their families.

Globally, the protection and sustainable management of mangroves have attracted heightened attention in international legal and regulatory frameworks (Chamberland-Fontaine et al., 2022). This is reflected in the increasing acknowledgement of mangrove ecosystems within climate change, biodiversity, and coastal governance initiatives. As of 2023, 97 nations have integrated coastal and marine ecosystems, including mangroves, into their Nationally Determined Contributions (NDCs) under the Paris Agreement, while 61 nations have explicitly included the conservation or restoration of blue carbon ecosystems within their climate mitigation and/or adaptation strategies. Mangroves are also highly relevant to achieving multiple objectives under the Kunming-Montreal Global Biodiversity Framework (GBF), especially those related to spatial planning, ecosystem restoration, protected areas, and climate resilience. Under the Ramsar Convention framework, 305 designated Ramsar sites worldwide include mangrove ecosystems, prompting member nations to formulate policies and regional efforts to conserve and restore coastal wetlands, including mangroves (International Union for Conservation of Nature, 2023).

Reflecting these international developments, a growing number of nations are incorporating their international obligations into their national policies by harmonising Nationally Determined Contributions (NDCs), National Biodiversity Strategies and Action Plans (NBSAPs), and national wetland inventories. This method is deemed effective for expediting mangrove conservation and restoration more synergistically. As of 2023, at least 75 jurisdictions had officially incorporated mangroves into their national legislation. Among these, 64 governments have explicit restrictions on mangrove clearence or conversion, while at least 19 have enacted comprehensive prohibitions on mangrove use, subject to limited exceptions for traditional or subsistence purposes. This tendency indicates that both international and national legal frameworks are progressively enhancing mangrove preservation through prohibitions, use limits, and sustainability-oriented governance strategies (International Union for Conservation of Nature, 2023).

Diverse approaches across several nations further demonstrate that enhancing mangrove policies can be achieved through a range of legal and policy instruments tailored to each country's circumstances. Belize has formulated a National Mangrove Action Plan, informed by land tenure analysis, to identify priority regions for mangrove restoration while aligning these initiatives with national NDC objectives concurrently. Costa Rica has initiated a Blue Carbon Strategy—deemed the first of its type worldwide—by employing economic incentives and market-based mechanisms to mitigate the loss and degradation of blue carbon habitats and wetlands. Ecuador has adopted a robust strategy through a constitutional court ruling that protect mangrove ecosystems from economic activities in coastal and bay regions, alongside agreements between the government and indigenous user communities for the sustainable utilisation of mangroves. Indonesia, meanwhile, has implemented the Bio-rights programme, a microfinance initiative that facilitates mangrove conservation and restoration while providing alternative livelihood opportunities for local communities, bolstered by substantial government support through programme allocation and funding for community groups.

Additional instances demonstrate that mangrove conservation can be enhanced through monitoring systems, carbon market regulation, and regional collaboration. Mexico has established a National Mangrove Monitoring System that provides baseline data and conducts periodic assessments, facilitating governments to support conservation through court decisions and new legislation reforms that enhance protection for endangered mangroves. Panama incorporates mangroves into its low-carbon economic strategy through the Blue Natural Heritage initiative, which encompasses the formulation of carbon footprint-reduction programs, the creation of a national carbon market framework, and executive decrees governing mangrove conservation, sustainable utilisation, and blue carbon projects. Likewise, countries have agreed to develop a Regional Mangrove Vision and a Regional Mangrove Action Plan to guide policy dialogue and strengthen sustainable mangrove management. These experiences demonstrate that effective mangrove conservation depends not only normative prohibitions, but also support for economic mechanisms, comprehensive monitoring data, recognition of community rights, and inter-regional collaboration.

Various legal policy alternatives has also proved effective in promoting mangrove protection, restoration, and sustainable utilisation. This encompasses the establishment of incentives and financial mechanisms, including blue carbon schemes, Payments for Ecosystem Services (PES), fiscal incentives, subsidies, and product certification. They also encompass community-based mangrove management via conservation agreements, community concessions, and acknowledgement of rights, tenure, and legal pluralism, as well as dedicated national mangrove strategies and the incorporation of mangrove protection into comprehensive planning laws and regulations. The evolution of mangrove law internationally demonstrates that safeguarding this ecosystem requires a comprehensive approach combining legal clarity, ecological sustainability, community engagement, and cross-sectoral policy support (Van Lavieren et al., 2012).

Within Indonesia's natural resource management framework, mangroves are among the ecosystems most susceptible to land conversion and exploitation, especially for prawn aquaculture. This vulnerability is closely correlated to the economic factors of the aquaculture industry, particularly given that Indonesia is the fifth-largest prawn exporter globally, with a 6.0% market share, trailing only Ecuador, India, Vietnam, and China (Harianto, 2025).

Indonesia's commitment to mangrove conservation is also reflected in its international collaboration. The governments of Germany and Indonesia have executed a grant agreement for EUR 20 million for the Mangrove Forest Protection Programme, including the establishment of the World Mangrove Centre in Indonesia. This arrangement is delineated in the Grant arrangement and Separate Agreement about Project Forest Programme VI, entitled "Protection of Mangrove Forest." Designed as an eight-year initiative, the programme seeks to protect and restrore mangrove forests in selected areas through socially, environmentally, and economically sustainable management led by forestry institutions in partnership with local populations. Project sites comprise the Karang Gading Wildlife Reserve in North Sumatra, the Mahakam and Berau Deltas in East Kalimantan, and Sorong City and South Sorong Regency in West Papua (Asmin, 2024).

Mangrove management is crucial from an ecological standpoint as it contributes directly to environmental quality and the preservation of green open spaces. Mangrove planting aids coastal restoration and enhances broader ecological functions, such as mitigating global warming. The principal objectives of mangrove ecosystem management encompass three facets: maximising production and ecological advantages via a sustainability-oriented ecosystem approach; restoring degraded mangrove forests; and enhancing institutional frameworks alongside scientific and technological capacities to facilitate effective mangrove management. Indonesia possesses exceptionally rich mangrove biodiversity, including *Rhizophora sp., Avicennia sp., Sonneratia sp., Bruguiera sp., Xylocarpus sp., Ceriops sp.,* and *Excoecaria sp.* This biological diversity highlights the importance of mangroves ecosystems not only for human well-being but also for maintaining broader ecological equilibrium (Fitria, 2021).

Mangrove ecosystems in Indonesia are protected through a comprehensive body of legislation governing environmental protection, forestry, natural resource conservation, and coastal and small island management safeguards mangrove ecosystems. Law enforcement efforts to address mangrove degradation are supported by administrative, civil, and criminal measures to preserve the ecosystem. The primary legal foundation is found in the 1945 Constitution of the Republic of Indonesia, specifically Article 28H(1), which guarantees every individual's right to a good and healthy environment, while Article 33(4) establishes sustainability, efficiency, equity, and environmental conservation as guiding principles in the management of the national economy and natural resources.

Law No. 41 of 1999 on Forestry provides the principal legal framework for the protection of mangrove habitats within forested areas. This law characterises forests as cohesive ecosystems comprising land with biological resources, predominantly trees, in their native habitats. In contrast, protected forests are defined as forested regions primarily intended to preserve life-support systems, such as water regulation, flood mitigation, erosion control, prevention of seawater intrusion, and maintenance of soil fertility. Mangroves are designated as an essential element of state-controlled protected forests managed for the greatest public benefit. The identification of mangrove forest zones is conducted via forest inventory and acknowledgement by the central government through forest management plans. Moreover, regulations regarding coastal buffer zones indicate that mangrove conservation objectives encompass both the vegetation itself and the natural habitat. The legislation acknowledges the involvement of indigenous and local populations via participatory methods in the oversight and execution of conservation initiatives. In instances of deterioration, it necessitates rehabilitation and reclamation by replanting and habitat restoration. Collectively, these regulations establish a relatively robust legal framework for mangrove protection.

The forestry system is complemented by Government Regulation No. 26 of 2008 on the National Spatial Plan, which further strengthens mangrove conservation through spatial planning. Under this Regulations, coastal mangrove areas are designated as essential components of national spatial preservation and classified as protected areas comparable to nature reserves, conservation areas, and cultural heritage sites. Therefore, the alteration or eradication of mangrove vegetation is fundamentally forbidden, save for strictly regulated purposes – such as education, scientific inquiry, or non-intrusive ecotourism. The law mandates comprehensive zoning for mangrove regions and designates protective corridors thar function as natural barriers against erosion and saltwater encroachment. By integrating mangrove conservation into national spatial planning, the Regulation seeks to balance ecological protection with social and and economic priorities.

The legal framework for mangrove conservation is further strengthened by Law No. 32 of 2009 about Environmental Protection and Management. This legislation prioritises sustainability, conservation, precaution, ecological equilibrium, and public engagement as fundamental tenets of environmental policy. In this context, mangrove ecosystems are considered as a vital element of the ecological system, serving as coastal buffers, protects erosion and saltwater intrusion, and habitats for many species. The legislation also delineates standards for evaluating damage to mangrove ecosystems by defining acceptable thresholds for physical and biological alterations. Moreover, the preparation of environmental inventories and the development of Environmental Protection and Management Plans at different levels of government provides the basis for the protection, restoration, and sustainable use of mangroves.

Presidential Regulation No. 73 of 2012 on the National Strategy for Mangrove Ecosystem Management underscores the necessity of intersectoral coordination among the Ministry of Environment and Forestry, the Ministry of Marine Affairs and Fisheries, and regional governments. This regulation recognises mangrove ecosystems as intricate systems comprising vegetation, fauna, and microbes within tidal zones, indicating that any disruption to a single component—such as land conversion for aquaculture or settlement—can affect overall biophysical processes (Banjade et al., 2017).

Further legislative developments are reflecteed in the 2024 amendment to the law on the Conservation of Biological Natural Resources and Ecosystems, together with Government Regulation No. 27 of 2025 on Mangrove Ecosystem Protection and Management. This law mandates the sustainable management of all mangrove habitats, both within and beyond forest zones, to preserve their ecological, social, and economic services upon which coastal populations depend (Satriawan & Satriawan, 2026).

Government Regulation No. 27 of 2025 establishes a more extensive framework formangrove protection and management. It requires conservation planning to be based on systemic inventories of mangrove extent, ecosystem type, condition, and legal status, together with assessment of ecosystem services, vulnerability to climate change, and the designation of protection or utilisation functions – serving as the foundation for planning at national and regional levels. The use of mangroves is confined to particular activities, including research, education, environmental services, non-timber forest products, aquaculture, ecotourism, and sustainable traditional practices. The regulation also prioritises control through preventive measures, including damage criteria standards, strategic environmental assessments, environmental permits, and public awareness initiatives, alongside mitigation strategies such as hydrological rehabilitation, protective structure construction, and bioremediation.

III. THE MARGINALISATION OF WOMEN IN MANGROVE MANAGEMENT: STRUCTURAL, EPISTEMIC INJUSTICE, AND GLOBAL POLICY CHALLENGES

Access to and control over natural resources, such as mangrove habitats, fundamentally shape patterns of resource use and governance. In reality, unequal power dynamics – frequently influenced by gender, ethnicity, socio-economic status, and age differences – result in inequities in access, management, and the benefits derived from these resources. In the context of mangroves, these differences affect men and women differently, influencing both livelihood opportunities and vulnerability to environmental destruction. Many studies show that women play a crucial role in managing coastal areas through activities such as collecting non-timber forest products, processing derivative products, providing food for the household, and maintaining the sustainable resources at the household and community levels. However, these contributions remain inadequately documented, limiting their visibility within both academics and policymakers. National policies continue to priotritise primary production while playing comparatively less attention to downstream stages of the resource-use chain, where women's participation is often more active. Consequently, women's contributions to the local economy, food security, and the health of mangrove ecosystems are frequently undervalued, while their role in implementing policies is often not given sufficient attention (Chuku et al., 2022; Harper et al., 2013).

A consistent gap in the literature is evident. Previous studies have tended to emphasise the ecological and economic value of mangroves at the primary production level, while women's contributions are concentrated in the post-harvest chain, ecosystem processing, and resource maintenance, all of which remain comparatively underexplored. Furthermore, the limited integration of local empirical findings into broader policy frameworks has resulted in women's contributions being overlooked in legal instruments and governance. This section attempts to address both of these gaps by linking evidence of women's participation in various case areas within institutional and legal analyses, while simultaneously placing them within the framework of feminist political ecology and environmental justice. In doing so, women's roles are eexamined not merely as descriptive observations but as manifestations of broader structural inequalities.

Women in coastal areas have a strong relationship to the mangrove ecosystem and the diverse livelihoods they support. Many undertake unpaid work or employment in the informal sector, such as gathering firewood, producing crafts from mangroves, collecting seafood, and securing household food supplies. When mangrove ecosystems are degraded through direct human activities or the impacts of climate change, the natural resources upon which women deepend for both household livelihoods and income generation are progressively diminished (Purwanti & Wijaningsih, 2025). Many of these jobs, such as small-scale fishing, gathering marine resources, and farming along the shore, are very vulnerable to climate change, including the increasing frequency and intensity of storms (Treviño, 2022).

The deterioration in mangrove quality also reduces their capacity to protect coastal communities from tidal waves and coastal erosion, disproportionately affecting women in vulnerable coastal areas. Restoring the health of mangrove ecosystems and biodiversity becomes an important step for strengthening climate resilience, more resilient, especially for households headed by women. Through their long-standing involvement in the informal management of mangrove resources, women have gained local knowledge and skills essential to the success of long-term restoration initiatives (Begum et al., 2025; Chapman et al., 2026).

Women are clearly involved in mangrove management at the local level, but they often face unfair treatment when it comes to receiving credit for their work in global conservation strategies (Begum et al., 2022; Cormier-Salem, 2017). Furthermore, their participation in mangrove management is intricately linked to their roles across diverse areas of natural resource management, as exemplified by the involvement of female fishermen in the sustainable production and processing of marine resources. In coastal environments, women contribute directly to the ecosystem stewardship through resource management, environmental monitoring, and community-based conservation initiatives, while also investing their time, knowledge, and, in many cases, financial resources in local environmental programmes. Despite their substantial involvement in mangrove management and ecotourism at the local level, these contributions remain insufficiently recognised within global environmental governance and conservation policy (Nisaa et al., 2025; Scheyvens, 2000).

In this environment, women are often not taken into account when decisions are being made about mangrove conservation measures even though their contributions are important (Purwanti, Wijaningsih, et al., 2022; Stone et al., 2008). This exclusion widens gender inequality and adds to the burden on women who must balance domestic affairs with resource management. Even in areas where they are active in mangrove rehabilitation and ecotourism, control over conservation policies remains in the hands of men, as demonstrated in Bangladesh and Indonesia(Begum et al., 2025; Purwanti et al., 2025).

Another problem concerns that mangrove management strategies typically do not account for women's roles in coastal ecosystems. Women are frequently underrepresented in formal decision-making processes concerning mangrove protection and governance. In Siak, Indonesia, for instance, women are involved in both ecological and economic activities, yet they still have little say in government (Nisaa et al., 2025). In Mozambique, women also play an important role in collecting shellfish, but they do not have a protection in policy decisions or resource management (Ramoni-Perazzi et al., 2026). Only 29% of women in Kenya help develop rules for mangrove conservation, highlighting the unequal gender roles in policy-making (Ahmed et al., 2023).

Table 1. Division of gender roles and women's participation level in mangrove management in a number of case areas

Case Area	Women's Main Role	Position in Decision Making
Lamu, Kenya (Ahmed et al., 2023)	Participation in the preparation of mangrove conservation regulations at the community level	Only about 29% of women are involved in formulating rules
Pulau Inhaca, Mozambik (Ramoni-Perazzi et al., 2026)	Subsistence shellfish collection (median about 3 kg/day), while men fish commercially (median about 15 kg/day)	Have no voice in policy decisions and resource management
Siak, Indonesia (Nisaa et al., 2025)	Involvement in ecological and economic activities through mangrove ecotourism	Voice in government and governance is still very limited
Teluk Youtefa, Papua, Indonesia (Tijjani et al., 2021; Senis et al., 2026)	Management of "Women's Forest" (Tonotwiyat) based on customary law, supports food security, and cultural preservation	Full control by women, but threatened by pollution and land conversion

Source: compiled by the author from Ahmed et al. (2023), Ramoni-Perazzi et al. (2026), Nisaa et al. (2025), Tijjani et al. (2021), and Senis et al. (2026).

Privatisation and economic reform have contributed to deepening gender inequality in the management of mangrove resources. As happened in Vietnam, the existing changes have narrowed and hindered women's access to mangrove resources, especially heads of households (Hue, 2006). They face a heightened risk of financial loss, which further exacerbates existing gender inequities. In Fiji, although women act as the main carers of mangroves, their work is frequently undervalued, while conventional gender roles continuously restrict their access to essential resources (Pearson et al., 2019).

The additional burden borne by women also constitute a major obstacle to effective mangrove management. Women engaged in mangroves often face a "double burden," balancing home and resource management duties (Mangroves for the Future (MFF) et al., 2018; Purwanti et al., 2025). This workload is further exacerbated by the limited recognition of their contributions within conservation policies, particularly in coastal areas. In Mozambique, women typically engage in subsistence work, such as gathering shellfish, while men engage in commercial fishing (Ramoni-Perazzi et al., 2026). This creates a gendered division of labour that makes it harder for women to find jobs.

Cultural conventions and patriarchal traditions frequently restrict women's involvement in mangrove management. For instance, in Cameroon, ingrained patriarchal norms constrains women's access to land and restrict their participation in community development programs (Lueong & Fogwe, 2022). The Enggros Tribe in Papua manages the "Women's Forest," providing a rare example of women exercising sole management of mangrove resources (Tijjani et al., 2021). However, this system is also at risk from external factors, such as pollution and land conversion. Moreover, conservation stories sometimes put hidden costs on women. In India, for instance, women are hesitant to adopt alternative energy sources because of cultural norms and societal obligations, which unfairly put them in charge of using resources in good for the environment (Hussain & Badola, 2010).

Despite growing international recognition of women's contributions to mangrove conservation, substantial challenges remain within global environmental governance. Women are frequently marginalised in the development and implementation of global policies, especially within international frameworks – such as REDD+ and other conservation initiatives (Larson et al., 2015; Westholm, 2016). This marginalisation limits their access not only to restoration projects but also to decision-making processes that directly affect their livelihoods (Sembiring & Garunja, 2026; Suherman et al., 2026). The Ramsar Convention is also an important international agreement that protects biodiversity and ecosystem services in mangroves (Batanjski et al., 2016; Kochoni et al., 2025). Nonetheless, this treaty does not specifically address gender or the role of women in mangrove management.

Miranda Fricker's (2007) idea of epistemic injustice describes that people suffer because they are denied knowledge, especially when this harms their credibility or ability to share knowledge. Fricker's (2007) thesis is pertinent in the discourse on gender and patriarchy, where women frequently encounter epistemic marginalisation. In patriarchal society, women's voices are often disregarded or discredited, resulting in what Fricker (2007) describes as "testimonial injustice." This happens when prejudice undermines the credibility in public life, including in scientific research, formulating policies, or discussing social issues because of gender bias. In these patriarchal systems, women's expertise is devalued, reinforcing societal hierarchies that keep them from making decisions and holding positions of power (Hasbi et al., 2026).

This issue relates to the ideas of Walby (1991), who developed the theory of patriarchy, which explains how gender inequality is built into the system. Walby (1991) contends that patriarchy constitutes a multifaceted set of social structures that perpetuate gender inequality across several domains – including economics, politics, and family. Walby (1991) stresses that patriarchy is not only a personal or cultural issue, but also a structural issue that is built into institutions and societal norms. Walby's (1991) study further elucidates how patriarchal regimes consistently diminish women's contributions to knowledge and decision-making within the framework of epistemic injustice. This system not only restricts women's access to positions of authority, but also views their experiences and knowledge as inferior, perpetuating their marginalisation within broader social narratives.

Building on Fricker (2007) and Walby's (1991) ideas, the epistemic injustice that women experience in patriarchal societies goes beyond the marginalisation of their voices in decision-making. It is also closely correlated to the fact that, even though international policies recognise the importance of women's roles, they often fail to incorporate gender equality principles. Seck (2024) contends that, despite the growing focus on gender equality in environmental governance, its implementation remains highly fragmented and varies across regions. Similarly, MacKinnon (1987) argues that legal systems frequently reproduce rather than remedy gender inequality because they are shaped by deeply embedded patriarchal institutions. In her view, the law is frequently employed as an instrument to perpetuate gender inequity rather than to rectify it. This institutional weakness is evident in international instruments such as the Rio Declaration and Agenda 21 clearly. They recognise women's role in managing ecosystems, but they still lack enforceable rules for gender equality. This indicates that there is a policy gap that does not meet women's requirements. As noted in the context of the SDGs, empowering women through inclusive policies and ongoing education is crucial to increasing their involvement in protecting ecosystems, thereby advancing gender equality in sustainable development.

IV. GENDER-RESPONSIVE APPROACHES IN ENVIRONMENTAL PRESERVATION: FOCUSING ON MANGROVE ECOSYSTEMS

Environmental concerns have emerged as a prominent topic of global discourse. It is closely associated with a variety of disasters, including floods, severe winters, and heavy rainfall accompanied by powerful winds (Adger & Brooks, 2003; Kemp, 2002). These concerns are being addressed by all nations worldwide. Almost every aspect of existence is impacted by these environmental issues. Although environmental matter affects society as a whole, its consequences are experienced unevenly, with women often bearing a disproportionate share of its social and economic impacts. Despite the long-standing advocacy for gender equality, women continue to experience injustice that heighten their vulnerability to environmental change (Natalis et al., 2023; Shiva, 2010; Shiva et al., 2014).

By gaining a deeper understanding of how a gender approach can bolster and reinforce global environmental preservation efforts, the role of women in environmental preservation can be better understood. Rocheleau et al. (2013) argue that gender issues are inextricably correlated to environmental issues in the context of global and local power dynamics. Through this method, environmental issues cannot be adequately addressed without considering the gender dimension. This feminist political ecology framework confirms the environmental justice perspective, which places the distribution of ecological burdens, benefits, and the recognition of previously marginalised actors at the heart of the issue. Recent evidence reinforces this foundation: the Intergovernmental Panel on Climate Change Working Group II report confirms that vulnerability to climate change is shaped by inequalities across gender and other social factors, resulting in unequal distribution of risks (IPCC, 2022). At the same time, UN Women emphasises that climate action and disaster risk reduction can only be effective when gender equality is truly mainstreamed into policy, rather than relegated to an afterthought (UN Women, 2022). By linking the discussion of mangroves to these two frameworks and evidence, women's roles are no longer limited to descriptive terms, but are positioned as a structural justice issue that determines the success of any conservation effort.

The patriarchal cultural system, which establishes an unequal relationship between men and women, is the structural foundation of the challenges encountered by women. This is evident in the state's diminished capacity to safeguard its citizens and in policies that are not gender-inclusive due to the pressures of globalisation. The state affords greater protection to capital owners in the economic sector than to women's welfare and the preservation of natural resources. As a result, women are systematically excluded from the public sphere and denied meaningful opportunities to participate in decisions that directly impact their survival. Their participation in public decision-making therefore remains constrained by institutional and structural barriers rather than individual choice.

Women are perceived as the primary users of natural resources, including water, land, and forests, accross developing regions of the world. This is due to their responsibility for securing food, fuel, and animal feed. Even though the majority of women in these countries are unable to own land or farmland, they devote most of their time to labouring on farms to provide for their families. The environment intersects with women's roles in both the domestic and public spheres. Within this context, ecofeminism provides an important perspective for understanding how women can make a significant contribution to environmental protection (Natalis et al., 2025; Saraswati, 2025).

Women's contributions significantly influence environmental management and development, both of which require active engagement and unwavering commitment (Abdulsalam et al., 2024; Oppong & Mathibe, 2024). Additionally, women are the primary and most significant educators for their children in their capacity as mothers. From an early age, children can develop environmental awareness and learn environmentally responsible behaviour through their mothers' influence. By adopting environmentally friendly practices within the family, children become acclimated to caring for their surroundings and develop a sense of environmental responsibility. Over time, these ptractices can foster a generation with stronger environmental awareness. According to Bangun (2020), women play a critical role in environmental protection by monitoring environmental harm and pollution, and developing a gender-responsive living environment by serving as "agents of change" and policymakers.

The relationship between the structural barriers experienced by women, the various roles they perform, and the policy interventions required to address these challenges is summarised in a conceptual framework presented in Figure 1. In this conceptual circle, it can be seen how gender mainstreaming and co-management become a concession that transforms women's ecological, economic, and socio-cultural roles into sustainable and equitable conservation achievements.

Figure 1. Conceptual framework of a gender-responsive approach to mangrove management

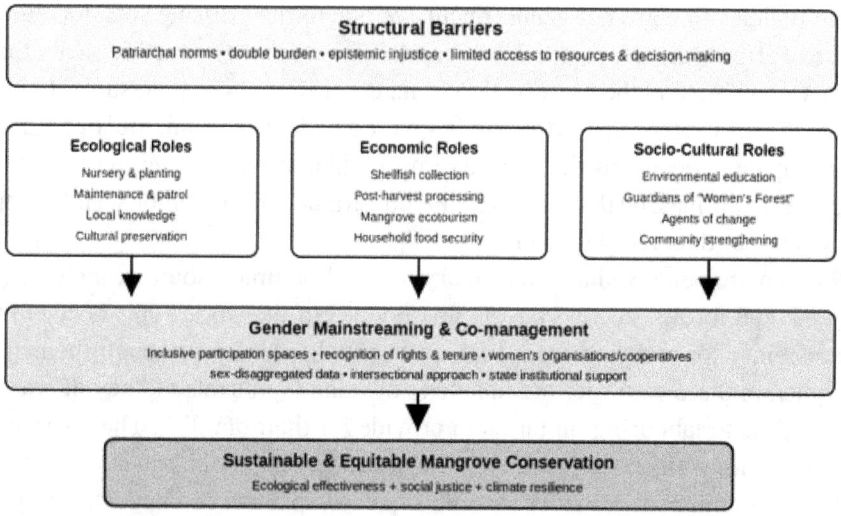

It is crucial to adopt a gender-responsive approach to mangrove management to promote more sustainable and economically valuable outcomes to mitigate social inequalities. An assessment of the long-term effects on resource and environmental sustainability can be facilitated by understanding the differences in women's and men's experiences, requirements, and roles in interacting with mangrove ecosystems. Women's economic contributions to the mangrove value chain are frequently overlooked, and pressures on coastal resources are not completely considered when their roles are disregarded. In a broader sense, prioritising gender equality in mangrove management policies and practices also creates an opportunity to address informal and vulnerable employment issues that women frequently encounter, strengthens their access to information and justice, as well as increases awareness of health and reproductive rights. Furthermore, this method underscores the significance of initiatives to mitigate and redistribute unpaid domestic labour, particularly in light of the challenges posed by climate change and the deterioration of coastal resource quality.

One of the principal instruments relevant to this issue is the 1979 Convention on the Elimination of All Forms of Discrimination against Women (CEDAW). This instrument emphasises gender equality by recognising women's right to be free from all forms of discrimination and by delineating the steps needed to achieve that objective. Additionally, CEDAW advocates incorporating equality principles into nations' constitutions and regulations. CEDAW has also included articles on environmental issues over time, thereby serving as a critical entry point for the mainstreaming of gender in the administration of coastal and marine resources in Asia. Furthermore, the 1995 Beijing Declaration and Platform for Action (BFA) has the objective of establishing priority actions in 12 critical areas, including the environment to promote equality and expand opportunities for women. This document explicitly emphasises the gender-based violence and sexual violence that impact women in both natural disasters and armed conflict scenarios.

Gender equaliry has likewise been incorporated into operational environmental governance through the Gender Action Plan (GAP) adopted under the Framework Convention on Climate Change (UNFCCC) and the Convention on Biological Diversity (CBD). The GAP outlines a range of gender-responsive priorities that provide a strategic foundation for developing gender-responsive priority actions that can serve as a strategic entry point for establishing a gender-responsive policy environment at both the national and local levels. Gender considerations are also embedded in numerous other international instruments, including Agenda 21, which promotes gender mainstreaming at the community level, the Convention to Combat Desertification (UNCCD), the Sendai Framework for Disaster Risk Reduction, the Sustainable Development Goals (SDGs), the Code of Conduct for Responsible Fisheries, and numerous indigenous community-related instruments are among the numerous international agreements that incorporate gender issues.

The dynamics of coastal communities' lives, which are heavily reliant on the sustainability of ecosystems, including mangrove forests, are fundamentally inseparable from the mainstreaming of gender in policies and actions to protect and preserve the sea. Women are frequently overlooked in the management of coastal resources, despite their significant contributions. This includes processing captures, collecting marine life in the intertidal zone, and preserving household economic sustainability during periods of resource depletion. Consequently, the enhancement of the gender perspective in marine conservation is not only a matter of justice; it also involves ensuring the effectiveness of environmental policies. The degradation of mangroves has a direct impact on women's livelihoods, health, workload, in addition to causing broader ecological damage through coastal abrasion and the loss of fish habitats.

The collection of empirical evidence that clearly demonstrates the distinct ways in which ocean degradation and mangrove damage affect women becomes a critical initial step. In this context, an intersectional approach is indispensable due to the heterogeneity of women's experiences. The disruption of mangrove ecosystems, which serve as natural buffers, may limit women's access to clean water and healthcare services in remote coastal areas. In contrast, women in more connected areas may experience economic pressure due to declining catch rates or a shortage of raw materials for processing. In addition, older women or those in more vulnerable socio-economic circumstances are frequently unable to adjust to environmental changes. Policies may fail to address genuine requirements on the ground if they lack comprehensive data sensitive to these differences.

In addition, marine conservation policy solutions must be developed with careful consideration of gender perspectives lens that address not only ecological aspects but also their wider social implications. In the context of mangroves, this entails comprehending that the ecosystem serves as a carbon sink and coastal protector, in addition to providing a source of income for women who collect crustaceans, shellfish, or other materials for consumption and sale. When mangroves are damaged, women are often required to travel greater distance to access the same resources, which ultimately increases their burden and reduces their time for other activities, such as education or social engagement. Consequently, mangrove rehabilitation policies must explicitly consider the economic benefits that women can derive from these efforts, their involvement in planning, and their access to rehabilitation sites.

The sustainability of marine and mangrove conservation efforts depends on empowering women. Such empowerment encompasses not only the provision of technical training, such as mangrove planting techniques or fisheries processing, but also the enhancement of women's negotiation and leadership abilities in decision-making environments. In numerous coastal communities, women continue to face structural barriers that prevent them from actively participating in formal forums. These barriers result from social conventions and the time constraints imposed by domestic responsibilities. Consequently, empowerment strategies must be contextually designed to reflect local contexts by creating inclusive participation spaces, adopting flexible meeting schedules, and supporting a more equitable division of domestic labour. When women participate meaningfully in mangrove governance, conservation decisions are more likely to reflect the needs and priorities of the community as a whole.

Additionally, the state's institutional support is equally critical. Women require organisational platforms that enable them to convene, exchange experiences, and advocate for their shared interests. The presence of women's organisations or cooperatives in coastal and mangrove management can enhance their bargaining power by enabling them to access government programs and engage with markets. Consequently, women's organisations in the fisheries, aquaculture, and littoral zone management sectors should be explicitly acknowledged and supported by policies and regulations. Such support can encompass a variety of services, such as facilitating the formation of organisations, providing technical assistance, and securing legal recognition to legitimise their activities.

In the context of gender mainstreaming, integrating women into the blue economy should also be a top priority. In practice, although the blue economy is widely regarded as a development approach that promotes the sustainable use of marin resources, women's contributions within it are frequently marginalised. Women nevertheless play a substantial role throughout the fisheries value chain, particularly in post-harvest phases, such as processing, packaging, and marketing. Women are also engaged in a variety of economic activities in mangrove contexts, including the production of food products from mangrove resources and handicrafts. Given this, blue economy policies must guarantee that women have equitable access to productive resources, such as credit, training, and markets. Furthermore, these policies must incorporate comprehensive occupational health and safety standards together with social protection that address the requirements of women.

Additionally, gender equality must be consistently prioritised in development cooperation that concentrates on sustainable littoral and marine management. It is crucial to support women-led initiatives that aim to protect mangroves and other coastal ecosystems – not only from a justice perspective, but also from an effectiveness perspective. Numerous community-based initiatives spearheaded by women have demonstrated their ability to enhance economic well-being while simultaneously preserving the environment. For instance, women's organisations that engage in mangrove rehabilitation frequently establish mangrove-based economic enterprises, which serve as an immediate incentive for ecosystem preservation. The success of such initiatives demonstrates that women are not only beneficiaries but also agents of change who drive the transformation toward more sustainable resource management.

Ultimately, a comprehensive and sustainable approach is necessary for gender mainstreaming in the protection and conservation of the oceans, particularly in mangrove ecosystems. It is insufficient to incorporate gender components into existing policies merely; a paradigm shift is required that recognises women as critical actors in the administration of natural resources. This encompasses the acknowledgement of the indigenous knowledge held by women, the enhancement of their access to resources and opportunities, and the reinforcement of their position within social and institutional structures. Such an approach not only enhances the effectiveness of mangrove conservation but also promotes broader social justice for coastal communities.

The significance of gender issues in addressing and comprehending environmental challenges is derived from the fact that the effects of environmental degradation are not experienced equally but are shaped by social relations, including gender constructs. In numerous communities, women are primarily responsible for the management of domestic necessities, including water, fuel, and food, which renders them more susceptible to environmental changes. The burdens of women are frequently disproportionately increased by the effects of climate change, land degradation, and natural disasters. For example, during droughts, women in certain regions may be required to travel farther to fetch water, exacerbating their burden and limiting their time for social engagement, economic activities, or education. Such situation demonstrates that gender inequality is not only the result of social structures; but it is also exacerbated by ecological pressures that are beyond the control of humans.

Conversely, women are the vulnerable demographic and significant players in the management of natural resources. They are frequently directly engaged in environmental-related activities – including agriculture, forest management, water conservation, and coastal zone management, in their daily lives. As a result of this engagement, women acquire profound local knowledge of the ecosystems surrounding them. This knowledge becomes a valuable asset for developing sustainable resource management strategies. Consequently, empowering women and recognising their contributions to environmental governance is not only a matter of justice, but also a practical necessity to improve the efficacy of conservation policies to empower women and acknowledge their contributions to environmental management. The presence of women in resource governance, including community-based forest management, has a positive impact on environmental sustainability, as studies indicate they tend to prioritise long-term sustainability and precautionary principles.

Furthermore, achieving environmental sustainability depends on the principle of gender justice. Gender inequality restricts women's access to resources, education, and decision-making spaces, thereby weakening broader conservation efforts. Marginalised women frequently are unable to contribute their knowledge, experiences, and needs to the formulation of policies, which leads to solutions that are less effective and inclusive. Alternatively, the meaningful participation of women in decision-making processes contributes to policies that are more responsive to the diverse needs of communities and more likely to achieve sustainable outcomes.

An intersectional approach is essential for understanding the relationship between gender and the environment, as it acknowledges that women's experiences are far from uniform. The manner in which an individual experiences environmental impacts is also influenced by several factors, such as social class, ethnicity, age, and economic status. For instance, women from minority groups or those residing in poverty frequently suffer from a dual burden: they are more susceptible to environmental degradation while simultanepisly having fever resources with which to resources for adaptation. Consequently, effective environmental policies and programs must account for this complexity addressing general issues and targeting the most vulnerable groups.

Furthermore, women are instrumental in efforts to adapt to and mitigate the effects of climate change. They are pioneers in developing sustainable practices across a range of contexts, including educating younger generations about environmental issues, community-based natural resource management, and environmentally friendly agricultural innovations. The participation of women in these activities demonstrates their potential as the agents of change who can facilitate the transition to more sustainable development models. Community capacity to confront the increasing complexity of climate change will be enhanced by promoting women's leadership in environmental issues through policies and empowerment programs.

In the final analysis, sustainability necessitates women's participation in environmental decision-making. Policies are at risk of neglecting the perspectives and requirements of the majority of community members when participation is unequal, thereby diminishing their effectiveness on the ground. Conversely, research across numerous regions indicates that the inclusion of women in natural resource management initiatives, including sustainable agriculture and forest management, not only enhances community well-being but also fortifies the long-term sustainability of the environment. Consequently, gender mainstreaming in environmental issues is a normative decision and a strategic necessity to ensure that environmental preservation efforts that are conducted in a sustainable, inclusive, and equitable manner.

V. CONCLUSION

Mangrove ecosystem protection becomes a strategic priority that connects ecological, social, and legal concerns and has a direct impact on sustainable development. Mangroves are not only important for protecting shorelines and storing carbon, but also has an important role as the main source of income for people living along the coast. However, several issues such as deforestation, land-use changes, and inadequate policy implementation reveal a gap between expectations and reality. Gender research reveals that women make significant contributions to mangrove management – both ecologically and economically. They continue to encounter structural marginalisation, restricted access, and epistemic unfairness within decision-making processes. This incident suggests that gender disparity hurts both social fairness and the efficiency of environmental policies. To mainstream gender within mangrove law, policy, and governance, it is essential to facilitate women's participation, recognise indigenous expertise, and strengthen women's empowerement. Thus, truly sustainable mangrove conservation can only be achieved through a holistic, inclusive, and equitable approach that positions women as key actors in coastal resource governance. Based on these findings, the following actionable recommendations for policymakers and practitioners can be synthesised:

1. Mandate quotas and inclusive participation spaces for women in mangrove management decision-making forums, along with flexible meeting schedules to avoid conflicts with domestic responsibilities.
2. Integrate gender mainstreaming principles explicitly into mangrove regulations, including those derived from Government Regulation No. 27 of 2025, anticipating women's roles can be acknowledged as legal subjects and not merely beneficiaries.
3. Develop a gender-disaggregated data collection system and an intersectional approach that concerns class, ethnicity, and age, so that policies target the most vulnerable groups.
4. Provide legal recognition and institutional support for coastal women's organisations and cooperatives, including access to credit, training, and markets within a blue economy framework.
5. Recognise and protect women's indigenous knowledge, such as the management practices of the "Women's Forest" in Youtefa Bay, from the threats of pollution and land conversion.
6. Direct development cooperation and climate finance to support women-led mangrove rehabilitation initiatives, in line with commitments to gender equality within the global climate framework.

REFERENCES

Abdulsalam, M., Elelu, S.-A., Innocent, M. O., Ibrahim, G. O., Livinus, M. U., Lateefat, S. O., & Muhammad, A. S. (2024). Women Empowerment in Environmental Conservation. In Aransiola, S. A., Babaniyi, B. R., Aransiola, A. B., & Maddela, N. R. (Eds.), *Prospects for Soil Regeneration and Its Impact on Environmental Protection* (pp. 351–370). Springer Nature Switzerland., DOI: 10.1007/978-3-031-53270-2_16

Adger, W. N., & Brooks, N. (2003). Does Global Environmental Change Cause Vulnerability to Disaster? In *Natural Disaster and Development in a Globalizing World* (pp. 19–42). Routledge.

Ahmed, J., Kathambi, B., & Kibugi, R. (2023). Policy Perspective on Governance Standards Setting Using Community Participation for Sustainable Mangrove Management in Lamu Kenya. *International Journal of Conservation Science, 14*(1), 295–306. DOI: 10.36868/IJCS.2023.01.20

Asante, F., Hugé, J., Asare, N. K., & Dahdouh-Guebas, F. (2023). Does Mangrove Vegetation Structure Reflect Human Utilization of Ecosystem Goods and Services? *iScience, 26*(6), 106858. DOI: 10.1016/j.isci.2023.106858 PMID: 37255662

Ashrafuzzaman, M. (2025). Mangroves as a Cornerstone of Nature-Based Solutions for Global Climate Resilience. In Nagy, G. J., & Ayal, D. Y. (Eds.), *Handbook of Nature-Based Solutions to Mitigation and Adaptation to Climate Change* (pp. 1071–1099). Springer Nature Switzerland., DOI: 10.1007/978-3-031-34967-6_18

Asmin, W. A. W. (2024). *Pengaturan Perlindungan Hutan Mangrove Dalam Perspektif Hukum Lingkungan Internasional. 9*(3), 642–661. https://snllb.ulm.ac.id/prosiding/index.php/snllb-lit/article/view/1103

Bangun, B. H. (2020). Ecofeminism and Enviromental Protection: A Legal Perspective. *Jambe Law Journal, 3*(1), 1–18. DOI: 10.22437/jlj.3.1.1-18

Banjade, M. R., Liswanti, N., Herawati, T., & Mwangi, E. (2017). *Governing Mangroves: Unique Challenges for Managing Indonesia's Coastal Forests.* Center for International Forestry Research (CIFOR). https://www.cifor-icraf.org/knowledge/publication/6595/

Batanjski, V., Batrićević, A., Purger, D., Alegro, A., Jovanović, S., & Joldžić, V. (2016). Critical Legal and Environmental View on the Ramsar Convention in Protection from Invasive Plant Species: An Example of the Southern Pannonia Region. *International Environmental Agreement: Politics, Law and Economics, 16*(6), 833–848. DOI: 10.1007/s10784-015-9307-4

Begum, F., Lobry de Bruyn, L., Kristiansen, P., & Islam, M. A. (2022). Forest Co-Management in the Sundarban Mangrove Forest: Impacts of Women's Participation on Their Livelihoods and Sustainable Forest Resource Conservation. *Environmental Development*, *43*, 100731. DOI: 10.1016/j.envdev.2022.100731

Begum, F., Lobry de Bruyn, L., Kristiansen, P., & Islam, M. A. (2025). What Factors Influence Women's Participation in Co-Management? A Case Study of Sundarban Mangrove Forest Management in Bangladesh. *Journal of Environmental Planning and Management*, *68*(14), 3337–3362. DOI: 10.1080/09640568.2024.2346600

Bunting, P., Rosenqvist, A., Hilarides, L., Lucas, R. M., Thomas, N., Tadono, T., Worthington, T. A., Spalding, M., Murray, N. J., & Rebelo, L.-M. (2022). Global Mangrove Extent Change 1996–2020: Global Mangrove Watch Version 3.0. *Remote Sensing (Basel)*, *14*(15), 3657. DOI: 10.3390/rs14153657

Cañada, M. C. B., Velasco, C. R., & Lota, M. M. (2022). Gender Roles in the Utilization and Challenges in the Management of Mangrove Forests in Casiguran, Aurora, Philippines. *Open Journal of Ecology*, *12*(4), 257–270. DOI: 10.4236/oje.2022.124015

Chamberland-Fontaine, S., Thomas Estrada, G., Heckadon-Moreno, S., & Hickey, G. M. (2022). Enhancing the Sustainable Management of Mangrove Forests: The Case of Punta Galeta, Panama. *Trees, Forests and People*, *8*, 100274. DOI: 10.1016/j.tfp.2022.100274

Chapman, M., Turner, N. J., & Kark, S. (2026). Women's Local Knowledge: An Important Factor in Effective Community-Based Wildlife Management. *Journal of Environmental Management*, *400*, 128559. DOI: 10.1016/j.jenvman.2026.128559 PMID: 41592469

Chuku, E. O., Effah, E., Adotey, J., Abrokwah, S., Adade, R., Okyere, I., Aheto, D. W., Kent, K., Osei, I. K., Omogbemi, E. D., Adité, A., Ahoedo, K., Sankoh, S. K., Soro, Y., Wélé, M., Saine, D. F., & Crawford, B. (2022). Spotlighting Women-Led Fisheries Livelihoods Toward Sustainable Coastal Governance: The Estuarine and Mangrove Ecosystem Shellfisheries of West Africa. *Frontiers in Marine Science*, *9*, 884715. DOI: 10.3389/fmars.2022.884715

Cormier-Salem, M.-C. (2017). Let the Women Harvest the Mangrove. Carbon Policy, and Environmental Injustice. *Sustainability (Basel)*, *9*(8), 1485. DOI: 10.3390/su9081485

Das, S. C., Thammineni, P., & Ashton, E. C. (2022). Mangroves: A Unique Ecosystem and Its Significance. In S. C. Das, Pullaiah, & E. C. Ashton (Eds.), *Mangroves: Biodiversity, Livelihoods and Conservation* (pp. 3–11). Springer Nature Singapore. DOI: 10.1007/978-981-19-0519-3_1

Dogeje, F. (2026). Empowering Coastal Economies: Gendered Insights into Women's Participation in Seaweed Aquaculture in Zanzibar. *Marine Policy*, *183*, 106900. DOI: 10.1016/j.marpol.2025.106900

Fitria, A. (2021). Ekosistem Mangrove Dan Mitigasi Pemanasan Global. *Jurnal Ekologi. Masyarakat Dan Sains*, *2*(1), 29–34. DOI: 10.55448/ems.v2i1.20

Fricker, M. (2007). *Epistemic Injustice: Power and the Ethics of Knowing*. Oxford University Press., DOI: 10.1093/acprof:oso/9780198237907.001.0001

Harianto, M. (2025, June 26). *KKP Pamerkan Potensi Udang Indonesia Di Shrimp Summit 2025 Bali*. https://www.antaranews.com/berita/4926773/kkp-pamerkan-potensi-udang-indonesia-di-shrimp-summit-2025-bali

Harper, S., Zeller, D., Hauzer, M., Pauly, D., & Sumaila, U. R. (2013). Women and fisheries: Contribution to food security and local economies. *Marine Policy*, *39*, 56–63. DOI: 10.1016/j.marpol.2012.10.018

Hasbi, Y., Saputra, F., Iskandar, H., & Asmara, R. (2026). Reproductive Justice and Women's Health Rights in Indonesia: Between Legal Protection and Structural Inequality. *Masalah-Masalah Hukum*, *55*(1), 47–81. DOI: 10.14710/mmh.55.1.2026.47-81

Hue, L. T. V. (2006). Gender, Doi Moi and Mangrove Management in Northern Vietnam. *Gender, Technology and Development*, *10*(1), 37–59. DOI: 10.1177/097185240501000103

Hussain, S. A., & Badola, R. (2010). Valuing Mangrove Benefits: Contribution of Mangrove Forests to Local Livelihoods in Bhitarkanika Conservation Area, East Coast of India. *Wetlands Ecology and Management*, *18*(3), 321–331. DOI: 10.1007/s11273-009-9173-3

Intergovernmental Panel on Climate Change (IPCC). (2022). *Climate Change 2022: Impacts, Adaptation and Vulnerability. Contribution of Working Group II to the Sixth Assessment Report of the Intergovernmental Panel on Climate Change*. Cambridge University Press., DOI: 10.1017/9781009325844

International Union for Conservation of Nature. (2023, December 8). *Legal and Policy Recommendations to Support International Mangrove Targets*. International Union for Conservation of Nature. https://iucn.org/story/202312/legal-and-policy-recommendations-support-international-mangrove-targets

Kemp, D. (2002). *Global Environmental Issues: A Climatological Approach*. Routledge. DOI: 10.4324/9780203425305

Kochoni, B. I., Salako, K. V., Danquah, J. A., Sinsin, C. B. L., Mensah, S., & Glèlè Kakaï, R. (2025). Contribution of the Ramsar Convention to the Conservation of West-African Mangroves: A Case Study in Benin. *Wetlands Ecology and Management*, *33*(1), 15. DOI: 10.1007/s11273-024-10026-z

Larson, A. M., Dokken, T., Duchelle, A. E., Atmadja, S., Resosudarmo, I. A. P., Cronkleton, P., Cromberg, M., Sunderlin, W., Awono, A., & Selaya, G. (2015). The role of women in early REDD+ implementation: Lessons for future engagement. *International Forestry Review*, *17*(1), 43–65. DOI: 10.1505/146554815814725031

Lueong, L. A., & Fogwe, Z. N. (2022). Gender Biases in the Resource Exploitation and Management of Cameroonian Inland Wetlands Within the Upper Nun Valley, Ndop, North West Region. In Dalei, N. N., & Gupta, A. (Eds.), *Economics and Policy of Energy and Environmental Sustainability* (pp. 257–270). Springer Nature Singapore., DOI: 10.1007/978-981-19-5061-2_13

MacKinnon, C. A. (1987). *Feminism Unmodified: Discourses on Life and Law*. Harvard University Press.

Mangroves for the Future (MFF), Stockholm Environment Institute (SEI), & Southeast Asian Fisheries Development Center (SEAFDEC). (2018). *Gender in Coastal and Fisheries Resource Management: A Regional Synthesis Report*. Mangroves for the Future. https://www.mangrovesforthefuture.org/assets/Repository/Documents/Regional-Synthesis-Report-Gender-in-coastal-and-fisheries-resource-management.pdf

Natalis, A., Purwanti, A., & Asmara, T. (2023). Anthropocentrism Vs Ecofeminism: How Should Modern Environmental Law Be Reformed? *Sortuz: Oñati Journal of Emergent Socio-Legal Studies, 13*(1), 38–68. https://opo.iisj.net/index.php/sortuz/article/view/1686

Natalis, A., Wibowo, S. L., Hardiyanti, M., & Putrijanti, A. (2025). Exploring Intersectionality: The Legal Challenges Faced by Indigenous Women in Palm Oil Land Conflicts. *Diponegoro Law Review*, *10*(2), 173–197. DOI: 10.14710/dilrev.10.2.2025.173-198

Nisaa, F., Rismawati, R., Azwar, W., & Zabua, B. H. (2025). Women's Environmental Governance Roles in Mangrove Ecotourism. *E3S Web Conf., 671*. DOI: 10.1051/e3sconf/202567105009

Nuraeni, E., & Kusuma, Y. W. C. (2023). The Role of Community-Based Tourism for Mangroves Conservation in Banten, Indonesia. *Jurnal Pengelolaan Sumberdaya Alam Dan Lingkungan, 13*(4), 606–612. DOI: 10.29244/jpsl.13.4.606-612

Oppong, C., & Mathibe, M. S. (2024). Women's Role in Promoting Sustainable Development Through Stakeholder Engagement and Environmental Policy: African Perspectives on the Circular Economy. *Business Strategy & Development*, *7*(4), e70051. DOI: 10.1002/bsd2.70051

Pearson, J., McNamara, K. E., & Nunn, P. D. (2019). Gender-Specific Perspectives of Mangrove Ecosystem Services: Case Study from Bua Province, Fiji Islands. *Ecosystem Services, 38*, 100970. DOI: 10.1016/j.ecoser.2019.100970

Purwanti, A., Mahfud, M. A., Wibawa, K. C. S., & Natalis, A. (2022). The Policy of Gender Mainstreaming in Fishing Communities for Poverty Reduction. *Journal of Governance & Regulation, 11*(4), 267–276. DOI: 10.22495/jgrv11i4siart7

Purwanti, A., & Wijaningsih, D. (2025). Feminist Climate Justice as a Framework for Inclusive Climate Governance. *IOP Conference Series. Earth and Environmental Science, 1537*(1), 012054. DOI: 10.1088/1755-1315/1537/1/012054

Purwanti, A., Wijaningsih, D., Mahfud, M. A., & Natalis, A. (2022). Gender Inequality Against Women Fishers in Indonesia. *Indonesia Law Review, 12*(3), 121–136. https://scholarhub.ui.ac.id/ilrev/vol12/iss3/5/

Ramoni-Perazzi, J., Orlandoni-Merli, G., Soto-Werschitz, A., Crescenzi, D., Munissa, D., Gonca, G., Mazuze, G., Alberto, M., Bernardini, N., Camale, N., Nanvonamuquitxo, S., Attorre, F., Nicosia, E., Mapanga, S. F., & Ramoni-Perazzi, P. (2026). Socioeconomics of Artisanal Fishery and Shellfish Collection in Mozambique: A Gender Perspective from Inhaca Island. *Sustainability (Basel), 18*(2), 578. DOI: 10.3390/su18020578

Rasquinha, D. N. (2024). Manning the Mangroves: Gender, Regional Identities, and Social History Shape Mangrove Forest Dependence and Governance. *Ecology and Society, 29*(1), 17. DOI: 10.5751/ES-14809-290117

Rocheleau, D., Thomas-Slayter, B., & Wangari, E. (2013). *Feminist Political Ecology: Global Issues and Local Experience*. Routledge. DOI: 10.4324/9780203352205

Samal, R., & Dash, M. (2024). Women's Participation and Gender Equality in Sustainable Ecotourism: A Bibliometric Analysis. *Women's Studies International Forum*, *107*, 102993. DOI: 10.1016/j.wsif.2024.102993

Saraswati, R. (2025). Indigenous Women in the Draft Law of Indigenous Peoples: An Approach Based on Legal Neutrality and Pancasila Values. *Masalah-Masalah Hukum*, *54*(3), 409–430. DOI: 10.14710/mmh.54.3.2025.409-430

Satriawan, N. P., & Satriawan, H. A. (2026). Pengaturan Mangrove Pasca Terbitnya Peraturan Pemerintah Nomor 27 Tahun 2025. *Private Law*, *6*(1), 38–46. DOI: 10.29303/tdmre621

Scheyvens, R. (2000). Promoting Women's Empowerment Through Involvement in Ecotourism: Experiences from the Third World. *Journal of Sustainable Tourism*, *8*(3), 232–249. DOI: 10.1080/09669580008667360

Seck, S. L. (2024). Women and the Marine Environment in International Law. *Environmental Policy and Law*, *53*(5–6), 321–332. DOI: 10.3233/EPL-239013

Sembiring, A. F., & Garunja, E. (2026). Transforming Energy Policy in Indonesia: Addressing Gender Bias and Promoting Women's Empowerment in the Renewable Energy Sector. *Jurnal Hukum Progresif*, *14*(1), 1–48. DOI: 10.14710/jhp.14.1.1-48

Senis, Y., Bless, A. E., Ungirwalu, A., Muzendi, A. S. M., Matulesi, M., & Yendri, Y., Salosa, & Munro, J. (2026). Women's Contributions to Mangrove Ecosystems: Indigenous Governance, Exploitation Practices, and Adaptation Strategies in Enggros and Tobati, Youtefa Bay. *BIO Web Conf.*, 213. DOI: 10.1051/bioconf/202621301007

Shiva, V. (2010). *Staying Alive: Women, Ecology and Survival in India*. Women Unlimited.

Shiva, V., Mies, M., & Salleh, A. (2014). *Ecofeminism*. Zed Books.

Stone, K., Bhat, M., Bhatta, R., & Mathews, A. (2008). Factors Influencing Community Participation in Mangroves Restoration: A Contingent Valuation Analysis. *Ocean and Coastal Management*, *51*(6), 476–484. DOI: 10.1016/j.ocecoaman.2008.02.001

Suhartanto, D., Kartikasari, A., Amalia, F. A., Novianti, S., Awaludin, I., & Tuan Ismal, T. A. (2025). Unravelling the Eco-Motivated Journey: The Role of Gender and Motivation in Shaping Loyalty in Mangrove Ecotourism. *Journal of Outdoor Recreation and Tourism*, *51*, 100915. DOI: 10.1016/j.jort.2025.100915

Suherman, H. A., Wulansari, F., Heryanto, B., & Lubis, H. F. (2026). Toward an Inclusive Papua: Harmonization of Indigenous Legal Values and Gender Equality Principles in Women's Empowerment. *Jurnal Hukum Progresif, 14*(1), 122–166. DOI: 10.14710/jhp.14.1.122-166

Tijjani, S., Mizuno, K., & Herdiansyah, H. (2021). The Loss of Ecosystem Services in Women's Forest at Youtefa Bay, Jayapura, Papua, Indonesia. *IOP Conference Series. Earth and Environmental Science, 940*(1), 012061. DOI: 10.1088/1755-1315/940/1/012061

Treviño, M. (2022). "The Mangrove is Like a Friend": Local Perspectives of Mangrove Cultural Ecosystem Services Among Mangrove Users in Northern Ecuador. *Human Ecology: an Interdisciplinary Journal, 50*(5), 863–878. DOI: 10.1007/s10745-022-00358-w

Van Lavieren, H., Spalding, M., Alongi, D. M., Kainuma, M., Clüsener-Godt, M., & Adeel, Z. (2012). *Securing the Future of Mangroves*. United Nations University.

Walby, S. (1991). *Theorizing Patriarchy*. Wiley.

Westholm, L. (2016). Fruits from the Forest and the Fields: Forest Conservation Policies and Intersecting Social Inequalities in Burkina Faso's REDD+ Program. *International Forestry Review, 18*(4), 511–521. DOI: 10.1505/146554816820127578

World Wildlife Fund. (2026, January 26). *Mangrove Conservation Around the World*. World Wildlife Fund. World Wildlife Fund. https://www.worldwildlife.org/news/stories/mangrove-conservation-around-the-world/

Chapter 12
Mageri Segoro as Local Wisdom in Mangrove Management:
Good Environmental Governance Based on Empowerment

Johan Bhimo Sukoco
http://orcid.org/0009-0004-8150-9710
Diponegoro University, Indonesia

Wulan Kinasih
http://orcid.org/0000-0003-2514-3473
Universitas Brawijaya, Indonesia

Yoga Aldi Saputra
http://orcid.org/0000-0001-7307-245X
Diponegoro University, Indonesia

Salma Tarek Eid Esmael
http://orcid.org/0009-0000-9822-3541
Salma Tarek Eid Law Firm, Egypt

ABSTRACT

Mangrove conservation is one of the main challenges in coastal area development. The Central Java Provincial Government, Indonesia, has a mangrove conservation policy agenda based on Good Environmental Governance, called Mageri Segoro. This approach emphasizes the empowerment of coastal communities. This program is imbued with local wisdom. In 2025, the government targeted the planting of 1.5–2 million mangrove seedlings in coastal areas through this program, but only 1.9

DOI: 10.4018/979-8-3373-9320-9.ch012

million mangrove seedlings were planted. This achievement also needs further study because the ecological success of mangroves is highly dependent on the level of post-planting care and maintenance. This chapter will explore the extent to which the Mageri Segoro program has been implemented. This study uses a qualitative method with a descriptive approach, which is carried out through a process of analysis, identification, and in-depth examination of various findings from literary sources.

INTRODUCTION

Mangrove areas tend to experience degradation every year due to natural factors and human activities, so their sustainability must be maintained. Mangrove conservation has recently become a concern for the Central Java Provincial Government, which has designed the Mageri Segoro Program. Mageri segoro, in Javanese, means fencing the sea, which literally does not involve concrete walls but rather a green belt of mangroves that serve as a natural barrier to protect coastal areas from environmental degradation. This program was launched on June 5, 2025, simultaneously in all cities/regencies in Central Java Province. DLHK Provinsi Jawa Tengah, (2025) states that 200,000 mangrove seedlings were planted simultaneously in the Mageri Segoro Program in 185 villages spread across 54 sub-districts and 17 districts/cities along the coast. The Mageri Segoro Program is a collaborative, cross-sectoral commitment in Central Java to strengthen coastal ecosystem resilience. This is crucial for maintaining environmental sustainability in coastal areas. Furthermore, the program is expected to promote sustainable improvements in the well-being of coastal communities.

Mangrove planting in this program was carried out along the northern and southern coastlines of Central Java. This activity was organized as part of a series of events commemorating World Environment Day in 2025. This demonstrates the government's commitment to protecting coastal ecosystems from various threats of erosion and other environmental damage. Coastal protection and restoration are the focus of the government in this program. Nature-based solutions form the basis of this program. Mageri Segoro, which means "fencing the sea" in Javanese, refers to the efforts made to build a green belt of mangroves as a natural barrier to protect coastal areas from abrasion and other environmental damage.

Based on the results of network, overlay, and density visualization using VOSviewer, previous studies show that the concepts of empowerment and governance approach are dominant themes and form the center of the global research network, with very high levels of density and connectivity. Network and density visualizations show that community empowerment is widely studied in the context of public policy, stakeholder participation, and sustainability in general. However, the integration of local wisdom with the Good Environmental Governance framework, particularly in the context of mangrove ecosystem management, has not emerged as a dominant node or temporal hotspot. This indicates a significant research gap. The approach to empowering local coastal communities and traditional values has not been systematically articulated in environmental governance models. Thus, this research occupies a strategic position as a bridge between modern institutional approaches and local socio-cultural practices, while expanding the discourse on governance from a normative perspective to contextual ecosystem-based practices.

Several previous researchers have examined environmental issues from various perspectives (Nirmala et al., 2025; Margono et al., 2024; ; Suryani et al., 2023 ; Devi et al., 2025; Sosunova & Porras, 2022; Kurniati et al., 2024). Several recent studies have also linked the environment to technological development (Lakhouit, 2025; Hussain et al., 2024). Other researchers have also conducted studies on communities in the context of disaster mitigation and the environment (Muktiali, Sudharto Hadi, Purnaweni, et al., 2023) (Khoirul Anas et al., 2023; Muktiali, Sudharto Hadi, & Purnaweni, 2023; Prabawani et al., 2024; Purnaweni et al., 2024; Kristanti et al., 2024; Mangkunegara et al., 2025; Hidayati et al., 2026). Previous studies on mangroves have also been examined by earlier researchers (Husna et al., 2023); (Khoirul Anas et al., 2023).

Unlike previous studies that generally emphasize institutional governance or general community empowerment, this study develops a contextual framework that places local wisdom as the core mechanism of governance in empowering coastal communities for sustainable mangrove management. The urgency of this research is even greater given the increasing ecological pressures in coastal areas and the need for adaptive, participatory, and sustainable mangrove governance. By filling the research gap identified through VOSviewer, this study not only contributes to the enrichment of academic literature but also offers a conceptual and empirical basis for the formulation of more inclusive and contextual mangrove management policies. This study will examine the extent to which the implementation of the Mageri Segoro Program as a mangrove conservation program can be viewed from the perspectives of Good Environmental Governance, Empowerment, and Local Wisdom.

RESEARCH METHODOLOGY

This study is a qualitative study employing a descriptive approach. It explores the extent to which the Mageri Segoro program, implemented in Central Java Province, emphasizes an empowerment-based approach to Good Environmental Governance. This qualitative study draws on both primary and secondary data. The study aims to explore the complex social interactions surrounding the research phenomenon under investigation (Feuer & Makarov, 2025; Kemparaj & Chavan, 2013). Primary data were collected through semi-structured interviews and observations. The interviews in this study were conducted with several stakeholders, such as the Regional Disaster Management Agency (BPBD), the Department of Environment and Forestry (DLHK), and community members. Meanwhile, secondary data in this study was obtained through document analysis derived from various local regulations, strategic planning documents, annual reports, and so on. This document analysis is crucial in helping researchers explore narratives and validate findings (Dalglish et al., 2021; Okello, 2023; Morgan, 2022).

In this study, informants were selected using purposive sampling, which was then expanded using snowball sampling (Johnson, 2005). In this study, interviews were conducted to gather in-depth information from stakeholders, while observations were conducted to examine the empirical context of the extent to which the Mageri Segoro Program, as a form of local wisdom, is implemented using an empowerment-based Good Environmental Governance approach in Central Java, Indonesia. These interviews and observations were conducted to strengthen the depiction of the phenomenon (Jimenez et al., 2019; Mahama & Khalifa, 2017; Rose & Grosvenor, 2013). Additionally, documentary analysis reinforced the data collection techniques from both observations and interviews. Source triangulation and methodological triangulation were used in this study to compare interview results with observations and documents, as well as to verify consistency and strengthen the interpretation (Meydan & Akkaş, 2024). Furthermore, this study was also supported by network analysis, overlay, and density visualization using VOSviewer.

This study analyzed the data using the qualitative analysis framework proposed by (Creswell, 2009), following six stages. The following are the six stages of data analysis. First, the researcher prepared the data and then organized it. Second, the researcher conducted a comprehensive review of the collected data. Third, the researcher analyzes the data and performs coding (assigning codes). Fourth, the researcher codes the data and categorizes it. Fifth, the researcher describes the data or information obtained. Sixth, the researcher interprets the data and provides meaning to it. These six stages of data analysis are used to describe and analyze the extent of the implementation of the Mageri Segoro program conducted in Central Java Province, which emphasizes an empowerment-based Good Environmental Governance approach. It is hoped that this study can provide a comprehensive overview of the phenomenon under study.

LITERATURE REVIEW

What is Good Environmental Governance?

Good Environmental Governance (GEG) is a form of environmental governance that ensures development proceeds in a manner that consistently safeguards the environment and the quality of life for communities (Bennett & Satterfield, 2018). GEG connects various actors (government, the private sector, and the public) in the responsible management, utilization, and protection of natural resources while upholding principles of sustainability (Fariz et al., 2024; Orazalin & Mahmood, 2021). This GEG constitutes environmental management by integrating various principles of good governance, with a focus on ecosystem sustainability. GEG does not merely emphasize bureaucratic order, but rather the extent to which the policies and practices of all stakeholders (government, private sector, community, and others) synergize in realizing sustainable environmental protection and management.

A transparent institutional framework for environmental governance, along with effective policies and programs, is essential to support the success of GEG (Purniawati et al., 2020). GEG is an environmental governance framework that promotes the achievement of effectiveness, equity, responsiveness, and resilience in maintaining environmental quality while utilizing natural resources based on sustainable principles (Bennett & Satterfield, 2018).

GEG aims to establish environmental governance based on principles such as effectiveness, equity, responsiveness, and robustness (Purniawati et al., 2020). Furthermore, GEG is expected to promote transparency and accountability in environmental management (Fariz et al., 2024). Additionally, GEG is expected, both in concept and practice, to encourage public participation and engagement in the formulation and implementation of environmental policies (Sokhatska, 2024).

GEG also aims to establish legal certainty, upholding the enforcement of clear environmental regulations (Orazalin & Mahmood, 2021). On the other hand, GEG is also essential for achieving decentralization and strong institutional frameworks, ensuring a clear division of authority and the effective functioning of institutions responsible for environmental issues (Dokuzoğlu & Güzel, 2024).

Effective interaction among various stakeholders in environmental governance is a key pillar of GEG's success. GEG practices can be described as follows: first, strengthening the government's role in formulating policies, laws, and programs related to environmental governance. Next, the business sector complies with various policies and regulations related to environmental governance, including implementing various environmentally friendly practices. On the other hand, the public actively participates in various practices that can promote the creation of

GEG, as well as conducts monitoring and oversight regarding deviant practices in the implementation of GEG (Azimi et al., 2023; Orazalin & Mahmood, 2021).

GEG is an extension of the concept of Good Governance (Ramadhiyanti, 2025). Building on this, Good Environmental Governance serves as a strategic approach to promoting environmental conservation, including mangrove conservation. Mangrove management is expected to implement the various principles outlined in GEG. The practice of GEG is expected to safeguard biodiversity along the coastline, a condition that is anticipated to support the development of the mangrove ecosystem.

Mangrove Management & SDGs

Mangrove management has become a critical focus within environmental conservation efforts due to the vital ecological, economic, and social functions mangrove ecosystems provide (Bindiya et al., 2023). Effective mangrove management integrates ecological restoration with sustainable use, emphasizing community involvement and adaptive governance to ensure long-term resilience (Lovelock et al., 2022).

The concept of Good Environmental Governance (GEG) is increasingly applied in mangrove management to promote transparency, accountability, inclusiveness, and sustainability. This governance framework aligns with local wisdom and traditional knowledge, recognizing the importance of empowering coastal communities who depend on mangrove resources (Golebie et al., 2022). Empowerment-based approaches facilitate active participation, capacity building, and shared responsibility among stakeholders, enhancing conservation outcomes. Programs like Mageri Segoro in Central Java exemplify this by integrating local cultural values and community engagement into mangrove restoration efforts.

Linking mangrove management with the SDGs highlights the multidimensional benefits of conserving these ecosystems. SDG 13 (Climate Action) directly benefits from mangrove restoration through carbon storage and resilience against climate-related hazards such as storms and sea-level rise (Chow, 2018). Additionally, SDG 14 (Life Below Water) is supported by mangroves as critical habitats that sustain marine biodiversity and fisheries productivity. Mangrove conservation also contributes to SDG 15 (Life on Land) by preserving coastal terrestrial ecosystems and preventing land degradation (Bindiya et al., 2023).

Furthermore, mangrove management addresses socio-economic dimensions aligned with SDG 1 (No Poverty) and SDG 2 (Zero Hunger) by supporting livelihoods through sustainable fisheries and ecotourism. It promotes SDG 6 (Clean Water and Sanitation) by filtering pollutants and maintaining water quality in coastal zones. The empowerment of local communities in managing mangroves fosters gender equality (SDG 5) and strengthens partnerships (SDG 17) between government agencies, NGOs, and local stakeholders, ensuring inclusive and collaborative conservation strategies.

Literature emphasizes that integrating local wisdom into mangrove management enhances the cultural relevance and acceptance of conservation initiatives, leading to better stewardship and sustainability (Maulidah, 2023). Combining these with scientific approaches and policy support creates a holistic framework for mangrove governance (T.T. et al., 2022).

Coastal Governance and Environmental Issues

Coastal governance is an integrated approach that regulates the utilization, conservation, and protection of coastal resources to achieve sustainable development (Partelow et al., 2020). Coastal governance balances economic, social, and ecological aspects to ensure holistic management (Bra oveanu, 2023; Ahlhorn, 2018). This approach also promotes active community participation in decision-making processes, recognizing the value of local knowledge and traditional wisdom in maintaining coastal sustainability (Gillgren et al., 2019).

Environmental issues frequently addressed in coastal governance include the degradation of natural habitats such as mangroves, coral reefs, and seagrass beds (Nichols et al., 2019). These ecosystems serve as natural buffers against erosion and flooding and provide critical habitats for biodiversity. Habitat destruction is often caused by human activities like coastal reclamation, unplanned development, land-based pollution, and climate change impacts, including sea-level rise and increased natural disasters (Sher, 2022). Consequently, coastal governance requires stringent regulations and continuous monitoring to preserve environmental quality and ecosystem functions.

Beyond ecological concerns, coastal governance closely relates to socio-economic issues such as land-use conflicts, community access to coastal resources, and resilience of coastal communities to environmental changes (Fathima Mafaziya Nijamdeen et al., 2025). Inclusive governance approaches grounded in community rights aim to enhance social welfare while ensuring the sustainability of natural resources. Strengthening local capacity and empowering communities are key factors for the successful implementation of coastal governance policies (Fathima Mafaziya Nijamdeen et al., 2025; Whitney et al., 2018).

National and international policies and regulations play a crucial role in supporting effective coastal governance. Clear and consistent legal frameworks, along with mechanisms for coordination among governmental agencies, strengthen coastal area management. Moreover, cross-border collaboration is essential in managing transboundary coastal zones, especially in addressing global challenges like climate change and marine pollution (Li et al., 2025).

Coastal Management: Concepts and Approaches

Coastal management is a process involved in the management of coastal areas, ranging from planning and implementation to the evaluation of programs or policies regarding the management of various resources in coastal areas (Kay & Alder, 2017). Coastal management requires an integrated approach, requires effective coordination between the central government, local governments, and local communities in coastal areas (Yuan & Chang, 2021). Some objectives of coastal management include (Ayyam et al., 2019): First, promoting the use and utilization of various coastal resources based on the principle of sustainability. Second, aiming to provide efforts to protect biodiversity in coastal areas. Third, to promote the control of pollution and damage to ecosystems and biodiversity in coastal areas. Fourth, to promote improved economic and social well-being for coastal communities. This integrated approach to coastal management requires effective collaboration among stakeholders, such as the government, the private sector, and the community (Yuan & Chang, 2021). Coastal management also requires effective coordination between land and marine areas (Gracia Prieto, 2022).

Coastal management practices must also take into account various environmental assessments, ranging from climate change, urbanization, pollution, and so on (R. K. Turner, 2000). This must be taken into account, as coastal management is expected to ensure that various disaster prevention efforts in coastal areas are carried out effectively and efficiently. Coastal management is vital for mitigating disaster risks and improving the well-being of coastal communities (Yuan & Chang, 2021).

The participation of local communities is crucial to the success of coastal management efforts. Cross-sectoral collaboration is essential in coastal management practices. Furthermore, harmonization and coordination among stakeholders are critical to the successful implementation of coastal management (Ayyam et al., 2019). Therefore, the implementation of coastal policies requires harmonious relationships among various parties, ranging from the government, the private sector, the community, and other stakeholders. Ideally, these coastal policies should align with the principles of cross-sectoral integration, community participation, social justice, and adaptation to climate change (Van Assche et al., 2020). Coastal management requires effective cross-sectoral coordination, as well as synergy among various stakeholders across different levels of government (Toledo et al., 2025). Building on this, coastal management is essential for efforts to conserve biodiversity in coastal areas. This aligns with this study, which examines the Mageri Segoro Program as one of the mangrove management initiatives implemented in Central Java, Indonesia.

Empowerment is Key

Empowerment is the key factor that determines the success of efforts to increase the capacity of coastal communities. The higher the level of empowerment, the higher the positive correlation with community participation in mangrove management. In addition, this level of empowerment determines the success of coastal environmental regeneration efforts themselves (Iskandar & Mussadun, 2024). Mangrove conservation involving community empowerment tends to have a high success rate, due to its significant impact on the income and health of coastal communities, and the protection of coastal areas (Basyuni, 2025).

Empowerment is always related to assistance. In relation to mangrove management, long-term assistance is expected to increase the sense of ownership of coastal resources by local communities. This is reinforced by the integration of scientific knowledge and local wisdom of the local community (Firman Ali Rahman et al., 2024). Therefore, the empowerment of coastal communities should ideally be promoted through various efforts to enhance their capabilities by making the most of their local knowledge.

Community empowerment is essential for coastal zone management. The involvement of local communities in the implementation of coastal policies, including mangrove management, is critical to the success of coastal management practices. Therefore, various coastal policies must accommodate the rights and needs of coastal communities themselves. Protection of vulnerable, disadvantaged, and marginalized groups is essential in coastal community empowerment efforts (Evans et al., 2023). Coastal community empowerment programs must be strengthened. This is crucial because coastal zone management is often hindered by the low capacity of coastal communities. This lack of community involvement, coupled with insufficient efforts to empower coastal communities, makes the successful management of coastal areas increasingly elusive (Sofianto, 2017).

Several previous studies show that empowerment in mangrove conservation can be narrowed down into three forms, namely strengthening institutional capacity, empowering the coastal community economically, and increasing the capacity and knowledge of the coastal community. Several forms of these empowerment activities can be seen in Table 1:

Table 1. Empowerment & mangrozve

No.	Forms of Empowerment	Form of activity	Researcher (years)
1.	Strengthening institutional capacity	Establishment of mangrove tourism villages, strengthening local regulations regarding mangrove conservation	(Angga & Labetubun, 2023); (Ramdhani & Rahaju, 2022); (Macamo et al., 2024).
2.	Economic empowerment of coastal communities	Additional income from the processing of pedada fruit jam products, ecotourism, MSMEs processing mangrove products	(Iskandar & Mussadun, 2024); (Puspitaloka et al., 2025); (Adriyanti et al., 2023); (Ropita et al., 2022).
3.	Increasing the capacity and knowledge of coastal communities	Rehabilitation, training in mangrove planting and selection of appropriate plant types	(Gómez-Ruiz et al., 2022); (Firman Ali Rahman et al., 2024).

Source: processed from research results (2026).

Several previous studies related to mangroves and empowerment, as shown in Table 1, indicate that these studies focused on institutional capacity building, the economic empowerment of coastal communities, and the enhancement of coastal communities' capacity and knowledge. However, few studies have focused on local wisdom within the context of community-based Good Environmental Governance (GEG). Therefore, this study focuses on describing and analyzing the Mageri Segoro Program from the perspective of local wisdom within community-based Good Environmental Governance (GEG).

RESULT & DISCUSSION

Portrait of mangroves in Central Java

Mangrove ecosystems in Central Java are widely distributed along the coastal areas, particularly on the northern coast. This region includes Demak Regency, Semarang City, Brebes Regency, Pemalang Regency, Kendal Regency, and Jepara Regency. Mangroves are also found on the southern coast, particularly in the Segara Anakan area of Cilacap Regency. This distribution indicates that mangroves in Central Java exhibit a spatial pattern influenced by coastal geomorphological conditions. In the northern region, human activity is very intense, whereas in the southern region, natural conditions remain relatively well-preserved.

Mangrove ecosystems play a vital role in the sustainability of coastal areas. Mangrove forests serve as a natural barrier against coastal erosion and ocean waves. Additionally, these forests are effective blue carbon sinks and help mitigate climate change. These ecosystems also provide vital habitats for various types of coastal wildlife. However, mangroves in Central Java face various pressures, particularly from human activities in coastal areas. Therefore, understanding the distribution, functions, and diversity of mangroves is crucial as a foundation for sustainable management and conservation. This can be seen in Figure 2. The map above illustrates changes in Central Java's mangrove forests from 2020 to 2025 (Figure 2). These changes are primarily concentrated along the northern coast facing the Java Sea.

Figure 1.

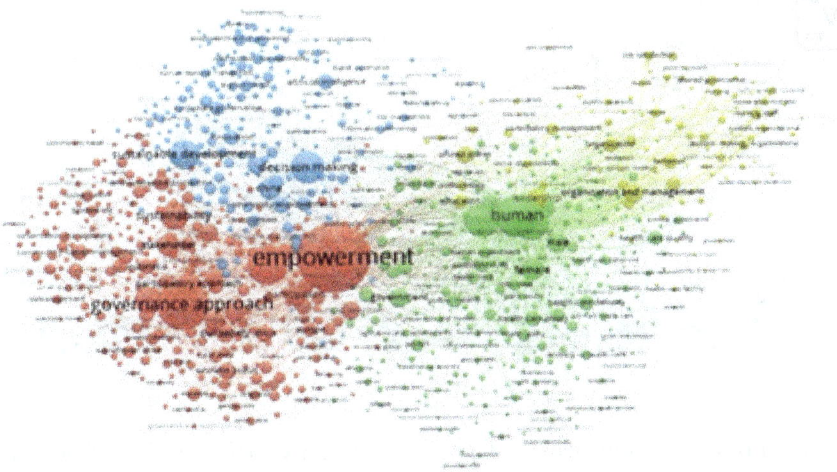

Figure 2.

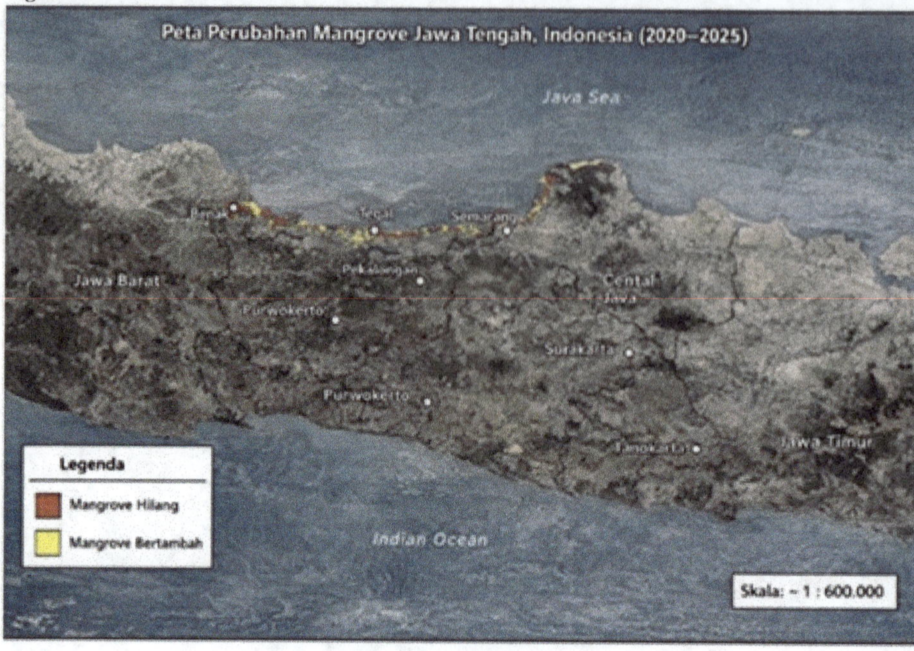

The condition of mangroves in Central Java can be described as fluctuating and dynamic. There is no upward trend, but no drastic decline has been observed either, suggesting that the mangroves are generally well-preserved. However, the predominance of red in some areas indicates that the ecosystem faces significant challenges. Without stronger and more coordinated management interventions, the potential loss of mangroves will continue. However, if rehabilitation is carried out systematically and in an ecosystem-based manner, the mangroves in this region can still recover and function as natural coastal protectors and as a foundation for the livelihoods of the surrounding communities.

Mangrove conservation efforts in Central Java involve various stakeholders, including the government, local communities, the private sector, and academics. Various programs being implemented demonstrate that mangrove conservation focuses not only on ecological aspects but also considers the social and economic aspects of coastal communities. The government plays a leading role in conservation through mangrove rehabilitation and planting programs in various coastal areas.

The government has also designated mangrove conservation areas, particularly in the southern region, namely Segara Anakan in Cilacap. The designation of these areas is accompanied by regulations that restrict exploitative activities, such as logging and land conversion, in order to preserve the ecosystem. This policy serves as the foundation for more targeted and sustainable mangrove management. Local communities also play an active role in mangrove conservation through community-based approaches. These activities are often driven by an awareness of the importance of mangroves in protecting their homes from coastal erosion. Communities are also developing sustainable mangrove management practices through silvofishery systems, which integrate aquaculture with mangrove conservation. This approach allows communities to derive economic benefits without damaging the ecosystem, thereby creating a balance between economic needs and environmental conservation.

The private sector and nongovernmental organizations also play a vital role in supporting mangrove conservation. Through corporate social responsibility (CSR) programs, various companies donate funds and assist in the implementation of mangrove rehabilitation. Additionally, collaborations between research institutions and nongovernmental organizations are established to expand rehabilitation efforts and enhance the effectiveness of these programs. Academics and researchers contribute through research, mapping, and community service activities.

Mageri Segoro: A Local Wisdom Based on GEG

The Mageri Segoro Program is an initiative organized by the Provincial Government of Central Java, Indonesia, aimed at combating coastal erosion and tidal flooding along the coastline through mangrove planting and maintenance. The Mageri Segoro Program is being implemented on a large scale along the northern and southern coasts of Central Java Province. The program's primary objective is to prevent various disasters along the coastal areas, such as coastal erosion. Additionally, the program is expected to mitigate the impacts of the climate crisis, protect coastal ecosystems, and secure the shoreline. In 2025, the Mageri Segoro Program successfully planted approximately 200,000 mangrove seedlings.

In 2025, the Mageri Segoro Program successfully planted more than 1 million mangrove seedlings simultaneously across 17 regencies and cities in Central Java Province, earning it a spot in the Indonesian Record Museum (MURI). The following is the number of mangrove plantings in the Mageri Segoro Program in 2025, as shown in Figure 3. Figure 3 shows significant variations in performance between districts in Central Java Province in mangrove planting under the Mageri Segoro Program. Pemalang District recorded the highest number of mangrove plantings, namely 143,000 seedlings. On the other hand, Demak District was only able to plant 6,000 seedlings. This condition indicates disparities in institutional capacity, resource support, and community participation in the implementation of the Mageri Segoro Program.

Figure 3.

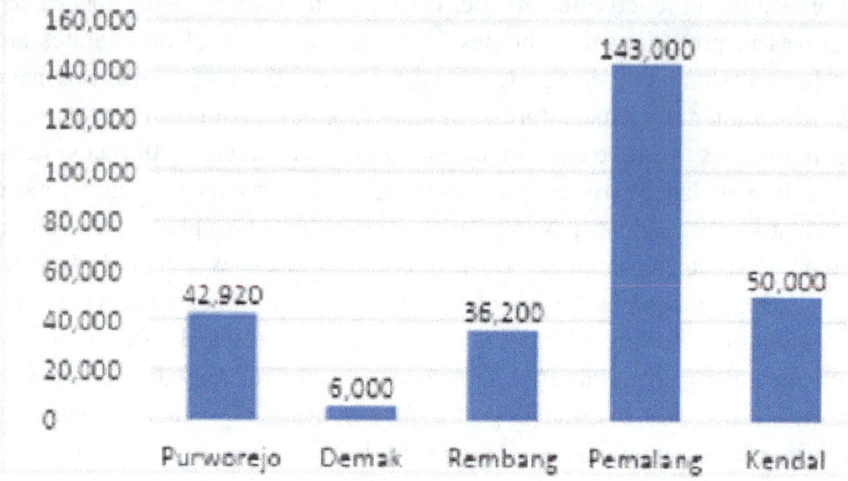

The disparity in achievements between regions is an interesting issue to examine in the implementation of the Mageri Segoro Program. The concentration of the program in certain areas is indicated to be caused by uneven performance achievements in other areas. However, even though Pemalang Regency has the highest number of mangrove plantings, the success of the Mageri Segiri Program is not only determined by the output of mangrove seedling planting, but also by the quality of environmental governance. Figure 2 indicates that Good Environmental Governance practices in Pemalang Regency are running well compared to other regions. Furthermore, this success is indicated to be driven by high levels of participation, sustainability, and accountability in the region. On the other hand, Figure 2 shows that the very low performance of the Mageri Segoro Program in Demak Regency is indicated to be caused not only by the low number of mangrove seedlings successfully planted, but also by the weak aspects of program sustainability and accountability.

The Mageri Segoro program is supported by strong local wisdom in the coastal areas of Central Java. Local cultures such as mutual cooperation, sea offerings, community discussions, and so on are strengths in the implementation of this program. Coastal communities are encouraged to actively participate in this program, from farmer groups and environmental communities to coastal communities, to be actively involved in planting mangrove seedlings. This is important in providing the community with an understanding that mangrove conservation is not only the responsibility of the government, but also the collective responsibility of all elements of society.

Sedekah laut, for example, is a local wisdom in coastal communities in Central Java that is indicated to encourage the success of the Mageri Segoro Program. The main values in sedekah laut, such as gratitude and respect for nature, are expected to encourage coastal communities to preserve the mangrove ecosystem. Sedekah laut is also expected to maintain environmental balance.

On the other hand, local values of coastal communities, such as mutual cooperation, are a strength in determining the success of mangrove conservation. This is in line with the importance of community participation and involvement in environmental development. Good Environmental Governance practices need to be supported by the involvement of various stakeholders, including community elements. Local values such as mutual cooperation are expected to be a strength in the implementation of empowerment-based Good Environmental Governance in mangrove conservation through the Mageri Segoro Program.

Other local values such as community discussions are effective social control instruments in promoting the success of the Mageri Segoro Program. Community discussions, for example, are realized through village deliberations in decision-making related to mangrove conservation efforts. In these community discussions, social sanctions that will be imposed in the event of mangrove destruction can be discussed.

How Important is Local Wisdom in Mangrove Management?

Mangrove management is not only closely linked to rehabilitation and government policy interventions, but is also strongly influenced by the local wisdom of coastal communities (Damastuti & de Groot, 2017). Local knowledge reflects the knowledge, values, and practices that have evolved over generations through communities' interactions with their environment, including an understanding of the ecological functions and utilization of mangrove resources (Galvão et al., 2024; Longépée et al., 2021). In the context of mangroves in Central Java, local wisdom is particularly important because communities are the actors closest to these ecosystems and possess direct experience in utilizing them sustainably (Damastuti & de Groot, 2019; Handayani & Suparno, 2023). Engaging local communities has proven to enhance the success of biodiversity conservation while supporting the livelihoods of coastal communities (Damastuti et al., 2022; Roslinda et al., 2021). Integrating local ecological knowledge with modern scientific approaches is viewed as the most effective strategy for sustainable mangrove restoration and management (Haq et al., 2023; Marquez & Olavides, 2024; Ravaoarinorotsihoarana et al., 2023).

Local knowledge serves as the foundation for environmental conservation, food security, and socio-ecological sustainability. Traditional practices and values are not merely symbolic; rather, they function as effective internal regulatory systems for maintaining environmental sustainability (Abas et al., 2022; Brondízio et al., 2021).

Indigenous knowledge has also been found to contribute significantly to efforts to protect biodiversity and manage natural resources sustainably (Gadgil et al., 2021; N. J. Turner et al., 2022). The most pressing issue today is the decline of traditional knowledge due to market expansion and modernization, as well as the lack of adequate recognition of local wisdom within policy frameworks (Az-Zahra & Guntur, 2025; Wheeler & Root-Bernstein, 2020). Future sustainability depends heavily on the willingness of all parties to incorporate and strengthen local wisdom within education, management, and development systems (Permana et al., 2025; Wirata, 2025). Thus, without the active participation of indigenous and local communities, sustainability policies and programs are likely to be ineffective and at risk of failure (Haq et al., 2023; Lushombo, 2025).

Local wisdom plays a major role in maintaining a balance between the use and conservation of mangrove ecosystems. The long-standing local wisdom of coastal communities generally encompasses knowledge about the optimal planting time, which plant species are suitable for specific soil conditions, and how to utilize available natural resources without damaging the ecosystem. For government programs such as Mageri Segoro, local wisdom is a key factor in the success of implementation on the ground. The program essentially aims to establish a "green belt" of mangroves that protects the coastline from erosion caused by waves and tides. However, success is determined not only by the number of seedlings planted, but also by the continuity of management after planting is complete

Local values in conservation programs such as Mageri Segoro are closely linked to the concept of community empowerment. Empowerment in this context does not merely involve positioning communities as field implementers but also provides them with opportunities to actively participate in decision-making, management, and the utilization of mangrove ecosystems (Ida Afidah, 2025; Dushkova & Ivlieva, 2024). This approach has proven more effective because communities are no longer merely the objects of the program but become the primary agents who feel a sense of ownership over the mangrove ecosystems in their region (Ravaoarinorotsihoarana et al., 2023; Sathiyamoorthy & Sakurai, 2024). Meaningful engagement can also ensure the long-term sustainability of the program, as the community has a direct stake in preserving these resources (Nasution et al., 2025; Ntibona et al., 2023).

The connection between local wisdom and community empowerment is clearly evident in the experience of community-based management. When communities are given the space and trust to manage mangrove ecosystems, they tend to be more committed to ensuring their sustainability because they will directly depend on the ecological and economic benefits generated (Jadin & Rousseau, 2022;Macamo et al., 2024). This approach can also foster various local innovations, one of which is the development of community-based mangrove ecotourism, which has proven capable of improving community well-being while supporting conservation (Budhi et al., 2025; Nuraeni & Kusum, 2023; TRIYANTI et al., 2025).In addition to eco-tourism, the diversification of mangrove-based processed products can also serve as an alternative strategy for sustainable community livelihoods (Lukman et al., 2025; (TJAHJONO et al., 2022).

One of the major obstacles in mangrove conservation efforts is the lack of integration between top-down (government policy) and bottom-up (local knowledge) approaches. Programs like Mageri Segoro can be highly successful if they can balance these two approaches through an adaptive co-management framework that involves all stakeholders (Muhl et al., 2023; Bown et al., 2013). If local knowledge is not incorporated, conservation risks becoming unsustainable because it fails to account for the local socio-ecological context (Djosetro & Behagel, 2024; Loch & Riechers, 2021). Conversely, local initiatives without policy support and resources from the government are often limited in scope and impact (Raha et al., 2024; Rivers et al., 2023). Therefore, synergy between scientific knowledge, formal policies, and local wisdom is a critical prerequisite for the success of long-term coastal management programs (d'Hont & Slinger, 2022; Muhl et al., 2023).

Local wisdom serves as the foundation for sustainable mangrove management. When integrated into government programs such as Mageri Segoro and supported by community empowerment strategies, mangrove management not only helps preserve the environment but also contributes to improved community well-being (Ida Afidah, 2025; Damastuti et al., 2022; Ridwan et al., 2025). Drawing on various cases in Indonesia, community-based approaches that combine local wisdom and formal policy support yield effectiveness and sustainability in conservation (Sasmito et al., 2023). In particular, in Central Java, this approach has successfully mitigated mangrove degradation in the region, thereby positioning local stakeholders as key contributors to achieving national coastal ecosystem restoration targets (Arifanti et al., 2022; (Basyuni, 2025) Building on this, a mangrove management approach that prioritizes local wisdom is an important step that must be considered by various stakeholders in the efforts, practices, and strategies of mangrove management.

Mangrove Management Through Empowerment-Based Good Environmental Governance: A Recommendation

Mangrove ecosystems are critical coastal environments that provide numerous ecological, economic, and social benefits (K et al., 2026). Their role extends beyond mere environmental functions; they serve as natural barriers against coastal erosion, storm surges, and tsunamis, protecting human settlements and infrastructure from natural disasters (Macamo et al., 2024). Additionally, mangroves act as nurseries for many marine species, supporting fisheries that are vital for local food security and economic livelihoods. Empowerment-based Good Environmental Governance

(GEG) offers a comprehensive framework to address these challenges by integrating community participation, local knowledge, and institutional accountability into mangrove conservation efforts.

The empowerment-based approach within GEG emphasizes the active involvement and capacity building of local communities, recognizing that sustainable environmental management cannot be achieved through top-down policies alone (Dushkova et al., 2025). Instead, it requires the meaningful participation of those who directly depend on and interact with the mangrove ecosystems. This approach aligns with the core principles of GEG—transparency, participation, accountability, and responsiveness ensuring that governance structures are both inclusive and effective. Empowering communities involves equipping them with the necessary knowledge, technical skills, institutional support, and access to information, enabling them to take ownership of mangrove management processes. This shift transforms community members from passive recipients of conservation policies into active agents of environmental stewardship (Salerno et al., 2021).

The community of Mageri Segoro exemplifies the successful integration of empowerment and local wisdom within the GEG framework. This coastal community has developed a rich repository of traditional ecological knowledge and sustainable practices that have been passed down through generations. Their local wisdom includes an understanding of seasonal patterns, species behaviors, and ecosystem interdependencies that are often overlooked by conventional scientific methods. These practices are embedded in cultural values and social norms that regulate resource use, ensuring equitable access and long-term sustainability. By incorporating local wisdom into formal governance mechanisms, the management of mangroves becomes more culturally relevant and socially accepted, fostering stronger community engagement and compliance (George et al., 2016).

Empowerment in the context of Mageri Segoro extends beyond knowledge sharing to strengthening institutional capacities and creating enabling environments for community-led management. This includes organizing training programs focused on mangrove restoration techniques, establishing community-based monitoring groups, and facilitating access to external resources such as technical assistance and funding. Empowered communities gain the confidence and legitimacy to engage in dialogue with government agencies, NGOs, and other stakeholders, advocating for their rights and interests (Christens, 2024). This dynamic fosters transparency and accountability, as empowered communities are better positioned to oversee governance processes and demand responsible management.

To operationalize empowerment-based GEG in mangrove management, several key recommendations emerge. Firstly, reinforcing the institutional capacity of communities like Mageri Segoro is paramount. This entails continuous, participatory training and mentoring programs that respect and build upon local wisdom. Capacity building should not only enhance technical skills, such as mangrove planting and monitoring, but also organizational competencies, including leadership, conflict resolution, and resource mobilization. Strengthened institutions enable communities to manage mangrove resources effectively and to participate confidently in multi-stakeholder governance platforms.

Secondly, fostering active community participation at every stage of mangrove management is crucial. Participation should encompass planning, implementation, monitoring, and evaluation phases, ensuring that management strategies are contextually appropriate and socially legitimate. Inclusive participation mechanisms such as community forums, participatory mapping, and joint decision-making bodies facilitate dialogue and consensus-building, empowering community members to contribute their knowledge and priorities.

Thirdly, establishing transparent and accountable governance mechanisms accessible to local communities is essential. Transparency tools such as public reporting, community scorecards, and grievance redress systems enable communities to monitor governance activities, identify irregularities, and hold authorities accountable. These mechanisms reduce risks of corruption, exclusion, and mismanagement, thereby enhancing trust among stakeholders.

Fourthly, integrating local wisdom with scientific knowledge and modern technology enhances the effectiveness and adaptability of mangrove conservation. Scientific approaches provide valuable data on ecosystem health, climate change impacts, and restoration methodologies. When combined with local wisdom—practical insights, cultural relevance, and long-standing ecological understanding—this integration yields innovative management strategies that are both evidence-based and community-driven.

Furthermore, empowerment-based GEG contributes not only to ecological sustainability but also to the socio-economic well-being of coastal communities (Mohdari et al., 2025). Sustainable mangrove management supports diverse livelihood activities such as fisheries, ecotourism, and sustainable harvesting of forest products. By positioning communities as equal partners in governance, empowerment-based GEG strengthens social cohesion and preserves cultural heritage, enabling intergenerational knowledge transfer and cultural continuity (Mustanir et al., 2025).

In summary, empowerment-based Good Environmental Governance represents a holistic and inclusive approach to mangrove management that balances ecological preservation with community welfare. The case of Mageri Segoro illustrates the potential of integrating local wisdom, capacity building, participatory governance, and transparent accountability to achieve resilient and sustainable mangrove ecosystems. Key recommendations including institutional capacity strengthening, inclusive participation, transparent governance, and knowledge integration provide a strategic roadmap for replicating this approach in other coastal regions. Implementing these recommendations will ensure that mangroves continue to deliver critical ecosystem services while empowering communities to safeguard their environment and livelihoods for the future.

CONCLUSION

The differences in performance achievements between regions in mangrove seedling planting in the Mageri Segoro Program in 2025 show the complexity of the driving and inhibiting factors. This is not only due to institutional support, but also influenced by the extent to which local wisdom encourages participation and sustainability of the program. Furthermore, regions with a high level of mutual cooperation tend to show better Good Environmental Governance scores.

Based on this, it can be concluded that the successful implementation of the Mageri Segoro Program does not only depend on program design and quantitative targets, but also on the capacity of local governments to accommodate the values, norms, and local knowledge of coastal communities. Therefore, local wisdom is a very strategic social capital to be developed in realizing mangrove conservation practices based on the principles of good environmental governance based on empowerment.

REFERENCES

Abas, A., Aziz, A., & Awang, A. (2022). A Systematic Review on the Local Wisdom of Indigenous People in Nature Conservation. *Sustainability (Basel), 14*(6), 3415. DOI: 10.3390/su14063415

Adriyanti, D. T., Putri, T. W. S., Mindaryani, A., Prasetya, A., & Petrus, H. T. B. M. (2023). Community Empowerment through Sustainable Utilisation of Mangrove Forests for the Production of Natural Dyes with An Environmentally Friendly System (Phase I: Increasing Awareness and Knowledge Enhancement). [Indonesian Journal of Community Engagement]. *Jurnal Pengabdian Kepada Masyarakat, 9*(1), 38. DOI: 10.22146/jpkm.70455

Afidah, I. (2025). Integrating Local Wisdom for Sustainable Forest Community Empowerment: Lessons From Indonesia and Malaysia. *Journal of Information Systems Engineering & Management, 10*(48s), 437–451. DOI: 10.52783/jisem.v10i48s.9553

Ahlhorn, F. (2018). *Status: Basics of ICZM* (pp. 35–48). DOI: 10.1007/978-3-658-17052-3_3

Al Fariz, R. D., Muis, R., Anggraini, N., Rachman, I., & Matsumoto, T. (2024). Good Environmental Governance Roles in Sustainable Solid Waste Management in Indonesia: A Review. *Journal of Community Based Environmental Engineering and Management, 8*(8), 45–56. DOI: 10.23969/jcbeem.v8i1.12035

Angga, L. O., & Labetubun, M. A. H. (2023). Pemberdayaan Masyarakat Hukum Adat Dalam Pelaksanaan Program Desa Wisata Mangrove. *AIWADTHU: Jurnal Pengabdian Hukum, 3*(1), 17–26. DOI: 10.47268/aiwadthu.v3i1.1191

Arifanti, V. B., Kauffman, J. B., Subarno, , Ilman, M., Tosiani, A., & Novita, N. (2022). Contributions of mangrove conservation and restoration to climate change mitigation in Indonesia. *Global Change Biology, 28*(15), 4523–4538. DOI: 10.1111/gcb.16216 PMID: 35470521

Ayyam, V., Palanivel, S., & Chandrakasan, S. (2019). Strategies and Collaborations for Management of Coastal Areas. In *Coastal Ecosystems of the Tropics - Adaptive Management* (pp. 261–286). Springer Singapore., DOI: 10.1007/978-981-13-8926-9_13

Az-Zahra, A. N., & Guntur, A. I. (2025). Indigenous knowledge and economic resistance: Local strategies against market penetration in Eastern Indonesia. *Edelweiss Applied Science and Technology, 9*(9), 206–214. DOI: 10.55214/2576-8484.v9i9.9785

Azimi, M. N., Rahman, M. M., & Nghiem, S. (2023). Linking governance with environmental quality: A global perspective. *Scientific Reports*, *13*(1), 15086. DOI: 10.1038/s41598-023-42221-y PMID: 37699950

Basyuni, M. (2025). Community service and empowerment of abandoned ponds through mangrove rehabilitation activities with an ecological and hydrological approach in Percut Sei Tuan, North Sumatra. *Journal of Saintech Transfer*, *8*(1), 86–93. DOI: 10.32734/jst.v8i1.18712

Bennett, N. J., & Satterfield, T. (2018). Environmental governance: A practical framework to guide design, evaluation, and analysis. *Conservation Letters*, *11*(6), e12600. Advance online publication. DOI: 10.1111/conl.12600

Bindiya, E. S., Sreekanth, P. M., & Bhat, S. G. (2023). *Conservation and Management of Mangrove Ecosystem in Diverse Perspectives.*, DOI: 10.1007/978-981-19-5841-0_13

Bown, N. K., Gray, T. S., & Stead, S. M. (2013). Co-management and adaptive co-management: Two modes of governance in a Honduran marine protected area. *Marine Policy*, *39*, 128–134. DOI: 10.1016/j.marpol.2012.09.005

Bra oveanu, F. (2023). Role and responsibilities of the national committee of the coastal zone in sustainable coastal zone development. *Ovidius University Annals of Constanta - Series. Civil Engineering*, *25*(1), 67–72. DOI: 10.2478/ouacsce-2023-0010

Brondízio, E. S., Aumeeruddy-Thomas, Y., Bates, P., Carino, J., Fernández-Llamazares, Á., Ferrari, M. F., Galvin, K., Reyes-García, V., McElwee, P., Molnár, Z., Samakov, A., & Shrestha, U. B. (2021). Locally Based, Regionally Manifested, and Globally Relevant: Indigenous and Local Knowledge, Values, and Practices for Nature. *Annual Review of Environment and Resources*, *46*(1), 481–509. DOI: 10.1146/annurev-environ-012220-012127

Budhi, M. K. S., Yasa, I. N. M., Saskara, I. A. N., Lestari, N. P. N. E., Suasih, N. N. R., Rustini, N. K. A., & Riani, N. L. T. (2025). Sustainability of community-based mangrove ecotourism in Bali, Indonesi. *Decision Science Letters*, *14*(3), 699–706. DOI: 10.5267/j.dsl.2025.3.010

Chow, J. (2018). Mangrove management for climate change adaptation and sustainable development in coastal zones. *Journal of Sustainable Forestry*, *37*(2), 139–156. DOI: 10.1080/10549811.2017.1339615

Christens, B. D. (2024). Building Community Power. In *The Cambridge Handbook of Community Empowerment* (pp. 1–32). Cambridge University Press., DOI: 10.1017/9781009153720.001

Creswell, J. W. (2009). *Research Design: Qualitative, Quantitative, and Mixed Methods Approaches* (3rd ed.). SAGE Publications.Inc.

d'Hont, F. M., & Slinger, J. H. (2022). Including local knowledge in coastal policy innovation: Comparing three Dutch case studies. *Local Environment*, *27*(7), 897–914. DOI: 10.1080/13549839.2022.2084722

Dalglish, S. L., Khalid, H., & McMahon, S. A. (2021). Document analysis in health policy research: The READ approach. *Health Policy and Planning*, *35*(10), 1424–1431. DOI: 10.1093/heapol/czaa064 PMID: 33175972

Damastuti, E., & de Groot, R. (2017). Effectiveness of community-based mangrove management for sustainable resource use and livelihood support: A case study of four villages in Central Java, Indonesia. *Journal of Environmental Management*, *203*, 510–521. DOI: 10.1016/j.jenvman.2017.07.025 PMID: 28841518

Damastuti, E., & de Groot, R. (2019). Participatory ecosystem service mapping to enhance community-based mangrove rehabilitation and management in Demak, Indonesia. *Regional Environmental Change*, *19*(1), 65–78. DOI: 10.1007/s10113-018-1378-7 PMID: 30872973

Damastuti, E., de Groot, R., Debrot, A. O., & Silvius, M. J. (2022). Effectiveness of community-based mangrove management for biodiversity conservation: A case study from Central Java, Indonesia. *Trees, Forests and People*, *7*, 100202. DOI: 10.1016/j.tfp.2022.100202

Devi, A., Singh, N., Thandra, N., Gupta, N., Salman, Z. N., & Natarajan, K. (2025). AI and IoT-Enabled Smart Urban Waste Management System for Efficient Collection, Segregation, and Disposal. *E3S Web of Conferences, 619*, 03002. DOI: 10.1051/e3sconf/202561903002

Djosetro, M., & Behagel, J. (2024). Including local knowledge in conservation planning: The case of the western coastal protected areas in Suriname. *Ecosystems and People (Abingdon, England)*, *20*(1), 2361683. Advance online publication. DOI: 10.1080/26395916.2024.2361683

Dokuzoğlu, S., & Güzel, A. E. (2024). Democracy, governance, and environmental policy effectiveness: A cross-country analysis with Sustainable Governance Indicators. *European Journal of Government and Economics*, *13*(1), 103–119. DOI: 10.17979/ejge.2024.13.1.9902

Dushkova, D., & Ivlieva, O. (2024). Empowering Communities to Act for a Change: A Review of the Community Empowerment Programs towards Sustainability and Resilience. *Sustainability (Basel)*, *16*(19), 8700. DOI: 10.3390/su16198700

Dushkova, D., Ivlieva, O., Pouget, C., & Vandewalle, M. (2025). How to support communities in the long-term sustainability transition: The tailored empowerment program. *Blue-Green Systems*, *7*(1), 210–237. DOI: 10.2166/bgs.2025.049

Evans, L. S., Buchan, P. M., Fortnam, M., Honig, M., & Heaps, L. (2023). Putting coastal communities at the center of a sustainable blue economy: A review of risks, opportunities, and strategies. *Frontiers in Political Science*, *4*, 1032204. Advance online publication. DOI: 10.3389/fpos.2022.1032204

Fathima Mafaziya Nijamdeen, T. W. G., Beunen, R., Löhr, A., & Van Assche, K. (2025). *Strategies for Transforming Coastal Governance: Addressing Barriers and Evolving Dependencies*. DOI: 10.5194/oos2025-933

Feuer, Z. S., & Makarov, D. V. (2025). Qualitative and mixed methods in urology. In *Translational Urology* (pp. 195–198). Elsevier., DOI: 10.1016/B978-0-323-90186-4.00083-3

Gadgil, M., Berkes, F., & Folke, C. (2021). Indigenous knowledge: From local to global. *Ambio*, *50*(5), 967–969. DOI: 10.1007/s13280-020-01478-7 PMID: 33566330

Galvão, M. L., Rodrigues, T. N. M., Santos, I. S., & Fernandes, M. E. B. (2024). Traditional ecological knowledge of mangrove wood use on the Brazilian Amazon coast. *Ethnobiology and Conservation*, *13*. Advance online publication. DOI: 10.15451/ec2024-01-13.03-1-19

George, A. S., Scott, K., Mehra, V., & Sriram, V. (2016). Synergies, strengths and challenges: Findings on community capability from a systematic health systems research literature review. *BMC Health Services Research*, *16*(S7), 623. DOI: 10.1186/s12913-016-1860-1 PMID: 28185589

Gillgren, C., Støttrup, J. G., Schumacher, J., & Dinesen, G. E. (2019). Working together: Collaborative decision making for sustainable Integrated Coastal Management (ICM). *Journal of Coastal Conservation*, *23*(5), 959–968. DOI: 10.1007/s11852-018-0631-z

Golebie, E. J., Aczel, M., Bukoski, J. J., Chau, S., Ramirez-Bullon, N., Gong, M., & Teller, N. (2022). A qualitative systematic review of governance principles for mangrove conservation. *Conservation Biology : The Journal of the Society for Conservation Biology*, *36*(1), e13850. Advance online publication. DOI: 10.1111/cobi.13850 PMID: 34668608

Gómez-Ruiz, P. A., Betancourth-Buitrago, R. A., Arteaga-Cote, M., Carbajal-Borges, J. P., Teutli-Hernández, C., & Laffon-Leal, S. (2022). Fostering a participatory process for ecological restoration of mangroves in Pantanos de Centla Biosphere Reserve (Tabasco, Mexico). *Ecosystems and People (Abingdon, England)*, *18*(1), 112–118. DOI: 10.1080/26395916.2022.2032358

Gracia Prieto, F. J. (2022). The complexity of studying coasts: From forms and processes to management. *Cuadernos de Investigación Geográfica*, *48*(2), 219–255. DOI: 10.18172/cig.5451

Handayani, E., & Suparno, S. (2023). The role of customary law in the governance of sustainable agrarian culture in local communities. *Corporate Law and Governance Review*, *5*(1), 29–37. DOI: 10.22495/clgrv5i1p3

Haq, S. M., Pieroni, A., Bussmann, R. W., Abd-ElGawad, A. M., & El-Ansary, H. O. (2023). Integrating traditional ecological knowledge into habitat restoration: Implications for meeting forest restoration challenges. *Journal of Ethnobiology and Ethnomedicine*, *19*(1), 33. DOI: 10.1186/s13002-023-00606-3 PMID: 37559120

Hidayati, I., Prabawani, B., Rahmadhony, A., Hayati, A. N., & Boddy, J. (2026). The Critical Role of Local Leaders in Managing the Impacts of Climate Change in Demak. *Indonesia*, •••, 119–133. DOI: 10.1007/978-3-032-15334-0_8

Husna, J., Wijayanti, I., Budiastusi Wiratmo, L., Indrahti, S., Naryoso, A., Edzan Che Nasir, N., Patria Ratna, M., Mulyaning Andini, B., & Raihan Putra Ratarno, D. (2023). Mapping the Scientific Literature on Mangrove Conservation in Indonesia: A Bibliometric Analysis to Environmental Research. *E3S Web of Conferences*, *448*, 03036. DOI: 10.1051/e3sconf/202344803036

Hussain, D., Elomri, D. A., Kerbache, D. L., & Omri, D. A. E. (2024). Smart city solutions: Comparative analysis of waste management models in IoT-enabled environments using multiagent simulation. *Sustainable Cities and Society*, *103*, 105247. DOI: 10.1016/j.scs.2024.105247

Iskandar, A. N. V., & Mussadun, . (2024). Pemberdayaan dan Partisipasi Masyarakat dalam Pengelolaan Ekosistem Mangrove: Strategi Regenerasi Lingkungan di Desa Bedono, Demak. *Journal of Regional and Rural Development Planning*, *8*(3), 259–271. DOI: 10.29244/jp2wd.2024.8.3.259-271

Jadin, J., & Rousseau, S. (2022). Local community attitudes towards mangrove forest conservation. *Journal for Nature Conservation*, *68*, 126232. DOI: 10.1016/j.jnc.2022.126232

Jimenez, M. E., Hudson, S. V., Lima, D., & Crabtree, B. F. (2019). Engaging a Community Leader to Enhance Preparation for In-Depth Interviews With Community Members. *Qualitative Health Research, 29*(2), 270–278. DOI: 10.1177/1049732318792848 PMID: 30101661

Johnson, T. P. (2005). Snowball Sampling. In *Encyclopedia of Biostatistics*. Wiley., DOI: 10.1002/0470011815.b2a16070

K, A., Prakash, P., KT, M., & PV, M. (2026). Ecological and Economic Significance of Mangroves: A Comprehensive Review. *Aquatic Life Sciences, 3*(1), 9–19. DOI: 10.58920/aqlis0301428

Kay, R., & Alder, J. (2017). *Coastal Planning and Management*. CRC Press., DOI: 10.1201/9781315272634

Kemparaj, U., & Chavan, S. (2013). Qualitative research: A brief description. *Indian Journal of Medical Sciences, 67*(3), 89. DOI: 10.4103/0019-5359.121127 PMID: 24231397

Khoirul Anas, A., Maryono, M., & Purnaweni, H. (2023). Concept of Waste Management in The Abrasion Puddle Area. *E3S Web of Conferences, 448*, 03049. DOI: 10.1051/e3sconf/202344803049

Kristanti, D., Purnaweni, H., Dwimawanti, I. H., & Yuwono, T. (2024). Harmonizing tourism and nature conservation: A governance model for sustainable tourism villages in Indonesia. *IOP Conference Series. Earth and Environmental Science, 1414*(1), 012078. DOI: 10.1088/1755-1315/1414/1/012078

Kurniati, R., Handayani, W., Prabawani, B., Nida, R. S., & Savitri, S. F. (2024). COMMUNITY PARTICIPATION, COMMUNITY READINESS, AND PREFERENCES TO PROMOTE EDU-TOURISM IN MANGROVE AREAS OF TELUK AWUR VILLAGE, JEPARA, INDONESIA. *PLANNING MALAYSIA, 22*. Advance online publication. DOI: 10.21837/pm.v22i33.1555

Lakhouit, A. (2025). Revolutionizing urban solid waste management with AI and IoT: A review of smart solutions for waste collection, sorting, and recycling. *Results in Engineering, 25*, 104018. DOI: 10.1016/j.rineng.2025.104018

Li, X., Teng, X., Zhang, X., & Sang, X. (2025). Transboundary Marine Spatial Planning: Theoretical Implications, Practical Analysis, and Future Prospects. *E3S Web of Conferences, 617*, 01022. DOI: 10.1051/e3sconf/202561701022

Loch, T. K., & Riechers, M. (2021). Integrating indigenous and local knowledge in management and research on coastal ecosystems in the Global South: A literature review. *Ocean and Coastal Management, 212*, 105821. DOI: 10.1016/j.ocecoaman.2021.105821

Longépée, E., Ahmed Abdallah, A., Jeanson, M., & Gollety, C. (2021). Local Ecological Knowledge on Mangroves in Mayotte Island (Indian Ocean) and Influencing Factors. *Forests, 12*(1), 53. DOI: 10.3390/f12010053

Lovelock, C. E., Barbier, E., & Duarte, C. M. (2022). Tackling the mangrove restoration challenge. *PLoS Biology, 20*(10), e3001836. DOI: 10.1371/journal.pbio.3001836 PMID: 36251664

Lukman, K. M., Quevedo, J. M. D., Rifai, H., Alifatri, L. O., Ulumuddin, Y. I., Sofue, Y., Uchiyama, Y., & Kohsaka, R. (2025). Mangrove forest food products as alternative livelihood measures: Mangrove conservation by community in Muara Gembong, Bekasi Regency, Indonesia. *Discover Sustainability, 6*(1), 237. DOI: 10.1007/s43621-025-01049-4

Lushombo, L. W. (2025). Ecological Sustainability for "Life on Land": Wellspring of Indigenous Knowledge. *Religions, 16*(3), 311. DOI: 10.3390/rel16030311

Macamo, C. da C. F., Inácio da Costa, F., Bandeira, S., Adams, J. B., & Balidy, H. J. (2024). Mangrove community-based management in Eastern Africa: Experiences from rural Mozambique. *Frontiers in Marine Science, 11*, 1337678. Advance online publication. DOI: 10.3389/fmars.2024.1337678

Mahama, H., & Khalifa, R. (2017). *Field interviews: Process and analysis The Routledge Companion to Qualitative Accounting Research Methods.*

Mangkunegara, R. A., Alfirdaus, L. K., & Fitriah, F. (2025). A Human Rights Approach to Examine Indonesia's Social Forestry Policies. *Journal of Southeast Asian Human Rights, 8*(2), 347. DOI: 10.19184/jseahr.v8i2.52858

Margono, A. B., Hadi, S. P., & Hidayat, J. W. (2024). Addressing ecosystem, sedimentation and collaborative approaches challenges in Jatibarang reservoir: A path towards sustainable water management. *Edelweiss Applied Science and Technology, 8*(6). Advance online publication. DOI: 10.55214/25768484.v8i6.3375

Marquez, G. P. B., & Olavides, R. D. (2024). Integrating science-based and local ecological knowledge: A case study of mangrove restoration and rehabilitation projects in the Philippines. *Disasters, 48*(S1), e12630. Advance online publication. DOI: 10.1111/disa.12630 PMID: 38840490

Maulidah, F. I. J. (2023). The Tangible and Intangible Benefits of Mangrove Forests as a Factor Affecting Community Participation in Mangrove Management. *Journal of Tropical Ethnobiology*, *6*(2), 112–125. DOI: 10.46359/jte.v6i2.174

Meydan, C. H., & Akkaş, H. (2024). *The Role of Triangulation in Qualitative Research.*, DOI: 10.4018/979-8-3693-3306-8.ch006

Mohdari, M., Noor, M., & Elliyana, E. (2025). Integrating Green Management and Economic Development: The Role of Esg Strategies in Regional Sustainability. *Journal of Production. Operations Management and Economics*, (51), 78–92. DOI: 10.55529/jpome.51.78.92

Morgan, H. (2022). Conducting a Qualitative Document Analysis. *The Qualitative Report.* Advance online publication. DOI: 10.46743/2160-3715/2022.5044

Muhl, E.-K., Armitage, D., Anderson, K., Boyko, C., Busilacchi, S., Butler, J., Cvitanovic, C., Faulkner, L., Hall, J., Martynuik, G., Paul-Burke, K., Swerdfager, T., Thorpe, H., & van Putten, I. (2023). Transitioning toward "deep" knowledge co-production in coastal and marine systems: Examining the interplay among governance, power, and knowledge. *Ecology and Society*, *28*(4), art17. DOI: 10.5751/ES-14443-280417

Muktiali, M., Sudharto Hadi, P., & Purnaweni, H. (2023). Micro and Macro Scale Policy Framework to Decrease Risk of Flood Disasters in Central Java Province. *IOP Conference Series. Earth and Environmental Science*, *1268*(1), 012071. DOI: 10.1088/1755-1315/1268/1/012071

Muktiali, M., Sudharto Hadi, P., Purnaweni, H., & Mussadun, M. (2023). Study of Green Infrastructure and Community Resilience to Flood Disasters in Central Java. *E3S Web of Conferences, 448*, 03021. DOI: 10.1051/e3sconf/202344803021

Mustanir, A., Sellang, K., Adnan, A. A., & Lubis, S. (2025). Tracing the Evolution of Community Empowerment Models in Development Planning. *Jurnal Ilmu Sosial Dan Ilmu Politik*, *28*(3), 400. DOI: 10.22146/jsp.99202

Nasution, F. J., Basyuni, M., Basyuni, M., Yunasfi, , Yunasfi, , Hartini, K. S., Patana, P., Mubaraq, A., & Wee, A. K. S. (2025). Community Participation in Mangrove Conservation and Ecosystem Assessment in Tanjung Rejo, Indonesia. *Online Journal of Biological Sciences*, *25*(2), 519–530. DOI: 10.3844/ojbsci.2025.519.530

Nichols, C. R., Zinnert, J., & Young, D. R. (2019). *Degradation of Coastal Ecosystems: Causes.* Impacts and Mitigation Efforts., DOI: 10.1007/978-3-319-75453-6_8

Nirmala, R. J., Sukoco, J. B., & Suwitri, S. (2025). From Policy to Practice: Analyzing Regional Waste Regulations Using Waste Governance Framework within the Island of Java. *IOP Conference Series. Earth and Environmental Science, 1537*(1), 012043. DOI: 10.1088/1755-1315/1537/1/012043

Ntibona, L. N., Shalli, M. S., & Mangora, M. M. (2023). Willingness and drivers of community participation in mangrove conservation in the Rufiji Delta, Tanzania. *Western Indian Ocean Journal of Marine Science, 22*(1), 31–45. DOI: 10.4314/wiojms.v22i1.4

Nuraeni, E., & Kusum, Y. W. C. (2023). The role of community-based tourism for mangroves conservation in Banten, Indonesia. [Journal of Natural Resources and Environmental Management]. *Jurnal Pengelolaan Sumberdaya Alam Dan Lingkungan, 13*(4), 606–612. DOI: 10.29244/jpsl.13.4.606-612

Okello, W. K. (2023). On the Recovery of Black Life. In *A Practical Guide to Teaching Research Methods in Education* (pp. 163–170). Routledge., DOI: 10.4324/b23320-28

Orazalin, N., & Mahmood, M. (2021). Toward sustainable development: Board characteristics, country governance quality, and environmental performance. *Business Strategy and the Environment, 30*(8), 3569–3588. DOI: 10.1002/bse.2820

Partelow, S., Schlüter, A., Armitage, D., Bavinck, M., Carlisle, K., Gruby, R. L., Hornidge, A.-K., Le Tissier, M., Pittman, J. B., Song, A. M., Sousa, L. P., Văidianu, N., & Van Assche, K. (2020). Environmental governance theories: A review and application to coastal systems. *Ecology and Society, 25*(4), art19. DOI: 10.5751/ES-12067-250419

Permana, D. Y., Imamulhadi, I., Idris, I., & Mariana, M. (2025). Integrating Indigenous Wisdom in Environmental Protection: Exploring Village Authority within the Framework of State Responsibility in Indonesia. *Jambura Law Review, 7*(2), 359–389. DOI: 10.33756/jlr.v7i2.29582

Prabawani, B., Hadi, S. P., Fisher, M. R., Warsono, H., Dewi, R. S., & Ainuddin, I. (2024). Socioeconomic perspective of agroforestry development in Central Java. *Environmental and Sustainability Indicators, 22*, 100354. DOI: 10.1016/j.indic.2024.100354

Provinsi Jawa Tengah, D. L. H. K. (2025, June). Aksi Hijau Terbesar Jawa Tengah [*Juta Mangrove untuk Masa Depan Pesisir.*]. *Mageri Segoro, 1*, 5.

Purnaweni, H., Djumiarti, T., Roziqin, A., & Santoso, B. (2024). How do local government strategies advance social accountability? The challenges from environmental management of Slaughterhouse in Semarang City, Indonesia. *Local Environment, 29*(2), 245–261. DOI: 10.1080/13549839.2023.2282094

Purniawati, P., Kasana, N., & Rodiyah, R. (2020). Good Environmental Governance in Indonesia (Perspective of Environmental Protection and Management). *The Indonesian Journal of International Clinical Legal Education, 2*(1), 43–56. DOI: 10.15294/iccle.v2i1.36381

Puspitaloka, D., Purnomo, H., Gasela Majid, R., Dyah Kusumadewi, S., Pebrial Ilham, Q., Akhfa Hapsari, F., & Nadhira, S. (2025). Advancing Sustainable Mangrove Restoration: A Community-Driven Edu-Ecotourism Business Model in Banyuasin. *Media Konservasi, 30*(2), 334. DOI: 10.29244/medkon.30.2.334

Raha, D., Davies-Vollum, K. S., Hemstock, S. L., Boateng, I., Islam, M. T., & Pierce, C. A. E. (2024). We need collaboration and co-creation to address challenges facing coastal communities. *Nature Human Behaviour, 8*(5), 814–822. DOI: 10.1038/s41562-024-01875-y PMID: 38789526

Firman Ali Rahman, Misdarti, Markum, Suyono, Reni Retnowathi, M. Barmawi, Hanapi, Wahyu Wijayanti, & Alfian Pujian Hadi. (2024). Pemberdayaan dan Pendampingan Masyarakat Pesisir Peduli Mangrove di Kawasan Mangrove Bangko-Bangko, Kabupaten Lombok Barat, Nusa Tenggara Barat. *AL-HAYAT: Jurnal Pengabdian Masyarakat, 2*(1), 13–18. DOI: 10.62588/ahjpm.2024.v2i1.0001

Ramadhiyanti, T. (2025). THE IMPLEMENTATION OF GOOD ENVIRONMENTAL GOVERNANCE IN WASTE MANAGEMENT. *JURNAL ILMU SOSIAL, 23*(2), 14–24. DOI: 10.14710/jis.23.2.2024.14-24

Ramdhani, D. S., & Rahaju, T. (2022). PEMBERDAYAAN MASYARAKAT PESISIR (STUDI PADA KOMUNITAS BANK SAMPAH BINTANG MANGROVE KELURAHAN GUNUNG ANYAR TAMBAK, KECAMATAN GUNUNG ANYAR, KOTA SURABAYA). *Publika*, 953–968. DOI: 10.26740/publika.v10n3.p953-968

Ravaoarinorotsihoarana, L. A., Ratefinjanahary, I., Aina, C., Rakotomahazo, C., Glass, L., Ranivoarivelo, L., & Lavitra, T. (2023). Combining Traditional Ecological Knowledge and Scientific Observations to Support Mangrove Restoration in Madagascar. *Forests, 14*(7), 1368. DOI: 10.3390/f14071368

Ridwan, R. E., Julianto, M. A., Marpaung, P. N., Utami, A. R., Ismail, T., Wijayanti, S. W., Subroto, S. T., Farizan F, D., Ma'ruf, K., & Setiawan, R. J. (2025). Coastal Community Empowerment Based on Local Wisdom and Technological Innovation for Climate Change Adaptation, Erosion Control, and Sustainable Ecotourism Development. *Sustainable Marine Structures*, 9–35. DOI: 10.36956/sms.v7i4.2701

Rivers, N., Strand, M., Fernandes, M., Metuge, D., Lemahieu, A., Nonyane, C. L., Benkenstein, A., & Snow, B. (2023). Pathways to integrate Indigenous and local knowledge in ocean governance processes: Lessons from the Algoa Bay Project, South Africa. *Frontiers in Marine Science*, *9*, 1084674. Advance online publication. DOI: 10.3389/fmars.2022.1084674

Ropita, R., Gazali, M., Anggraini, D., Mardalena, S., Salsabila, F., Syafitri, R., Zulfadhli, Z., Zuriat, Z., Al Farisi, I., Zurba, N., Rahma, E. A., Rinawati, R., & Wahyuni, S. (2022). PEMBERDAYAAN MASYARAKAT PESISIR PEMANFAATAN MANGROVE Sonneratia alba SEBAGAI SELAI BUAH PEDADA DI DESA LHOK BUBON, ACEH BARAT. *Marine Kreatif*, *6*(2), 136. DOI: 10.35308/jmk.v6i2.6552

Rose, R., & Grosvenor, I. (2013). *Interviews. Doing Research in Special Education: Ideas into Practice*, .

Roslinda, E., Listiyawati, L., Ayyub, A., & Al Fikri, F. (2021). The Involvement of Local Community in Mangrove Forest Conservation in West Kalimantan. *Jurnal Sylva Lestari*, *9*(2), 291–301. DOI: 10.23960/jsl29291-301

Salerno, J., Romulo, C., Galvin, K. A., Brooks, J., Mupeta-Muyamwa, P., & Glew, L. (2021). Adaptation and evolution of institutions and governance in community-based conservation. *Conservation Science and Practice*, *3*(1), e355. Advance online publication. DOI: 10.1111/csp2.355

Sasmito, S. D., Basyuni, M., Kridalaksana, A., Saragi-Sasmito, M. F., Lovelock, C. E., & Murdiyarso, D. (2023). Challenges and opportunities for achieving Sustainable Development Goals through restoration of Indonesia's mangroves. *Nature Ecology & Evolution*, *7*(1), 62–70. DOI: 10.1038/s41559-022-01926-5 PMID: 36593293

Sathiyamoorthy, S., & Sakurai, T. (2024). Effectiveness of community participation in Mangrove restoration: The evidence from northern Sri Lanka. *Environmental Economics and Policy Studies*, *26*(4), 759–779. DOI: 10.1007/s10018-024-00397-1

Sher, A. A. (2022). Threats to Biodiversity: Habitat Change. In *An Introduction to Conservation Biology*. Oxford University Press., DOI: 10.1093/hesc/9780197564370.003.0004

Sofianto, A. (2017). PRINSIP-PRINSIP PENANGGULANGAN KEMISKINAN DI WILAYAH PESISIR UTARA JAWA TENGAH. *Buletin Ilmiah Marina Sosial Ekonomi Kelautan Dan Perikanan*, *2*(2), 81. DOI: 10.15578/marina.v2i2.4679

Sokhatska, O. (2024). ENVIRONMENTAL POLICY IN THE PUBLIC ADMINISTRATION SYSTEM: CHALLENGES OF SUSTAINABLE DEVELOPMENT. *Philosophy and Governance*, (2), 31–41. DOI: 10.70651/3041-248X/2024.2.04

Sosunova, I., & Porras, J. (2022). IoT-Enabled Smart Waste Management Systems for Smart Cities: A Systematic Review. *IEEE Access : Practical Innovations, Open Solutions*, *10*, 73326–73363. DOI: 10.1109/ACCESS.2022.3188308

Suryani, D. A., Purwanto, P., Purnaweni, H., & Yuniningsih, T. (2023). Sustainability of Rural Ecotourism. *E3S Web of Conferences, 448*, 03048. DOI: 10.1051/e3sconf/202344803048

Tjahjono, A., Adi Intyas, C., & Fattah, M.TJAHJONO. (2022). MANGROVE MANAGEMENT STRATEGY FOR SUSTAINABLE BUSINESS BASED ON INDONESIAN ECOLOGICAL PRODUCTS. *Geo Journal of Tourism and Geosites*, *43*(3), 1045–1055. DOI: 10.30892/gtg.43325-919

Toledo, I., Laino, E., Iglesias, G., Palazón, A., & Aragonés, L. (2025). Local authorities or national frameworks? A global review on coastal protection policies. *Environmental Development*, *53*, 101119. DOI: 10.1016/j.envdev.2024.101119

Triyanti, R., Wijaya, R. A., Priyatmoko, R., Rahmawati, E., Asih, T. S., Nugraha, R. B. A., & Zamroni, A.TRIYANTI. (2025). The Transformation of Community-Based Mangrove Conservation to Coastal Tourism on the Northern Coast of Java, Indonesia. *International Journal of Conservation Science*, *16*(1), 207–226. DOI: 10.36868/IJCS.2025.01.14

T.T., P., E., M., & T.P., V. (2022). *Understanding mangrove governance: Research methods and guidelines*. Center for International Forestry Research (CIFOR). DOI: 10.17528/cifor/008478

Turner, N. J., Cuerrier, A., & Joseph, L. (2022). Well grounded: Indigenous Peoples' knowledge, ethnobiology and sustainability. *People and Nature*, *4*(3), 627–651. DOI: 10.1002/pan3.10321

Turner, R. K. (2000). Integrating natural and socio-economic science in coastal management. *Journal of Marine Systems*, *25*(3–4), 447–460. DOI: 10.1016/S0924-7963(00)00033-6

Van Assche, K., Hornidge, A.-K., Schlüter, A., & Vaidianu, N. (2020). Governance and the coastal condition: Towards new modes of observation, adaptation and integration. *Marine Policy*, *112*, 103413. DOI: 10.1016/j.marpol.2019.01.002

Wheeler, H. C., & Root-Bernstein, M. (2020). Informing decision-making with Indigenous and local knowledge and science. *Journal of Applied Ecology*, *57*(9), 1634–1643. DOI: 10.1111/1365-2664.13734

Whitney, C. K., Bennett, N., Ban, N., Allison, E. H., Armitage, D., Blythe, J. L., Burt, J. M., Cheung, W. W. L., Finkbeiner, E. M., Kaplan-Hallam, M., Perry, I., Turner, N. J., & Yumagulova, L. (2018). *Adaptive capacity: from assessment to action in coastal social-ecological systems*. DOI: 10.31230/osf.io/sxcb9

Wirata, G. (2025). Integrating Traditional Knowledge and Local Wisdom into Public Policy for Addressing Social and Ecological Crises. *Journal of Hunan University Natural Sciences*, *52*(8). Advance online publication. DOI: 10.55463/issn.1674-2974.52.8.6

Yuan, W., & Chang, Y.-C. (2021). Land and Sea Coordination: Revisiting Integrated Coastal Management in the Context of Community Interests. *Sustainability (Basel)*, *13*(15), 8183. DOI: 10.3390/su13158183

Chapter 13
Pathways to Sustainable Mangrove Governance:
Bridging Policy and Practice in a Changing Climate

Niamjit Das
http://orcid.org/0000-0001-7619-8306
Shahjalal University of Science and Technology, Bangladesh

ABSTRACT

This chapter examines mangrove governance as a pathway for bridging policy and practice in climate resilience. It explores national and local structures, highlighting strengths and weaknesses in policies, decentralization challenges, and institutional fragmentation. Case studies from Bangladesh, Indonesia, Kenya, and the Philippines illustrate diverse approaches, from disaster risk reduction and large-scale restoration to community stewardship and carbon finance. Persistent barriers include conflicts between development and conservation, weak enforcement, equity gaps, and political-economic complexities. Future directions emphasize resilient governance systems, multi-level and transboundary cooperation, scaling community-based models, and embedding mangroves into sustainability agendas. Governance is reaffirmed as the cornerstone of sustainable mangrove management, requiring adaptive, inclusive, and collaborative frameworks to ensure ecological resilience and social equity.

INTRODUCTION

Mangroves are globally significant ecosystems that provide a wide range of ecological, social, and economic benefits. They act as natural buffers against coastal hazards, sequester significant amounts of carbon, sustain biodiversity, and support

DOI: 10.4018/979-8-3373-9320-9.ch013

the livelihoods of millions of people living in coastal zones (Alongi, 2014; Donato et al., 2011). Despite their recognized importance, mangroves remain highly vulnerable, with global losses estimated at 20–35% over the past half-century due to aquaculture expansion, urbanization, pollution, and weak governance structures (Friess et al., 2019; Hamilton & Casey, 2016). This paradox—where mangroves are celebrated in international policy frameworks yet continue to decline locally—underscores a persistent governance gap between policy ambition and practical implementation.

International frameworks have increasingly acknowledged the role of mangroves in addressing climate change and biodiversity loss. The Paris Climate Accord emphasizes the importance of nature-based solutions, including coastal blue carbon ecosystems, in achieving nationally determined contributions (NDCs) (Herr & Landis, 2016). Similarly, the Convention on Biological Diversity (CBD) recognizes mangroves as priority ecosystems for conservation and restoration (CBD, 2020). The Ramsar Convention on Wetlands also highlights mangroves as critical habitats for migratory species and global biodiversity (Ramsar Secretariat, 2018). Yet, despite these commitments, translating global frameworks into effective national and local action remains uneven. Many countries lack coherent strategies, adequate financing, or institutional capacity to implement mangrove-related policies (Herr, Himes-Cornell, & Laffoley, 2016).

At the national level, mangrove governance often suffers from fragmented institutional arrangements. In the Philippines, overlapping bureaucracies and weak enforcement have undermined conservation efforts, with responsibilities divided among multiple agencies such as the Department of Environment and Natural Resources and local government units (Primavera, 2000). Indonesia has made strides in adopting multilevel governance approaches to align national climate strategies with local realities, yet challenges remain in ensuring coherence across scales and sectors (Nainggolan, Suroso, Argo, & Sagala, 2024). In Africa, governance approaches vary widely, with some countries experimenting with community-based restoration while others struggle with institutional weaknesses and competing development priorities (Quenum et al., 2024). These cases illustrate that while policies exist, their implementation is often hindered by institutional fragmentation, limited resources, and lack of community empowerment.

Bridging the governance gap requires reimagining mangrove management beyond formal institutions. Community-based management and indigenous knowledge have proven effective in strengthening conservation outcomes. In the Philippines, local stewardship has been central to mangrove rehabilitation, with communities engaging in reforestation and sustainable use practices (Walters, 2004). In Kenya, the Mikoko Pamoja project demonstrates how linking carbon markets with community governance can generate both ecological and social benefits (Wylie, Sutton-Grier, & Moore, 2016). Indigenous knowledge systems, which often emphasize holistic and adaptive management, provide valuable insights for designing governance frameworks that are culturally appropriate and ecologically sustainable (Kusmana, 2011). These examples highlight the importance of participatory approaches that empower local communities as co-managers of mangrove ecosystems.

New tools and financing mechanisms offer pathways to strengthen mangrove governance. Participatory mapping and remote sensing technologies enhance monitoring and accountability, enabling communities and policymakers to track changes in mangrove cover and ecosystem health (Friess et al., 2012). Ecosystem service valuation provides a framework for integrating mangroves into economic decision-making, highlighting their contributions to climate mitigation, disaster risk reduction, and livelihoods (Barbier, 2016). Financing innovations—including climate funds, payment for ecosystem services, and blended finance—create opportunities to scale conservation efforts and embed mangroves into broader sustainability agendas (Goldstein et al., 2020). These tools not only improve governance capacity but also help bridge the divide between policy frameworks and practical implementation.

Ultimately, sustainable mangrove management requires governance systems that are adaptive, inclusive, and multi-level. Adaptive governance emphasizes flexibility and learning, enabling institutions to respond to changing ecological and social conditions (Folke et al., 2005). Inclusive governance ensures that marginalized groups, including women and indigenous peoples, have a voice in decision-making processes (Agrawal & Gibson, 1999). Multi-level governance connects global frameworks with national strategies and local realities, fostering coherence across scales and sectors (Ostrom, 2010). By weaving together these principles, mangrove governance can move beyond fragmented policies toward integrated systems that deliver tangible outcomes for ecosystems and communities.

The governance gap in mangrove management is not merely a technical challenge but a political and social imperative. Bridging policy and practice requires collaboration across scales, sectors, and borders, as well as recognition of the central role of communities and indigenous knowledge. This chapter situates mangrove governance within the broader climate and sustainability discourse, examining how international commitments can be translated into effective local action. By drawing on case studies from Asia and Africa, it highlights both the pitfalls of fragmented governance and the promise of participatory, community-driven approaches. In doing so, it argues that mangroves must be governed not only as ecological assets but as vital foundations for climate resilience and human prosperity.

GLOBAL POLICY CONTEXT

International Conventions and Agreements

Mangroves have become increasingly visible in international environmental governance frameworks because of their dual role in climate mitigation and adaptation. The Paris Climate Accord (2015) was a landmark agreement that emphasized the importance of nature-based solutions in achieving nationally determined contributions (NDCs). Coastal blue carbon ecosystems, including mangroves, were highlighted as critical assets for reducing greenhouse gas emissions and enhancing resilience against climate impacts (Herr & Landis, 2016). Mangroves' ability to store carbon at densities far exceeding most terrestrial forests (Donato et al., 2011) positions them as indispensable in global mitigation strategies. At the same time, their role in buffering shorelines against storms and sea-level rise underscores their adaptation value (Alongi, 2014). Despite this recognition, many countries have yet to fully integrate mangroves into their NDCs, reflecting a gap between global commitments and national implementation.

The Convention on Biological Diversity (CBD) has also elevated mangroves within its frameworks. Under the Aichi Biodiversity Targets and the post-2020 Global Biodiversity Framework, mangroves are recognized as priority ecosystems for restoration and conservation (CBD, 2020). Mangroves are recognized as priority ecosystems for restoration and conservation under the CBD, reflecting their biodiversity value and ecological functions (CBD, 2020; Das, 2025). They are valued not only for biodiversity protection but also for their contributions to ecosystem services such as fisheries, water purification, and cultural heritage. However, implementation remains uneven. Many countries struggle to reconcile biodiversity goals with economic development priorities, leading to continued mangrove loss despite CBD commitments (Friess et al., 2019). This reflects a broader challenge in biodiversity governance: strong normative commitments but weak enforcement and financing.

The Ramsar Convention on Wetlands (1971) provides another important framework, designating mangroves as wetlands of international importance. Ramsar emphasizes their ecological functions, including habitat provision for migratory species, and their socio-economic value for coastal communities (Ramsar Secretariat, 2018). Yet, despite Ramsar's recognition, enforcement mechanisms remain weak. Many Ramsar sites suffer from inadequate management, insufficient funding, and limited monitoring capacity (Davidson, 2014). This reflects a broader challenge in global wetland governance: strong normative commitments but weak practical enforcement. Together, these international conventions provide a robust normative foundation for mangrove conservation. However, the persistent gap between commitments and implementation highlights the need for stronger institutional coherence, financing mechanisms, and integration across scales.

Regional Frameworks and Their Limitations

Regional frameworks complement global agreements by addressing mangrove governance within specific geographic contexts. For example, the Coral Triangle Initiative on Coral Reefs, Fisheries and Food Security (CTI-CFF), involving six Southeast Asian nations, integrates mangroves into its marine biodiversity conservation agenda (CTI-CFF, 2009). Similarly, regional fisheries management organizations acknowledge mangroves as critical nursery grounds for commercially important species (FAO, 2018). These frameworks recognize mangroves' ecological and economic importance and provide platforms for cooperation among neighboring states.

Despite these efforts, regional frameworks often face limitations. They may lack binding enforcement mechanisms, rely heavily on donor funding, and struggle with coordination across diverse political and institutional contexts (Friess et al., 2019). Regional frameworks often prioritize coral reefs, while mangroves receive secondary attention despite their biodiversity significance (Das, 2025; Friess et al., 2019). In many cases, regional initiatives emphasize biodiversity but fail to fully integrate mangroves into climate adaptation and mitigation strategies. This creates a gap between ecological recognition and policy coherence, limiting the effectiveness of regional governance. Moreover, regional frameworks often prioritize marine ecosystems such as coral reefs, with mangroves receiving secondary attention. This imbalance reflects broader challenges in coastal governance, where mangroves are undervalued compared to more visible ecosystems. Addressing these limitations requires greater integration of mangroves into regional climate and biodiversity agendas, supported by stronger institutional mechanisms and financing.

Blue Carbon Initiatives and Integration into Global Climate Strategies

The concept of blue carbon has emerged as a transformative framework for integrating mangroves into global climate strategies. Blue carbon refers to the carbon captured and stored by coastal ecosystems such as mangroves, seagrasses, and salt marshes (Murray et al., 2011). Mangroves are particularly significant due to their high carbon density and long-term storage capacity (Donato et al., 2011). Their ability to store "irrecoverable carbon" makes them critical for avoiding emissions that cannot be offset elsewhere (Goldstein et al., 2020).

Initiatives led by organizations such as the Blue Carbon Initiative and IUCN have advanced methodologies for measuring, reporting, and verifying mangrove carbon stocks (Herr, Himes-Cornell, & Laffoley, 2016). These efforts have enabled countries to incorporate mangroves into their NDCs, linking local conservation with global climate commitments. For example, Indonesia has integrated mangrove restoration into its climate strategies, while Kenya's *Mikoko Pamoja* project demonstrates how community-based conservation can be linked to carbon finance (Wylie, Sutton-Grier, & Moore, 2016).

Blue carbon initiatives also highlight opportunities for innovative financing. Payment for ecosystem services (PES), climate funds, and blended finance mechanisms can support mangrove conservation at scale (Murray et al., 2011). By monetizing the carbon value of mangroves, these initiatives create incentives for governments and communities to invest in restoration and sustainable management. However, challenges remain in scaling blue carbon initiatives. Technical barriers in monitoring, valuation, and verification persist, while institutional fragmentation often hampers integration into national climate policies (Friess et al., 2012). Moreover, there is a risk that blue carbon could be treated as a technical add-on rather than a transformative pathway for governance. Ensuring that blue carbon initiatives empower communities and align with broader sustainability agendas is essential for their long-term success.

NATIONAL AND LOCAL GOVERNANCE STRUCTURES

National Coastal and Mangrove Policies: Strengths and Weaknesses

Many countries have adopted national policies to protect mangroves as part of broader coastal management frameworks. For example, Bangladesh has enacted legislation to conserve the Sundarbans mangrove forest, recognizing its ecological and socio-economic importance. However, studies show that despite these policies, threats such as deforestation, aquaculture expansion, and industrial development persist due to weak institutional capacity and enforcement gaps (Islam, 2023). Similarly, in Southeast Asia, national coastal laws often include mangroves within protected areas or designate them as special forest zones, but implementation remains inconsistent (GLF, 2022).

Strengths of national policies include formal recognition of mangroves as critical ecosystems, integration into biodiversity and climate strategies, and establishment of protected areas. Weaknesses lie in limited funding, lack of coordination among agencies, and insufficient incorporation of local community knowledge and participation (Van Lavieren et al. 2012). This mismatch between policy ambition and ground realities highlights the need for adaptive and inclusive governance. Table 1 summarizes comparative case studies of mangrove governance approaches across four countries, highlighting both strengths and persistent challenges.

Table 1. Comparative case studies of mangrove governance approaches

Country	Governance Focus	Strengths	Challenges	References
Bangladesh	Disaster risk reduction & resilience	Integration into disaster management policies; community reforestation	Shrimp aquaculture pressures; weak enforcement	Iftekhar & Islam, 2004; Rahman et al. 2014; Islam, 2023; Mancheño et al., 2025
Indonesia	Large-scale restoration & multilevel governance	National climate commitments; ambitious restoration targets	Institutional fragmentation; financing gaps	Kusmana, 2011; Nainggolan et al., 2024
Philippines	Community-based stewardship & institutional reform	Local reforestation success; NGO partnerships	Overlapping mandates; weak enforcement	Primavera, 2000; Walters, 2004
Kenya	Carbon markets & community governance	Innovative PES model (*Mikoko Pamoja*); community benefits	Scaling challenges; monitoring & verification	Wylie et al., 2016; Herr & Landis, 2016

Decentralization and Local Governance Challenges

Decentralization has been promoted in many countries to empower local governments in managing mangrove resources. In theory, this allows for more context-specific decision-making and community participation. In practice, however, decentralization often leads to uneven capacity across local institutions. For instance, local governments may lack technical expertise, financial resources, or political authority to enforce mangrove conservation measures (GLF, 2022). In Bangladesh, local authorities in the Sundarbans region face difficulties in balancing conservation with livelihood needs, leading to conflicts between national directives and local realities (Islam, 2023).

Community-based management initiatives, such as participatory reforestation programs, have shown promise in enhancing local stewardship. Importantly, community perceptions shape governance outcomes; for example, research in Luwu, Indonesia, demonstrates that local attitudes toward mangroves strongly influence conservation success (Sulaiman et al., 2019). Yet, without adequate support from national institutions, these efforts remain vulnerable to external pressures such as commercial aquaculture or infrastructure development.

Institutional Fragmentation and Overlapping Bureaucracies

Mangrove governance is often undermined by fragmented institutional arrangements, with responsibilities spread across ministries of environment, forestry, fisheries, and local government. This overlapping authority leads to confusion, duplication of efforts, and diluted accountability (GLF, 2022). Local-level experiences, such as mangrove management in Rembang Regency, Indonesia, reveal how fragmented policies hinder effective implementation (Roziqin, 2018). At the national scale, the Sundarbans in Bangladesh are managed by the Forest Department, while fisheries and tourism fall under other agencies, creating conflicting mandates and weak coordination (Islam, 2023). Similar challenges are observed across Southeast Asia, where mangroves are subject to both forestry and marine laws without clear delineation of responsibilities (Van Lavieren et al., 2012). Addressing these challenges requires integrated governance frameworks that clarify mandates and foster inter-agency collaboration.

Role of Enforcement and Accountability Mechanisms

Even where strong policies exist, enforcement remains a major challenge. Illegal logging, land conversion for shrimp farming, and industrial encroachment continue to degrade mangroves due to weak monitoring and limited accountability (Islam, 2023). Enforcement agencies often lack resources, personnel, and political support to implement regulations effectively. In some cases, corruption and vested interests further undermine enforcement efforts.

Accountability mechanisms, such as transparent reporting, community monitoring, and judicial oversight, are critical for strengthening governance. Participatory monitoring programs, where local communities are involved in tracking mangrove health and reporting violations, have proven effective in enhancing compliance (Van Lavieren et al. 2012). However, these mechanisms require institutional support and integration into national governance structures to be sustainable.

National and local governance structures for mangroves reveal a paradox: while policies exist and decentralization is promoted, institutional fragmentation and weak enforcement undermine effectiveness. Regional authority interventions can also influence governance outcomes; for example, in West Java, local government involvement shaped mangrove ecotourism initiatives, highlighting both opportunities and challenges for sustainable management (Feti et al., 2020). Bridging this gap requires integrated governance frameworks, capacity-building for local institutions, and robust accountability mechanisms.

BRIDGING POLICY AND PRACTICE

Mechanisms for Translating Policy into Action

The persistent gap between policy ambition and practical implementation in mangrove governance underscores the need for effective mechanisms that translate commitments into tangible outcomes. Policies often articulate broad conservation goals, but without clear pathways for implementation, they remain aspirational. Mechanisms such as integrated coastal zone management (ICZM), ecosystem-based management, and adaptive co-management have been proposed to operationalize policy frameworks (Cicin-Sain & Belfiore, 2005; McLeod & Leslie, 2009). These approaches emphasize cross-sectoral coordination, stakeholder participation, and iterative learning, enabling policies to adapt to changing ecological and social conditions. For example, ICZM frameworks in South and Southeast Asia have attempted to align mangrove conservation with fisheries, tourism, and disaster risk reduction, though success varies depending on institutional capacity and political will (Islam, 2023). Innovative tools such as geospatial modeling have proven useful in identifying suitable restoration sites, as demonstrated in Yogyakarta, Indonesia (Suraningsih et al., 2021). These approaches complement ICZM and adaptive co-management by providing evidence-based guidance for local implementation.

Financial mechanisms also play a crucial role in translating policy into practice. Payment for ecosystem services (PES), carbon finance, and blended funding models provide incentives for communities and governments to invest in mangrove restoration (Murray et al., 2011). The integration of mangroves into national climate strategies through blue carbon initiatives demonstrates how global policy commitments can be localized through financial innovation (Herr & Landis, 2016). However, ensuring equitable distribution of benefits and avoiding elite capture remain critical challenges.

Role of Institutions, NGOs, and Civil Society

Institutions at national and local levels are central to bridging policy and practice. Government agencies provide the legal and regulatory frameworks, but their effectiveness depends on coordination, capacity, and accountability. In many countries, weak institutional capacity undermines enforcement, while overlapping mandates create confusion (Primavera, 2000). Strengthening institutions requires not only resources but also reforms that clarify responsibilities and foster collaboration across sectors.

Non-governmental organizations (NGOs) and civil society actors have emerged as vital partners in mangrove governance. NGOs often fill gaps left by state institutions, providing technical expertise, funding, and advocacy. For instance, Conservation International and The Nature Conservancy have supported mangrove restoration projects that integrate local communities into decision-making (Friess et al., 2019). Civil society organizations also play a watchdog role, holding governments accountable for policy commitments and exposing illegal activities such as logging or land conversion. Their involvement enhances transparency and ensures that governance processes remain inclusive.

Community-based organizations are particularly important in contexts where local livelihoods depend on mangroves. By mobilizing grassroots participation, they foster stewardship and build social capital. Successful examples include community forestry initiatives in the Philippines, where local groups manage mangrove areas under formal agreements with government agencies (Walters, 2004). These initiatives demonstrate that bridging policy and practice requires not only top-down institutional frameworks but also bottom-up engagement.

Indigenous Knowledge and Participatory Governance Models

Indigenous and local knowledge systems provide valuable insights for sustainable mangrove management. Traditional practices often emphasize holistic and adaptive approaches, balancing resource use with conservation. In Indonesia, customary laws (adat) regulate access to mangrove resources, ensuring sustainable harvesting and protection of sacred sites (Kusmana, 2011). Similarly, in East Africa, local communities have long managed mangroves through informal governance systems that emphasize collective responsibility (Wylie, Sutton-Grier, & Moore, 2016).

Participatory governance models that integrate indigenous knowledge with scientific expertise offer promising pathways for bridging policy and practice. Co-management arrangements, where communities share decision-making authority with government agencies, enhance legitimacy and effectiveness (Berkes, 2009). Participatory mapping and monitoring tools empower communities to document ecological changes and advocate for policy interventions (Friess et al., 2012). These approaches not only improve ecological outcomes but also strengthen social equity by giving marginalized groups a voice in governance.

However, integrating indigenous knowledge into formal policy frameworks requires overcoming institutional biases and power asymmetries. Governments and international organizations must recognize the legitimacy of local practices and ensure that participatory processes are genuinely inclusive rather than tokenistic. Building trust between communities and institutions is essential for sustaining participatory governance.

Building Coherence Between Global Frameworks and Local Realities

One of the greatest challenges in mangrove governance is aligning global frameworks with local realities. International conventions such as the Paris Accord, CBD, and Ramsar provide overarching commitments, but their translation into national and local contexts often encounters obstacles. Global frameworks tend to emphasize standardized metrics and reporting, while local realities are shaped by diverse socio-economic, cultural, and ecological conditions (Davidson, 2014). Bridging this gap requires adaptive governance that connects scales and fosters coherence.

Multi-level governance frameworks offer a pathway for building coherence. By linking global commitments with national strategies and local implementation, they create vertical integration across scales (Ostrom, 2010). For example, Indonesia's efforts to integrate mangrove restoration into its NDCs demonstrate how global climate commitments can be localized through national policies and community-based projects (Nainggolan et al., 2024). Similarly, Kenya's Mikoko Pamoja project illustrates how local initiatives can contribute to global climate goals while addressing community needs (Wylie et al., 2016).

Horizontal coherence is equally important, requiring coordination across sectors such as forestry, fisheries, tourism, and disaster management. Integrated governance frameworks that align sectoral policies reduce conflicts and enhance synergies. For instance, aligning mangrove conservation with disaster risk reduction strategies ensures that restoration projects also contribute to coastal protection (Friess et al., 2019). Building coherence also involves fostering dialogue between stakeholders, ensuring that global frameworks are not imposed top-down but adapted to local contexts through participatory processes.

Bridging policy and practice in mangrove governance requires mechanisms that translate commitments into action, institutions that coordinate effectively, NGOs and civil society that advocate and implement, and participatory models that integrate indigenous knowledge. Recent systematic reviews of global mangrove management highlight recurring concerns such as institutional fragmentation and financing gaps, reinforcing the importance of integrated governance frameworks that build coherence across scales (Purnaweni et al., 2026). Building coherence between global frameworks and local realities is essential for ensuring that policies are not only ambitious but also practical and inclusive. Ultimately, sustainable mangrove governance depends on multi-level, adaptive, and participatory systems that align global commitments with local needs and empower communities as co-managers of these vital ecosystems.

CASE STUDIES

Bangladesh: Mangroves in Disaster Risk Reduction and Resilience

The Sundarbans mangrove forest in Bangladesh is the largest contiguous mangrove ecosystem in the world and plays a critical role in disaster risk reduction. Mangroves act as natural barriers against cyclones and storm surges, reducing wave energy and protecting coastal communities (Mancheño et al., 2025). Following Cyclone Sidr in 2007 and Cyclone Aila in 2009, research demonstrated that villages protected by mangrove belts experienced significantly less damage compared to those without such protection (Iftekhar & Islam, 2004). National policies have recognized this role, integrating mangrove conservation into disaster risk reduction strategies. However, challenges remain due to population pressure, illegal logging, and aquaculture expansion (Islam, 2023). Despite these pressures, community-based reforestation programs have shown promise in enhancing resilience, highlighting the importance of aligning disaster management policies with local stewardship.

Indonesia: Large-Scale Restoration and Multilevel Governance Reforms

Indonesia has one of the largest mangrove areas globally, but decades of deforestation for aquaculture and coastal development have degraded vast tracts. In recent years, Indonesia has embarked on ambitious restoration programs, pledging to rehabilitate over 600,000 hectares of mangroves by 2024 as part of its climate commitments (Nainggolan et al., 2024). These efforts are embedded within multilevel governance reforms that seek to align national climate strategies with local implementation. The Ministry of Environment and Forestry collaborates with provincial and district governments, NGOs, and local communities to coordinate restoration activities. While progress has been made, challenges include ensuring coherence across scales, avoiding duplication of efforts, and securing long-term financing (Friess et al., 2019). Indonesia's experience demonstrates both the potential and complexity of large-scale restoration, underscoring the need for adaptive governance that integrates national ambitions with local realities. Ecological studies in Papua further highlight the rich mangrove diversity and community structure that must be considered in shaping Indonesia's conservation priorities (Setyadi et al., 2021).

Philippines: Institutional Challenges and Community-Based Solutions

The Philippines has a long history of mangrove degradation due to aquaculture expansion, urbanization, and weak enforcement of conservation laws. Institutional fragmentation has been a major challenge, with overlapping responsibilities among the Department of Environment and Natural Resources, Bureau of Fisheries and Aquatic Resources, and local government units (Primavera, 2000). This has led to inconsistent enforcement and policy incoherence. However, community-based solutions have emerged as effective alternatives. Local stewardship programs, where communities engage in mangrove reforestation and sustainable harvesting, have demonstrated success in restoring degraded areas (Walters, 2004). These initiatives often rely on partnerships with NGOs and academic institutions, which provide technical support and advocacy. The Philippines illustrates how bottom-up approaches can complement weak institutional frameworks, highlighting the importance of empowering communities as co-managers of mangrove ecosystems.

Kenya (Mikoko Pamoja): Linking Carbon Markets with Community Governance

Kenya's *Mikoko Pamoja* project is a pioneering example of linking carbon markets with community governance. Located in Gazi Bay, the project involves local communities in mangrove conservation and restoration, generating verified carbon credits that are sold in voluntary markets (Wylie, Sutton-Grier, & Moore, 2016). Revenues are reinvested in community development projects such as education, water supply, and health services, creating tangible benefits for local residents. Mikoko Pamoja demonstrates how innovative financing mechanisms can align global climate goals with local needs, fostering both ecological and social resilience. The project has gained international recognition as a model for community-based blue carbon initiatives, though scaling such efforts requires overcoming technical and institutional barriers in monitoring and verification (Herr & Landis, 2016). Kenya's experience highlights the potential of carbon finance to bridge policy and practice when grounded in strong community governance.

Comparative Insights Across Regions

Comparing these case studies reveals both commonalities and differences in mangrove governance. In Bangladesh, mangroves are framed primarily as disaster risk reduction assets, while in Indonesia they are integrated into large-scale climate commitments. The Philippines illustrates the challenges of institutional fragmentation but also the promise of community-based solutions. Kenya's Mikoko Pamoja highlights the role of innovative financing and carbon markets in empowering communities. Across these contexts, several insights emerge:

1. **Community engagement is critical.** Whether in Bangladesh's reforestation programs, the Philippines' stewardship initiatives, or Kenya's carbon projects, local participation enhances legitimacy and effectiveness.
2. **Institutional coherence matters.** Indonesia's multilevel governance reforms demonstrate the importance of aligning national strategies with local implementation, while the Philippines shows the pitfalls of fragmented institutions.
3. **Innovative financing can bridge gaps.** Carbon markets and PES schemes provide incentives for conservation, but equitable benefit-sharing is essential to avoid reinforcing inequalities.
4. **Global frameworks must adapt to local realities.** While international conventions provide overarching commitments, their success depends on translation into context-specific strategies that reflect local socio-economic and ecological conditions.

Together, these case studies underscore the central argument of this chapter: bridging policy and practice requires adaptive, participatory, and multi-level governance that connects global ambitions with local realities.

TOOLS AND PATHWAYS FOR EFFECTIVE GOVERNANCE

Monitoring and Evaluation Frameworks

Effective governance of mangroves requires robust monitoring and evaluation (M&E) systems to track ecological health, policy implementation, and socio-economic outcomes. Traditional monitoring approaches often relied on field surveys, which, while valuable, are resource-intensive and limited in scale. Recent advances in remote sensing, GIS, and ecological indicators have transformed M&E frameworks, enabling large-scale and cost-effective monitoring (Friess et al., 2019). For example, satellite imagery has been used to assess mangrove cover changes globally, providing critical data for policy evaluation (Hamilton & Casey, 2016).

M&E frameworks also need to incorporate social dimensions, such as community participation, livelihood impacts, and equity considerations. Adaptive management approaches emphasize iterative learning, where monitoring results inform policy adjustments (Folke et al., 2005). Local studies on growth analysis and modeling (Das, 2015; Das, 2018) provide methodological insights that strengthen ecological baselines, highlighting how scientific data supports the design of effective monitoring frameworks. This ensures that governance systems remain responsive to ecological changes and socio-economic dynamics. However, challenges remain in ensuring data accessibility, integrating local knowledge, and securing long-term funding for monitoring programs.

Figure 1 illustrates the diverse ecosystem services provided by mangroves and their integration into global and national climate agendas. At the center, mangrove forests are shown as multifunctional ecosystems delivering four major services: carbon sequestration, coastal protection, livelihoods and biodiversity, and integration into climate and sustainability strategies.

- *Carbon Sequestration:* Mangroves act as blue carbon sinks, storing large amounts of carbon and contributing to climate mitigation (Donato et al., 2011; Goldstein et al., 2020).
- *Coastal Protection:* They buffer storm surges, reduce erosion, and safeguard coastal infrastructure (Spalding et al., 2014; Barbier, 2016).
- *Livelihoods & Biodiversity:* Mangroves support fisheries, ecotourism, and biodiversity conservation, sustaining millions of coastal livelihoods (Friess et al., 2019).
- *Integrated Climate & Sustainability Strategies:* Embedding mangroves into national climate policies, disaster risk reduction, and sustainable development frameworks ensures long-term resilience (Herr & Landis, 2016; Van Lavieren et al. 2012).

Together, these interconnected services position mangroves as vital components of both mitigation and adaptation strategies in global climate governance. The following figure synthesizes these ecosystem services and their policy linkages.

Figure 1. Ecosystem services of mangroves in climate agendas

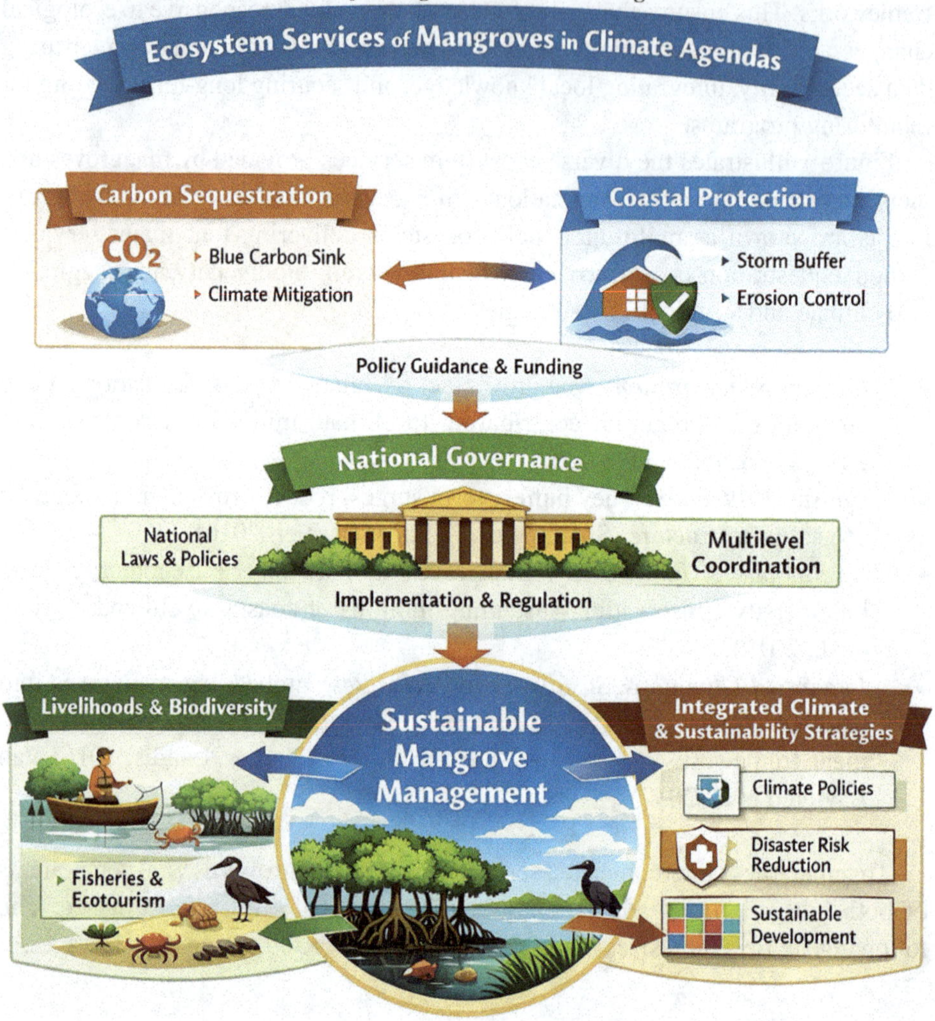

Participatory Mapping and Remote Sensing

Participatory mapping has emerged as a powerful tool for bridging policy and practice by involving communities directly in documenting and managing mangrove ecosystems. Through participatory GIS, local stakeholders can map resource use, identify threats, and prioritize conservation areas (Corbett, 2009). This enhances legitimacy and empowers communities to advocate for policy interventions.

Remote sensing technologies complement participatory approaches by providing large-scale data on mangrove distribution, health, and change. High-resolution satellite imagery and drone-based surveys allow for precise mapping of mangrove cover and biomass (Friess et al., 2012). When combined with participatory mapping, these tools create hybrid governance models that integrate scientific data with local knowledge. For instance, in Indonesia, participatory mapping has been used to align national restoration targets with community priorities, ensuring coherence across scales (Nainggolan et al., 2024).

The integration of participatory mapping and remote sensing strengthens accountability, enhances transparency, and fosters collaboration between institutions and communities. Table 2 summarizes the key tools and pathways that operationalize effective mangrove governance.

Table 2. Tools and pathways for effective governance

Tool/Pathway	Description	Example Application	References
Monitoring & Evaluation	Tracking ecological and social outcomes	Global mangrove cover database (CGMFC-21)	Hamilton & Casey, 2016; Friess et al., 2019
Participatory Mapping & Remote Sensing	Combining local knowledge with geospatial data	Indonesia's community mapping for restoration	Corbett, 2009; Nainggolan et al., 2024
Ecosystem Service Valuation	Quantifying mangrove benefits in monetary terms	Coastal protection valuation in Southeast Asia	Barbier, 2016; Spalding et al., 2014
Financing Mechanisms	Climate funds, PES, blended finance	Mikoko Pamoja carbon credits	Murray et al., 2011; Wylie et al., 2016
Legal Innovations	Adaptive laws, community forestry, zoning	Philippines' community forestry agreements	Camacho, 2010; Walters, 2004

Ecosystem Service Valuation and Integration into Policy

Mangroves provide a wide range of ecosystem services, including carbon sequestration, coastal protection, fisheries support, and cultural values. Valuing these services is essential for integrating mangroves into economic and policy decision-making. Ecosystem service valuation frameworks quantify the benefits of mangroves in monetary terms, making them visible in policy debates and budget allocations (Barbier, 2016).

For example, studies have shown that mangroves reduce storm damage costs by billions of dollars annually, highlighting their role in disaster risk reduction (Spalding et al., 2014). Similarly, their carbon storage capacity positions them as critical assets in climate mitigation strategies (Donato et al., 2011). Recent studies demonstrate that climate and salinity jointly regulate above-ground carbon accumulation in the Sundarbans, reinforcing mangroves' critical role in blue carbon initiatives and highlighting the need for valuation frameworks that account for ecological variability (Chowdhury et al., 2024). Integrating ecosystem service valuation into policy enables governments to justify investments in conservation and restoration, while also creating opportunities for innovative financing mechanisms.

However, valuation approaches must be applied carefully to avoid commodifying ecosystems and overlooking non-market values such as cultural heritage and spiritual significance. Inclusive valuation frameworks that combine economic, ecological, and social dimensions are essential for equitable governance.

Financing Mechanisms: Climate Funds, PES, Blended Finance

Financing remains a major barrier to effective mangrove governance. Traditional funding sources, such as government budgets and donor aid, are often insufficient and unsustainable. Innovative financing mechanisms offer pathways to scale conservation efforts.

Climate funds, such as the Green Climate Fund, provide resources for countries to integrate mangroves into climate adaptation and mitigation projects (Herr & Landis, 2016). Payment for ecosystem services (PES) schemes create direct incentives for communities to conserve mangroves by compensating them for the ecosystem services they provide (Murray et al., 2011). For example, carbon finance projects like Kenya's *Mikoko Pamoja* demonstrate how PES can link global climate markets with local governance (Wylie, Sutton-Grier, & Moore, 2016).

Blended finance, which combines public, private, and philanthropic funding, is increasingly recognized as a way to mobilize large-scale investments in mangrove restoration (Goldstein et al., 2020). By leveraging multiple funding sources, blended finance reduces risks and enhances sustainability. However, ensuring equitable benefit-sharing and avoiding elite capture are critical challenges that must be addressed through transparent governance mechanisms.

Legal Innovations and Policy Instruments

Legal frameworks and policy instruments are essential for institutionalizing mangrove governance. Traditional legal approaches often focus on protected area designation, but innovative instruments are expanding the toolkit. For example, community forestry laws in the Philippines allow local groups to manage mangroves under formal agreements, enhancing stewardship and accountability (Walters, 2004).

Other innovations include integrating mangroves into national climate legislation, embedding ecosystem service valuation into environmental impact assessments, and adopting adaptive legal frameworks that respond to ecological changes (Camacho, 2010). Internationally, legal instruments such as the Ramsar Convention provide normative commitments, but national implementation requires context-specific laws and regulations.

Policy instruments such as zoning regulations, land tenure reforms, and incentive schemes also play critical roles. Secure land tenure for communities enhances stewardship, while zoning laws prevent destructive activities such as aquaculture expansion in sensitive areas (Primavera, 2000). Legal innovations must balance conservation goals with livelihood needs, ensuring that governance frameworks are both effective and equitable.

Tools and pathways for effective mangrove governance encompass monitoring and evaluation frameworks, participatory mapping, ecosystem service valuation, financing mechanisms, and legal innovations. Together, they provide the means to translate policy into practice, enhance accountability, and empower communities. However, their effectiveness depends on integration across scales, inclusivity, and adaptive governance. By combining scientific tools with participatory approaches, innovative financing, and legal reforms, mangrove governance can move beyond fragmented policies toward coherent systems that deliver ecological resilience and social equity.

CHALLENGES AND BARRIERS

Conflicts Between Development and Conservation Priorities

Mangrove governance is frequently undermined by conflicts between conservation objectives and development priorities. Coastal zones are highly sought after for aquaculture, tourism, and urban expansion, leading to widespread mangrove conversion (Friess et al., 2019). In Bangladesh, shrimp aquaculture has been a major driver of mangrove degradation, despite national policies emphasizing conservation (Iftekhar & Islam, 2004). Long-term monitoring reveals that the Sundarbans mangrove ecosystem has become increasingly homogeneous since the 1980s, reflecting governance gaps and unsustainable development pressures (Sarker et al., 2019). In the Philippines, rapid urbanization and infrastructure development have placed immense pressure on mangrove ecosystems, often prioritizing short-term economic gains over long-term ecological sustainability (Primavera, 2000).

These conflicts highlight the tension between economic development and environmental protection. While mangroves are recognized globally for their role in climate mitigation and adaptation, local and national development agendas often prioritize industrial growth and coastal infrastructure. These conflicts are further intensified by broader drivers such as deforestation, climate change, and socio-cultural disruption, which reshape governance priorities and community resilience (Das, 2026). Bridging this divide requires integrated planning frameworks that align conservation with sustainable development, but political and economic incentives frequently favor development, sidelining conservation efforts. These challenges are not only ecological but also institutional and socio-cultural, as highlighted in recent studies (Sarker et al., 2016; Das, 2026). Table 3 summarizes the key challenges in mangrove governance, which are elaborated in the following subsections. The following subsections elaborate on each of these challenges, drawing on case studies from Bangladesh, the Philippines, and beyond.

Table 3. Key challenges in mangrove governance

Challenge	Manifestation	Example Context	References
Development vs. Conservation	Conversion for aquaculture, tourism, urbanization	Bangladesh shrimp farming; Philippine coastal development	Primavera, 2000; Iftekhar & Islam, 2004; Das, 2026
Weak Enforcement	Illegal logging, limited monitoring capacity	Sundarbans wildlife protection gaps	Islam, 2023
Institutional Fragmentation	Overlapping mandates across agencies	Philippines' DENR vs. fisheries agencies	Walters, 2004
Equity & Inclusion	Marginalized groups excluded from decision-making	Gender inequities in resource use	Agrawal & Gibson, 1999; Das, 2014; Van Lavieren et al. 2012
Political Economy	Elite capture of benefits; community costs	Aquaculture expansion in Southeast Asia	Primavera, 2000; Wylie et al., 2016

Weak Enforcement and Institutional Fragmentation

Even where strong policies exist, enforcement remains a persistent challenge. Illegal logging, land conversion, and industrial encroachment continue to degrade mangroves due to weak monitoring and limited accountability (Islam, 2023). Enforcement agencies often lack resources, personnel, and political support to implement regulations effectively. In some cases, corruption and vested interests further undermine enforcement efforts (Walters, 2004). Evidence from the Sundarbans World Heritage ecosystem demonstrates that despite formal protection, enforcement failures continue to undermine mangrove conservation (Sarker et al., 2016).

Institutional fragmentation exacerbates these challenges. Mangrove governance is often divided among multiple ministries and agencies, including environment, forestry, fisheries, and local government. This overlapping authority leads to confusion, duplication of efforts, and diluted accountability (Primavera, 2000). In Indonesia, despite ambitious restoration targets, coordination across national, provincial, and local levels remains inconsistent (Nainggolan et al., 2024). Weak enforcement is not only a product of institutional fragmentation but is also undermined by broader socio-cultural disruptions and climate pressures (Das, 2026). Addressing these challenges requires integrated governance frameworks that clarify mandates, foster inter-agency collaboration, and strengthen enforcement capacity.

Equity and Inclusion Issues in Governance

Equity and inclusion are critical dimensions of mangrove governance but are often overlooked. Marginalized groups, including women, indigenous peoples, and small-scale fishers, frequently lack representation in decision-making processes (Agrawal & Gibson, 1999). This exclusion undermines the legitimacy of governance frameworks and perpetuates inequalities. For example, in the Philippines, community-based initiatives have demonstrated the importance of local participation, yet national policies often fail to institutionalize inclusive governance (Walters, 2004).

Gender dynamics also play a significant role. Women are often key users of mangrove resources, particularly for fuelwood and small-scale fisheries, but their voices are rarely heard in policy discussions (Van Lavieren et al. 2012). Ensuring equity requires participatory governance models that empower marginalized groups, recognize indigenous knowledge, and integrate social justice into conservation strategies. Without inclusion, governance risks reinforcing existing inequalities and undermining sustainability.

Political Economy of Mangrove Management

The political economy of mangrove management shapes governance outcomes by influencing who benefits from conservation and who bears the costs. Mangroves are often caught in power struggles between state institutions, private investors, and local communities. In many cases, elites capture the benefits of mangrove conversion for aquaculture or tourism, while local communities bear the ecological and social costs (Primavera, 2000).

International financing mechanisms, such as carbon markets and climate funds, introduce new dynamics into the political economy. Projects like Kenya's *Mikoko Pamoja* demonstrate how carbon finance can empower communities, but scaling such initiatives requires navigating complex institutional and market structures (Wylie, Sutton-Grier, & Moore, 2016). At the same time, global climate commitments may prioritize carbon sequestration over local livelihood needs, creating tensions between international agendas and community realities (Herr & Landis, 2016).

Understanding the political economy of mangrove management is essential for designing governance frameworks that balance ecological sustainability with social equity. This requires transparency, accountability, and mechanisms to ensure that benefits are equitably distributed.

Mangrove governance faces multiple challenges and barriers, including conflicts between development and conservation, weak enforcement, institutional fragmentation, equity and inclusion issues, and complex political economy dynamics. Addressing these challenges requires integrated, participatory, and adaptive governance frameworks that align global commitments with local realities. Only by confronting these barriers directly can mangrove governance move beyond fragmented policies toward coherent systems that deliver ecological resilience and social justice.

FUTURE DIRECTIONS

Building Resilient Governance Systems

The future of mangrove governance lies in building resilient systems that can withstand ecological, social, and political shocks. Climate change, with its associated sea-level rise and intensifying storms, will continue to test the resilience of mangrove ecosystems and the communities that depend on them. Governance systems must therefore be adaptive, integrating scientific monitoring with local knowledge to respond to changing conditions (Folke et al., 2005). Resilient governance also requires institutional coherence, reducing fragmentation by aligning mandates across ministries and agencies. By embedding flexibility and learning into governance structures, mangrove management can evolve in response to uncertainty and complexity, ensuring long-term sustainability (Ostrom, 2010).

Figure 2 depicts the multi-level governance pathways that connect global policy frameworks, national governance systems, and local practices to achieve sustainable mangrove management. At the top, international agreements such as the *Paris Agreement*, *Convention on Biological Diversity (CBD)*, and *Ramsar Convention* provide overarching policy guidance and funding (CBD, 2020; Ramsar Secretariat, 2018). These commitments inform national governance structures, which establish laws, climate strategies, and mechanisms for multilevel coordination (Nainggolan et al., 2024; Kusmana, 2011). At the local level, community-based co-management, restoration projects, and carbon finance initiatives operationalize these policies through participatory and inclusive approaches (Walters, 2004; Wylie et al., 2016). The central element—*Sustainable Mangrove Management*—is reinforced by four pillars: participatory mapping and monitoring, ecosystem service valuation, equity and inclusion, and adaptive governance (Folke et al., 2005; Barbier, 2016).

Together, these interconnected components demonstrate how bridging policy and practice across scales fosters resilient, equitable, and climate-compatible mangrove governance systems. The following figure illustrates how global, national, and local governance systems interact to achieve resilient mangrove management.

Figure 2. Integrated framework for sustainable mangrove governance

Strengthening Multi-Level and Transboundary Cooperation

Mangrove ecosystems often span administrative boundaries, requiring cooperation across local, national, and regional scales. Strengthening multi-level governance is essential for aligning global commitments with national strategies and local implementation (Nainggolan et al., 2024). Transboundary cooperation is particularly important in regions such as the Sundarbans, shared by Bangladesh and India, where coordinated management is critical for biodiversity conservation and disaster risk reduction (Mancheño et al., 2025). Regional frameworks like the Coral Triangle Initiative demonstrate the potential of collaborative governance, but future efforts must enhance enforcement, financing, and integration of mangroves into broader marine and coastal agendas (CTI-CFF, 2009). Building stronger multi-level and transboundary cooperation will ensure that mangrove governance is not fragmented but harmonized across scales and borders.

Scaling Successful Community-Based Models Globally

Community-based governance models have proven effective in many contexts, demonstrating that local stewardship enhances legitimacy and sustainability. Projects such as *Mikoko Pamoja* in Kenya illustrate how community-led conservation can be linked to innovative financing mechanisms like carbon markets, delivering both ecological and social benefits (Wylie, Sutton-Grier, & Moore, 2016). Similarly, community forestry initiatives in the Philippines show how local groups can successfully manage mangroves under formal agreements with government agencies (Walters, 2004). Scaling these models globally requires adapting them to diverse socio-political contexts, ensuring that local communities are empowered and supported by national institutions and international frameworks. Future directions should focus on creating enabling environments for community-based governance, including secure land tenure, capacity-building, and equitable benefit-sharing (Berkes, 2009).

Embedding Mangroves into Broader Sustainability and Climate Agendas

Mangroves must be embedded into broader sustainability and climate agendas to ensure their long-term protection. Their role in carbon sequestration positions them as critical assets in global climate mitigation strategies (Donato et al., 2011). At the same time, their capacity to reduce storm damage and protect coastlines highlights their importance in climate adaptation (Spalding et al., 2014). Embedding mangroves into national climate policies, disaster risk reduction strategies, and sustainable development plans will elevate their status from niche ecosystems to central pillars of resilience. Ecosystem service valuation frameworks can further integrate mangroves into economic planning, making their benefits visible in policy debates and budget allocations (Barbier, 2016). Future governance must therefore move beyond isolated conservation projects to mainstream mangroves within holistic sustainability agendas. Table 4 summarizes the strategic directions for future mangrove governance, consolidating the pathways discussed above.

Table 4. Future directions for mangrove governance

Direction	Strategic Focus	Expected Outcomes	References
Resilient Governance Systems	Adaptive, learning-based institutions	Flexibility under climate uncertainty	Folke et al., 2005; Ostrom, 2010
Multi-level Cooperation	Aligning global, national, local governance	Harmonized strategies across scales	Nainggolan et al., 2024; CTI-CFF, 2009
Transboundary Collaboration	Joint management of shared ecosystems	Coordinated Sundarbans management	Mancheño et al., 2025
Scaling Community Models	Empowering local stewardship globally	Enhanced legitimacy, equity	Walters, 2004; Wylie et al., 2016
Embedding in Climate Agendas	Integration into NDCs, DRR, SDGs	Mainstreamed mangrove protection	Donato et al., 2011; Barbier, 2016

Future directions in mangrove governance emphasize resilience, cooperation, community empowerment, and integration into broader sustainability frameworks. Building resilient governance systems will ensure adaptability in the face of climate change. Strengthening multi-level and transboundary cooperation will harmonize efforts across scales and borders. Scaling successful community-based models globally will empower local stakeholders and enhance legitimacy. Embedding mangroves into broader sustainability and climate agendas will secure their role as vital ecosystems for both people and planet. Together, these pathways chart a course toward bridging policy and practice, ensuring that mangrove governance evolves into a coherent, adaptive, and inclusive system capable of meeting the challenges of the 21st century.

CONCLUSION

Mangroves stand at the intersection of global climate commitments, biodiversity conservation, and local livelihood needs. Throughout this chapter, we have seen how international conventions, national policies, and community-based initiatives recognize the importance of mangroves, yet a persistent gap remains between policy ambition and practical implementation. Bridging this gap is not simply a technical exercise—it is a governance challenge that requires aligning diverse actors, scales, and priorities. Reaffirming the need to bridge policy and practice is essential if mangroves are to fulfill their potential as both ecological lifelines and social safety nets in a changing climate.

Governance emerges as the central linchpin for sustainable mangrove management. Strong governance frameworks provide the institutional coherence, enforcement capacity, and participatory mechanisms needed to translate commitments into action. Without effective governance, even the most ambitious policies risk remaining aspirational. Conversely, when governance is adaptive, inclusive, and accountable, mangrove conservation can deliver multiple benefits—from climate resilience and biodiversity protection to livelihood security and cultural preservation. Governance is therefore not a peripheral concern but the foundation upon which sustainable mangrove management rests.

Looking ahead, the path to sustainable mangrove governance lies in adaptive, inclusive, and collaborative frameworks. Adaptive governance ensures that policies remain responsive to ecological changes and socio-economic dynamics, embracing uncertainty as an opportunity for learning. Inclusive governance empowers marginalized groups, integrates indigenous knowledge, and ensures that equity and justice are embedded in conservation strategies. Collaborative governance fosters partnerships among governments, NGOs, civil society, and communities, creating synergies across scales and sectors.

Ultimately, bridging policy and practice requires a paradigm shift: from fragmented, top-down approaches to integrated, participatory, and multi-level systems. By embracing adaptive, inclusive, and collaborative governance, we can ensure that mangroves continue to thrive as vital ecosystems, safeguarding both people and planet in the face of climate change.

REFERENCES

Agrawal, A., & Gibson, C. C. (1999). Enchantment and disenchantment: The role of community in natural resource conservation. *World Development, 27*(4), 629–649. DOI: 10.1016/S0305-750X(98)00161-2

Alongi, D. M. (2014). Carbon cycling and storage in mangrove forests. *Annual Review of Marine Science, 6*(1), 195–219. DOI: 10.1146/annurev-marine-010213-135020 PMID: 24405426

Barbier, E. B. (2016). The protective service of mangrove ecosystems: A review of valuation methods. *Marine Pollution Bulletin, 109*(2), 676–681. DOI: 10.1016/j.marpolbul.2016.01.033 PMID: 26851868

Berkes, F. (2009). Evolution of co-management: Role of knowledge generation, bridging organizations and social learning. *Journal of Environmental Management, 90*(5), 1692–1702. DOI: 10.1016/j.jenvman.2008.12.001 PMID: 19110363

Camacho, A. E. (2010). Adapting governance to climate change: Managing uncertainty through a learning infrastructure. *Emory Law Journal, 59*(1), 1–77.

CBD. (2020). *Global Biodiversity Outlook 5*. Secretariat of the Convention on Biological Diversity.

Chowdhury, M., Sarker, S. K., Marma, M., Rahman, M. S., & Datta, A. (2024). Climate and salinity together control above ground carbon accumulation in the Sundarbans mangrove ecosystem. *Ocean and Coastal Management, 255*, 107242. DOI: 10.1016/j.ocecoaman.2024.107242

Cicin-Sain, B., & Belfiore, S. (2005). Linking marine protected areas to integrated coastal and ocean management: A review of theory and practice. *Ocean and Coastal Management, 48*(11–12), 847–868. DOI: 10.1016/j.ocecoaman.2006.01.001

Coral Triangle Initiative on Coral Reefs. Fisheries and Food Security (CTI-CFF). (2009). *Regional Plan of Action*. Jakarta: CTI-CFF Secretariat.

Corbett, J. (2009). *Good practices in participatory mapping. International Fund for Agricultural Development*. IFAD.

Das, N. (2014). Assessment of dependency levels of the forest community people livelihoods through non-timber forest products in the north-eastern region of Bangladesh. *International Journal of Forest Usufructs Management, 15*(1), 61–69.

Das, N. (2015). Comparative growth analysis and yield performance of *Glycine max* under *Jatropha curcas* based agrisilviculture system of agroforestry in the northern part of Bangladesh. *Journal of Forests*, *2*(2), 14–23. DOI: 10.18488/journal.101/2015.2.2/101.2.14.23

Das, N., Mahapatra, C. K., Biswas, S. K., Das, P., & Majumdar, A. (2018). Suitability of the normal, log normal and Weibull distributions for modeling diameter and height distributions of Swietenia mahagoni plantations in Bangladesh. *Journal of Biology and Nature*, *8*(4), 146–155.

Das, N. (2025). *Biodiversity in forest ecosystems: Patterns, drivers, and conservation strategies*. Nova Science Publishers., DOI: 10.52305/AZJY2712

Das, N. (2026). Threats and challenges of deforestation, climate change, and socio-cultural disruption. In Islam, M. (Ed.), *Protecting biodiversity, culture, and communities through forest conservation* (pp. 65–116). IGI Global Scientific Publishing., DOI: 10.4018/979-8-3373-3992-4.ch003

Davidson, N. C. (2014). How much wetland has the world lost? Long-term and recent trends in global wetland area. *Marine and Freshwater Research*, *65*(10), 934–941. DOI: 10.1071/MF14173

Donato, D. C., Kauffman, J. B., Murdiyarso, D., Kurnianto, S., Stidham, M., & Kanninen, M. (2011). Mangroves among the most carbon-rich forests in the tropics. *Nature Geoscience*, *4*(5), 293–297. DOI: 10.1038/ngeo1123

FAO. (2018). *The State of World Fisheries and Aquaculture 2018*. Food and Agriculture Organization of the United Nations.

Feti, F., Hadi, S. P., Purnaweni, H., & Sudarno, . (2020). Does the intervention of regional authorities contribute to sustainable mangrove ecotourism? Case study on mangrove management at Karangsong, West Java, Indonesia. *Ecological Questions*, *31*(3), 7–14. DOI: 10.12775/EQ.2020.017

Folke, C., Hahn, T., Olsson, P., & Norberg, J. (2005). Adaptive governance of social-ecological systems. *Annual Review of Environment and Resources*, *30*(1), 441–473. DOI: 10.1146/annurev.energy.30.050504.144511

Friess, D. A., Krauss, K. W., Horstman, E. M., Balke, T., Bouma, T. J., Galli, D., & Webb, E. L. (2012). Are all intertidal wetlands naturally created equal? Bottlenecks, thresholds and knowledge gaps to mangrove and saltmarsh ecosystems. *Biological Reviews of the Cambridge Philosophical Society*, *87*(2), 346–366. DOI: 10.1111/j.1469-185X.2011.00198.x PMID: 21923637

Friess, D. A., Rogers, K., Lovelock, C. E., Krauss, K. W., Hamilton, S. E., Lee, S. Y., Lucas, R., Primavera, J., Rajkaran, A., & Shi, S. (2019). The state of the world's mangrove forests: Past, present, and future. *Annual Review of Environment and Resources*, *44*(1), 89–115. DOI: 10.1146/annurev-environ-101718-033302

Goldstein, A., Turner, W. R., Spawn, S. A., Anderson-Teixeira, K. J., Cook-Patton, S., Fargione, J., Gibbs, H. K., Griscom, B., Hewson, J. H., Howard, J. F., Ledezma, J. C., Page, S., Koh, L. P., Rockström, J., Sanderman, J., & Hole, D. G. (2020). Protecting irrecoverable carbon in Earth's ecosystems. *Nature Climate Change*, *10*(4), 287–295. DOI: 10.1038/s41558-020-0738-8

Hamilton, S. E., & Casey, D. (2016). Creation of a global database of continuous mangrove forest cover for the 21st century (CGMFC-21). *Global Ecology and Biogeography*, *25*(6), 729–738. DOI: 10.1111/geb.12449

Herr, D., Himes-Cornell, A., & Laffoley, D. (2016). *National Blue Carbon Policy Assessment Framework*. IUCN.

Herr, D., & Landis, E. (2016). *Coastal blue carbon ecosystems: Opportunities for nationally determined contributions. Policy Brief. Gland, Switzerland: IUCN*. The Nature Conservancy.

Iftekhar, M. S., & Islam, M. R. (2004). Managing mangroves in Bangladesh: A strategy analysis. *Journal of Coastal Conservation*, *10*(1), 139. DOI: 10.1652/1400-0350(2004)010[0139:MMIBAS]2.0.CO;2

Islam, M. (2023). Are the Sundarbans mangrove forest management policies adequate to conserve wildlife species? *Biodiversity and Conservation*, *32*(14), 4529–4550. DOI: 10.1007/s10531-023-02708-6

Kusmana, C. (2011). Management of mangrove ecosystem in Indonesia. *Jurnal Pengelolaan Sumberdaya Alam dan Lingkungan (Journal of Natural Resources and Environmental Management)*, *1*(2), 152–157.

Mancheño, A. G., Brander, L., Jafino, B. A., Urrutia, I., & Kazi, S. (2025). *Mangroves for coastal resilience in Bangladesh: Identification of potential locations, assessment of triple dividends, and cost estimation* (Policy Research Working Paper No. 11175). World Bank. worldbank.org

McLeod, K., & Leslie, H. (2009). *Ecosystem-based management for the oceans*. Island Press.

Murray, B. C., Pendleton, L., Jenkins, W. A., & Sifleet, S. (2011). *Green payments for blue carbon: Economic incentives for protecting threatened coastal habitats* (Report No. NI R 11-04). Nicholas Institute for Environmental Policy Solutions, Duke University.

Nainggolan, R. R. E., Suroso, D. S. A., Argo, T. A., & Sagala, S. A. H. (2024). Multilevel governance for climate-compatible development: A case study of mangrove governance in Indonesia. *Journal of Infrastructure. Policy and Development, 8*(6), 3740. DOI: 10.24294/jipd.v8i6.3740

Ostrom, E. (2010). Polycentric systems for coping with collective action and global environmental change. *Global Environmental Change, 20*(4), 550–557. DOI: 10.1016/j.gloenvcha.2010.07.004

Primavera, J. H. (2000). Development and conservation of Philippine mangroves: Institutional issues. *Ecological Economics, 35*(1), 91–106. DOI: 10.1016/S0921-8009(00)00170-1

Purnaweni, H., Hadi, S. P., Soeprobowati, T. R., Roziqin, A., & Rakhma, N. A. (2026). *What are the main concerns for global mangrove management research? Systematic literature review and bibliometric analysis*. Research in Ecology., DOI: 10.30564/re.v8i2.12149

Rahman, H. M. T., Sarker, S. K., Hickey, G. M., Haque, M. M., & Das, N. (2014). Informal institutional responses to government interventions: Lessons from Madhupur National Park, Bangladesh. *Environmental Management, 54*(5), 1175–1189. DOI: 10.1007/s00267-014-0325-8 PMID: 25034753

Ramsar Secretariat. (2018). *Global Wetland Outlook: State of the World's Wetlands and their Services to People*. Ramsar Convention Secretariat.

Roziqin, A. (2018). Environmental policy of mangroves management in Rembang Regency. In *E3S Web of Conferences* (Vol. 31, p. 09002). EDP Sciences., DOI: 10.1051/e3sconf/20183109002

Sarker, S. K., Reeve, R., Thompson, J., Nirmal, P. K., & Matthiopoulos, J. (2016). Are we failing to protect threatened mangroves in the Sundarbans world heritage ecosystem? *Scientific Reports, 6*(1), 21234. Advance online publication. DOI: 10.1038/srep21234 PMID: 26878801

Setyadi, G., Pribadi, R., Wijayanti, D. P., & Sugianto, D. N. (2021). Mangrove diversity and community structure of mimika district, papua, indonesia. *Biodiversitas (Surakarta), 22*(8). Advance online publication. DOI: 10.13057/biodiv/d220857

Spalding, M. D., McIvor, A. L., Tonneijck, F. H., Tol, S., & van Eijk, P. (2014). *Mangroves for coastal defence: Guidelines for coastal managers and policy makers*. Wetlands International and The Nature Conservancy.

Sulaiman, B., Bambang, A., Purnaweni, H., Lutfi, M., & Mohammed, E. (2019). Coastal community perception of mangroves in Suli Subdistrict, Luwu. *Jurnal Pendidikan IPA Indonesia, 8*(4), 561–569. DOI: 10.15294/jpii.v8i4.21396

Suraningsih, K., Purnaweni, H., & Helmi, M. (2021). Geospatial modeling to assess the land suitability of mangrove restoration at Bantul Regency, Yogyakarta, Indonesia. In *E3S Web of Conferences* (Vol. 317, p. 05025). EDP Sciences., DOI: 10.1051/e3sconf/202131705025

Van Lavieren, H., Spalding, M., Alongi, D. M., Kainuma, M., Clüsener-Godt, M., & Adeel, Z. (2012). *Securing the future of mangroves: A policy brief*. UNU-INWEH, UNESCO-MAB, ITTO, ISME, FAO, UNEP-WCMC and TNC. iucn.org

Walters, B. B. (2004). Local management of mangrove forests in the philippines: Successful conservation or efficient resource exploitation? *Human Ecology: an Interdisciplinary Journal, 32*(2), 177–195. DOI: 10.1023/B:HUEC.0000019762.36361.48

KEY TERMS AND DEFINITIONS

Mangrove Governance: The systems, policies, institutions, and practices that regulate the conservation, restoration, and sustainable use of mangrove ecosystems across global, national, and local scales.

Resilient Governance Systems: Adaptive, learning-based institutional frameworks designed to respond flexibly to ecological, social, and political uncertainties, ensuring long-term sustainability.

Monitoring and Evaluation (M&E): Frameworks for tracking ecological health, policy implementation, and socio-economic outcomes of mangrove management, often combining field surveys, remote sensing, and participatory approaches.

Participatory Mapping: A governance tool that engages local communities in documenting resource use, threats, and conservation priorities through participatory GIS, enhancing legitimacy and accountability.

Remote Sensing: The use of satellite imagery, drones, and GIS technologies to monitor mangrove distribution, biomass, and health at large scales.

Ecosystem Service Valuation: The process of quantifying the economic, ecological, and social benefits of mangroves, including carbon sequestration, coastal protection, fisheries support, and cultural values.

Payment for Ecosystem Services (PES): Financial mechanisms that compensate communities for conserving ecosystems, such as carbon credit projects like *Mikoko Pamoja* in Kenya.

Institutional Fragmentation: The division of governance responsibilities across multiple ministries or agencies, leading to overlapping mandates, duplication of efforts, and weak accountability.

Equity and Inclusion: Ensuring marginalized groups—including women, Indigenous peoples, and small-scale fishers—are represented in decision-making processes, thereby enhancing legitimacy and fairness in governance.

Political Economy of Mangrove Management: The power relations and economic interests that shape governance outcomes, often resulting in elite capture of benefits and community-level ecological and social costs.

Transboundary Cooperation: Collaborative governance across national borders for shared ecosystems, such as the Sundarbans between Bangladesh and India, requiring harmonized management strategies.

Community-Based Governance Models: Local stewardship initiatives where communities manage mangroves under formal agreements or through participatory conservation projects, often linked to innovative financing mechanisms.

Chapter 14
Overcoming the Gap Between Global Commitments and Local Reality:
Mangrove Governance and Legal Fragmentation

Muhammad Ansy Alghasi
http://orcid.org/0009-0006-4435-825X
Lensa Garuda Nusantara, Indonesia

Muhammad Ansy Althafzufar
https://orcid.org/0009-0006-3572-2398
Universitas Dr. Soetomo Surabaya, Indonesia

Raja Iqbal Islamy
Perhimpunan Advokat Indonesia Gresik dan Yayasan Lembaga Bantuan Hukum Kabupaten Gresik, Indonesia

ABSTRACT

This study analyzes the disconnect between Indonesia's climate pledges, like the 2022 G20 mangrove rehabilitation goal, and its fragmented legal framework. Field research in Mengare, Gresik, reveals that legislative overlap creates a legal identity crisis. Conflicting regulations under forestry and coastal management laws fuel jurisdictional disputes between the KLHK and KKP, effectively paralyzing enforcement. Furthermore, a performative compliance trap persists; CSR programs often prioritize symbolic planting over ecological survival, resulting in negligible success

DOI: 10.4018/979-8-3373-9320-9.ch014

rates. Discrepancies in government reporting further highlight a system detached from reality. While Indonesia adopts global climate norms, it fails to integrate them into enforceable domestic policy. We recommend legislative reconstruction via lex specialis and centralized authority to align diplomacy with genuine conservation.

INTRODUCTION

The adoption of the Paris Agreement in 2015 represented a pivotal moment in how the global community views the link between ecosystems and climate responsibilities. For the first time, signatory nations were broadly expected to include ecosystem-based carbon sequestration in their nationally determined contributions (NDCs). This move transitioned conservation from a secondary concern to a primary element of climate diplomacy. Mangroves became a central focus of this transition. Current estimates suggest that global mangrove ecosystems hold approximately 6.4 billion tonnes of carbon. Furthermore, the emissions produced when these forests are cleared or degraded are generally comparable to those from inland tropical deforestation. Consequently, mangrove governance carries a legal and political significance that reaches far beyond coastal ecology (Bertram et al., 2021). Follow-up multilateral agreements have only strengthened this trend. Both the 2022 Kunming-Montreal Global Biodiversity Framework and the UNFCCC Glasgow Climate Pact reaffirmed that coastal blue carbon ecosystems are vital for keeping the 1.5°C warming target attainable. These developments have placed coastal developing nations, which manage most of the world's remaining mangroves, under consistent pressure to turn international promises into meaningful domestic results.

With approximately 3.4 million hectares of mangrove coverage, representing more than 23 percent of the world's remaining mangrove forests, Indonesia is the most significant mangrove nation on Earth (Humas Kementerian Kehutanan Republik Indonesia, 2025). The immense scale of Indonesia's mangrove holdings creates substantial international legal expectations. Indonesia's mangroves alone are estimated to store about 3 billion tons of blue carbon. This figure ranks the country among the top three global repositories alongside Australia and Cuba, making the preservation of these forests a matter of global atmospheric importance rather than just a local environmental issue (Bertram et al., 2021).

These ecological facts have shaped Indonesia's role within the international climate regime in specific and binding ways. Under its Paris Agreement obligations, Indonesia has committed to reducing national greenhouse gas emissions by 29 percent through its own initiatives and by up to 41 percent with international assistance. The forestry and land-use sector, which officially includes mangroves, is expected to handle roughly 60 percent of this reduction target. This makes the effectiveness of mangrove governance a primary factor in whether Indonesia achieves its internationally stated goals (IESR, 2024). To signal its commitment to these duties, Indonesia launched a major mangrove rehabilitation program aimed at 600,000 hectares. This pledge was publicly confirmed before world leaders at the 2022 G20 Bali Summit. Presidential Regulation No. 82 of 2020 provided the institutional framework for this goal by creating the National Mangrove Rehabilitation and Management Agency as the lead organization. Together, these steps presented Indonesia as a proactive leader in climate action, willing to support its diplomatic words with regulatory structures and clear targets. This chapter examines whether that framework has actually been capable of meeting these challenges.

Most research on mangrove rehabilitation in Indonesia measures success in ecological terms, such as seedling survival or biomass recovery, rather than examining the social and political sides of restoration (Sitthi et al., 2025). Governance scholarship, by contrast, tends to stop at national-level institutional design and legal fragmentation. It rarely traces what happens once that fragmentation reaches local implementation: whether it gets reproduced, contested, or renegotiated on the ground (Tamanaha, 2021). A recent bibliometric review confirms this gap, finding that the social, political, and institutional dimensions of mangrove management remain understudied (Purnaweni et al., 2026). This chapter treats governance fragmentation itself as the central object of inquiry rather than background context. Using Koh's (1997) transnational legal process framework, it identifies the specific stage at which Indonesia's internalization of its international mangrove commitments breaks down, showing not just whether a compliance gap exists, but where and how it forms.

This chapter found that the disconnect between Indonesia's international commitments and its domestic legal reality is a structural problem created by the very design of Indonesian environmental law. At the heart of this issue is a type of regulatory fragmentation that has made consistent governance of mangrove ecosystems very difficult to achieve. This challenge stems from a fundamental legal ambiguity where mangroves are managed under two conflicting legal systems. According to Law No. 41 of 1999 on Forestry, mangroves are classified as part of the national forest estate and fall under the authority of the Ministry of Environment and Forestry (KLHK). However, under Law No. 27 of 2007 in conjunction with Law No. 1 of 2014 on the Management of Coastal Areas and Small Islands, these same ecosystems are defined as part of the coastal zone and are regulated by the Ministry of Maritime Affairs and Fisheries (KKP).

The results of this overlap are clear in the data. Only about 31 percent of Indonesia's mangroves are currently in good condition, while degradation continues at a rate of approximately 26,100 hectares per year (Walhi Jawa Timur, 2023). The national rehabilitation program managed only 33,000 hectares in its first year. This accounts for just 5.5 percent of the four-year target and suggests the program must speed up by five times its current rate to avoid failing on an international commitment (Embassy of Denmark to Indonesia, 2022). Fragmentation is also evident in the unreliability of national environmental data. The KLHK, the KKP, and the Geospatial Information Agency frequently publish mangrove coverage statistics that differ by hundreds of thousands of hectares. These discrepancies hurt Indonesia's credibility in multilateral carbon markets, which depend on standardized and verifiable monitoring. Using the framework of transnational legal process proposed by Koh (1997) Indonesia demonstrates a failure to internalize norms at every stage: from including international goals in policy discussions to embedding them in regulatory agencies and finally turning them into enforceable laws with real penalties.

This lack of structural coherence has been made worse by recent legislative changes. The passage of Law No. 11 of 2020 on Job Creation introduced rules that allow for more industrial and geothermal use of aquatic and coastal areas, which puts mangrove habitats at risk (Sidik et al., 2023). The end result is a legal environment where specific regulations work against each other. There is no single law that functions as a lex specialis for mangroves. Consequently, ecosystems that are claimed by several different legal regimes often end up being protected by none of them.

To understand how this regulatory fragmentation functions, this study focuses on Gresik Regency in East Java as its primary research site. Gresik is a location where the tensions mentioned earlier are concentrated and easily observed. Located on the northern coast of Java, Gresik has historically maintained about 5,828 hectares of mangroves across six coastal sub-districts, as noted in the Regency's Regional Spatial Plan (RTRW) 2010–2030 (Pemerintah Kabupaten Gresik, 2010). The economic value of these ecosystems is quite high. A study in Banyuurip village estimated the total economic value of its mangroves at Rp 16 billion. This figure highlights how much local coastal livelihoods depend on ecological services that the current legal framework is struggling to protect (Yunitasari et al., 2020).

The degradation patterns along the Gresik coast serve as a physical record of governance failure. In Manyar Subdistrict, roughly 238 hectares of mangroves have been converted into industrial sites and aquaculture ponds. In Ujungpangkah, another 84 hectares have been lost to illegal logging (Prasetyo et al., 2017). These numbers, however, only tell part of the story. Field research in the Mengare area of Bungah Subdistricts provides a human perspective that data alone cannot capture. Local residents described corporate mangrove planting, which is required by law, as a symbolic gesture. These efforts are often done to show compliance on paper without the necessary care to ensure the seedlings survive. One community leader in Mengare was very direct: "The Gresik government keeps records of how many hundreds of thousands of seedlings were planted, but the actual survival rate is zero percent." The Chairman of the local fishermen's association shared a similar view, noting that nearly 70 percent of the mangroves in Tanjung Widoro have disappeared due to coastal erosion. He pointed out that this erosion continues because the state has not built wave-breaking infrastructure, which is a necessary step for mangroves to grow, yet no current law clearly requires the government to provide it.

Perhaps the most revealing insight from the Mengare fieldwork is not the degradation itself, but the institutional reaction to it. Local participants described a jurisdictional battle between the regency and provincial governments over who gets to take credit for mangrove planting in coastal areas. When a company or NGO plants in the intertidal zone, the province claims the achievement. When planting happens on land, the regency takes the credit. This dispute over reporting numbers reflects the core problem of Indonesia's mangrove governance. Multiple authorities claim jurisdiction over the same areas, which spreads accountability so thin that no single institution is truly responsible for what happens to the forests.

METHOD

1. Type of Research

This study runs on two legs, normative legal analysis and empirical fieldwork, because neither one alone tells the full story, treating legal research as a structured process for finding and solving legal problems through library sources plus observation and interviews on the ground (Hamzani et al., 2024). One leg goes through Indonesia's mangrove governance laws sector by sector to flag where they overlap, contradict each other, or leave responsibilities undefined, then traces how that mess ends up weakening the country's international environmental commitments; the other asks a different question altogether, namely what happens once those same rules run into the capacity, incentives, and daily realities of an actual coastal community, which is the jump from law in books to law in action that no amount of reading statutes in a library will ever get you (Banakar, 2019). Holding both pieces together takes a theory made for exactly this problem, legal regimes that are each legitimate on their own yet end up working against one another once they overlap, and that is what normative legal pluralism supplies (Tamanaha, 2021), which is why the findings from Bungah Subdistrict, Gresik carry the weight they do.

2. Research Approach

This study relies on three approaches, statutory, conceptual, and comparative, each pulling a different thread out of the same problem. The statutory approach reads Indonesia's mangrove governance mess as, at root, a conflict-of-laws problem: it checks the Forestry Law (41/1999), the Coastal Areas laws (27/2007 and 1/2014), the Environmental Protection Law (32/2009), the Job Creation Law (11/2020), Presidential Regulation 82/2020, and ratified treaties like the Paris Agreement and the CBD against each other to find the antinomies, the places where two provisions hand out conflicting authority over the same thing (Noor, 2023). The conceptual approach leans on doctrinal scholarship to sharpen the ideas doing the analytical work here: Raustiala and Victor's (2004) idea of regime complexity, Koh's (1997) stages of how international norms get internalized domestically, and Lemos and Agrawal's (2006) point that regulatory outcomes depend on the surrounding political and institutional setting, not just the rules on paper (Hamzani et al., 2024). The comparative piece looks sideways at Thailand and Australia, both of which faced a similar coordination problem but solved it differently. Thailand centralized coastal authority under one agency, the Department of Marine and Coastal Resources, while Australia built a constitutionally grounded federal system that lets multiple levels of government coordinate without flattening them into one. The comparison follows Reimann and Zimmermann's (2019) functional method, which just asks how different legal systems end up solving the same problem (Al Abiad & Masadeh, 2024).

3. Research Location

This study looks at how regulatory fragmentation actually plays out on the ground, so the fieldwork site had to be somewhere that shows the fragmentation in practice, not just on paper, and all data collection happened in Mengare, a coastal village in Bungah Subdistrict, Gresik Regency, East Java, where none of the three government bodies involved, the Ministry of Environment and Forestry (KLHK), the Ministry of Maritime Affairs and Fisheries (KKP), and the Gresik Regional Government, were interviewed directly, so what they are said to do comes from regulatory documents rather than from anything collected at their offices. Mengare qualifies as a critical case because the jurisdictional overlap written into the regulations shows up locally as unclear permits and no one willing to take responsibility when things go wrong, because the mangroves are visibly eroding with no state-built wave breaker anywhere in sight, and because in the absence of any real state presence, local leaders and community groups have simply built their own arrangements to manage the coast (Creswell & Poth, 2018). Interviews also turned up a dispute between the regency and the provincial government over who gets credit for planting the mangroves, something residents reported secondhand rather than anything the researcher saw firsthand, and it lines up with the same fragmentation pattern documented at the national level (Irsadi et al., 2025).

4. Sources

This study draws on two kinds of material: legal sources ranked by normative authority, and empirical data gathered directly from the people and institutions involved in mangrove management in Mengare. The legal sources follow Noor's (2023) three-tier approach to Indonesian legal research. At the top sit binding primary materials: the 1945 Constitution, national mangrove legislation, government regulations, presidential decrees, ratified treaties like the Paris Agreement, and relevant court decisions. Below that are secondary materials, not binding but still carrying real weight, including academic writing, peer-reviewed research, UNFCCC and UNEP reports, and policy documents from KLHK and BRGM. Tertiary materials, mostly legal dictionaries and encyclopedias, only get pulled in when a term is left ambiguous elsewhere. The empirical data comes from three people in Mengare. Zaini, a community leader, has watched local mangrove management play out over the years and can say exactly where state promises never materialized. Atmudjin chairs the Mengare Fishermen's Association and represents those whose livelihoods take the direct hit from environmental degradation and missing state infrastructure. Lukman works with a local NGO, and his visibility across different organizations lets him speak to how both government agencies and companies actually operate on the ground. Read side by side, these three views expose how regulatory fragmentation plays out for Mengare's residents, though all three sit on the community or civil society side, with no one from a government regulator or company in the area, a gap the conclusion takes up (Yin, 2017).

5. Data Collection

This study combined three methods, library research, field observation, and interviews, working through them in sequence at first and then circling back once the findings raised new questions. Library research came first: content analysis of the relevant legal texts showed where the rules conflict and how international commitments had made their way into domestic law, giving us a baseline for what the law is supposed to require. Observation came next, recording mangrove governance as it stands on the ground, the ecological condition of the mangroves, whether wave breaking structures were in place, and what corporate CSR planting programs had produced (Clark et al., 2021), and this is where a persistent gap between the regulations and what actually gets delivered showed up. Observation can tell you a gap exists, but not why, or how people living with it manage, so we turned to interviews, selecting informants through purposive sampling (Ahmad & Wilkins, 2025), an approach suited to topics that can only be understood through people with direct, specialized experience of them (Pahwa et al., 2023). We settled on three informants because each holds a distinct, complementary position in the local governance system (Yin, 2017).

6. Data Analysis

This study leans on three methods at once, doctrinal analysis, thematic analysis, and triangulation, and uses each to check the others. Doctrinal analysis asks whether Indonesia's mangrove regulations actually hang together as a coherent legal hierarchy. Using Koh's (1997) transnational legal process framework, which follows a norm from public debate into institutional practice and eventually into a binding obligation, it doesn't just confirm that implementation failed; it points to the specific stage where things stall. Thematic analysis follows the standard path, familiarization, coding, building themes, interpretation (Naeem et al., 2023), with coding guided by Howlett, Ramesh, and Perl's (2020) implementation theory so the three informants' accounts connect back to the structural problems the doctrinal reading already turned up. Triangulation asks a plainer question: does a conclusion that looks solid from one angle survive the other two? Statutes alone can be dismissed as theoretical, interviews alone as anecdotal, but law on paper, actual agency practice, and what communities live through rarely line up by accident, and when they do, the claim that fragmented regulation produces real, felt implementation failures gets a lot harder to wave away (Denzin, 2009). All three together are meant to land on recommendations for regulatory harmonization that fit the actual institutional and ecological conditions of Indonesia's coastal governance, not an idealized version of it.

FINDINGS AND DISCUSSIONS

1. The Legal Identity Problem

The fundamental challenge underlying the governance failures in Indonesian mangrove management is an overabundance of conflicting regulations that govern the same environmental space. As a coastal ecosystem, mangroves face what can be described in juridical terms as a crisis of legal identity. They are simultaneously claimed by two separate legislative regimes, each possessing its own authority, distinct enforcement mechanisms, and incompatible management philosophies.

According to Law No. 41 of 1999 on Forestry, mangroves are classified as part of the national forest estate and therefore fall under the jurisdiction of the Ministry of Environment and Forestry (KLHK). At the same time, Law No. 27 of 2007 in conjunction with Law No. 1 of 2014 on the Management of Coastal Areas and Small Islands places the physically identical ecosystem within the coastal zone, subjecting it to the authority of the Ministry of Maritime Affairs and Fisheries (KKP). This dual arrangement has created significant confusion among members of the public and businesses seeking to utilize mangrove land. Furthermore, it has generated a persistent potential for conflict between government departments (Ervita & Makhas, 2021).

This dualism constitutes a juridical antinomy in the most substantive sense: two normatively valid legal provisions establish obligations and authorities that cannot be simultaneously fulfilled without violating one of them. Koh (1997) and Nurhati and Murdiyarso (2022) argue that an ecosystem cannot function as an effective legal object unless three conditions are in place. Its spatial boundaries need to be clearly defined, authority over it needs to be clearly assigned, and there needs to be a mechanism to hold someone accountable. The simultaneous absence of all three preconditions in the Indonesian mangrove legal framework is a direct consequence of this non-unified juridical construction. This dualism is further complicated by the enactment of Law No. 11 of 2020 on Job Creation, which added yet another layer of regulatory conflict. The omnibus law, particularly in its provisions regarding geothermal resources, legalizes geothermal mining in waterborne areas. This is a deregulatory step with the potential to devastate mangrove forests (Sidik et al., 2023). Rather than consolidating the fragmented legal regimes, this legislation deepens normative uncertainty by introducing area utilization imperatives that are incompatible with existing ecosystem protection obligations.

Field research in Mengare, Bungah Subdistricts, Gresik, confirms that this dualism operates as a lived institutional reality. Within the mangrove ecosystem of that area, the jurisdictional claims of KLHK under the forestry regime, the claims of KKP under the coastal management regime, and the authority of the Gresik Regional Government under the regional autonomy framework converge simultaneously without an adequate resolution mechanism. Consequently, no single authority possesses a mandate sufficiently clear and binding to take decisive enforcement action when violations occur. The synchronization between non-governmental organizations, corporations, and regional government that community leader Zaini of Mengare identifies as an urgent necessity reflects the absence of a coordination mechanism that is legally mandated and enforceable by an authority holding jurisdictional supremacy.

2. Regulatory Fragmentation in Practice

If dual legal classification represents the structural root of the problem, then regulatory fragmentation is the specific mechanism through which that root produces visible failure on the ground. Raustiala and Victor (2004) define fragmentation as a condition where a policy domain is governed by multiple distinct legal instruments, issued by various authorities, that are not integrated into a single, coherent normative system. What we see in Gresik, and specifically in the Mengare area, is a precise empirical manifestation of that definition.

The overlapping authority of KLHK, KKP, Perhutani, and the Gresik regional government is a real, active problem, not just a theoretical one, and it directly decides who protects, rehabilitates, and enforces the law in the area's mangrove zones. Zaini, a community leader in Mengare, named KLHK and Perhutani as the main managers, yet Gresik's Fisheries Office and Environmental Agency claim the same territory. Lukman, from a local NGO, said even basic land ownership gets disputed: the Environmental Agency claimed the land around the Mangrove Restoration and Learning Center even though PT Petrokimia Gresik built and funded it. A similar standoff played out in Karangsong, West Java, where the regional government pushed aside a local community group's mangrove management rights (Feti et al., 2020).

Figure 1. Mangrove nursery sign ("Pembibitan Mangrove") at the Mengare planting site

Figure 2. Bamboo and mangrove seedlings growing near a field marker in Mengare

Figure 3. Identification tag for Lumnitzera racemosa, locally known as "Kedukduk," at the Mengare research site

This fragmentation does more than cause administrative confusion; it produces measurable and documented enforcement paralysis. Between 2009 and 2019, Indonesia lost 182,091 hectares of mangroves to land conversion for aquaculture ponds, agriculture, oil palm plantations, and infrastructure projects, which indicates that supervision and law enforcement must be significantly optimized (Arifanti et al., 2021). At the local level in Gresik Regency, at least 238 hectares of mangroves in the Manyar sub-district have been converted into industrial estates and fishponds, while in Ujungpangkah, a minimum of 84 hectares have been lost to illegal logging (Prasetyo et al., 2017).

Atmudjin, the Chairman of the Mengare Fishermen's Association, provided perhaps the most incisive confirmation of this breakdown. He stated that the Fisheries Office had never actually implemented a mangrove program, noting that even the Community Supervisory Group established by that office was non-functional. This institutional failure signals the absence of a single entity legally designated as the sole accountable authority with the necessary capacity, budget, and enforcement power. As Sidik et al. (2023) confirm, problems related to blue carbon, including mangrove conversion, are frequently driven by incoherent policies, weak law enforcement, and the constant pressure to stimulate economic growth at the expense of ecology.

The situation in Gresik fails nearly every test of sustainable resource management. Ostrom (1990) famously established that the effective management of common-pool resources requires clearly defined boundaries, agreed-upon rules of use, robust monitoring frameworks, and tiered sanctions. In the mangrove governance of Gresik Regency, not one of these requirements is adequately met. Jurisdictional boundaries are blurry and imprecise, rules of use are subject to conflicting interpretations under different legal regimes, monitoring is almost entirely absent, and sanctions cannot be imposed consistently because no single institution holds a clear, undisputed mandate for enforcement. In short, when everyone is in charge, no one is actually accountable.

Fragmented authority does not, by itself, doom environmental outcomes. Ostrom's later work on polycentric systems makes that case directly: multiple, overlapping centers of authority can improve environmental governance, but only if coordination mechanisms and information-sharing arrangements tie them together (Ostrom, 2010). Seen this way, the Gresik case looks like fragmentation without coordination, not evidence that multiple authorities are inherently incompatible with effective mangrove governance. That distinction matters for the reform options discussed later in this chapter.

This chapter treats legal and jurisdictional fragmentation as the primary explanation for Mengare's governance failures, though other factors often cited in the literature, such as local politics, administrative capacity, budget constraints, and enforcement incentives, were not tested here and may well matter too. Since the fieldwork cannot separate their effects from fragmentation's, that remains an open question for later work. Worth noting: comparable research on household hazardous waste in Indonesia found public knowledge, not administrative or economic factors, was the strongest predictor of proper environmental practice (Nasarani et al., 2025).

Figure 4. Actual condition of degraded mangrove ecosystem in Gresik Regency

3. The Performative Compliance Trap

Among all the field findings of this study, none more profoundly reveals the depth of mangrove governance failure in Mengare than the phenomenon that may be described as the performative compliance trap. This phenomenon refers to a condition in which formal legal obligations are fulfilled in documentary terms yet produce none of the ecological outcomes that those obligations are designed to secure.

Zaini summarized this condition with a directness that no regulatory audit has managed to capture: "The Gresik government keeps records of how many hundreds of thousands of seedlings were planted, but the actual survival rate is zero percent." This statement refers to the reality that pressure from the regional government on corporations to plant mangroves through corporate social responsibility (CSR) mechanisms has produced documented records of hundreds of thousands of seedlings, but the survival rate of those seedlings is zero. Lukman confirmed that the companies concerned fulfilled their formal obligations by erecting labeled plaques showing each company's planting quota, but there was no follow-through in the form of post-planting maintenance.

Figure 5. Physical evidence of mangrove conversion

This phenomenon can be analyzed through the framework of decoupling between regulatory compliance and ecological performance. This is a familiar pattern in sociology: organizations often decouple their formal policies from what actually happens in practice (Meyer & Rowan, 1977). Although regulations mandate CSR, many companies treat it as a discretionary social activity aimed primarily at improving corporate image, which indicates weak internalization of legal norms into corporate governance structures (Fitrianggraeni et al., 2023). This condition is exacerbated by the absence of a standardized audit mechanism for CSR activities, which further limits the ability of regulators and stakeholders to assess corporate performance in fulfilling social obligations.

Although Government Regulation No. 47 of 2012 requires all Indonesian companies to implement CSR, this regulation is silent on limitations or restrictions regarding how companies may implement their CSR programs. In practice, companies may allocate their funds or resources to various social and environmental development programs at their own discretion. In the context of mangrove planting, the absence of minimum technical standards and measurable post-planting monitoring requirements means that a single day of planting is followed by abandonment, without any legal obligation burdening the company with responsibility for seedling survival.

Zaini identified the logic of the problem precisely: "The company's main concern is planting, not maintaining." This is an accurate description of the incentive structure created by the existing regulatory framework. When legal obligations end at the moment of planting and no binding survival verification mechanism exists, corporate economic rationality naturally gravitates towards minimum-cost compliance, which in this context means planting without maintenance.

This problem is also closely related to the ecologically inappropriate selection of seedlings. Zaini and Atmudjin consistently emphasized that the seedlings used in CSR planting programs are often brought in from outside the area, such as from Tuban, Pasuruan, or Madura, even though locally sourced seedlings are far better suited to the soil structure of the area and have already adapted to the environmental conditions of Mengare. This practice is unregulated in the existing CSR rules, so no legal mechanism mandates the use of local seedlings as an enforceable best practice.

Figure 6. Natural regeneration of mangrove seedlings (propagules) from parent trees in the intertidal zone

4. The Infrastructure Gap as a Legal Failure

One of the most legally significant findings from the field research in Mengare is that the failure of mangrove planting is a consequence of the state's failure to fulfill the obligations that constitute the ecological preconditions for all restoration efforts. Those obligations concern the construction of wave-breaking infrastructure.

Atmudjin stated with certainty that the primary solution is wave-breakers, noting that no planting activity can succeed before wave-breakers are in place. Zaini reinforced this with a more detailed analysis, explaining that planting a million trees without wave-breakers is equally futile. These statements are consistently supported by mangrove ecology science, which establishes that the successful transplantation of mangroves in intertidal zones exposed to high wave energy depends heavily on the availability of physical protective structures that reduce wave energy before propagules or seedlings can take root (Irsadi et al., 2025).

The absence of wave-breaking infrastructure is not a neutral technical gap. It is a failure by the state to meet its own environmental protection obligations. Law No. 32 of 2009 requires the state to prevent ecosystem damage and restore what has already been degraded, yet the government keeps pushing corporations to plant mangroves through CSR while ignoring the infrastructure those mangroves actually need to survive. You cannot demand a result from the private sector while withholding the conditions that make it possible. The Semarang-Demak coast tells the same story, where sea dikes and toll roads have cut through mangrove habitat and made local environmental problems worse (Widjonarko et al., 2025).

The destruction of nearly 70 percent of the mangrove area in Desa Tanjung Widoro as reported by Atmudjin constitutes the principal empirical evidence of this failure. Multi-level governance challenges can impede the realization of coastal ecosystem restoration agendas, with empirical evidence demonstrating how international inculcation and agenda-setting frequently cannot be implemented administratively or in practice at national and sub-national governance levels (Dong et al., 2025). In Mengare, this condition manifests in a very concrete form: despite the declaration of a national mangrove rehabilitation target at international forums, not a single regulatory instrument designates wave-breaking structure construction as a measurable and enforceable government obligation within restoration programs.

Figure 7. Bamboo wave-breaking infrastructure as a physical precondition for mangrove restoration

Zaini also introduced the dimension of land ownership conflict that further complicates this problem. Under regulations dating to 1912, fishponds lost to coastal erosion legally revert to state ownership. In practice, however, this ownership transfer mechanism is not operationalized in a manner that would allow the state to subsequently plant and manage those areas. As a result, there are expanses of land that are legally available for state planting programs, yet no institution operationally assumes responsibility for doing so.

5. The Credit-Claiming Dispute

One of the most analytically revealing field findings concerning the systemic nature of mangrove governance failure is the dispute over planting credit between the regency and provincial governments. Atmudjin reported an unwritten understanding among officials: mangrove planting conducted in the coastal intertidal area is claimed by the provincial government, while planting on land remains the regency claims. Zaini added with a tone reflecting deep frustration that, meanwhile, the province never actually planted anything.

This dispute carries analytical significance that far exceeds its scale as a local anecdote. It reveals a condition in which the accountability system has become entirely disconnected from its ecological purpose. When governing authorities are more preoccupied with contesting the right to report planting figures than with ensuring the survival of the reported plants, the logic driving government action is the logic of institutional performance reporting that is severed from ecological reality.

This phenomenon constitutes a precise micro-level manifestation of the failure of the Monitoring, Reporting, and Verification (MRV) system that has been identified as a systemic national problem. Indonesia currently lacks fully accredited MRV verifiers and mechanisms aligned with Article 6.4 of the Paris Agreement, which impedes transparency and weakens potential participation in international carbon markets (Jonathan, 2025). When planting claims between the regency and province are not based on an independent and standardized verification system, the figures entering the national reporting system will reflect institutional claim interests rather than the actual condition of the ecosystem.

Nurhati and Murdiyarso (2022) argue that MRV works only when three preconditions are met:: harmonized methodological standards, adequate institutional capacity, and independent verification. Not one of these three preconditions can be fulfilled in a condition where authorities that should be working in synergy over the management of the same ecosystem are instead competing to claim credit for the same achievement.

The causal link between this credit-claiming dispute and actual conservation failure can be traced directly. When Zaini mentioned that 15,000 mangrove seedlings previously planted in the Mengare area are now entirely gone due to coastal erosion, no authority, whether at the regency or provincial level, acknowledged that failure as its own responsibility. Planting figures enter reporting documents, but seedling mortality rates do not make it. This is precisely what is meant by an accountability system that has become fully disconnected from its ecological purpose.

6. The Implementation Gap of International Commitments

It is necessary to draw an analytical line connecting the condition of the Tanjung Widoro mangrove area, which has suffered nearly 70 percent degradation, directly to Indonesia's international commitments under the Paris Agreement. Using the transnational legal process framework developed by Koh (1997), the process of domesticating international norms into binding national law must pass through three distinct stages. Failure at any of these stages produces what is commonly termed an implementation gap.

As of 2023, 97 countries had included coastal and marine ecosystems, including mangroves, in their Nationally Determined Contributions (NDCs) under the Paris Agreement, and 61 countries had included the conservation or restoration of blue carbon ecosystems as mitigation or adaptation measures (Sun et al., 2024). Indonesia is among those countries and bears a responsibility considerably greater than most, given its stewardship of 23 percent of the world's mangroves.

At the first stage, discursive internalization, Indonesia demonstrates relatively strong performance. The 600,000-hectare rehabilitation commitment was publicly declared at the G20 Bali Summit in 2022, Presidential Regulation No. 82 of 2020 established the Peat and Mangrove Restoration Agency (BRGM) as the primary delivery mechanism, and mangroves are formally embedded in Indonesia's NDC calculations.

At the second stage, institutionalization, fragmentation becomes clearly visible. The BRGM, the Ministry of Environment and Forestry (KLHK), and the Ministry of Maritime Affairs and Fisheries (KKP) operate under mandates that are not integrated, and the jurisdictional conflict between the regency and provincial governments in Mengare reproduces the same tensions at the sub-national level. First-year achievement of the national mangrove rehabilitation program in 2021 was recorded at only 29,500 hectares across nine priority provinces, which is far below the pace required to meet the target (Walhi Jawa Timur, 2023).

At the third stage, legal imperatives, the most complete failure occurs. No legal instrument establishes enforceable post-planting maintenance obligations, no meaningful sanction exists for corporations that plant without ensuring seedling survival, and no accountability mechanism links reporting figures to the actual condition of the ecosystem. The challenges that obstruct the development of mangrove-based carbon projects include methodology standardization and accreditation, Monitoring, Reporting, and Verification (MRV) processes, transparency, stakeholder participation, and overall governance (Matta et al., 2025). In Mengare, all of these challenges are present simultaneously, making the area a sandbox for mangrove governance problems in their most comprehensive dimensions.

7. Toward Regulatory Reconstruction

A deep understanding of why Indonesian mangrove governance fails structurally requires a comparison with the experiences of other jurisdictions that have confronted different versions of the same fundamental institutional problem. Thailand and Australia are selected as comparators because each has addressed, through different institutional means, a version of the same core governance challenge, which is how to integrate fragmented authority in the management of coastal ecosystems.

Thailand has consolidated coastal and mangrove management authority under a single institution, namely the Department of Marine and Coastal Resources (DMCR) within the Ministry of Natural Resources and Environment. The Thailand National Coordinating Body, approved by the Permanent Secretary of the Ministry of Natural Resources and Environment, is hosted by the DMCR and functions as a cross-stakeholder coordination center comprising six government members, two civil society members, two academics, three international organizations, and five independent experts (Mangroves for the Future: Investing in Coastal Ecosystems, n.d.). This consolidation of authority has enabled Thailand to resolve the jurisdictional ambiguity that has encumbered Indonesia.

The results are empirically measurable. The government's decision to halt the allocation of coastal concessions in the late 1990s, combined with various tree-planting efforts and community-led conservation initiatives, effectively reversed Thailand's mangrove decline. Thailand now has 248,400 hectares of mangrove cover, representing approximately a 50 percent increase since 1996 (Cowan, 2024). The Thai experience also furnishes a relevant warning for Indonesia: a state-led, corporately funded mass-planting model without multi-year budgeting and monitoring produces results that stand in sharp contrast to community-based conservation efforts.

Thailand has also issued regulations on carbon credit sharing derived from community-managed mangrove planting and maintenance, with the aim of encouraging community participation in the conservation and restoration of marine and coastal resources while creating national wealth and ecosystem balance (Thailand, 2022). This constitutes a highly relevant legal precedent for Indonesia, which is the formalization of a mangrove carbon credit-sharing mechanism that positions local communities as active stakeholders and beneficiaries.

Australia offers a different but equally instructive model, namely federally coordinated environmental governance under the Environment Protection and Biodiversity Conservation Act. Following the Samuel Review, the Australian government responded with the Nature Positive Plan in December 2022, identifying priorities to guide its environmental reform agenda, including strengthened coordination between the federal and state governments in seven priority regions (McCredie et al., 2024). The Australian model demonstrates that in a multi-level governance system, regulatory fragmentation is resolved by establishing nationally binding standards across all levels of government, clear bilateral accreditation mechanisms, and an independent body overseeing compliance.

Australia has also developed its first national ocean ecosystem account focused on blue carbon ecosystems and their climate mitigation and resilience benefits, encompassing information on mangroves, seagrass, and tidal saltmarsh, as well as the carbon stocks they sequester and store (Department of Climate Change, Energy, the Environment and Water, 2024). This represents the operational implementation of an integrated, independent, and internationally reportable Monitoring, Reporting, and Verification (MRV) system, which directly addresses Indonesia's weakness in the fragmentation of data across the Ministry of Environment and Forestry (KLHK), the Ministry of Maritime Affairs and Fisheries (KKP), and the Geospatial Information Agency.

To sharpen the comparative analysis above, the following table presents a systematic mapping across six governance dimensions that are determinative, which are legal structure, institutional arrangements, rehabilitation obligations, MRV systems, integration of international commitments, and lessons available for adoption by Indonesia.

Table 1. Comparative analysis of mangrove governance: Indonesia, Thailand, and Australia

Dimension of Comparison	Indonesia	Thailand	Australia
I. Legal Structure and Ecosystem Classification			
Legal identity of mangroves	Dual classification: forest estate under Law No. 41/1999 and coastal component under Law No. 27/2007 in conjunction with Law No. 1/2014, applied simultaneously to the same physical ecosystem. This produces an unresolved juridical antinomy.	Mangroves are classified as marine and coastal resources under the Marine and Coastal Resources Management Act B.E. 2565. A single legal regime eliminates classificatory ambiguity.	Protected as a matter of national environmental significance under the Environment Protection and Biodiversity Conservation Act 1999 (1999), with nationally binding standards applicable across all levels of government.
Existence of lex specialis	None. No dedicated mangrove legislation exists that carries supremacy over either the forestry or the coastal management regime. This absence is the primary source of jurisdictional conflict.	The Marine and Coastal Resources Management Act 2015 functions de facto as a lex specialis for coastal ecosystems including mangroves.	The EPBC Act, the Nature Positive Plan (2022), and the Nature Repair Act 2023 together constitute a hierarchical and integrated legislative framework.
II. Institutional Structure and Authority			
Model of governing authority	Fragmented: KLHK, KKP, BRGM, and regional governments operate in parallel without a clear hierarchy of authority. Field research in Mengare documents enforcement paralysis as a direct consequence.	Consolidated under a single institution: the Department of Marine and Coastal Resources (DMCR) within the Ministry of Natural Resources and Environment. This model eliminates inter-ministerial conflict.	Federal-state coordination: the Commonwealth (EPBC Act) sets binding standards; the newly established Environment Protection Australia (EPA) functions as an independent federal regulator.
Cross-sectoral coordination mechanism	No legal mechanism mandates synchronisation between KLHK, KKP, and regional governments. Coordination is ad hoc and dependent on the initiative of individual officials.	DMCR coordinates six government members, two civil society members, two academics, three international organisations, and five independent experts within a single national coordinating body.	Bilateral approval frameworks between the federal government and states or territories, reinforced by National Environmental Standards and regional planning

continued on following page

Table 1. Continued

Dimension of Comparison	Indonesia	Thailand	Australia
Resolution of jurisdictional conflict	No mechanism exists. The dispute between the regency and provincial governments over credit for mangrove planting in Mengare is left unresolved by any formal system.	DMCR as a single authority eliminates jurisdictional conflict between ministries in the management of mangroves and coastal resources.	The Samuel Review (2020) identified duplication between federal and state laws. The Nature Positive Plan 2022 designs bilateral accreditation to eliminate overlap.
III. Rehabilitation and Maintenance Obligations			
Planting obligations	Mandated through CSR mechanisms under Law No. 40/2007 as amended by Law No. 6/2023. However, implementation is discretionary, lacking minimum technical standards and relying on pressure from regional government.	Integrated into long-term national DMCR programmes. The Thailand Mangrove Alliance (2024) engages more than 90 organizations across 24 coastal provinces.	The Nature Repair Market (2023) is a national voluntary incentive scheme that certifies conservation and restoration actions through tradeable biodiversity certificates.
Post-planting maintenance obligations	No legally enforceable obligation exists. CSR regulation sets no seedling survival standards or post-planting monitoring requirements. Field research in Mengare records a seedling survival rate of zero percent.	The Thailand Mangrove Alliance emphasises that mangrove rehabilitation must be a long-term commitment with multi-year budgeting and continuous monitoring.	The EPBC Act requires approval conditions that include ongoing compliance and audits enforceable by the EPA.

continued on following page

Table 1. Continued

Dimension of Comparison	Indonesia	Thailand	Australia
Infrastructure preconditions for restoration	No regulatory instrument designates wave-breaking structure construction as an enforceable state obligation and precondition for mangrove restoration programs.	Breakwater infrastructures such as bamboo fences and nearshore breakwaters are essential preconditions for mangrove restoration as they reduce wave energy by up to 70% and facilitate the sedimentation necessary for seedling survival (Thieu Quang & Mai Trong, 2020). These structures create stable substrates for natural recolonization, with nearshore designs achieving success rates over 65% by shielding young plants from hydro-morphological stress (Sitthi et al., 2025). Although cost-effective and eco-friendly, these temporary measures require consistent maintenance to withstand monsoon impacts until the forest becomes self-sustaining (Friedrich, 2017).	Breakwater infrastructures are frequently utilized as critical preconditions for mangrove restoration in high-energy coastal environments. These structures protect vulnerable seedlings from wave stress and facilitate the sediment accretion necessary for survival (Saunders et al., 2024). While Australian strategies emphasize nature-based solutions, hybrid approaches involving offshore breakwaters are trialed in areas like Port Phillip Bay to overcome hydrodynamic barriers (Kamali & Hashim, 2011). Such interventions are vital in erosion-prone regions to establish stable intertidal conditions, although they are typically viewed as temporary measures within broader Ecological Mangrove Restoration (EMR) frameworks (Tonneijck, 2016).
IV. MRV Systems and Accountability			
Mangrove MRV system	Mangrove coverage data among KLHK, KKP, and BIG frequently diverge by hundreds of thousands of hectares. Indonesia lacks fully accredited MRV mechanisms aligned with Article 6.4 of the Paris Agreement.	DMCR manages a centralised mangrove monitoring system. The Thailand Mangrove Alliance (2024) is developing multi-year monitoring across 24 coastal provinces with standardized data.	Australia has developed its first national ocean ecosystem account covering mangrove, seagrass, and tidal saltmarsh carbon stocks. Environment Information Australia (EIA) provides high-quality environmental data as the basis for policy decisions.
Planting credit accountability mechanism	The jurisdictional dispute between regency and provincial governments over planting credit, documented in Mengare, is unresolved by any system. Planting figures are reported without verification of seedling survival.	DMCR Regulation 2022 on Carbon Credit Sharing establishes a mechanism for distributing carbon credits from community-managed mangrove planting and maintenance, with clear attribution.	Environment Information Australia is mandated to report on progress towards nature-positive targets every two years, based on independently established baselines.

continued on following page

Table 1. Continued

Dimension of Comparison	Indonesia	Thailand	Australia
V. Integration of International Commitments			
Internalization of NDC into domestic policy	The 600,000-hectare target was declared at the G20 Bali Summit in 2022. First-year achievement totalled only 33,000 hectares, representing 5.5 percent of the target. No legal imperatives mechanism ensures implementation consistent with the commitment.	Thailand's NDC explicitly integrates blue carbon targets. The Carbon Credit Sharing Regulation 2022 operationalizes a payment mechanism based on ecological performance.	The Nature Positive Plan (2022) explicitly aligns EPBC Act reforms with the Kunming-Montreal Global Biodiversity Framework. Australia has led the development of Ramsar Convention resolutions on blue carbon wetlands.
Participation of local communities	Dependent on the voluntary commitment of local groups without adequate budget support or legal obligations. The Mengare community manages conservation without sustained government accompaniment.	The Marine and Coastal Resources Management Act 2015 legally guarantees the participation of communities, local governments, and other groups in mangrove restoration and conservation.	First Nations peoples, conservation groups, and farmers may participate in the Nature Repair Market as producers of biodiversity certificates with direct financial benefits.
VI. Key Lessons for Indonesian Reform			
Principal lessons	Subject of reform: legislative reconstruction through a mangrove lex specialis, institutional consolidation, integrated MRV standards, and legally enforceable post-planting maintenance obligations are all required.	Consolidation of authority within a single institution (DMCR model); formalization of carbon credit-sharing mechanisms for local communities; multi-year monitoring as a mandatory condition of planting programmes.	Nationally binding standards across all levels of government; an independent verification body; the Nature Repair Market as a market-based incentive for ecosystem conservation and restoration.

CONCLUSION

The foundational source of Indonesia's mangrove governance failure is a juridical antinomy embedded in its own legislation. The simultaneous classification of mangroves as forest estate under Law No. 41 of 1999 and as coastal ecosystem components under Law No. 27 of 2007 in conjunction with Law No. 1 of 2014 assigns contradictory authorities to the Ministry of Environment and Forestry (KLHK) and the Ministry of Maritime Affairs and Fisheries (KKP) over the same physical ecosystem. In the absence of a lex specialis capable of resolving this dualism, no institution can exercise clear, unchallenged jurisdiction, and this chapter argues that effective ecosystem governance rests on three conditions: clear spatial boundaries, a defined authority to enforce them, and accountability mechanisms with real teeth. So far, none of the three has been met.

Regulatory fragmentation translates this legislative antinomy into measurable enforcement collapse at the field level. In Gresik Regency, the parallel and uncoordinated mandates of the KLHK, the KKP, and the regional government have produced a governance void in which at least 238 hectares of mangrove in Manyar Subdistricts have been converted to industrial and aquaculture uses, and a minimum of 85 hectares in Ujungpangkah have been lost to illegal logging, with no single institution held accountable for either outcome. As Ostrom (1990) establishes, the absence of defined boundaries, agreed rules, monitoring, and tiered sanctions renders the protection of any common-pool resource structurally impossible.

The corporate CSR planting mechanism, which the regional government relies upon as its primary instrument of mangrove rehabilitation, has been reduced to a form of performative compliance that generates documentary evidence of fulfillment while producing zero ecological result. The testimony of community leader Zaini that the recorded planting of hundreds of thousands of seedlings yielded a survival rate of zero percent exposes a structural legal gap. Government Regulation No. 47 of 2012 imposes no post-planting maintenance obligation, sets no minimum seedling survival standard, and provides no audit mechanism through which regulators can verify whether corporate CSR activities have achieved their intended ecological purpose.

The jurisdictional contest between the regency and provincial governments over which level of government is entitled to claim credit for mangrove planting in the coastal zone reveals that the accountability system has become entirely decoupled from its ecological purpose. When institutional competition over reportable planting numbers displaces any concern for seedling survival, the Monitoring, Reporting, and Verification (MRV) system ceases to function as a mechanism of genuine environmental accountability and becomes instead an instrument of institutional self-promotion. Indonesia's lack of fully accredited MRV verifiers aligned with Article 6.4 of the Paris Agreement means that this decoupling propagates upward into Indonesia's national and international climate reporting obligations.

Applying Koh's (1997) transnational legal process framework to the Indonesian mangrove governance context reveals that the implementation gap between the Paris Agreement commitments and the conditions documented in Mengare reflects failure at all three stages of norm domestication. Discursive internalization has occurred, as evidenced by the 600,000-hectare rehabilitation target declared at the G20 Bali Summit. Institutionalization has been attempted through the establishment of the Peat and Mangrove Restoration Agency (BRGM). However, legal imperatives has failed entirely: no enforceable obligations, no meaningful sanctions, and no operational accountability mechanisms have been created to convert diplomatic commitments into the ecological outcomes that those commitments were designed to produce.

One of the notorious examples lies in infrastructure. The absence of state-constructed wave-breaking infrastructure constitutes a legally cognizable failure of the state's environmental protection obligations. Field testimony from Atmudjin and Zaini establishes that no mangrove planting program can survive in the Mengare intertidal zone without prior installation of physical wave-energy reduction structures. Since no regulatory instrument designates this infrastructure as an enforceable governmental precondition for restoration programs, the state requires ecological outcomes from private actors while withholding the physical conditions necessary to achieve them. The destruction of nearly 70 percent of the mangrove area in Desa Tanjung Widoro stands as the direct empirical consequence of this normative gap.

The comparative analysis of Thailand and Australia identifies three levels of reform that are both theoretically coherent and operationally grounded in Indonesia's specific institutional conditions. At the legislative level, Indonesia requires a mangrove lex specialis that terminates the jurisdictional dualism and establishes a single legal identity for this sui generis ecosystem. At the institutional level, the BRGM requires a strengthened mandate, adequate budget, and consolidated authority modeled on Thailand's Department of Marine and Coastal Resources (DMCR) to function as a genuine single-authority governance body. At the operational level, Indonesia requires harmonized inter-ministerial MRV standards, legally enforceable post-planting maintenance conditions attached to all CSR planting permits, and a transparent responsibility-sharing mechanism between central, provincial, and regency governments that replaces the competition for planting credit with accountability grounded in verifiable ecological outcomes. Without reform at all three levels simultaneously, the gap between Indonesia's climate diplomacy and the ecological reality of its coastal communities will continue to widen.

Beyond its policy recommendations, this chapter contributes to governance scholarship by showing empirically how a national-level juridical antinomy is reproduced, not resolved, once it reaches the level of local implementation. Existing literature on legal fragmentation and transnational legal process has largely stayed at the level of national policy design, leaving open the question of what happens once such policies actually touch the ground. The Mengare case answers that question by tracing three mechanisms: jurisdictional credit-claiming, performative CSR compliance, and the absence of enforceable post-planting obligations. Together, these show how fragmentation is not just experienced but actively entrenched by the actors closest to the resource. This gives a more granular account of Koh's (1997) internalization failure than a purely doctrinal analysis could offer, and it places the chapter's findings inside, rather than merely next to, ongoing debates on institutional fragmentation and symbolic compliance.

This study has some clear limitations. The empirical material rests on three informants in Mengare, all from the community and civil-society side; no one from KLHK, KKP, BRGM, regional government agencies, or the corporate actors behind CSR planting was interviewed for this chapter. That means the account of institutional motivation and inter-agency conflict given here reflects the perspective of those affected by it, not confirmation from the institutions themselves. It also means the weight placed on legal fragmentation, relative to other possible factors such as administrative capacity, budget constraints, or local political interests, has not been tested independently. Readers should keep this evidentiary scope in mind when weighing the chapter's institutional claims.

Future research should draw directly on the perspectives of government agencies and corporate actors, whether through interviews, freedom-of-information requests, or analysis of internal reporting documents. This would test whether the jurisdictional conflict and performative compliance described by community informants look the same from inside the institutions, or whether officials and companies tell a different story. It would also be worth comparing Indonesia to other Southeast Asian jurisdictions dealing with similar blue carbon governance problems, not just Thailand and Australia, since that comparison would show which of the reforms proposed here are particular to Indonesia's legal system and which are just symptoms of fragmented environmental governance found across developing countries generally.

REFERENCES

Ahmad, M., & Wilkins, S. (2025). Purposive sampling in qualitative research: A framework for the entire journey. *Quality & Quantity*, *59*(2), 1461–1479. DOI: 10.1007/s11135-024-02022-5

Al Abiad, H., & Masadeh, A. (2024). *Law Comparison as a Research Method in Legal Studies, and Its Importance in Promoting Uniformity in Legal Systems.*, DOI: 10.1007/978-3-031-56121-4_42

Arifanti, V. B., Novita, N., Subarno, , & Tosiani, A. (2021). Mangrove deforestation and CO 2 emissions in Indonesia. *IOP Conference Series. Earth and Environmental Science*, *874*(1), 012006. DOI: 10.1088/1755-1315/874/1/012006

Banakar, R. (2019). On Socio-Legal Design. SSRN *Electronic Journal.* DOI: 10.2139/ssrn.3463028

Bertram, C., Quaas, M., Reusch, T. B. H., Vafeidis, A. T., Wolff, C., & Rickels, W. (2021). The Blue Carbon Wealth of Nations. *Nature Climate Change*, *11*(8), 704–709. DOI: 10.1038/s41558-021-01089-4

Clark, T., Foster, L., Sloan, L., & Bryman, A. (2021). *Bryman`s Social Research Methods* (Sixth). Oxford University Press.

Cowan, C. (2024). *Thailand's Budding Mangrove Restoration Plans Spark Both Hope and Concern.* Mongabay. https://news.mongabay.com/2024/10/thailands-budding-mangrove-restoration-plans-spark-both-hope-and-concern/

Creswell, J. W., & Poth, C. N. (2018). *Qualitative Inquiry and Research Design: Choosing Among Five Approaches* (Fourth). Sage Publication Inc.

Denzin, N. K. (2009). *The Research Act: A Theoretical Introduction to Sociological Methods.* Transaction Publishers.

Department of Climate Change. Energy, The Environment, and W. (1999). *Environment Protection and Biodiversity Conservation Act 1999.* Department of Climate Change, Energy, The Environment, and Water.

Dong, K., Zhao, C., Nepal, R., & Zander, K. K. (2025). Are natural disasters stumbling blocks to carbon inequality mitigation? A global perspective. *Ecological Economics*, *227*, 108392. DOI: 10.1016/j.ecolecon.2024.108392

Embassy of Denmark to Indonesia. (2022). *Danish Support to The Indonesia Mangroves for Coastal Resilience Project.*

Ervita, K., & Makhas, K. A. (2021). *Understanding Spatial Planning to Protect Mangrove Ecosystems*. World Resources Indonesia. https://wri-indonesia.org/en/insights/understanding-spatial-planning-protect-mangrove-ecosystems

Feti, F., Hadi, S. P., Purnaweni, H., & Sudarno, . (2020). Does the intervention of regional authorities contribute to sustainable mangrove eotourism? Case Study on Mangrove Management at Karangsong, West Java, Indonesia. *Ecological Questions*, *31*(3), 7. DOI: 10.12775/EQ.2020.017

Fitrianggraeni, S., Avianti, F. N., & Purnama, S. (2023). *Redifining Corporate (Social) Responsibility: The Indonesian Perspective in Indonesia 2023*. Anggraeni and Partners. https://www.ap-lawsolution.com/actio/redifining-corporate-social-responsibility-the-indonesian-perspective-in-indonesia-2023

Friedrich, H. (2017). *Bamboo for Coastal Protection*. LinkedIn. https://www.linkedin.com/pulse/bamboo-coastal-protection-hans-friederich

Hamzani, A. I., Widyastuti, T. V., Khasanah, N., & Mohd Rusli, M. H. (2024). Implementation approach in legal research. *International Journal of Advances in Applied Sciences*, *13*(2), 380. DOI: 10.11591/ijaas.v13.i2.pp380-388

Howlett, M., Ramesh, M., & Perl, A. (2020). *Studying Public Policy: Principles and Processes* (Fourth). Oxford University Press.

Humas Kementerian Kehutanan Republik Indonesia. (2025). *Mangrove Festival 2025: Indonesia Perkuat Konsolidasi Nasional Rehabilitasi Ekosistem Mangrove*. Kementerian Kehutanan Republik Indonesia. https://www.kehutanan.go.id/news/mangrove-festival-2025-indonesia-perkuat-konsolidasi-nasional-rehabilitasi-ekosistem-mangrove

IESR. (2024). *Indonesia Energy Transition Outlook 2024: Peaking Indonesia Energy Sector Emission by 2030 The Beginning or The End of Energy Transition Promise*.

Irsadi, A., Jabbar, A., Dewi, N. K., Somantri, D., Khair, A. S. E., Sidiq, W. A. B. N., Mutiatari, D. P., & Martuti, N. K. T. (2025). Coastal Erosion Reduces Resilience and Disrupts Compositional Dynamics of The Mangrove Ecosystem. *Biosaintifika*, *17*(1), 128–142. DOI: 10.15294/biosaintifika.v17i1.17292

Jonathan, P. (2025). *Indonesia has Launched its International Carbon Trading Platform*. Mt. Stonegate. https://www.mtstonegate.com/post/indonesia-has-launched-its-international-carbon-trading-platform

Kamali, B., & Hashim, R. (2011). Mangrove restoration without planting. *Ecological Engineering*, *37*(2), 387–391. DOI: 10.1016/j.ecoleng.2010.11.025

Koh, H. H., Chayes, A., Chayes, A. H., & Franck, T. M. (1997). Why Do Nations Obey International Law? *The Yale Law Journal, 106*(8), 2599. DOI: 10.2307/797228

Lemos, M. C., & Agrawal, A. (2006). Environmental Governance. *Annual Review of Environment and Resources, 31*(1), 297–325. DOI: 10.1146/annurev.energy.31.042605.135621

Mangroves for the Future: Investing in Coastal Ecosystems. (n.d.). *Thailand.* Mangroves for the Future: Investing in Coastal Ecosystems. https://www.mangrovesforthefuture.org/countries/members/thailand/

Matta, T. C., Pereira, L. S., Belmonte, Y. C. B., Chaves, F. de O., & Soares, M. L. G. (2025). Evaluating terrestrial laser scanning for structural characterization of mangrove forests in Southeastern Brazil. *Forest Ecology and Management, 583,* 122567. DOI: 10.1016/j.foreco.2025.122567

McCredie, B., Pleming, R., & Johnstone, E. (2024). *Second Stage of EPBC Act Reforms Announced: Draft legislation expected shortly.* Allens. https://www.allens.com.au/insights-news/insights/2024/04/second-stage-of-epbc-act-reforms-announced/

Meyer, J. W., & Rowan, B. (1977). Institutionalized Organizations: Formal Structure as Myth and Ceremony. *AJS; American Journal of Sociology, 83*(2), 340–363. DOI: 10.1086/226550

Naeem, M., Ozuem, W., Howell, K., & Ranfagni, S. (2023). A Step-by-Step Process of Thematic Analysis to Develop a Conceptual Model in Qualitative Research. *International Journal of Qualitative Methods, 22,* 16094069231205789. Advance online publication. DOI: 10.1177/16094069231205789

Nasarani, G., Purnaweni, H., Maryono, M., Rachmawati, S., Suryadi, I., Surahman, N., & Fitriani, N. (2025). Factors That Influence the Management of Household Hazardous Waste in Semarang City, Central Java, Indonesia: A Path Analysis. *International Journal of Environmental Impacts, 8*(5), 888–898. DOI: 10.56578/ijei080505

Noor, A. (2023). Socio-Legal Research: Integration of Normative and Empirical Juridical Research in Legal Research. *Jurnal Ilmiah Dunia Hukum, 7*(2), 94. DOI: 10.56444/jidh.v7i2.3154

Nurhati, I. S., & Murdiyarso, D. (2022). *Strategi Nasional Pengelolaan Ekosistem Mangrove: Sebagai Rujukan Konservasi dan Rehabilitasi Kawasan Berkelanjutan dan Pembangunan Rendah Karbon.* Center for International Forestry Research (CIFOR). DOI: 10.17528/cifor-icraf/008790

Ostrom, E. (1990). *Governing the Commons*. Cambridge University Press., DOI: 10.1017/CBO9780511807763

Ostrom, E. (2010). Polycentric systems for coping with collective action and global environmental change. *Global Environmental Change*, *20*(4), 550–557. DOI: 10.1016/j.gloenvcha.2010.07.004

Pahwa, M., Cavanagh, A., & Vanstone, M. (2023). Key Informants in Applied Qualitative Health Research. *Qualitative Health Research*, *33*(14), 1251–1261. DOI: 10.1177/10497323231198796 PMID: 37902082

Pemerintah Kabupaten Gresik. (2010). *Pemutakhiran dan Penyerasian Analisis dan Perencanaan Rencana Tata Ruang Wilayah (RTRW)*. Kabupaten Gresik.

Prasetyo, A., Santoso, N., & Prasetyo, L. B. (2017). Kerusakan Ekosistem Mangrove di Kecamatan Ujung Pangkah Kabupaten Gresik Provinsi Jawa Timur. *Jurnal Silvikultur Tropika*, *8*(2), 130–133.

Purnaweni, H., Hadi, S. P., Soeprobowati, T. R., Roziqin, A., & Rakhma, N. A. (2026). *What are the Main Concerns for Global Mangrove Management Research? Systematic Literature Review and Bibliometric Analysis*. Research in Ecology., DOI: 10.30564/re.v8i2.12149

Raustiala, K., & Victor, D. G. (2004). The Regime Complex for Plant Genetic Resources. *International Organization*, *58*(02). Advance online publication. DOI: 10.1017/S0020818304582036

Reiman, M., & Zimmermann, R. (Eds.). (2019). *The Oxford Handbook of Comparative Law* (Second). Oxford University Press.

Saunders, M. I., Cannard, T., Fischer, M., Sheppard, M., Twomey, A., Morris, R., Bishop, M. J., Mayer-Pinto, M., Malcolm, F., Vozzo, M., Steven, A., Swearer, S. E., Lovelock, C. E., Pomeroy, A. W. M., McLeod, I., & Waltham, N. J. (2024). A roadmap to coastal and marine ecological restoration in Australia. *Environmental Science & Policy*, *159*, 103808. DOI: 10.1016/j.envsci.2024.103808

Sidik, F., Lawrence, A., Wagey, T., Zamzani, F., & Lovelock, C. E. (2023). Blue carbon: A new paradigm of mangrove conservation and management in Indonesia. *Marine Policy*, *147*, 105388. DOI: 10.1016/j.marpol.2022.105388

Sitthi, A., Pimple, U., Piponiot, C., & Gond, V. (2025). Assessing the Effectiveness of Mangrove Rehabilitation Using Above-Ground Biomass and Structural Diversity. *Scientific Reports*. 10.1038/s41598-025-92514-7?urlappend=%3Futm_source%3Dresearchgate.net%26utm_medium%3Darticle

Sun, Y. Y., Faturay, F., Lenzen, M., Gössling, S., & Higham, J. (2024). Drivers of global tourism carbon emissions. *Nature Communications*, *15*(1), 10384. Advance online publication. DOI: 10.1038/s41467-024-54582-7 PMID: 39658580

Tamanaha, B. Z. (2021). *Legal Pluralism Explained*. Oxford University PressNew York., DOI: 10.1093/oso/9780190861551.001.0001

Thailand. (2022). *Regulation of the Department of Marine and Coastal Resources on the Carbon Credit Trading and Benefit Sharing Obtained from Planting and Maintaining Mangrove Forest for the Community B.E.2565*. Thailand.

Thieu Quang, T., & Mai Trong, L. (2020). Monsoon wave transmission at bamboo fences protecting mangroves in the lower mekong delta. *Applied Ocean Research*, *101*, 102259. DOI: 10.1016/j.apor.2020.102259

Tonneijck, F. (2016). *Mangrove Restoration: To Plant or not to plant?* Wetlands International. https://www.wetlands.org/publication/mangrove-restoration-to-plant-or-not-to-plant/

Walhi Jawa Timur. (2023). *Hutan Mangrove di Jawa Timur dalam Ancaman*. Walhi Jawa Timur. https://walhijatim.org/2023/07/26/hutan-mangrove-di-jawa-timur-dalam-ancaman/

Widjonarko, W., Purnaweni, H., Maryono, M., & Soeprobowati, T. R. (2025). Modelling Environmental Impact of Sea Dike and Toll Road in Semarang-Demak Indonesia Based on Satellite Imagery Data. *Geoplanning: Journal of Geomatics and Planning*, *12*(1), 31–44. DOI: 10.14710/geoplanning.12.1.31-44

Yin, R. K. (2017). *Case Study Research and Applications: Design and Methods* (Sixth). Sage Publication Inc.

Yunitasari, D., Zainuri, Z., & Masfufah, K. (2020). Analisis Valuasi Ekonomi Berdasarkan Perhitungan Total Economic Value Ekosistem Mangrove Di Desa Banyuurip Kabupaten Gresik. *Media Trend*, *15*(2), 345–358. DOI: 10.21107/mediatrend.v15i2.5415

Chapter 15

Marketing Sustainable Mangroves Value Creation, Green Branding, and Policy Support in Global Coastal Management

Robetmi Jumpakita Pinem
http://orcid.org/0000-0003-2704-1219
Diponegoro University, Indonesia

Sari Listyorini
Diponegoro University, Indonesia

ABSTRACT

This study examines sustainable marketing practices in mangrove ecotourism by focusing on value creation, green branding, and policy support within the context of coastal management. Using a quantitative approach, data were collected through structured questionnaires administered to 25 managers and 100 visitors at mangrove ecotourism destinations in Indonesia. Descriptive and inferential analyses were conducted to compare perceptions of sustainable marketing implementation across dimensions such as sustainability communication, stakeholder engagement, community participation, environmental education, and destination competitiveness. The findings reveal significant perception gaps between managers and visitors. While managers perceive branding, collaboration, and sustainability communication as effective, visitors report limited information clarity, inadequate environmental ed-

DOI: 10.4018/979-8-3373-9320-9.ch015

ucation, and low involvement in conservation activities. These results highlight the need for stronger communication, inclusive stakeholder collaboration, and clearer sustainability narratives.

1. INTRODUCTION

Recently, sustainability has become a major concern, both for natural resource management and tourism development worldwide. Mangroves, which thrive along Indonesia's coasts, play a key role in curbing climate change and protecting the coastline (Sutanto et al., 2022) . However, their role goes beyond that. Mangroves also serve as a source of economic and social livelihoods for many local communities (Rahman et al., 2022; Arifin et al., 2023). When mangroves are developed into ecotourism destinations, it is not only about environmental preservation but also about promoting community well-being and socio-economic benefits for surrounding communities (Fatimatuzzahro et al., 2020).

In ecotourism development, sustainable marketing strategies are increasingly important. This means that marketing is not just about profit, but also integrates economic, social, and environmental values into every process, from planning to evaluation (Font & McCabe, 2017; Kumar & Christodoulopoulou, 2022). This approach is believed to create shared value for all parties: managers, visitors, local communities, and other stakeholders (Mihalic, 2020).

Indonesia has considerable potential for mangrove ecotourism development, yet its long-term success depends on effective management, stakeholder collaboration, and sustainable development strategies (Nugraha et al., 2024; Purnaweni et al., 2026). On the one hand, many tourists now seek natural experiences and want to learn about the environment, so mangrove management must be creative and innovative. However, on the other hand, there remains a gap in understanding between the marketing strategies developed by managers and the expectations or perceptions of visitors and the surrounding community (Putra et al., 2020; Lee & Chen, 2021). This gap can reduce program effectiveness, dampen community participation, and ultimately impact economic and environmental sustainability.

In reality, active community participation and the involvement of various stakeholders across sectors are key to creating truly adaptive and inclusive ecotourism destinations (Junaidi et al., 2021; Rasoolimanesh et al., 2022). Managers must ensure that their marketing strategies meet market needs, provide tangible benefits to the community, and effectively convey sustainable values to visitors. Communication must also be open and responsive to environmental changes to ensure these destinations remain competitive (Björk et al., 2021; Damayanti et al., 2021).

Unfortunately, research on sustainable marketing in the mangrove ecotourism sector in Indonesia is still relatively limited, especially research that comprehensively compares the perceptions of managers and visitors. Most studies focus on a single group or only highlight environmental and economic aspects, without addressing the social dimensions and sustainability communication in depth (Putra et al., 2020; Mihalic, 2020).

The study will show that sustainable marketing should be looked at through a lens that is more than simply utilitarian. It will also examine how managers communicate to achieve sustainability objectives and how they can incorporate the input of a variety of stakeholders in developing a sound sustainable marketing strategy, which is critical for long-term value creation and sustainable coastal governance. By incorporating these dimensions of sustainable marketing, this chapter offers additional perspectives on what constitutes ecotourism in a mangrove setting that go beyond the traditional definitions currently used.

Therefore, this research aims to fill this gap. By comparing the perceptions of managers and visitors regarding sustainable marketing strategies in mangrove programs, this study aims not only to broaden academic knowledge but also to provide direct recommendations useful for ecotourism managers and policymakers. It is hoped that future mangrove marketing strategies can be more adaptive, participatory, and, of course, sustainable.

The broad objectives of this study are:

1. To analyze the perceptions of managers and visitors regarding the implementation of sustainable marketing strategies in mangrove programs.
2. To identify the strengths and weaknesses of marketing strategies from the perspectives of both groups.
3. To formulate concrete recommendations based on field findings to improve the effectiveness of sustainable marketing in the mangrove ecotourism sector in Indonesia.

2. LITERATURE REVIEW

2.1. Sustainable Marketing and Ecotourism

Sustainable marketing is more than just a strategy. It's about incorporating economic, social, and environmental considerations into every stage of marketing (Kumar & Christodoulopoulou, 2022). In the world of ecotourism, this effort is not just about increasing the number of tourists. The goal is much bigger: to ensure that tourism provides real benefits for local communities, while still preserving nature (Font & McCabe, 2017; Mihalic, 2020).

The sustainable development of mangrove ecotourism in Indonesia requires strategic planning, government support, and community participation (Baiti et al., 2026). The foundation of communities and their dependence on fragile ecosystems makes sustainable marketing key, both for nature conservation and economic empowerment (Rahman et al., 2022; Kurniati et al., 2024).

2.2. Economic, Social and Environmental Dimensions in Sustainable Marketing

Recent research shows that sustainable marketing in ecotourism must not forget three important things: economic, social and environmental benefits (Björk et al., 2021; Rasoolimanesh et al., 2022). Economically, this means that people's incomes increase, job opportunities open up, and the regional economy becomes stronger. From a social perspective, regional identity and pride increases, environmental education becomes better, and the community is more involved and empowered (Junaidi et al., 2021). Meanwhile, from the environmental aspect, conservation, ecosystem restoration and reducing the negative impacts of tourism continue to be pursued (Arifin et al., 2023; Turisno et al. 2021).

These three dimensions are closely connected. Economic success often arises when the community is truly involved in tourism management and promotion, and environmental protection is more guaranteed if the community has received direct economic and social benefits (Putra et al., 2020). In essence, the balance and synergy between these pillars determines whether the program runs well and is sustainable or not.

2.3. Stakeholders play a role, the community is involved, sustainability values are communicated

In fact, the success of a sustainable marketing strategy is greatly influenced by the roles of various stakeholders. Stakeholders include managers, government, local communities, tourists and the private sector (Rasoolimanesh et al., 2022; Purnaweni et al., 2025). Local community involvement gives a sense of ownership to the mangrove program. Local community involvement also increases the effectiveness of communication and promotion of sustainability (Junaidi et al., 2021; Srifitriani et al., 2025).

Communication of sustainability values to visitors is key. This communication makes tourists not only enjoy the destination, but also understand and support environmental conservation (Font & McCabe, 2017). Effective communication strengthens visitors' positive perceptions of the program. This strong perception encourages responsible tourism behavior and increases loyalty to the destination (Lee & Chen, 2021; Awang et al., 2026).

A communication gap between managers and visitors can contribute to the development of negative perceptions about sustainability efforts. This negative perception reduces community participation and reduces interest in visiting (Mihalic, 2020; Zhang et al., 2020). On the other hand, when stakeholders actively collaborate and innovate in marketing strategies, economic benefits, social benefits and environmental benefits can be achieved simultaneously and sustainably.

2.4. Relationship and Influence of Dimensions in the Context of Mangrove Ecotourism

There is a dynamic interplay between managers' and visitors' perceptions regarding the various dimensions of sustainable marketing. Positive manager perceptions of economic, social and environmental benefits encourage management to be more innovative and collaborative (Arifin et al., 2023). Conversely, visitors' perceptions of sustainability value and communication quality increase satisfaction, revisit intention, and support for conservation (Lee & Chen, 2021; Agustina & Artanti, 2020).

Communication and stakeholder involvement strengthens those relationships. Managers who can explain the value of sustainability clearly and involve local communities will foster trust and visitor participation (Rahman et al., 2022; Kristanti et al. 2024). So, the relationship between dimensions is dynamic and mutually reinforcing. Each aspect can influence and be influenced by other aspects.

A comparative study of managers' perceptions and visitors' perceptions is needed. This study identifies gaps and opportunities for improvement in sustainable marketing strategies for mangrove programs. Managers understand the relationships and influences between dimensions. Managers design strategies that are more effective, more adaptive, and have a long-term impact. This strategy supports the sustainability of mangrove ecotourism in Indonesia.

2.5. Stakeholder Theory, Value Co-Creation, and Relational Governance.

Ecotourism can better understand for sustainable marketing differs depending on the theory used to analyze the situation. For example, stakeholder theory states that an organization should create value for all stakeholders both directly and indirectly affected by its actions including customers, local communities, government, environmental organizations, and all other stakeholders affected by the ecotourism experience. Therefore, mangrove ecotourism's ability to create sustainable outcomes depends upon the ability of stakeholders to balance their interests with each other.

Value co-creation theory elaborates on stakeholder theory by stating that, in order to achieve sustainability, all stakeholders must actively engage in creating both the tourism experience and the associated conservation initiatives. Thus, the visitor becomes an active participant versus a passive consumer and utilizes environmental education, environmental conservation activities, providing feedback and exhibiting responsible tourism behavior to generate value. Similarly, local communities become strategic partners in generating economic, social, and environmental value.

Relational governance supports both of these theories by emphasizing that successful long-term sustainability is reliant upon collaboration, trust, a shared understanding, and continued communication between stakeholders. Relational governance does not rely on formal regulations as the primary form of governance but instead emphasizes cooperation, transparency, and mutual commitment as effective approaches to sustainable destination management.

Thus, this study views sustainable marketing by way of being a relational communication process whereby stakeholder engagement and sustainability communication function as strategic mechanisms linking environmental protection, local community development and increased visitor engagement in ecotourism experiences.

3. RESEARCH METHODS

3.1. Research Design

This research uses a quantitative approach with a comparative survey design. So, this research wants to examine the differences and relationships between managers' and visitors' perceptions of the implementation of sustainable marketing strategies in mangrove programs in Central Java. So, this research chooses a quantitative approach in order to get an objective picture, disseminate results, and carry out clear statistical analysis (Creswell & Creswell, 2018).

3.2. Research Location and Time

Researchers visited several locations of the mangrove ecotourism program on the coast of Central Java, Indonesia. The location is in an area that has implemented a sustainable marketing program. Researchers collected data from February to April 2026.

3.3. Population and Sample

There are two main groups in the population in this study, namely:

1. Mangrove program manager, internal stakeholders, including group leaders, daily administrators and field officers
2. Mangrove program visitors (local, domestic and educational tourists who directly interact with mangrove destinations)

Sampling used purposive sampling technique. Sampling followed the following inclusion criteria:

a. For managers: actively involved for at least one year in the mangrove program
b. Visitors: must take part in at least one mangrove tourism or education program

The respondents collected were 25 managers. The respondents collected were 100 visitors.

3.4. Research Instrument

We collected data using a structured questionnaire using a 1–5 Likert scale; 1 indicates strongly disagree, and 5 indicates strongly agree. The manager questionnaire contained 12 statements. The visitor questionnaire contained 8 statements. The questionnaire measured perceptions of economic benefits, social benefits, environmental benefits, local community involvement, stakeholder involvement, responsiveness to market needs, and communication of sustainability values.

Furthermore, the initial section of the questionnaire contained demographic questions. These included age, gender, educational level, and occupation.

3.5. Respondent Description

3.5.1. Respondent Profile

Table 1. Demographic distribution of mangrove program manager respondents (N=25)

Category	Subcategory	Amount (n)	Percentage (%)
Age	< 30 yo	3	12,0
	31-40 yo	10	40,0
	41-50 yo	9	36,0
	> 50 yo	3	12,0
Gender	Male	19	76,0
	Female	6	24,0
Education	Junior High School/Equivalent	1	4,0
	High School/Equivalent	7	28,0
	Diploma/Bachelor's Degree	16	64,0
	Master's/PhD	1	4,0
Occupation	Core management	13	52,0
	Field staff	7	28,0
	Others	5	20,0

3.5.2. Respondent Profile

Table 2. Demographic distribution of respondents visiting the mangrove program (N=100)

Category	Subcategory	Amount (n)	Percentage (%)
Age	< 20 yo	8	8,0
	20-30 yo	38	38,0
	31-40 yo	34	34,0
	41-50 yo	13	13,0
	> 50 yo	7	7,0
Gender	Male	56	56,0
	Female	44	44,0

continued on following page

Table 2. Continued

Category	Subcategory	Amount (n)	Percentage (%)
Education	Junior High School/Equivalent	5	5,0
	High School/Equivalent	31	31,0
	Diploma/Bachelor's Degree	58	58,0
	Master's/PhD	6	6,0
Occupation	Students	32	32,0
	Private/Professionals	25	25,0
	Civil Servants	12	12,0
	Self-Employed	15	15,0
	Others	16	16,0

3.6. Data Collection Procedures

The research team collected data directly at the research site. Data collection used structured interviews and questionnaires. The team explained the research objectives and secured informed consent from all respondents. The team gave respondents sufficient time to read the questionnaire, then completed it independently. If necessary, the team provided assistance for clarification. The team checked the questionnaire for completeness. The team randomly rechecked a subset of respondents to ensure data validity. Data were collected and encrypted to maintain participant confidentiality.

3.7. Validity and Reliability Testing

Content validity testing was conducted by consulting with experts in sustainable marketing and ecotourism. The instrument's reliability was tested using Cronbach's Alpha coefficient. The results of the reliability test for all questionnaire items showed an alpha value >0.7, indicating high reliability (Taber, 2018).

3.8. Data Analysis Techniques

3.8.1. Descriptive Analysis

Descriptive analysis displays the demographic characteristics of respondents and the distribution of perception scores for each dimension. Statistics used include the mean, standard deviation, median, and frequency distribution. Descriptive analysis aims to show general perception trends in the management and visitor groups.

3.8.2. Inferential Analysis

The researcher analyzed differences in perceptions between managers and visitors using independent samples t-tests for each main dimension of the questionnaire. The researcher chose a two-sample independent sample t-test because the data came from two unrelated groups (Field, 2018). The researcher conducted the test at a 5% significance level ($\alpha=0.05$, $\alpha=0.05$). Before conducting the t-test, the assumption of normality was checked using the Kolmogorov-Smirnov or Shapiro-Wilk test, and the homogeneity of variance was checked using Levene's test to allow for parametric testing. If the assumptions were not met, an alternative non-parametric test, namely the Mann-Whitney-U test, was used.

3.8.3. Analysis of Relationships Between Variables

In addition to the difference test, researchers conducted a Pearson correlation analysis. Pearson correlation analysis identified the relationship between perception scores for each dimension. Pearson correlation analysis was applied to both the management and visitor groups. Pearson correlation examined the relationship between the economic benefit, social, environmental, communication, and stakeholder engagement dimensions.

3.8.4 Additional Analysis: If data permits, researchers conducted a simple linear regression analysis. Researchers explored the influence of the sustainability communication and stakeholder engagement dimensions on visitor satisfaction and loyalty.
All analyses used statistical software (SPSS v26, Microsoft Excel, or R) to ensure accurate and replicable results.

4. RESULTS AND DISCUSSION

The following is user feedback:

4.1. Respondent Characteristics

4.1.1. Mangrove Program Managers

This study involved 25 mangrove program managers in Central Java. Demographic data shows that 76% of managers are aged 31–50, indicating that the majority are of productive age and experienced. Of the total, 76% of managers are male and 24% are female, indicating that the mangrove program is still dominated by men, although female participation is starting to emerge. Managers have diverse educational backgrounds. 64% have a diploma/bachelor's degree. 28% have a high school/equivalent degree. The remainder have a junior high school degree and a master's/doctoral degree. This indicates that the majority of managers have secondary or higher education. Education enables managers to adopt knowledge-based sustainable management and marketing practices. In their work, more than half of respondents (52%) are core managers, for example, group leaders or coordinators. Field staff comprise 28%. The remaining 20% fill supporting or other roles. This composition indicates a relatively stable organizational structure in managing the mangrove program.

4.1.2. Mangrove Program Visitors

Researchers recruited 100 visitors. Respondents were evenly distributed in age. The largest age group was between 20 and 40 years old, accounting for 72% of the total. The results showed that mangrove tourism attracts young people and young adults, who are more responsive to sustainability and ecotourism issues. Fifty-six percent of visitors were male and 44% female, indicating a nearly equal gender balance. The educational level of visitors was dominated by diploma/bachelor's degree graduates (58%). This was followed by high school/equivalent graduates (31%), master's/doctoral graduates (6%), and junior high school/equivalent graduates (5%). This indicates that most visitors are highly educated. Therefore, they tend to be aware of and interested in educational and environmental tourism. Visitors' occupations were diverse: 32% were students, 25% were private employees or professionals, 15% were self-employed, 12% were civil servants, and 16% were others (e.g., housewives, informal workers). This diversity indicates that the mangrove program appeals to all levels of society. Following is user feedback: Conclusion Respondent Characteristics: From the description above, it can be concluded that managers and visitors to mangrove programs in Central Java are generally of productive age and have secondary or higher education. Managers and visitors provide strong social capital for developing sustainable marketing programs based on participation, innovation, and education.

4.2. Managers' and Visitors' Perceptions of Sustainable Marketing Strategies

4.2.1. Average Perception Scores

Table 3. Average (mean) and Standard Deviation (SD) of managers' and visitors' perception scores for each dimension

Dimension	Manager (Mean ± SD)	Visitors (Mean ± SD)
Economic Benefits	4,52 ± 0,64	4,45 ± 0,70
Social Benefits	4,36 ± 0,61	4,30 ± 0,67
Environmental Benefits	4,48 ± 0,59	4,40 ± 0,65
Community Engagement	4,28 ± 0,68	4,22 ± 0,74
Stakeholder Engagement	4,32 ± 0,66	4,10 ± 0,78
Adapting to Market Needs	4,24 ± 0,70	4,18 ± 0,73
Sustainability Communication	4,10 ± 0,71	3,78 ± 0,82

4.2.2. Interpretation of Mean and Standard Deviation

The following is user feedback: 1. Economic Benefits: Based on the data, managers (mean = 4.52) rated the mangrove program very positively. Visitors (mean = 4.45) also rated the mangrove program very positively. The standard deviation (SD) was small across both groups, 0.64 for managers and 0.70 for visitors, indicating consistent opinions. These data indicate that the mangrove program increases community income, creates jobs, and adds economic value to the local community. 2. Social Benefits: Managers rated the social benefits dimension highly (mean = 4.36). Visitors rated the social benefits dimension highly (mean = 4.30). The SD was relatively low (0.61 and 0.67). High ratings indicate that the mangrove program increases social cohesion, strengthens local identity, and expands the social networks of residents and visitors. 3. Environmental Benefits: Both groups rated the environmental benefits highly (mean = 4.48 for managers; mean = 4.40 for visitors; SD = 0.59 and 0.65). This indicates a shared awareness that mangrove programs are important for coastal conservation, carbon sequestration, and biodiversity preservation (Setyadi et al. 2021; Halim et al., 2023). 4. Community Involvement: Managers rated community involvement as good (mean = 4.28; SD = 0.68). Visitors gave similar scores (mean = 4.22; SD = 0.74). High scores and moderate SDs indicate that local communities are active in management and promotion. However, some visitors still see room for improvement in participation. 5. Stakeholder Involvement: Differing views are emerging on this dimension. Managers gave a higher average score (4.32; SD = 0.66) than visitors (4.10; SD = 0.78). This indicates that managers feel they have engaged external parties (government, NGOs, and the private sector) quite well. However, visitors feel that cross-stakeholder collaboration is still not optimal or the benefits have not been felt directly. 6. Adaptation to Market Needs: The average score on this dimension is high and nearly balanced (managers = 4.24; visitors = 4.18; SD = 0.70 and 0.73). This indicates that the mangrove program is quite responsive and adaptive to changing visitor preferences and ecotourism market trends. 7. Sustainability Communication This dimension received the lowest score, especially from visitors (mean = 3.78; SD = 0.82). Managers gave it a higher score (mean = 4.10; SD = 0.71). This indicates a difference in perception: managers felt they were communicating sustainability values quite well, while visitors felt they still found the information they received unclear, less interactive, or less personal.

4.2.3. Discussion of Findings

The results above show that managers' and visitors' perceptions of the sustainable marketing strategy of the mangrove program in Central Java are very positive. Positive perceptions emerged across economic, social, and environmental aspects. This finding aligns with research by Rahman et al. (2022) and Arifin et al. (2023). These studies stated that mangrove-based ecotourism programs improve community welfare and conserve ecosystems when managed in a participatory and adaptive manner. The main differences lie in the dimensions of stakeholder engagement and sustainability communication. Managers assessed collaboration and communication as optimal. However, visitors still identified shortcomings. They considered information unclear, education inadequate, and visitor engagement in sustainability activities low. The data showed slightly higher SD scores for the communication dimension among visitor groups. This demonstrates variations in perceptions and experiences. In my opinion, this perception gap presents a challenge in implementing sustainable marketing. According to studies by Lee & Chen (2021) and Rasoolimanesh et al. (2022), the effectiveness of sustainable marketing is greatly influenced by inclusive two-way communication and the active participation of all stakeholders, including visitors. Therefore, I believe management must develop a more interactive communication strategy, utilize a variety of media, and involve visitors in education and concrete conservation actions. Furthermore, the high score for adapting to market needs indicates that management and visitors recognize flexibility and innovation in developing mangrove tourism services. This is crucial for maintaining the competitiveness of ecotourism destinations amidst the rapidly changing dynamics of the tourism market.

The theoretical implications associated with this research, and more specifically with the areas of communication and stakeholder engagement as relational constructs that link and define each of the three main dimensions of sustainability, have shown that a narrow view of sustainability as an outcome of the three pillars of economic, social, and environmental conditions is limiting. By nature of their position, managers/owners as well as visitors to ecotourism sites have created a perception gap towards sustainability initiatives. While there are tangible benefits related to sustainability initiatives, they will not always be accepted as effective if communication is poorly established and stakeholder participation is limited. It is important to view sustainable marketing in the context of mangrove ecotourism as a relational process in which value is co-created through ongoing interaction, dialogue and cooperation between managers, communities, visitors, government agencies and all other stakeholders. This is an extension of the traditional sustainability framework by placing an emphasis on relational governance as a way to achieve successful long-term coastal sustainability.

The finding offers new insight into the cause of the differences in people's perception of sustainable marketing between tourists and managers. For example, tourists rely on their own experiences while visiting to assess the sustainability of a site. This is in contrast to managers, who evaluate the effectiveness of any sustainability initiative based on their ability to plan, implement, and evaluate its effectiveness through their own organization's practices. If a tourist has limited access to sustainability information, has inadequate opportunities to participate in sustainability initiatives, or is not treated consistently in terms of interacting with local communities, he/she will evaluate that site as being less sustainable than he/she would if he/she had better access.

Sustainability is not just about the presence of environmental, economic, or social programs; it is also about how effectively stakeholders are connected through the relational processes of those stakeholders. Communication provides the mechanism through which sustainability initiatives will be translated into understanding on behalf of stakeholders. Likewise, stakeholder engagement provides the mechanism through which sustainability initiatives can create value together with the stakeholder and build shared ownership of the initiative. As such, the existence of gaps in perception reflects the difference in the experiences of stakeholders and does not merely reflect the difference in opinions. By providing this new understanding of sustainable marketing theory, the study has emphasized the role of relational governance in connecting sustainability programs to stakeholder support and ultimately, to the competitive position of the destination.

The research results reveal that ecotourism services generate substantial economic, social, and environmental benefits to both manager and visitor respondents. However, the visitor sample displayed lower levels of perceived sustainability communication and stakeholder involvement than the manager sample. These findings indicate that the perceived success of sustainable marketing is contingent not only upon the type of benefits realized but also the manner in which sustainability values are conveyed and the means by which stakeholders are involved in all aspects of the tourism experience. Given that the results are based on perception data, it is important to interpret the results as stakeholder assessments rather than as objective indicators of programme outcomes.

4.3. Conclusions and Implications

The above presentation concludes that:

1. The mangrove program in Central Java provides tangible economic, social, and environmental benefits. Management and visitors acknowledge the economic, social, and environmental benefits provided by the mangrove program in Central Java.
2. Local community involvement is good, but collaboration and sustainability communication are still lacking. Improve both to achieve a broader impact and be felt by all parties.
3. The perception gap between management and visitors, especially in communicating sustainability values, must be bridged. This can be done through innovative communication strategies, the use of digital technology, and active visitor involvement in environmental education and action programs.
4. The authors assess that the demographic profile of the respondents indicates a majority of productive age and secondary education. This demographic profile provides an opportunity to develop mangrove ecotourism. Mangrove ecotourism must be more creative, more participatory, and more adaptive to current developments. The practical implications of this finding require strengthening two-way communication. Strengthening two-way communication means inviting visitors to participate in sustainability actions. Strengthening two-way communication also means increasing transparency and increasing innovation in cross-sector stakeholder engagement. In this way, sustainable marketing strategies are no longer mere jargon. They become a tangible part of the practices and experiences of all parties involved.

5. RESEARCH LIMITATIONS

This study has several limitations that must be considered when interpreting the results and developing further studies. First, we conducted the study in only one area, a mangrove area in Central Java. Therefore, we cannot generalize the findings to mangrove programs in other regions of Indonesia or to other countries with different socio-ecological conditions. This generalization requires further testing (Font & McCabe, 2017; Rasoolimanesh et al., 2022). Second, the cross-sectional study design only examined perceptions at a single point in time. Therefore, changes in perceptions influenced by marketing programs or external events cannot be monitored longitudinally. Third, data were collected via self-report questionnaires. This data can introduce biases in respondents, such as social desirability bias or response bias (Field, 2018). Fourth, the sample of managers and visitors did show demographic variation, but the sample size of managers was small. Therefore, further statistical analysis, such as multivariate regression, is limited. Fifth, the dimensions of sustainable marketing measured were macro-scale. Micro aspects such as individual motivation, risk perception, and dynamics between stakeholders have yet to be explored in depth (Junaidi et al., 2021).

6. PRACTICAL IMPLICATIONS

This research provides practical suggestions for ecotourism managers, policymakers, and sustainable marketers in nature tourism:

1. Improve sustainability communication strategies

The gap in perception between managers and visitors regarding the communication of sustainability values indicates that communication strategies need innovation. Therefore, managers must develop more interactive educational materials. Furthermore, managers must utilize digital technologies, such as social media and interactive applications. Finally, managers must also adopt a storytelling approach to strengthen sustainability messages (Font & McCabe, 2017; Mihalic, 2020).

2. More inclusive stakeholder collaboration

Managers, local communities, government, NGOs, and the private sector must strengthen partnerships to achieve greater program impact and expand marketing networks (Rasoolimanesh et al., 2022). Co-creation through discussion forums, workshops, or CSR programs can foster a sense of ownership and strengthen long-term sustainability (Junaidi et al., 2021).

3. Empowering Local Communities

The findings reveal a high perception of economic and social benefits. The findings provide the basis for expanding training, broadening market access, and developing local mangrove-based products (Arifin et al., 2023). Managers can design community engagement programs for tourism product design, online marketing, and sustainability monitoring.

4. Evolving tourism services

Visitors have different needs. Managers must continuously innovate in services, facilities, and educational and environmental conservation tourism (Lee & Chen, 2021). Managers should utilize feedback systems to capture visitor desires and satisfaction in real time.

7. THEORETICAL IMPLICATIONS

This research makes theoretical contributions to the literature on sustainable marketing and ecotourism:

1. Strengthening the multi-dimensional synergy model

This research emphasizes that economic, social, environmental, communication, and stakeholder engagement benefits are interconnected in sustainable marketing strategies (Kumar & Christodoulopoulou, 2022). Empirical data supports the integrative approach in sustainable marketing theory. We found that achieving sustainability is inseparable from cross-dimensional collaboration and communication (Björk et al., 2021).

2. Contribution of a Comparative Perception Study

This research compares the perceptions of managers and visitors. This research adds to our understanding of the factors influencing the success of sustainable marketing strategies in mangrove ecotourism. This research emphasizes the importance of dialogue and adaptation between stakeholders (Putra et al., 2020; Rasoolimanesh et al., 2022). The following user input is provided:

3. Basis for Developing an Evaluation and Measurement Model

Perception indicators across multiple dimensions guide the development of sustainability measurement instruments. Sustainability measurement instruments are becoming more holistic and applicable in community-based ecotourism (Arifin et al., 2023; Mihalic, 2020).

4. Support for Participation and Co-creation Theory

Research shows that active participation by local communities is crucial. Value co-creation between managers and visitors is crucial to supporting the sustainability of nature tourism destinations (Junaidi et al., 2021).

REFERENCES

Agustina, R. P., & Artanti, Y. (2020). The role of satisfaction as a mediating variable on the effects of novelty seeking and familiarity on tourist revisit intention. *Diponegoro International Journal of Business*, *3*(2), 88–96. DOI: 10.14710/dijb.3.2.2020.88-96

Arief Rahman Halim, M., Soeprobowati, T. R., & Hadiyanto, H. (2023). ECOLOGICAL ASSESSMENT OF THE ROLE OF MANGROVE TREES IN CARBON SEQUESTRATION AND BIODIVERSITY IN KARIMUNJAWA NATIONAL PARK INDONESIA. *Geography, Environment, Sustainability*, *16*(3), 32–42. DOI: 10.24057/2071-9388-2022-2565

Arifin, Z., Rahman, M., & Ridwan, M. (2023). The role of community participation in mangrove ecotourism management in Indonesia. *Ocean and Coastal Management*, *236*, 106235. DOI: 10.1016/j.ocecoaman.2023.106235

Awang, M. Y., Farida, N., Listyorini, S., & Nugraha, H. S. (2026). Revisit Intention in Coastal Tourism: The Interaction of Experience Quality Destination Image, Perceived Value, and Destination Trust. *Institutions and Economies*, *18*(2), 227–253. DOI: 10.22452/IJIE.vol18no2.9

Baiti, M. W. A., Purwanti, F., & Haeruddin, H. (2026). Sustainable Ecotourism Development Strategies in the Mangrove Center Tuban, East Java, Indonesia. *Egyptian Journal of Aquatic Biology and Fisheries*, *30*(2), 433–450. DOI: 10.21608/ejabf.2026.467428.7537

Björk, P., Rahimi, R., & Nemati, R. (2021). Sustainability in tourism marketing: A systematic review. *Sustainability*, *13*(5), 2537. DOI: 10.3390/su13052537

Creswell, J. W., & Creswell, J. D. (2018). *Research Design: Qualitative, Quantitative, and Mixed Methods Approaches* (5th ed.). Sage.

Damayanti, M., Tyas, W. P., & Aswad, W. O. S. J. (2021). Conceptualizing multi-actors' collaboration in smart tourism destination. *IOP Conference Series. Earth and Environmental Science*, *673*(1), 012027. DOI: 10.1088/1755-1315/673/1/012027

Fatimatuzzahroh, F., Hadi, S. P., & Purnaweni, H. (2020). Does the intervention of regional authorities contribute to sustainable mangrove ecotourism? Case study on mangrove management at Karansong, West Java, Indonesia. *Ecological Questions*, *31*(3), 7–14. DOI: 10.12775/EQ.2020.017

Field, A. (2018). *Discovering Statistics Using IBM SPSS Statistics* (5th ed.). Sage.

Font, X., & McCabe, S. (2017). Sustainability and marketing in tourism: Its contexts, paradoxes, approaches, challenges and potential. *Journal of Sustainable Tourism*, *25*(7), 869–883. DOI: 10.1080/09669582.2017.1301721

Junaidi, J., Susilo, D., & Dewi, A. (2021). Stakeholder collaboration in sustainable tourism development: A case study in Indonesia. *Sustainability*, *13*(10), 5555. DOI: 10.3390/su13105555

Kristanti, D., Purnaweni, H., Dwimawanti, I. H., & Yuwono, T. (2024). Harmonizing tourism and nature conservation: A governance model for sustainable tourism villages in Indonesia. *IOP Conference Series. Earth and Environmental Science*, *1414*(1), 012078. Advance online publication. DOI: 10.1088/1755-1315/1414/1/012078

Kumar, V., & Christodoulopoulou, A. (2022). Sustainability and Marketing: A Review of Recent Advances and Future Directions. *Journal of the Academy of Marketing Science*, *50*(3), 542–566. DOI: 10.1007/s11747-021-00830-2

Kurniati, R., Handayani, W., Prabawani, B., Nida, R. S., & Savitri, S. F. (2024). COMMUNITY PARTICIPATION, COMMUNITY READINESS, AND PREFERENCES TO PROMOTE EDU-TOURISM IN MANGROVE AREAS OF TELUK AWUR VILLAGE, JEPARA, INDONESIA. *PLANNING MALAYSIA*, *22*(33). Advance online publication. DOI: 10.21837/pm.v22i33.1555

Lee, T. H., & Chen, C. C. (2021). Stakeholder perspectives on sustainable tourism development. *Tourism Management*, *82*, 104197. DOI: 10.1016/j.tourman.2020.104197

Mihalic, T. (2020). Conceptualising overtourism: A sustainability approach. *Annals of Tourism Research*, *84*, 102969. DOI: 10.1016/j.annals.2020.103025

Nugraha, W. A., Ambariyanto, A., & Insafitri, I. (2024). The mangrove ecosystems on the north coast of Bangkalan, Indonesia, and its potential for mangrove ecotourism. *Biodiversitas (Surakarta)*, *25*(2), 859–868. DOI: 10.13057/biodiv/d250246

Purnaweni, H., Soeprobowati, T. R., Chatterjee, U., Kiswoyo, G. P., Rakhma, N. A., & Manik, J. S. T. (2025). The role of stakeholders in community-based mangrove management in Tapak Area of Semarang City. E3S Web of Conferences, 650, 02047. DOI: 10.1051/e3sconf/202565002047

Putra, A., Dewi, M. H. U., & Saputra, I. (2020). Community-based mangrove ecotourism management. *IOP Conference Series. Earth and Environmental Science*, *485*, 012047. DOI: 10.1088/1755-1315/485/1/012047

Rahman, M., Ridwan, M., & Arifin, Z. (2022). Economic and environmental impacts of mangrove ecotourism in Indonesia. *Environmental Science and Pollution Research International*, *29*, 46789–46799. DOI: 10.1007/s11356-022-19974-5

Rasoolimanesh, S. M., Seyfi, S., Rastegar, R., & Hall, C. M. (2022). A systematic review of sustainable tourism indicators in protected areas. *Journal of Hospitality and Tourism Management, 50*, 250–261. DOI: 10.1016/j.jhtm.2021.12.001

Setyadi, G., Pribadi, R., Wijayanti, D. P., & Sugianto, D. N. (2021). Mangrove diversity and community structure of mimika District, Papua, Indonesia. *Biodiversitas (Surakarta), 22*(8), 3562–3570. DOI: 10.13057/biodiv/d220857

Srifitriani, A., Soeprobowati, T. R., & Puryono, S. (2025). Empowering communities in mangrove ecotourism: a pathway to sustainability and climate resilience in Bengkulu City. E3S Web of Conferences, 605, 03039. DOI: 10.1051/e3sconf/202560503039

Sutanto, H. A., Susilowati, I., Iskandar, D. D., & Waridin, . (2022). Mitigation and adaptation to climate change through sustainable mangrove management on the coast of Rembang Regency. *IOP Conference Series. Earth and Environmental Science, 1036*(1), 012014. Advance online publication. DOI: 10.1088/1755-1315/1036/1/012014

Taber, K. S. (2018). The Use of Cronbach's Alpha When Developing and Reporting Research Instruments in Science Education. *Research in Science Education, 48*(6), 1273–1296. DOI: 10.1007/s11165-016-9602-2

Turisno, B. E., Dewi, I. G. S., Mahmudah, S., & Soemarmi, A. (2021). Recovery policy and proper management of mangrove forests to preserve environmental sustainability and ecotourism in bangkalan Indonesia. *Journal of Environmental Management and Tourism, 12*(8), 2188–2193.

Zhang, Y., Song, H., & Wen, L. (2020). Sustainability in Tourism: A Review of Recent Developments. *Sustainability, 12*(6), 2505. DOI: 10.3390/su12062505

Chapter 16
Transforming Coastal Governance From the Grassroots: Evidence From Brebes, Indonesia

Junjung Sahala Tua Manik
http://orcid.org/0009-0009-6089-2761
Diponegoro University, Indonesia

Amni Zarkasyi Rahman
https://orcid.org/0000-0001-5862-7496
Diponegoro University, Indonesia

Maesaroh Maesaroh
Diponegoro University, Indonesia

ABSTRACT

Mangrove ecosystem degradation in coastal areas is not only an ecological problem, but also a governance, institutional, and livelihood sustainability issue. This article examines the transformation of Pandansari Hamlet, Kaliwlingi Village, Brebes Regency, from a coastal area affected by abrasion, tidal flooding, aquaculture decline, and mangrove degradation into a model of community-based mangrove governance. Using a qualitative case study approach, this study shows how coastal crisis generated ecological awareness, collective action, local institutional strengthening, the formation of Satgas Jaga Segara or mangrove police, culturally grounded environmental campaigns, multi-actor collaboration, and mangrove ecotourism development. The findings suggest that Pandansari's success cannot be reduced to mangrove planting alone, but reflects a broader transformation of

DOI: 10.4018/979-8-3373-9320-9.ch016

coastal governance. Sustainable mangrove restoration requires social legitimacy, local rules, community monitoring, benefit distribution, and adaptive capacity to connect ecological recovery with livelihood resilience.

1. INTRODUCTION

Mangrove ecosystems occupy a strategic position in coastal sustainability because they function as a green belt, a habitat for biota, a carbon sink, and a socio-economic space for coastal communities. Global literature indicates that mangroves are not only ecologically valuable but also provide ecosystem services that support coastal protection, fisheries, adaptation to climate change, and the well-being of local communities (Bimrah et al., 2022; Friess et al., 2019). In the language of environmental policy, mangroves are a nature-based solution: they act as a natural defense that dampens waves, traps sediment, and reduces the risk of coastal disasters at a social cost that is often lower than that of hard infrastructure (Narayan et al., 2016). However, experiences in many coastal regions of Indonesia show that the mangrove crisis is rarely purely ecological in nature. Mangrove degradation is usually intertwined with land-use conversion, the expansion of aquaculture ponds, weak oversight, conflicts of authority, limited community capacity, and development models that place too much emphasis on short-term gains. A study on mangroves in Karangsong, West Java, for example, shows that local government intervention in mangrove ecotourism can create tension if it reduces the space for community participation—which had previously been the driving force behind rehabilitation efforts (Fatimatuzzahroh et al., 2020). The future of mangroves is largely determined by governance: who makes decisions, who protects the areas, who reaps the benefits, and how local regulations are institutionalized.

Dukuh Pandansari, Kaliwlingi Village, Brebes Regency, is a key case for understanding this transformation. This area once faced significant pressure due to the clearing of mangrove land for shrimp and milkfish ponds. Initially, the ponds offered economic hope. In the long term, unsustainable aquaculture practices and the loss of the mangrove buffer zone weakened the co 's resilience against coastal erosion and tidal flooding. The aquaculture ponds were damaged, settlements were threatened, and the community began to experience a decline in their standard of living. This crisis then became a socio-ecological turning point: the community launched a mangrove rehabilitation movement, established local institutions, conducted patrols, developed cultural campaigns, and linked conservation with ecotourism. A similar experience was observed in the management of the Tapak Mangrove in Semarang, which demonstrates that successful mangrove rehabilitation develops through collaboration among various stakeholders and the strengthening of coastal community institutions, although the configuration of actors differs from that in Pandansari (Purnaweni et al., 2025).

This article argues that Pandansari's success cannot be understood merely as a success in mangrove planting. Pandansari is more accurately interpreted as a community-based transformation of coastal governance—a process in which experiences of crisis give rise to ecological awareness, collective action, local institutions, multi-stakeholder collaboration, and livelihood diversification. Within the framework of social-ecological system sustainability, sustainability does not depend on a single variable but rather on the interactions among resources, actors, rules, and social context (Ostrom, 2009). Therefore, the central question of this article is: how did the Pandansari community transform a coastal crisis into community-based mangrove governance that supports social-ecological resilience?

Although the literature on mangrove conservation has grown rapidly, most research still focuses on ecological rehabilitation, ecosystem services, blue carbon, and the effectiveness of vegetation restoration. Studies explaining how governance transformations emerge from community initiatives, evolve into local institutions, and subsequently attract collaboration among various actors remain relatively limited. As a result, the process of transition from community initiatives toward institutionalized collaborative governance remains poorly understood in the literature on coastal governance. This article aims to fill this gap by demonstrating how crisis experiences trigger institutional transformation and give rise to "collaborative governance from below" as the foundation for successful mangrove restoration.

This article's contribution lies in three aspects. First, it positions mangrove restoration as a governance process, not merely as a technical planting project. Second, it introduces the concept of "collaborative governance from below" to explain collaboration that emerges from community initiatives, rather than solely from formal government designs. Third, this article demonstrates that local institutions such as the Jaga Segara Task Force or the "mangrove police" can serve as institutional innovations that bridge conservation, monitoring, education, and social legitimacy.

2. LITERATURE REVIEW AND THEORETICAL FRAMEWORK

2.1 Mangroves as a Socio-Ecological System

Mangroves need to be understood as a socio-ecological system, not merely a collection of coastal vegetation. As a socio-ecological system, mangroves connect ecological dynamics, livelihoods, local norms, institutions, policies, and power relations. Sustainability requires attention to the interconnections between resource systems, resource units, users, and governance systems (Ostrom, 2009). In the context of mangroves, vegetation degradation is often a symptom of deeper institutional problems: weak utilization regulations, a lack of oversight, or imbalances in the distribution of benefits.

Ecological studies show that mangroves serve to protect coastlines, store carbon, provide habitats for biota, and support fisheries (Bimrah et al., 2022; Friess et al., 2019). On the other hand, social studies indicate that mangrove sustainability requires community participation, social learning, and livelihood incentives. In the case of Awur Bay, Jepara, the community's readiness to support mangrove eco-tourism remains heavily dependent on their capacity, participation, awareness of sustainability, and utilization of local resources (Kurniati et al., 2024). This indicates that mangrove eco-tourism cannot be developed solely through physical infrastructure but must also be fostered through social institutions and community learning.

Several studies indicate that coastal areas in Indonesia face overlapping vulnerabilities. Research on coastal settlements in Semarang shows that community resilience to water-related disasters is significantly influenced by adaptive capacity, social networks, and institutional support (Dewi et al., 2021; Muktiali, 2020). Therefore, mangrove restoration in Pandansari should be viewed as an effort to strengthen socio-ecological resilience, not merely as a means of restoring vegetation cover.

2.2 Community-Based Conservation

Community-based conservation positions local communities as the primary actors in natural resource management. Conservation efforts are stronger when communities are given opportunities for participation, receive benefits, and feel a sense of ownership over the resources being protected (Berkes, 2004). This approach rejects the view that communities are merely objects in conservation programs. Instead, communities are recognized as holders of local knowledge, historical experience, social norms, and adaptive capacity. Pandansari embodies this logic. Mangrove rehabilitation did not arise solely from technocratic directives but also from residents' experiences in coping with coastal erosion, tidal flooding, damage to fish ponds, and threats to their homes. The community has an existential reason to protect the mangroves: mangroves are not merely distant ecological entities but living bulwarks that directly safeguard their livelihoods. This is what gives the Pandansari movement its social legitimacy.

Experiences from other coastal areas reinforce this argument. Community participation in the management of Karimunjawa National Park has been a key factor in the effectiveness of conservation area management (Wibowo et al., 2018). A study at Maron Beach, Semarang, also shows that opening recreational areas without careful management can reduce the post-rehabilitation functions of mangroves, including coastal protection and environmental quality (Fithor et al., 2019). Thus, community-based conservation does not mean leaving communities to fend for themselves, but rather ensuring that conservation has social roots, rules, and benefits that residents can tangibly experience.

2.3 Collaborative Governance from Below

Collaborative governance describes the process by which governments, communities, the private sector, universities, and nongovernmental organizations work together to address public issues that cannot be handled by a single actor alone. In the deliberative forums of collaborative governance, the involvement of non-state actors, consensus, and formal processes are considered crucial (Ansell & Gash, 2008). Additionally, three other essential dimensions are principled engagement, shared motivation, and the capacity for joint action (Emerson et al., 2012). The Pandansari case is notable because the collaboration did not originate entirely from the government. The initial initiative actually emerged from the community, which was directly experiencing an ecological crisis that threatened their living space. After the local movement showed results, external actors such as the local government, NGOs, universities, the media, and the private sector began to get involved. Therefore, this article uses the term "collaborative governance from below"—that is, collaboration driven by the legitimacy and commitment of grassroots communities.

The Indonesian context shows that environmental collaboration often faces policy fragmentation, conflicts of interest, and tensions between formal authorities and local practices. In environmental management, the importance of coordination among actors and clarity of responsibilities in responding to environmental damage are the most influential factors (Purnaweni et al., 2022). This occurs because the environmental legislative process at the local level involves advocacy coalitions that bring diverse beliefs, interests, and resources (Alamsyah et al., 2019). In the context of Pandansari, the challenge is to ensure that external collaboration strengthens—rather than replaces—community leadership. A recent study in Tapak, Semarang, shows that the Quadruple Helix model successfully strengthened mangrove conservation by assigning roles to the government as regulator, the community as conservation implementer, academics as knowledge providers, and the private sector as resource providers. However, Pandansari exhibits a different configuration because collaboration was built on community legitimacy before external support was secured (Purnaweni et al., 2025).

2.4 Adaptive Governance and Socio-Ecological Resilience

Adaptive governance is used to understand how communities and institutions respond to ecological uncertainty. Adaptive capacity in community-based resource management is determined by local knowledge, networks, leadership, learning, and access to technical and financial support (Armitage, 2005). Resilience is also crucial. Resilience does not merely mean the ability to return to the original state, but also the ability to absorb disturbances, adapt, and transform the system when old ways are no longer sufficient (Folke, 2006). In a coastal context, community resilience is shaped by the interplay between ecological protection, social preparedness, infrastructure, and policy capacity. For example, the COVID-19 restrictions implemented some time ago demonstrated that adaptive public policies must account for spatial variations and local social contexts (Handayani et al., 2022). Meanwhile, studies on community resilience in Padang and Sukorejo indicate that community resilience is largely determined by a combination of social, economic, and institutional capacities, as well as preparedness for risks (Ardinugroho & Handayani, 2020; Mulia & Handayani, 2024).

Pandansari exemplifies resilience as transformation. The community not only survives coastal erosion and tidal flooding but also transforms these crises into initiatives for conservation, institutional oversight, ecotourism, education, and local economic products. Thus, Pandansari's resilience is not merely the ability to withstand disasters but also the capacity to create a new model of coastal governance.

2.5 Mangrove Restoration, Ecotourism, and Nature-Based Solutions

Mangrove restoration is often narrowly understood as the planting of seedlings. Recent literature indicates that such an approach risks failure if it does not account for hydrology, sedimentation, site suitability, and natural regeneration (van Bijsterveldt et al., 2022). Many mangrove rehabilitation programs fail due to inappropriate sites, an overemphasis on quantitative seedling targets, and inadequate post-planting maintenance (Primavera & Esteban, 2008). In other words, the question of restoration is not merely how many seedlings are planted, but also how ecosystems and communities can sustain the recovery process. From the perspective of nature-based solutions, mangroves serve as a form of coastal defense that integrates ecological and social benefits (Narayan et al., 2016). However, when mangroves are developed for ecotourism, new challenges arise: how to ensure that tourism activities do not transform conservation into a commodity that actually puts pressure on the ecosystem. A study on mangrove edu-tourism in Awur Bay highlights the importance of community readiness and local capacity (Kurniati et al., 2024), while a study on the creative economy in a tourist village in Central Java emphasizes that the creative economy can strengthen community-based tourism when supported by local organization and innovation (Oktavilia et al., 2024).

Thus, Pandansari mangrove ecotourism must be understood as an instrument of conservation and learning, not merely an attraction. Ecotourism becomes sustainable when economic benefits reinvest in mangrove protection, expand community participation, and foster community pride in the coastal landscape. Based on a synthesis of various theoretical perspectives, this study views the transformation of coastal governance as a gradual process that begins with a socio-ecological crisis, then gives rise to ecological awareness among the community, evolves into collective action, shapes local institutions, expands collaborative governance, and ultimately results in socio-ecological resilience. This framework is used to understand how the relationships between ecological change, institutional dynamics, and multistakeholder collaboration develop simultaneously in the Pandansari case.

Figure 1. Analytical pathway from coastal crisis to social-ecological resilience

Figure 1 illustrates the analytical framework used in this study. The framework views the transformation of coastal governance as a gradual process initiated by *a coastal crisis* as the primary trigger for change. This crisis spurred the growth of ecological awareness among the community, which then evolved into collective action to restore the coastal ecosystem. As community participation strengthened, this collective action underwent a process of institutionalization through the formation of local organizations, monitoring mechanisms, community-based rules, and various social practices that strengthened the community's capacity to manage coastal resources.

This institutionalization process subsequently became the foundation for the formation of collaborative governance from below—a model of collaboration that originates from community initiatives and is then strengthened by the involvement of the government, universities, nongovernmental organizations, the media, and the private sector. This collaboration not only supports the success of mangrove rehabilitation but also promotes conservation-based economic diversification through the development of ecotourism and community-based enterprises. Ultimately, the interplay between ecosystem restoration, institutional strengthening, multi-stakeholder collaboration, and local economic development fosters socio-ecological resilience that enhances communities' capacity to address various coastal challenges in a sustainable manner.

This analytical framework is based on a synthesis of the theories of Social-Ecological Systems, Common-Pool Resources, Collaborative Governance, and Community-Based Conservation. The framework is not intended as a linear causal model but rather as an analytical guide to understanding the interrelationships between ecological changes, institutional dynamics, and multi-stakeholder collaboration in explaining the process of coastal governance transformation in Pandansari.

3. RESEARCH METHODOLOGY

This study employs a qualitative case study approach to gain an in-depth understanding of the transformation of mangrove governance in Pandansari Hamlet, Brebes Regency. This approach was chosen because the study focuses on community processes in responding to coastal crises, building collective action, forming local institutions, developing multi-stakeholder collaboration, and integrating conservation with community livelihoods (R. Yin, 2017; R. K. Yin, 2018). Data were collected through semi-structured interviews, field observations, community documents, policy documents, academic publications, and online media. Interviews were conducted with key actors who understood the history of mangrove rehabilitation, ecotourism management, the Jaga Segara Task Force, and the relationships among these actors. Observations were used to assess the ecological conditions of the area, conservation activities, and the interactions between mangrove areas, fish ponds, settlements, and community economic activities. Online media were utilized as secondary data to reconstruct the development of the conservation movement, including environmental campaigns, the formation of the Jaga Segara Task Force, the development of ecotourism, and the various challenges faced. All data sources were analyzed through triangulation to enhance the credibility of the findings (Houghton et al., 2012).

Data analysis was conducted using thematic analysis. The analysis process began with deductive coding based on the study's conceptual framework, namely coastal crisis, ecological awareness, collective action, local institutions, collaborative governance, ecotourism, and socio-ecological resilience. This was followed by inductive coding to identify themes emerging from the field data, such as the Jaga Segara Task Force, cultural campaigns through calung, local regulations, livelihood transformations, and *collaborative governance from below*. This approach allows the research to remain grounded in theory while capturing empirical findings that emerge from the local context (Braun & Clarke, 2006).

4. RESEARCH FINDINGS

4.1 The Coastal Crisis as a Socio-Ecological Turning Point

The transformation of coastal governance in Pandansari began with an ecological crisis that subsequently evolved into a social and economic crisis. The clearing of mangrove areas for the expansion of aquaculture ponds has altered the community's relationship with the coastal ecosystem. In the short term, mangrove conversion provided economic benefits through increased aquaculture production and the expansion of cultivation areas. However, in the long term, the loss of the mangrove buffer zone reduced the coast's natural function as a barrier against erosion, waves, and seawater intrusion. As a result, the shoreline continues to retreat, erosion is eroding aquaculture ponds and farmland, tidal flooding is increasingly inundating settlements, and aquaculture productivity is declining. This situation highlights the paradox of coastal development, where efforts to improve well-being through resource exploitation actually weaken the ecological foundation that supports the sustainability of people's livelihoods.

This crisis not only causes material losses but also changes how communities view the function of mangroves. Previously, mangroves were often perceived as vegetation that hindered the expansion of fish ponds and economic activities. However, the experience of land loss, the increasing frequency of tidal flooding, and declining aquaculture yields have led communities to realize that the presence of mangroves is a prerequisite for coastal security and the sustainability of their livelihoods. In other words, the ecological crisis has evolved into a process of *social learning* that has transformed communities' perceptions of the relationship between ecosystems and well-being.

This shift in perspective became the turning point for the emergence of collective action. Ecological awareness did not arise primarily through public education or government intervention, but rather through direct experience in facing the impacts of environmental degradation. These shared experiences fostered *a shared sense of urgency* that motivated communities to begin rehabilitating mangroves on their own. In this context, the crisis serves as a catalyst for governance transformation—a trigger that drives a shift from exploitative patterns of resource use toward a more collaborative and sustainability-oriented governance model. Thus, the starting point of the transformation in Pandansari was not the introduction of a government program, but rather the growth of a collective awareness among the community that the sustainability of coastal life can only be maintained through ecosystem restoration supported by institutional changes and collective action.

4.2 The Mangrove Rehabilitation Movement and Collective Action

The mangrove rehabilitation movement in Pandansari evolved from community-led planting efforts into an increasingly organized socio-ecological movement. In the early stages, the community planted mangroves as a means of protecting their fish ponds, settlements, and living spaces from the ever-increasing threat of coastal erosion. This initiative arose from the community's own awareness after directly experiencing coastal degradation that threatened their livelihoods. As the mangroves began to grow and their benefits became increasingly evident—whether in reducing coastal erosion, improving environmental conditions, or enhancing the community's sense of security—public support expanded. Mangrove planting is no longer viewed as a temporary project-based activity, but rather as a collective investment to ensure the village's sustainability and the well-being of future generations.

This development demonstrates that collective action does not form instantly but grows through shared experiences in facing environmental crises. The initial success of the rehabilitation efforts fostered confidence that conservation efforts can yield tangible results, thereby encouraging more community members to volunteer. Over time, the planting activities evolved into a collective movement that strengthened mutual aid, trust, and collective responsibility in protecting the mangrove areas. Thus, conservation is no longer viewed merely as a rehabilitation activity, but as part of the community's strategy to maintain the sustainability of their living environment.

This collective action demonstrates that the success of conservation is greatly influenced by the direct connection between environmental issues and the affected communities. The residents of Pandansari took action not because of incentives or formal obligations, but because they directly experienced the consequences of mangrove loss on their daily lives. This experience fostered *a sense of ownership* that motivated the community to protect and manage the mangroves sustainably. Community-based conservation is stronger when communities have social ownership of resources, are involved in decision-making processes, and derive tangible benefits from their management (Berkes, 2004). In the context of Pandansari, this social ownership serves as the primary asset that strengthens community participation and forms the foundation for the establishment of local institutions and the development of *collaborative governance from below*.

Figure 2. Mangrove rehabilitation by the Pandansari community

4.3 Cultural Campaign: Calung as the Language of Conservation

One of the key innovations in the mangrove conservation movement in Pandansari is the use of calung as a medium for environmental campaigns. Conservation messages are conveyed not only through meetings, information boards, or technical outreach, but also through artistic performances that are closely tied to community life. Schoolchildren, youth, and residents are involved in parades, outreach events, and mangrove planting activities. This approach makes conservation not merely an ecosystem rehabilitation effort, but also an integral part of the social and cultural life of coastal communities. The use of the calung demonstrates that environmental messages are more readily accepted when conveyed through a medium that holds emotional resonance for the community. The calung serves as a simple, engaging, and easily understandable means of public communication for people of all ages. The involvement of children also broadens the process of social learning, as conservation values are not only absorbed through formal activities but also disseminated within family and community settings.

From an environmental governance perspective, these findings indicate that governance occurs not only through regulations and bureaucratic forums but also through narratives, symbols, education, art, and everyday practices. In the context of Pandansari, culture functions as a "soft institution" that fosters awareness, a sense of ownership, social legitimacy, and collective responsibility toward mangroves. While regulations govern behavior, culture shapes the reasons why communities are willing to protect the environment voluntarily.

This culture-based strategy also strengthens community-based conservation. Mangroves are no longer viewed merely as coastal vegetation, but as protectors of the village, a source of livelihood, and a legacy for future generations. The involvement of schools, arts groups, families, and community leaders broadens the base of participation while supporting the regeneration of the conservation movement. Furthermore, cultural campaigns enhance the movement's legitimacy in the eyes of the government, universities, nongovernmental organizations, the media, and the private sector. Thus, the calung is not merely a traditional art form but also serves as a language of conservation that connects ecological knowledge, community identity, and cross-sector collaboration. While regulations govern behavior, culture preserves meaning.

Figure 3. Utilizing culture as a medium for environmental campaigns

4.4 The Jaga Segara Task Force, or "Mangrove Police," as an Institutional Innovation

The formation of **the Jaga Segara Task Force**, known to the community as *the "mangrove police,"* is an institutional innovation that marks a pivotal point in the transformation of mangrove governance in Pandansari. The presence of this task force demonstrates that the community does not stop at planting activities but also builds institutional mechanisms to ensure mangroves remain protected after the rehabilitation process. The Jaga Segara Task Force performs various functions, ranging from patrolling the area, monitoring activities that could potentially damage mangroves, educating the community and visitors, to resolving issues through persuasion when violations are found. Thus, mangrove rehabilitation is not viewed as a one-time activity, but rather as a long-term process requiring monitoring, maintenance, and collective commitment.

The existence of the Jaga Segara Task Force demonstrates that conservation success is determined not only by the number of seedlings planted but also by the institutional capacity to ensure the sustainability of rehabilitation outcomes. Many mangrove rehabilitation programs stop at the planting stage, resulting in restored areas suffering damage again due to weak monitoring and the absence of institutions responsible for maintenance. In the context of Pandansari, the community has filled this gap by establishing a community-based monitoring mechanism that operates routinely and sustainably. Monitoring is carried out by residents who live and work around the mangrove area, enabling them to detect potential disturbances more quickly than formal monitoring mechanisms.

This institutional framework also plays a role in establishing local norms and rules regarding mangrove use. Utilization of the area is not entirely prohibited but is regulated through a collective agreement, which emphasizes responsibility, permitting, and the obligation to replant vegetation if it is utilized. This approach demonstrates that community governance does not rely solely on formal sanctions but develops through trust, social reputation, a sense of ownership, and a shared commitment to the area's sustainability. Rules that emerge from the community are easier to comply with because they are built on shared experiences in addressing coastal crises and are viewed as mechanisms to protect collective interests.

These findings indicate that the Jaga Segara Task Force functions as a community-based monitoring institution that strengthens the governance of shared resources. This mechanism aligns with the principles of *common-pool resource* management, which emphasize the importance of rules tailored to local conditions, monitoring by resource users, and compliance built through social norms and shared responsibility (Ostrom, 2009b). In the case of Pandansari, monitoring is not perceived as a form of government control, but rather as a protective mechanism established and operated by the community itself. It is this condition that strengthens the legitimacy of local institutions while also serving as the foundation for the development of **collaborative governance from below**, in which community institutions serve as the starting point that subsequently attracts support from the government, universities, nongovernmental organizations, the media, and the private sector to strengthen sustainable mangrove conservation.

Figure 4. The "Jaga Segara" task force on patrol in the mangrove area

4.5 Multi-Stakeholder Collaboration and Local Capacity Building

Pandansari's transformation cannot be separated from the support of various parties. Local governments, NGOs, universities, the media, environmental communities, and the private sector provided support in the form of seedlings, mentoring, training, promotion, infrastructure, research, and program assistance. This collaboration expanded the capacity of the community, which was initially limited in terms of resources. However, the nature of Pandansari's collaboration differs from a top-down model. The community had already established the initiative and demonstrated its commitment. External support emerged only after the local movement gained legitimacy. In this regard, Pandansari exemplifies collaborative governance from the bottom up. External actors do not take ownership of the movement but rather serve to strengthen the community's capacity. This is crucial because collaboration that is insensitive to local leadership can undermine participation (Fatimatuzzahroh et al., 2020).

Collaboration also carries risks. As the number of stakeholders increases, coordination becomes more complex. The interests of conservation, tourism, corporate social responsibility (CSR), local policies, and the local economy can support one another, but they can also conflict. Therefore, the sustainability of collaboration requires defined roles, communication forums, accountability mechanisms, and the community's capacity to negotiate with external stakeholders.

4.6 Mangrove Ecotourism and Livelihood Diversification

Mangrove growth opens up new economic opportunities. Areas once viewed solely as coastal protectors are now evolving into spaces for education, tourism, and local production. Communities are developing mangrove trekking tours, homestays, culinary experiences, processed fishery products, naturally dyed batik, and the cultivation of crabs, shellfish, and seaweed, as well as environmental education services. This diversification is important because it can reduce dependence on fish ponds, which are vulnerable to erosion, tidal flooding, and changes in environmental quality.

Ecotourism in Pandansari not only generates income but also expands the social value of mangroves. Visitors, schoolchildren, college students, and environmental groups learn that mangroves serve as both ecological infrastructure and living spaces for local communities. Thus, ecotourism acts as an educational tool for the community. However, tourism development must still be carefully managed. Without regulations on visitor capacity, waste management, zoning, and the equitable distribution of benefits, ecotourism can place new pressures on the ecosystem.

Figure 5. Successful mangrove ecotourism developed by the community

5. DISCUSSION

5.1 From Tree Planting to Governance Transformation

Findings in Pandansari confirm that mangrove restoration cannot be viewed merely as a technical activity focused solely on tree planting. Planting is only the initial stage of the recovery process, while the sustainability of restoration depends heavily on the ability to establish a governance framework capable of safeguarding the ecosystem in the long term. The experience in Pandansari demonstrates that the success of rehabilitation stems from the formation of a mutually supportive institutional ecosystem, encompassing mangrove management groups, the Jaga Segara Task Force (mangrove police), local regulations, cultural campaigns, multi-stakeholder collaboration, and conservation-based economic development. The combination of these various elements creates social mechanisms that not only protect mangroves from further damage but also strengthen the community's commitment to making conservation an integral part of daily life. Thus, the process restores not only coastal vegetation but also social and institutional relationships, as well as the community's sense of ownership of the coastal environment.

These findings reinforce the view that successful mangrove restoration must consider the integration of ecological suitability and community social conditions (van Bijsterveldt et al., 2022). Various rehabilitation programs have shown that a high number of seedlings planted does not always correlate directly with restoration success if it is not accompanied by appropriate site selection, sustainable maintenance, and community participation in management (Primavera & Esteban, 2008). In the context of Pandansari, these social dimensions are, in fact, the distinguishing factors that determine the success of restoration. The community serves not only as the implementers of the planting efforts but also as monitors, educators, and beneficiaries of the mangroves' presence. Therefore, the success of restoration in Pandansari demonstrates that coastal ecosystem recovery is a process of governance transformation that integrates ecological, institutional, cultural, and economic aspects into a single conservation system that reinforces one another and is capable of long-term sustainability.

5.2 Mangrove Police and the Governance of the Commons

The existence of the Jaga Segara Task Force, or *"mangrove police,"* can be understood as a tangible form of co-governance in mangrove management. Mangroves are a *common-pool resource* whose benefits can be enjoyed by the entire community, but they are also vulnerable to degradation if their use is not regulated. In such circumstances, the existence of local monitoring mechanisms becomes crucial for maintaining a balance between utilization and conservation. The Jaga Segara Task Force fulfills this role through community-based monitoring rooted in local norms and supported by strong social legitimacy. This approach ensures that monitoring is not perceived as external control but rather as a collective responsibility to protect a shared asset. These findings indicate that the effectiveness of oversight is not determined solely by the severity of formal sanctions, but also by the social closeness between the monitors and the community being monitored.

Members of the Jaga Segara Task Force come from the local community and thus have a deeper understanding of the area's characteristics, mangrove utilization patterns, and the community's social dynamics. These conditions allow issues to be resolved persuasively through education, warnings, and deliberation before they escalate into larger conflicts. Thus, the monitoring function is not only repressive but also preventive and educational, thereby fostering compliance that stems from a shared sense of responsibility.

This finding reinforces the principles of shared resource management, which emphasize the importance of rules tailored to local conditions, monitoring mechanisms, graduated sanctions, and collective conflict resolution. Although the Jaga Segara Task Force does not explicitly adopt this terminology, the practices that have developed on the ground reflect the implementation of these principles through routine monitoring, issuing warnings, educating the community, and controlling activities that have the potential to damage mangroves. Thus, the success of mangrove management in Pandansari demonstrates that local institutions that have earned the community's trust can serve as effective governance mechanisms for safeguarding the sustainability of shared resources, even when government oversight capacity is limited.

5.3 "Collaborative Governance from Below" as a Theoretical Contribution

The concept of "collaborative governance from below" is the main contribution of this study. Most of the literature on collaborative governance explains that collaboration is generally initiated by the government, which establishes forums, invites stakeholders, and facilitates the process of joint decision-making (Ansell & Gash, 2008; Emerson et al., 2012). In contrast, the experience in Pandansari reveals a different pattern. Collaboration did not arise from a government initiative but rather from the community's response to a coastal erosion crisis that directly threatened their living space. The community first established local leadership, developed shared rules, created oversight mechanisms, and demonstrated the success of conservation efforts before securing support from the government, universities, nongovernmental organizations, the media, and the private sector. Thus, the legitimacy of the collaboration was not built through formal mandates but through the experience of crisis, social trust, and the success of collective action.

These findings indicate that communities are not merely beneficiaries or implementers of rehabilitation programs, but are also capable of acting as agenda-setters who determine the direction of coastal management. Once communities possess institutional capacity, trusted leadership, and evidence of success on the ground, external actors will find it easier to build equitable partnerships. In this context, the government does not lose its role but undergoes a transformation in function—shifting from the primary controller to a facilitator, provider of regulatory frameworks, resource connector, and supporter of sustainable collaboration. This shift results in more adaptive relationships because collaboration evolves based on community needs, rather than merely following top-down program designs.

The implications of these findings are crucial for coastal rehabilitation policies in Indonesia. Many conservation programs remain unsustainable because they still treat communities as mere objects of implementation, while planning, decision-making, and resource management processes remain dominated by the government. The experience in Pandansari demonstrates that sustainability is actually easier to achieve when communities are given the space to develop their own initiatives, institutions, and leadership, which are then strengthened through policy support and cross-actor collaboration. Such an approach fosters a stronger *sense of ownership*, ensuring that conservation efforts do not cease when a project or funding ends.

Conceptually, these findings expand the discussion on collaborative governance by demonstrating that collaboration does not always have to originate from the government, but can also grow organically from the community's capacity to address shared challenges. These findings also broaden the direction of mangrove research in Indonesia—which has traditionally focused more on ecological conservation—toward an understanding of institutional transformation, community leadership, and community-based governance as key factors in the success of long-term conservation (Husna et al., 2023).

5.4 Culture, Narratives, and the Legitimacy of Conservation

Findings in Pandansari indicate that conservation success depends not only on technical interventions but also on the ability to translate environmental issues into social language that communities can understand. Issues such as coastal erosion, climate change, blue carbon, and biodiversity often feel abstract to coastal communities when communicated solely through scientific or administrative approaches. Through a cultural campaign using the calung, these messages are translated into experiences closely tied to people's daily lives, so that conservation is no longer viewed merely as an environmental agenda but as an effort to protect their village, livelihoods, and the future of the next generation. In this context, culture serves as a bridge connecting scientific knowledge with the local values that have long been part of the community.

This finding aligns with research showing that public narratives play a crucial role in shaping how communities understand environmental issues while also determining how a solution gains social legitimacy (Hasfi & Fisher, 2025). In Pandansari, the narrative of mangroves as *a bulwark of life* has successfully built a collective identity while strengthening community support for the conservation movement. Through cultural symbols, the involvement of children, and community participation in various conservation activities, conservation values have been transformed into social norms passed down across generations.

Consequently, environmental governance cannot rely solely on regulations, incentives, or formal oversight. Conservation success also requires narrative tools capable of fostering awareness, *a sense of ownership*, and collective commitment to the environment. Regulations can foster compliance, but narratives build conviction; regulations control behavior, while narratives shape the meaning behind that behavior. Therefore, the Pandansari experience demonstrates that the integration of policy instruments and cultural approaches results in governance that is more adaptive, participatory, and sustainable, as conservation evolves into a social movement rather than merely the implementation of a government program.

5.5 Socio-Ecological Resilience and Sustainability Risks

The Pandansari experience demonstrates that socio-ecological resilience is built not only through ecosystem restoration but also through the strengthening of local institutions, social capital, and the community's economic sustainability. Mangrove rehabilitation has successfully reduced coastal vulnerability to erosion while simultaneously restoring the area's ecological functions. At the same time, the conservation movement has strengthened mutual aid, trust, and the community's collective identity as coastal guardians. This resilience is further reinforced by the presence of local institutions, such as the Jaga Segara Task Force, community-based regulations, and the development of ecotourism and mangrove-based products that create diversified livelihoods. Thus, Pandansari's success is measured not only by the increase in mangrove cover but also by the community's growing capacity to adapt, collaborate, and maintain the sustainability of conservation efforts in the face of various changes. However, resilience is not a permanent state. Pandansari still faces various challenges, including ongoing coastal erosion and tidal flooding, the impacts of climate change, the potential for illegal logging, dependence on key community leaders, and the risk of ecotourism commercialization that could undermine conservation goals. These conditions indicate that resilience must be maintained through a continuous process of adaptation, not merely through the success of rehabilitation efforts during a specific period.

Strengthening resilience requires more comprehensive strategies, such as regenerating community leadership, ensuring equitable distribution of economic benefits, managing tourism carrying capacity, integrating conservation programs with local government policies, and conducting continuous monitoring of ecological conditions. These findings reinforce the view that mitigation efforts in coastal areas require a combination of complementary structural and nestructural approaches (Wungo et al., 2024). In the context of Pandansari, physical coastal protection through mangrove rehabilitation will only be sustainable if supported by strong local institutions, consistent community participation, and collaborative governance capable of adapting to future challenges. Thus, socio-ecological resilience is not merely the end result of conservation, but rather the collective capacity of the community to continuously learn, adapt, and maintain the sustainability of coastal ecosystems.

5.6 Why Did Pandansari Succeed?

Research findings indicate that Pandansari's success was not due to a single factor, but rather to a combination of several mutually reinforcing conditions. First, the erosion crisis created a shared sense of urgency, leading to conservation being viewed as a collective need rather than merely a government program. Second, local leadership emerged that was capable of organizing collective action and building community trust. Third, the formation of local institutions such as the Jaga Segara Task Force and community rules ensured the sustainability of conservation efforts after planting was completed. Fourth, the initial success of the rehabilitation efforts enhanced social legitimacy, encouraging the government, universities, nongovernmental organizations, the media, and the private sector to provide support. Fifth, the integration of conservation with ecotourism and local businesses creates economic benefits that strengthen community incentives to continue protecting mangroves. The combination of these five factors demonstrates that conservation success is the result of a governance transformation process, not merely the success of ecological rehabilitation.

This finding also opens opportunities for the development of collaborative governance theory, particularly in the context of coastal communities in developing countries. Until now, collaborative governance has generally been understood as a process initiated by the government through formal forums. The Pandansari experience shows that collaboration can also develop organically from the community's capacity to respond to crises, subsequently gaining legitimacy and support from external actors. Thus, "collaborative governance from below" can be understood as an alternative pathway for establishing collaborative governance that warrants further testing in various other socio-ecological contexts.

6. CONCLUSION

This article demonstrates that Pandansari's transformation from a coastal area experiencing erosion, tidal flooding, and the decline of fish ponds into a community-based mangrove governance system is a complex socio-ecological process. Pandansari's success lies not solely in the number of mangroves planted, but in the community's ability to foster ecological awareness, collective action, institutional oversight, local regulations, multi-stakeholder collaboration, and a conservation economy.

The main findings of this article include five points. First, the coastal crisis served as a catalyst for ecological awareness and community mobilization. Second, the mangrove rehabilitation movement evolved into collective action involving residents, schoolchildren, local groups, and community leaders. Third, the Satgas Jaga Segara, or "mangrove police," emerged as a crucial institutional innovation for safeguarding mangroves after planting. Fourth, multi-stakeholder collaboration strengthens local capacity, but its success depends on the ability to maintain community leadership. Fifth, mangrove ecotourism and local products broaden economic benefits, thereby giving conservation stronger social legitimacy.

Theoretically, this article contributes to the literature on community-based conservation, collaborative governance, and adaptive governance by demonstrating that sustainable mangrove restoration requires the integration of ecology, institutions, culture, and the local economy. The concept of "collaborative governance from below" is a key contribution because it explains collaboration that arises from community initiatives and is subsequently strengthened by external support. Practically, the Pandansari experience teaches that mangrove rehabilitation programs must be designed as long-term governance initiatives: strengthening community institutions, local regulations, community oversight, environmental education, livelihood integration, and collaborative networks. This study has limitations because it focuses on a single case. Further research could compare Pandansari with other mangrove villages in Indonesia, quantitatively measure the economic impact on households, evaluate the sustainability of mangrove policies, and analyze the role of gender in mangrove ecotourism.

More broadly, the Pandansari experience demonstrates that the future of mangrove restoration is not primarily determined by the number of seedlings planted, but rather by the ability to build local institutions that earn community trust, strengthen multistakeholder collaboration, and link conservation with community well-being. Thus, governance transformation is a key prerequisite for the sustainability of coastal restoration.

REFERENCES

Alamsyah, A., Suwitri, S., Kismartini, K., & Yuwanto, Y. (2019). Applying the Advocacy Coalition Framework to Understand the Process of Local Legislation in Indonesia. *Sriwijaya Law Review*, 34–47. DOI: 10.28946/slrev.Vol3.Iss1.136.pp34-47

Ansell, C., & Gash, A. (2008). Collaborative governance in theory and practice. *Journal of Public Administration: Research and Theory*, *18*(4), 543–571. DOI: 10.1093/jopart/mum032

Ardinugroho, N. S., & Handayani, W. (2020). Landslide community resilience: An examination of six neighborhoods in Sukorejo, Semarang. *IOP Conference Series. Earth and Environmental Science*, *447*(1), 012015. Advance online publication. DOI: 10.1088/1755-1315/447/1/012015

Armitage, D. (2005). Adaptive capacity and community-based natural resource management. In *Environmental Management* (Vol. 35, No. 6, pp. 703–715). DOI: 10.1007/s00267-004-0076-z

Berkes, F. (2004). Rethinking community-based conservation. In *Conservation Biology* (Vol. 18, No. 3, pp. 621–630). DOI: 10.1111/j.1523-1739.2004.00077.x

Bimrah, K., Dasgupta, R., Hashimoto, S., Saizen, I., & Dhyani, S. (2022). Ecosystem Services of Mangroves: A Systematic Review and Synthesis of Contemporary Scientific Literature. In *Sustainability (Switzerland)* (Vol. 14, No. 19). MDPI. DOI: 10.3390/su141912051

Braun, V., & Clarke, V. (2006). Using thematic analysis in psychology. *Qualitative Research in Psychology*, *3*(2), 77–101. DOI: 10.1191/1478088706qp063oa

Dewi, S. P., Ristianti, N. S., & Kurniati, R. (2021). Coastal settlement resilience to water-related disasters in Semarang City. *IOP Conference Series. Earth and Environmental Science*, *623*(1), 012067. Advance online publication. DOI: 10.1088/1755-1315/623/1/012067

Emerson, K., Nabatchi, T., & Balogh, S. (2012). An integrative framework for collaborative governance. *Journal of Public Administration: Research and Theory*, *22*(1), 1–29. DOI: 10.1093/jopart/mur011

Fatimatuzzahroh, F., Hadi, S. P., & Purnaweni, H. (2020). Does the intervention of regional authorities contribute to sustainable mangrove ecotourism? A case study on mangrove management at Karansong, West Java, Indonesia. *Ecological Questions*, *31*(3), 7–14. DOI: 10.12775/EQ.2020.017

Fithor, A., Sutrisno, J., & Indarjo, A. (2019). Mangrove Ecosystem Management Strategy at Maron Beach, Semarang. *Ilmu Kelautan : Indonesian Journal of Marine Sciences, 23*(4), 156–162. DOI: 10.14710/ik.ijms.23.4.156-162

Folke, C. (2006). Resilience: The emergence of a perspective for social-ecological systems analyses. *Global Environmental Change, 16*(3), 253–267. DOI: 10.1016/j.gloenvcha.2006.04.002

Friess, D. A., Rogers, K., Lovelock, C. E., Krauss, K. W., Hamilton, S. E., Lee, S. Y., Lucas, R., Primavera, J., Rajkaran, A., & Shi, S. (2019).. . *The State of the World's Mangrove Forests: Past, Present, and Future., 7*, 16. DOI: 10.1146/annurev-environ-101718

Handayani, W., Insani, T. D., Fisher, M., Gim, T. H. T., Mardhotillah, S., & Adam, U. E. (2022). Effects of COVID-19 restriction measures in Indonesia: A comparative spatial and policy analysis of selected urban agglomerations. *International Journal of Disaster Risk Reduction, 76*, 103015. Advance online publication. DOI: 10.1016/j.ijdrr.2022.103015 PMID: 35601393

Hasfi, N., & Fisher, M. R. (2025). Narrating Vulnerability and Reshaping Resilience: Media Discourse and Political Coalitions on the Jakarta Floods. *Environmental Communication, 19*(4), 688–706. DOI: 10.1080/17524032.2024.2445565

Houghton, C., Casey, D., Shaw, D., & Murphy, K. (2012). Rigour in qualitative case-study research. *Nurse Researcher*. www.nurseresearcher.co.uk

Husna, J., Wijayanti, I., Budiastusi Wiratmo, L., Indrahti, S., Naryoso, A., Edzan Che Nasir, N., Patria Ratna, M., Mulyaning Andini, B., & Raihan Putra Ratarno, D. (2023). Mapping the Scientific Literature on Mangrove Conservation in Indonesia: A Bibliometric Analysis for Environmental Research. *E3S Web of Conferences, 448*. DOI: 10.1051/e3sconf/202344803036

Kurniati, R., Handayani, W., Prabawani, B., Shofa Nida, R., & Savitri, S. F. (2024). Community Participation, Community Readiness, and Preferences to Promote Edu-Tourism in Mangrove Areas of Teluk Awur Village, Jepara, Indonesia. In *PLANNING MALAYSIA Journal of the Malaysia Institute of Planners* (Vol. 22).

Muktiali, M. (2020). The Resilience of the Coastal Communities of Semarang City in the Face of Floods and Tides Due to the Effects of Climate Change. *IOP Conference Series. Earth and Environmental Science, 448*(1), 012093. Advance online publication. DOI: 10.1088/1755-1315/448/1/012093

Mulia, F. A., & Handayani, W. (2024). Assessment and Comparison of Community Resilience to Floods and Tsunamis in Padang, Indonesia. *Journal of Integrated Disaster Risk Management, 14*(1), 74–97. DOI: 10.5595/001c.115826

Narayan, S., Beck, M. W., Reguero, B. G., Losada, I. J., Van Wesenbeeck, B., Pontee, N., Sanchirico, J. N., Ingram, J. C., Lange, G. M., & Burks-Copes, K. A. (2016). The effectiveness, costs, and coastal protection benefits of natural and nature-based defenses. *PLoS One*, *11*(5), e0154735. Advance online publication. DOI: 10.1371/journal.pone.0154735 PMID: 27135247

Oktavilia, S., Firmansyah, Suharno, Fafurida, Rahman, Y. A., & Adzim, F. (2024). How Does the Development of the Creative Economy Serve as a Driving Force in Community-Based Tourism Village Development?: An Empirical Study in Central Java Province. *Journal of Ecohumanism*, *3*(7), 5008–5018. DOI: 10.62754/joe.v3i7.4609

Ostrom, E. (2009a). A General Framework for Analyzing Sustainability of Social-Ecological Systems. In *Science* (Vol. 325, No. 5939, pp. 416–419). DOI: 10.1126/science.1172133

Ostrom, E. (2009b). A general framework for analyzing sustainability of social-ecological systems. In *Science* (Vol. 325, No. 5939, pp. 419–422). DOI: 10.1126/science.1172133

Primavera, J. H., & Esteban, J. M. A. (2008). A review of mangrove rehabilitation in the Philippines: Successes, failures, and future prospects. *Wetlands Ecology and Management*, *16*(5), 345–358. DOI: 10.1007/s11273-008-9101-y

Purnaweni, H., Saputra, J., Roziqin, A., Kismartini, K., Djumiarti, T., & Seitz, T. (2022). Oil Spill Governance: Evidence from Bintan Island, Indonesia. *Sustainability (Basel)*, *14*(3), 1603. Advance online publication. DOI: 10.3390/su14031603

Purnaweni, H., Soeprobowati, T. R., Chatterjee, U., Kiswoyo, G. P., Rakhma, N. A., & Manik, J. S. T. (2025). The Role of Stakeholders in Community-Based Mangrove Management in the Tapak Area of Semarang City. *E3S Web of Conferences, 650*. DOI: 10.1051/e3sconf/202565002047

van Bijsterveldt, C. E. J., Debrot, A. O., Bouma, T. J., Maulana, M. B., Pribadi, R., Schop, J., Tonneijck, F. H., & van Wesenbeeck, B. K. (2022). To Plant or Not to Plant: When Can Planting Facilitate Mangrove Restoration? *Frontiers in Environmental Science*, *9*, 690011. Advance online publication. DOI: 10.3389/fenvs.2021.690011

Wibowo, B. A., Aditomo, A. B., & Prihantoko, K. E. (2018). Community Participation in the Management of the National Park in the Coastal Area of Karimunjawa Island. *IOP Conference Series. Earth and Environmental Science*, *116*(1). Advance online publication. DOI: 10.1088/1755-1315/116/1/012051

Wungo, G. L., Dewi, S. P., Mussadun, , & Riswananda, A. H. (2024). The Integration of Structural and Non-Structural Mitigation Based on a Collaborative Governance Process in Tanjung Mas Subdistrict. *IOP Conference Series. Earth and Environmental Science*, *1404*(1), 012039. Advance online publication. DOI: 10.1088/1755-1315/1404/1/012039

Yin, R. (2017). *Case Study Research and Applications: Design and Methods*. https://consensus.app/papers/case-study-research-and-applications-design-and-methods-yin/9a06266cf63e5e80a397f7ba9629a5f7/

Yin, R. K. (2018). *Case Study Research and Applications* (6th ed.). Sage Publications.

About the Contributors

Hartuti Purnaweni is a Professor at the Faculty of Social and Political Sciences, Universitas Diponegoro, Indonesia. Her research interests focus on environmental policy, public administration, and governance, with a strong emphasis on sustainable development and natural resource management. She has published extensively in national and international journals, contributed to book chapters, and actively engaged in research collaborations addressing environmental governance challenges in Indonesia and beyond. In addition to her academic work, she has served as a reviewer and editorial board member in several scholarly journals and is frequently invited as a resource person in policy forums and conferences on environmental issues.

Ali Roziqin is a Lecturer in the Department of Government Studies, Universitas Muhammadiyah Malang, Indonesia. He obtained his bachelor's degree from the Department of Public Administration at Universitas Diponegoro, Indonesia. He holds a Master of Public Administration from the Universitas Gadjah Mada, Indonesia. Hisresearch interests are in bureaucracy, public policy and administration, and public sector innovation. He has published papers in the reputable international journals like Sustainability (MDPI), Local Environment (Taylor and Francis), Tourism Critiques, Journal of Science and Technology Policy Management, Journal of the Knowledge Economy, and Public Administration and Policy: An Asia-Pacific Journal, as well as in the IGI Global

Nilam Adini Rakma is lecturer in the Department of Public Administration, Diponegoro University. She completed his Bachelor of Public Administration education at the Faculty of Social and Political Sciences (FISIP) Diponegoro University, graduating in 2015. She got Masters degree in the Master of Public Administration Program at the Faculty of Social and Political Sciences (FISIP) Diponegoro University, graduating in 2021. She has experience working as an HR consultant and Supervisor of the Learning & Development section of the Human

Resource division for more than 5 years. Seh has expertise in the field of human resource management and has started to actively write national and international books and journals

Setyo Budi Kurniawan is listed as World's Top 2% Scientist by Elsevier in the subject of Environmental Sciences (2023-2024). He is interested in the topic of environmental remediation, wastewater treatment, and circular economy. He is currently a postdoctoral research at Department of Chemical and Process Engineering, The National University of Malaysia, as well as researcher in Research Centre for Environment and Clean Technology, National Research and Innovation Agency of Indonesia.

***

Nur Ain Amani Abdul Mubin is a researcher in marine science with a focus on coral reef ecology and photophysiology. Her work centers on understanding how environmental stressors, particularly light and temperature effect on coral health, with an emphasis on turbid reef systems in Malaysia. She has conducted experimental and field-based studies on coral photoacclimation and resilience, contributing to broader insights on how reefs respond to multiple stressors. She has been involved in interdisciplinary initiatives linking ocean observation, data integration, and policy development, and has participated in international training on statistical analysis of oceanographic data. Her research interests also extend to blue carbon ecosystems and nature-based solutions for climate mitigation.

Alfiyah is an undergraduate student in the Department of Public Administration, Faculty of Social and Political Sciences, Diponegoro University, Indonesia. She is currently pursuing her bachelor's degree in Public Administration and has shown academic interest in governance and public policy issues. As a student, she has been involved in academic writing and research activities related to public administration. Through her academic engagement, she continues to develop her interest in the study and practice of governance in Indonesia.

Muhammad Ansy Alghasi is an experienced strategy consultant and researcher with a strong track record at leading institutions such as Boston Consulting Group, Kearney, and Redseer. A master's graduate from the University of New South Wales and a bachelor's graduate from the Bandung Institute of Technology, he possesses deep expertise in data analytics, business strategy, public policy, and petroleum engineering. In addition to his corporate career, he is actively involved as a cofounder of Lensa Garuda Nusantara, which focuses on public policy research and social projects, and has authored various international scholarly publications.

Muhammad Ansy Althafzufar is a Master of Laws student at Dr. Soetomo University in Surabaya who is actively engaged in legal and public policy research. His areas of focus include criminal law, civil law, renewable energy, and social justice. Through his various contributions to international journals and conferences, he consistently critiques the criminalization of vulnerable groups. He strives to formulate legal policy recommendations that are adaptive, equitable, and uphold human rights.

Muhamad Nur Chozin is an assistant lecturer in the Department of Public Administration at Universitas Negeri Yogyakarta. His expertise lies at the intersection of public policy and digital governance, with a primary focus on social and educational issues. He earned his Master's degree in Public Policy and Management from Universitas Gadjah Mada. In addition to his academic role, he works as a policy consultant at the Center for Education and Policy Research. You can reach him at muhamad.nur.chozin@uny.ac.id.

Niamjit Das is a Research Fellow at the Center for Natural Resource Studies in Dhaka. He completed his B.Sc. and M.Sc. in Forestry from the Department of Forestry & Environmental Science at Shahjalal University of Science & Technology, Bangladesh. With over 12 years of research experience, he has led and contributed to national and international funded projects, authored fifteen peer-reviewed journal articles, multiple book chapters, delivered numerous oral and poster presentations, and various newspaper and magazine pieces. His expertise spans forest and community ecology, biodiversity conservation, and climate change adaptation, employing computational tools and interdisciplinary strategies to deepen understanding of forest dynamics and ecosystem function.

Salma Tarek Eid Esmael is a lawyer at Salma Tarek Eid Law Firm in Egypt. She was born in Egypt on February 24, 1993. She completed her doctoral studies in Public Administration at the Faculty of Administrative Sciences, Universitas Brawijaya, Indonesia. Prior to that, she completed her bachelor's and master's degrees at the Faculty of Law, Cairo University. She has also worked as a lawyer for Emtelak (Nevada).

Yonis Gulzar is an Associate Professor at King Faisal University, Saudi Arabia. He received his Ph.D. in Computer Science from the International Islamic University Malaysia. His research focuses on artificial intelligence, deep learning, database systems, and AI applications in healthcare and agriculture. He has published extensively in journals, conferences, and books and has led multiple funded research projects.

Agung Wahyu Handaru is a Associate Professor and senior academic at Universitas Negeri Jakarta, Indonesia, specializing in management, organizational behavior, and strategic marketing. He has established a strong academic reputation through extensive research contributions in leadership, organizational commitment, employee performance, entrepreneurship, and digital transformation. He has published numerous articles in reputable national and international journals, covering diverse topics such as transformational leadership, job satisfaction, work engagement, entrepreneurial intention, and organizational communication. His research also extends to emerging themes including digital marketing, sustainability, tourism marketing, and organizational innovation. Several of his publications have received substantial citations, particularly in leadership, employee performance, and organizational behavior research, highlighting his contribution to advancing management knowledge and practice.

Maesaroh Herawati is a lecturer in Public Service Management at the Department of Public Administration, Faculty of Social and Political Sciences, Universitas Diponegoro, Indonesia. She is a member of the Public Governance Expertise Group. Her academic and research interests include public service management, public governance, service innovation, and public sector performance. Her work contributes to the development of effective, accountable, responsive, and citizen-oriented public administration.

Rizki Hidayati is a lecturer in the Department of Public Administration, Faculty of Social and Political Sciences, Universitas Negeri Yogyakarta. She completed her Bachelor's and Master's degrees in Public Administration at Universitas Diponegoro. Her academic and research interests focus on governance, community-based development, and innovation and collaboration in sustainable development, with particular emphasis on the tourism, social, and gender sectors.

Raja Iqbal Islamy is a Bachelor of Law graduate from the Islamic University of Indonesia with professional expertise as an advocate and legal consultant. Actively involved in legal advocacy, research, and legal assistance across civil law, criminal law, constitutional law, and human rights law. Actively involved in social and cultural research while consistently contributing written works to both online and print media and currently serving as a board member of the Legal Aid Center of the Indonesian Advocates Association (PERADI), Gresik Branch, for the 2025–2028 term, and as Vice Chair of the Advocates Division of the PERADI Gresik Branch Executive Board for the 2025–2028 term.

Wulan Kinasih is a lecturer at Universitas 17 Agustus 1945 Semarang. She completed her doctoral studies in Public Administration at the Faculty of Administrative Sciences, Universitas Brawijaya, Indonesia. She is a recipient of the Indonesian Education Scholarship (BPI) from the Indonesian Ministry of Higher Education, Science, and Technology. Her research interests focus on sustainable tourism governance and local community empowerment. He wrote a book titled "Rahayuning Reyog Ponorogo: Khazanah Budaya Indonesia yang Mendunia".

Kismartini is a Professor in the Department of Public Administration, Faculty of Social and Political Sciences, Diponegoro University, Indonesia. She has extensive experience in teaching, research, and academic development in the field of public administration. Her scholarly interests include collaborative governance, coastal and environmental governance, public policy, and sustainable development. She has published widely in national and international journals, books, and other academic publications. Through her academic and research contributions, she has actively advanced the study and practice of public administration, particularly in relation to governance innovation, stakeholder collaboration, and sustainability-oriented policy.

Gandisa Pangestuti Kiswoyo is a lecturer and researcher at the Department of Public Administration, Faculty of Social and Political Sciences, Yogyakarta State University. She earned her Bachelor's and Master's degrees in Public Administration from the Faculty of Social and Political Sciences, Diponegoro University. Her research interests focus on disaster management and environmental governance in the public sector.

Asih Widi Lestari is a Lecturer in the Department of Public Administration, Faculty of Social and Political Sciences, Diponegoro University, Indonesia. She holds a doctoral degree and has academic expertise in public policy. She is actively engaged in teaching, research, and community service in the field of public administration, with a particular focus on public policy studies. Through her academic work, she contributes to the development of public administration scholarship and policy-related research in Indonesia.

Sari Listyorini is a lecturer and researcher in the Department of Business Administration, Faculty of Social and Political Sciences, Universitas Diponegoro, Indonesia. Her teaching and research focus on marketing management, consumer behavior, digital marketing, and business strategy. She has actively participated in numerous national and international research collaborations and has published scholarly articles in reputable academic journals. In addition to her research activities, Dr. Listyorini is committed to teaching, supervising undergraduate and postgraduate students, and contributing to community engagement programs. Her research emphasizes the application of marketing theories to contemporary business challenges, including consumer decision-making, digital transformation, and sustainable business practices. Through academic collaboration and interdisciplinary research, she continues to contribute to the advancement of business administration and marketing knowledge while supporting the development of innovative and socially responsible business practices.

Sumaira is a Lecturer of Economics at the Forman Christian College (A Chartered University)'s Department of Economics. She has more than 10 years of teaching and research experience in Consumer psychology, Tourism economics, Environmental economics, Religious extremism, and Research Methods fields. She has also worked as a Lecturer of Economics at Qarshi University, Lahore, School of International Studies in Sciences and Arts (SISA) and Bahaudin Zakriya University, Multan. She has published research articles in well-reputed international and national journals of economics and social sciences. Moreover, she has expertise in data analytical softwares including STATA, EViews, QGIS, Python and SPSS.

Junjung Sahala Tua Manik, S.Sos., M.A.P., is a lecturer in the Department of Public Administration, Faculty of Social and Political Sciences, Universitas Diponegoro, Indonesia. He is affiliated with the Public Governance Expertise Group. His research interests include environmental governance, indigenous community governance, environmental health policy, and community-based environmental management. His academic work focuses on collaborative governance, environmental sustainability, and the role of local communities in addressing public and ecological challenges.

Ruth Stany Melisa works at the intersection of media, power, and digital public life. She is currently completing her Master's degree in Political Communication at Universitas Paramadina, with a focus on how people encounter, interpret, and respond to political realities through everyday communication. Her work grows from a simple curiosity: why certain narratives stay, while others disappear. This curiosity has led her to explore digital discourse, symbolic resistance, and the ways popular culture quietly shapes political meaning. Rather than treating communication as a neutral process, she approaches it as a space where meanings are continuously negotiated, contested, and lived. Alongside her academic path, she is involved in communication work with public institutions and civil society initiatives, particularly on issues related to environmental narratives and digital media. These experiences keep her research grounded, while also opening new questions about how communication operates beyond theory. She is interested in producing work that does not stop at explanation, but invites reflection, work that remains attentive to the complexities of power, while still connected to the realities people navigate every day.

Mahadi Mohammad is the Director of the Centre for Marine and Coastal Studies (CEMACS), Universiti Sains Malaysia. A marine biologist with more than 20 years of experience, he focuses on coral reef ecology, marine ecosystem resilience, and the responses of marine organisms to environmental stress, particularly within turbid, human-influenced tropical coastal waters. He leads multidisciplinary research on coral reefs, seagrass ecosystems and the sustainable blue economy, translating scientific evidence into practical conservation, policy and coastal management solutions.

Norhanis Razalli is a Research Officer and marine environmental consultant based in Malaysia. Her work focuses on the development and long-term sustainability of marine reserves, particularly around the islands and proposed marine sanctuaries of Penang, a coastal state in northwestern Peninsular Malaysia. Her research centres on seagrass ecology and blue carbon, supporting science-based conservation and marine management in line with the United Nations Sustainable Development Goals.(SDGs) for 2030

Aga Natalis is a lecturer and researcher at the Faculty of Law, Diponegoro University, Indonesia. He earned his bachelor's, master's, and doctoral degrees in law from the same university. His scholarship focuses on feminist legal theory, ecofeminism, law and gender, prostitution policy, socio-legal studies, and progressive law. He also serves in several editorial roles for legal journals and reviews manuscripts for international publications. His work combines critical legal inquiry with a strong commitment to gender justice and social transformation.

P. Selvakumar currently serves as an Associate Professor in the Department of Science and Humanities at Nehru Institute of Technology, Coimbatore, Tamil Nadu, India. With a teaching career spanning over a decade, he previously held positions as an Assistant Professor in the Department of Science and Humanities at various engineering colleges in Coimbatore from 2011 to 2023. Dr. P. Selvakumar's primary area of academic and research interest is phytochemistry, where he has made substantial contributions. His research credentials include 30 journal publications, 14 patents, and 25 conference papers. He has actively promoted faculty and student development by organizing 5 Faculty Development Programs (FDPs) and completing 3 NPTEL courses along with 60 Coursera certifications. His dedication to student success is evident in his role as a committed mentor, consistently guiding and supporting his students academically and personally. Recognized for his excellence in teaching, Dr. P. Selvakumar has received numerous accolades, including Best

Teacher Awards. He is also an active member of MISTE (The Indian Society for Technical Education) and continues to play a vital role in both acade

Robetmi Jumpakita Pinem, S.AB., MBA., Ph.D. is an Assistant Professor in the Department of Business Administration, Faculty of Social and Political Sciences, Universitas Diponegoro, Indonesia. He earned a Ph.D. in International Trade from Kangwon National University, South Korea, an MBA from National Taiwan University of Science and Technology, Taiwan, and a Bachelor of Business Administration from the University of Sumatera Utara, Indonesia. His research interests include international business, digital marketing, green marketing, entrepreneurship, SMEs, consumer behavior, and artificial intelligence in business. He has published extensively in reputable international journals, including SSCI- and Scopus-indexed publications, and has led numerous nationally and internationally funded research projects. Dr. Pinem actively teaches undergraduate and postgraduate students, supervises research, and promotes international academic collaboration. His academic achievements include the Outstanding Academic Award from Kangwon National University in 2025 and several prestigious scholarships recognizing his academic excellence.

Ani Purwanti is a Professor of Law at Diponegoro University, Indonesia. She earned her bachelor's and master's degrees from Diponegoro University and completed her doctorate at Universitas Indonesia. Her academic work focuses on gender and law, women's empowerment, women fishers, migrant workers, reproductive rights, and sustainable development. She has led research projects and community engagement programs addressing gender inequality, coastal communities, and legal protection. Her scholarship reflects a strong commitment to gender justice and socially responsive legal development.

Ramya Raghavan is currently serving as faculty at Sri Sathya Sai University for Human Excellence, India. Dr Raghavan's research interests include Genetics, Biostatistics, Biophotons, Epigenetic factors, Intellectual disability, and Assistive Technologies. Her recent editorial work on a book dedicated to advancing cancer treatment through spatial omics approaches underscores her commitment to innovation. As an experienced editor specializing in medical and scientific publications, Dr. Raghavan possesses a unique ability to bridge complex scientific concepts with broader societal impact. Her expertise extends to the critical intersection of the exposome and Social Determinants of Health (SDOH). Dr. Raghavan's work underscores the importance of a holistic understanding of environmental exposures, from the molecular level to population-wide influences. Currently working on various early developmental programming projects, aligning with the Developmental Origins of Health and Disease (DOHaD) paradigm, and also addressing health disparities and an equitable approach to disease prevention.

Amni Zarkasyi Rahman is an Associate Professor and the Head of the Doctoral Program in Public Administration at Universitas Diponegoro. Specializing extensively in public governance, his scholarly work covers institutional capacity, government-community collaboration, public policy, and local sustainable development. Bridging theory and practice, he actively serves as a policy advisor and expert assessing development planning documents and government governance in Indonesia. He contributes significantly to the discipline through numerous books, academic monographs, and peer-reviewed articles in prestigious national and international journals.

Rakshitha Ramesh, Student in the Department of Life Sciences, Sri Sathya Sai University for Human Excellence, India. 585313.

Sazlina Salleh is an Associate Professor at Universiti Sains Malaysia with over 15 years of experience in marine ecology and conservation. Her research focuses on the impacts of climate change on marine photosynthetic organisms, including microalgae, corals, and seagrasses. She has led numerous nationally and internationally funded projects and published extensively in reputable journals. As Deputy Director of the Centre for Policy Research, she promotes interdisciplinary collaboration to produce research that supports effective policymaking. She is also committed to teaching, mentoring future marine ecologists, and raising public awareness of marine conservation through community outreach.

Yoga Aldi Saputra is a Lecturer in the Department of Public Administration, Faculty of Social and Political Sciences, Universitas Diponegoro, Indonesia. He obtained his Master's degree in Public Policy and Management from Universitas Gadjah Mada under the Indonesia Endowment Fund for Education (LPDP) Scholarship. His research focuses on environmental governance, climate change adaptation, coastal governance, ecosystem restoration, and sustainable tourism, with particular attention to how governance shapes resilience and community livelihoods in coastal regions of Southeast Asia. His current research examines the governance of coastal dune restoration and ecosystem-based climate adaptation in Indonesia. His work has been published in international peer-reviewed journals, including the Journal of Tourism Futures (Q1, Emerald Publishing), and presented at international academic forums, including the Asia Research Institute, National University of Singapore.

Doni Sugianto Sihotang is a doctoral student in Marketing at Jakarta State University, Indonesia. His academic background combines a Bachelor's degree in Social Anthropology from Universitas Sumatera Utara and a Master's degree in Marketing Management from Jakarta State University. His research interests focus on consumer behavior, digital marketing, tourism marketing, and behavioral analytics. He has actively contributed to several peer-reviewed publications, exploring topics such as consumer decision-making, destination branding, and electric vehicle adoption. He also serves as a Graduate Teaching Assistant and Research Assistant at Jakarta State University, supporting research in marketing and behavioral sciences. Doni has received several national awards, including the BIMA DRTPM Research Grant from the Indonesian Ministry of Education and the PMDSU Fast-Track Scholarship for outstanding young researchers. His interdisciplinary background in anthropology and marketing enables him to integrate cultural and behavioral insights into the study of consumer decision processes and sustainable tourism development.

Dina Miza Suhaimi is a Lecturer in Social Policy (Multicultural Policy) at the Centre for Policy Research (CPR), Universiti Sains Malaysia (USM), Penang, Malaysia.

Johan Bhimo Sukoco is a lecturer at Diponegoro University in Semarang, Indonesia. He completed his doctoral studies in Public Administration at the Faculty of Administrative Sciences, Universitas Brawijaya, Indonesia. He was an awardee of the Indonesian Education Scholarship (BPI) from the Indonesian Ministry of Higher Education, Science, and Technology, having previously completed his Diploma, Bachelor's, and Master's degrees at Universitas Sebelas Maret (UNS) in Surakarta, Indonesia. His research interests focus on community empowerment.

Antonia Eka Sari Widyastuti is a Lecturer at the Department of Public Administration, Faculty of Social and Political Science, Yogyakarta State University. She received her Master's degree in Public Administration from Gadjah Mada University and is currently pursuing a doctoral degree in Management and Public Policy at the same university. Her research interests include public sector management and human resource management.

Tri Yuniningsih is an Associate Professor in the Department of Public Administration, Faculty of Social and Political Sciences, Diponegoro University, Indonesia. She has extensive academic experience in teaching, research, and community engagement in the field of public administration. Her scholarly work has contributed to the development of public administration studies, particularly in governance, innovation, and public sector issues. She has been actively involved in academic forums, research activities, and publications at both national and international levels. Through her academic contributions, she continues to support the advancement of public administration scholarship and practice in Indonesia.

Teguh Yuwono is an associate professor at the Department of Political and Government Science. He currently serves as the Dean of the Faculty of Social and Political Sciences (FISIP) at Universitas Diponegoro (Undip) for the 2024–2029 term. He has a diverse range of research interests primarily focused on public policy, regional autonomy, and governance.

Index

A

B

C

E

F

G

I

L

M

P

V

W

www.ingramcontent.com/pod-product-compliance
Lightning Source LLC
LaVergne TN
LVHW061528290926
849162LV00004B/705